A Clinically-Oriented Approach to
ANATOMY AND PHYSIOLOGY

INCLUDES ACCESS TO THE KHQ STUDY APP

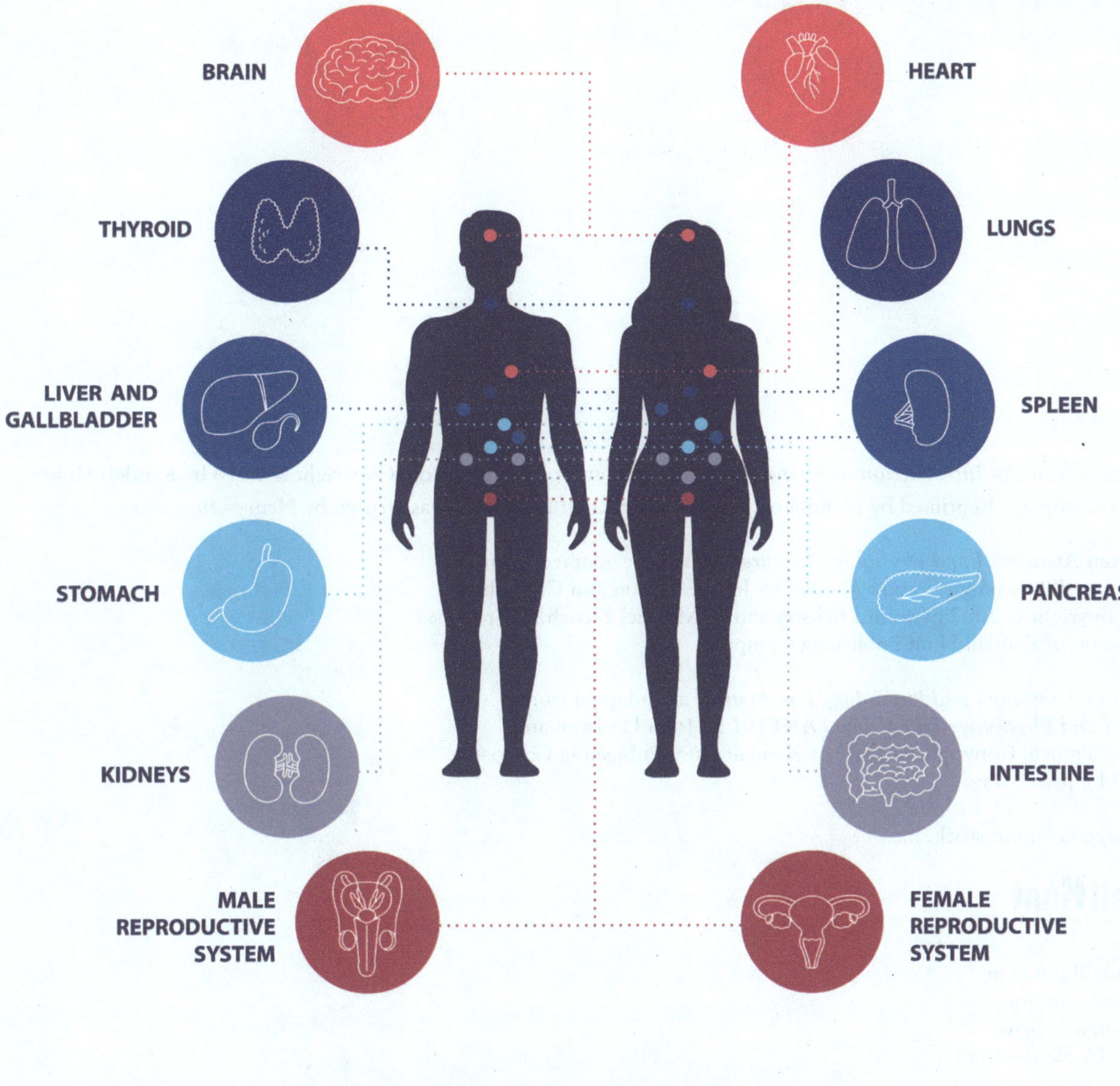

- BRAIN
- THYROID
- LIVER AND GALLBLADDER
- STOMACH
- KIDNEYS
- MALE REPRODUCTIVE SYSTEM

- HEART
- LUNGS
- SPLEEN
- PANCREAS
- INTESTINE
- FEMALE REPRODUCTIVE SYSTEM

AN ADAPTATION OF JOHN ERICKSON'S PUBLICATION

HENRY OH
Idaho State University

Kendall Hunt publishing company

All content is from *An Introduction to Anatomy and Physiology* by John Erickson. Copyright © 2010 by Kendall Hunt Publishing Company. Reprinted by permission. Text between ⌊half brackets⌋ was written by Henry Oh.

Images from *Anatomy I and Physiology Lecture Manual* are adapted from *Anatomy I and Physiology Lecture Manual* by John Erickson and C. Michael French. Copyright © 2007 by John Erickson and C. Michael French. Reproduced by permission of Kendall Hunt Publishing Company.

Images from *Anatomy I and Physiology Lab Manual* are adapted from *Anatomy I and Physiology Lab Manual ANP101* by John Erickson and C. Michael French. Copyright © 2007 by Kendall Hunt Publishing Company. Reprinted by permission.

Cover image © Shutterstock, Inc.

Kendall Hunt
publishing company

www.kendallhunt.com
Send all inquiries to:
4050 Westmark Drive
Dubuque, IA 52004-1840

Copyright © 2021 by Kendall Hunt Publishing Company

PAK ISBN: 978-1-7924-4018-2
Text alone ISBN: 978-1-7924-4019-9

All rights reserved. No part of this publication may be reproduced, stored in a retrieval system, or transmitted, in any form or by any means, electronic, mechanical, photocopying, recording, or otherwise, without the prior written permission of the copyright owner.

Published in the United States of America

This book is dedicated to
all my students,
professional colleagues, mentors and professors,
my friends and family,
and to everyone who cares about the
education of our students and
health needs of the community.

Contents

List of Clinical Approaches vii

Preface xi

About the Author xiii

Chapter 1 Introduction to Anatomy and Physiology 1
Chapter 2 Cells, Tissues, and Organ Systems 21
Chapter 3 The Integumentary System 47
Chapter 4 The Skeletal System 67
Chapter 5 The Muscular System 103
Chapter 6 The Nervous System and the Senses 129
Chapter 7 The Endocrine System 179
Chapter 8 The Cardiovascular System and Blood 203
Chapter 9 The Respiratory System 261
Chapter 10 The Lymphatic System and Immunity 287
Chapter 11 The Digestive System and Nutrition 315
Chapter 12 The Urinary System 363
Chapter 13 The Reproductive Systems 385

Appendix I *Physical Examination, History, and Interview* 417
Appendix II *Complete Blood Count (CBC), Basic Metabolic Panel (BMP), Kidney Panel, and Comprehensive Metabolic Panel (CMP), Liver Panel* 419
Appendix III *Cardiac Panel, Lipid Panel, and Urinalysis* 421
Appendix IV *12-Lead ECG: Chest Electrodes and Placement* 423

List of Clinical Approaches

CHAPTER 1
Clinical Introduction   2
Anatomical Position   5
Abdominal Region and Body Cavity   6
Carbohydrates   13
Cell Injury   18
Abdominal Region   18
Cell Physiology   18
COPD and Anatomical Position   19
Oxygen Therapy   19

CHAPTER 2
Clinical Introduction   22
Bacterial Infection   22
Tissue Repair   22
Oxygen and Brain Cells   38
General Assessment of a Patient   42
Pre-Op Patient Assessment (Assessing a Patient before Surgery)   43

CHAPTER 3
Clinical Introduction   48
Handwashing and Hand Sanitizers   59
Coronavirus   60
Patient with Thermal Burns on the Skin   64

CHAPTER 4
Clinical Introduction   68
Initial Assessment of a Patient with Chest Trauma (Fractured Ribs)   82
Knee Injury (Knee Cap or Patella)   88

Gait Rehabilitation Program 99
Simple Gait Evaluation 100

CHAPTER 5

Clinical Introduction 104
Muscle Strain 121
Manual Muscle Testing 127

CHAPTER 6

Clinical Introduction 130
Oxygen and Brain Cells 130
Sympathetic and Parasympathetic Systems 134
Stroke 149
Myotome Testing 154
Dermatome Testing 155
Cervical Fracture 162

CHAPTER 7

Clinical Introduction 180
Blood Glucose Testing 193
Hypoglycemia 199

CHAPTER 8

Clinical Introduction 204
EKG 222
Hypertension 235
Complete Blood Count (CBC) 248
ABO and Rh Blood Type 256
ABO and Rh Blood Typing 257
Chest Pain 258

CHAPTER 9

Clinical Introduction 262
Reading a Chest X-Ray 274
Pulmonary Function Testing 280
Patient with Asthma 284

CHAPTER 10

Clinical Introduction   288
Lymph Nodes   298
Vaccines, Antibodies, and Antigens   312

CHAPTER 11

Clinical Introduction   316
Gastroesophageal Reflux Disease (GERD)   331
Peptic Ulcer   332
Patient Assessment in Appendicitis   338
Weight Control   359

CHAPTER 12

Clinical Introduction   364
Kidney Stones   371
Urinalysis by Dipstick   379
Dialysis   380
Renal Failure   382

CHAPTER 13

Clinical Introduction   386
Male Reproductive Issues   399
Female Reproductive Issues   410

Preface

Anatomy and physiology can be taught and learned in many different ways depending on the background of the instructor, the type of the program, the various experiences of the students, the format or strategies of teaching and learning, and the mission and vision of the institution.

This is an introductory textbook intended for instructors and students who are in the health science or allied health occupation programs. Instructors and students of other disciplines may also find this text interesting and informative for use in their programs. As what the title of this book suggests, it has a clinically oriented approach. It is expected that the instructor has a clinical background to effectively teach the topics or lessons by relating them to their own clinical experience. For instructors whose background are other than health science, inviting guest speakers with clinical background to their class would be helpful and highly recommended. The learning process becomes more meaningful to the students by connecting or associating their learning to relevant clinical training. There are four clinical approaches in this book—clinical introduction, clinical notes, clinical skills, and clinical scenarios.

The clinical introduction presents the importance of the body system at the beginning of each chapter. It provides some examples in the clinical setting that relate to the specific body system. The clinical notes are additional information that pertain to specific topics found in the chapter. These clinical notes add clinical relevance to the lessons. The clinical skills are intended for hands-on or laboratory sessions to provide the students some clinical practice or simulation. The lab sessions reinforce their learning and provide them the opportunity to work in pairs or in groups. They start to learn the importance of teamwork early in the program to help them prepare for future clinical rotations or fieldwork where they would interact or work together with other healthcare providers. The clinical scenario is used to stimulate the students' critical thinking and decision-making skills and creativity to visualize themselves working in the clinical setting. This visualization reinforces their goals and the reasons why they are in this class and in the program. The Clinical Terms to Review found at the end of each chapter serves as a guide or a checklist for reviewing.

The overall purpose of this book is to build a strong, clinically oriented foundation in anatomy and physiology to help the students prepare for advanced classes in their respective programs. The theoretical framework of this book is based on competency education.

ORGANIZATION OF THE BOOK

The chapters are organized in four sections:

1. Structural support and movement—integumentary, skeletal, and muscular systems
2. Communication, coordination, and control—nervous and endocrine systems
3. Life support and defense—cardiovascular, respiratory, and lymphatic systems
4. Processing, regulation, and lifecycle—digestive, urinary, and reproductive systems

ACKNOWLEDGMENT

I wish to acknowledge the support and encouragement of several individuals who helped to make this textbook a reality. I am expressing my sincere thanks to the contributors of this book, to the staff from Kendall Hunt, and especially to Noelle Henneman, senior publishing specialist, and Caleb Smith, acquisitions editor, both from Kendall Hunt Publishing.

I also wish to thank Eugene Demekhin, MBA, RRT; Max Eskelson, MS, RRT; David Flint, MPH; and Dave Smith, MS, PTA, for their clinical contributions to certain chapters in this book.

About the Author

Source: Henry Oh

Dr. Henry Oh is a multi-credentialed healthcare professional with clinical practice, teaching, and management experience. He has served as a speaker on a variety of topics in seminars and faculty in-services. He is a certified medical technologist (MT), a registered respiratory therapist (RRT) with a neonatal pediatric specialty (NPS), a chartered biologist (CBiol), and a certified allied health instructor (AHI). He is a fellow of the Association of Clinical Scientists, the Human Biology Association, and the Royal Society of Biology in the United Kingdom.

Henry has been recognized with numerous awards which include: U.S. Professor of the Year in Health Sciences in 2020 from Extraordinary People's Award, Master Teacher of Honor Award in 2013 from Kappa Delta Pi International Honor Society in Education, Asia Pacific Excellence Award for Outstanding Achiever in Medicine and Allied Sciences in 2017, Exceptional Merit Award in 2019, and Distinguished Achievement Award in 2009 from the American Medical Technologists, the Outstanding Atenean Award in Academics in 2008, Editor of the Year Award in 2014 from the American Medical Technologists, National Teacher of the Year Honorable Mention in 2007 from the Career Colleges Association, and many other awards.

Henry has served in several organizations and in various capacities: president of Lambda Beta Society, which is the national honor society for the profession of respiratory care; vice-president of Utah State Society of American Medical Technologists; president of New Mexico State Society of American Medical Technologists; Governor-appointed Board Chairman of the New Mexico Respiratory Care Advisory Board, which is the state licensure board for respiratory care practitioners. He earned a doctorate degree in instructional leadership in health sciences, and an honorary doctorate in humanities from the International Institute of Leaders.

Dr. Oh currently serves as the department chair with a faculty rank of full clinical professor of Health Occupations at Idaho State University in the College of Technology.

Chapter 1

Introduction to Anatomy and Physiology

LEARNING OBJECTIVES

Upon completion of this chapter, you will be able to:

1. Use anatomical terms to identify major body regions, sections, and cavities.
2. Describe the importance of inorganic and organic compounds in the human body.
3. Define kinetic energy and potential energy and describe how energy is transferred in the human body.
4. Identify the criteria for life, and the requirements for life.
5. Describe homeostasis.

CHAPTER OUTLINE

Introduction

Anatomical Landmarks
- Anatomical Planes and Positions
- Abdominal Regions and Body Cavities

Biological Chemistry
- Inorganic Compounds
- Organic Compounds

Energy
- Kinetic and Potential Energy
- Transfer of Energy

Criteria for Life
- Organization
- Response to Stimuli
- Growth
- Reproduction
- Movement
- Metabolism and Excretion
- Requirements to Sustain Life

Homeostasis
- Negative Feedback
- Positive Feedback
- Intrinsic and Extrinsic Regulation

INTRODUCTION

Why do I need to study anatomy and physiology? What is the importance of anatomy and physiology in my program? These are some of the questions that may pop into your mind which can be answered in several ways. Let us define first what is anatomy and what is physiology. Anatomy is the study of the different structures of the human body, and physiology is the study of the functions of the different organs and systems of the human body.

As a future health care provider, you will be called upon to assess the patient or find out what is affecting the patient or what the patient is complaining about, such as chest pain, injury, cough, fever, weakness in some body parts, or any other signs and symptoms. A foundational knowledge of anatomy and physiology is needed in order to make accurate assessments. It would help you to communicate more effectively about your findings to other members of the health care team.

Students who pursue a career in the health sciences will study anatomy and physiology as part of their program's curriculum. Basically, this course serves as the most important foundation for all other courses in a program, whether it is a certificate, diploma, or degree program. All other health science courses are built upon this fundamental course—anatomy and physiology.

This section introduces you to anatomical terms, body cavities, biological chemistry, criteria for life, and homeostasis. This chapter, and all the other chapters in this book, are clinically oriented. What does this mean? How would this benefit you? The general concepts of anatomy and physiology will be discussed with a focus on clinical applications. These clinical applications refer to those scenarios, skills, or competencies that you may encounter when you work in the clinical setting. These skills or competencies are some of the basic applications commonly performed when you meet your patient.

Many of the clinical skills are basic, but some of them may be intermediate or advance. The purpose is to give you the "feeling" of doing patient assessment or patient care in a simulated clinical environment. This is to provide you and your class a "real-world" meaningful learning experience to build upon when you move or advance to other courses in your program.

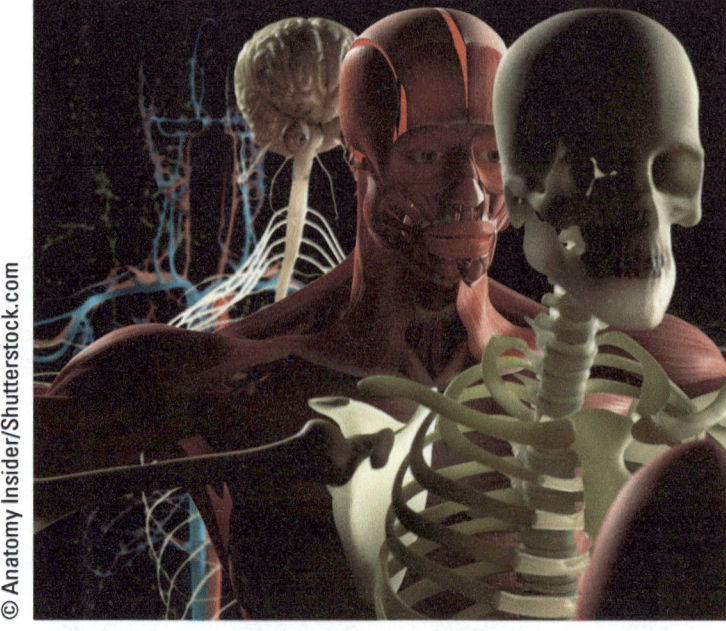

Figure 1.1

ANATOMICAL LANDMARKS

One of the important aspects of studying anatomy and physiology is to be able to identify anatomical positions and the landmarks or locations of specific structures and organs of the human body. To do this, you must learn about anatomical planes, anatomical positions, abdominal quadrants, and body cavities.

Anatomical Planes

- Frontal or coronal plane—divides the body into anterior and posterior sections; view is from the front
- Sagittal plane—divides the body into right and left sections; view is from the side or lateral
- Transverse or horizontal plane—divides the body or organ into superior and inferior sections

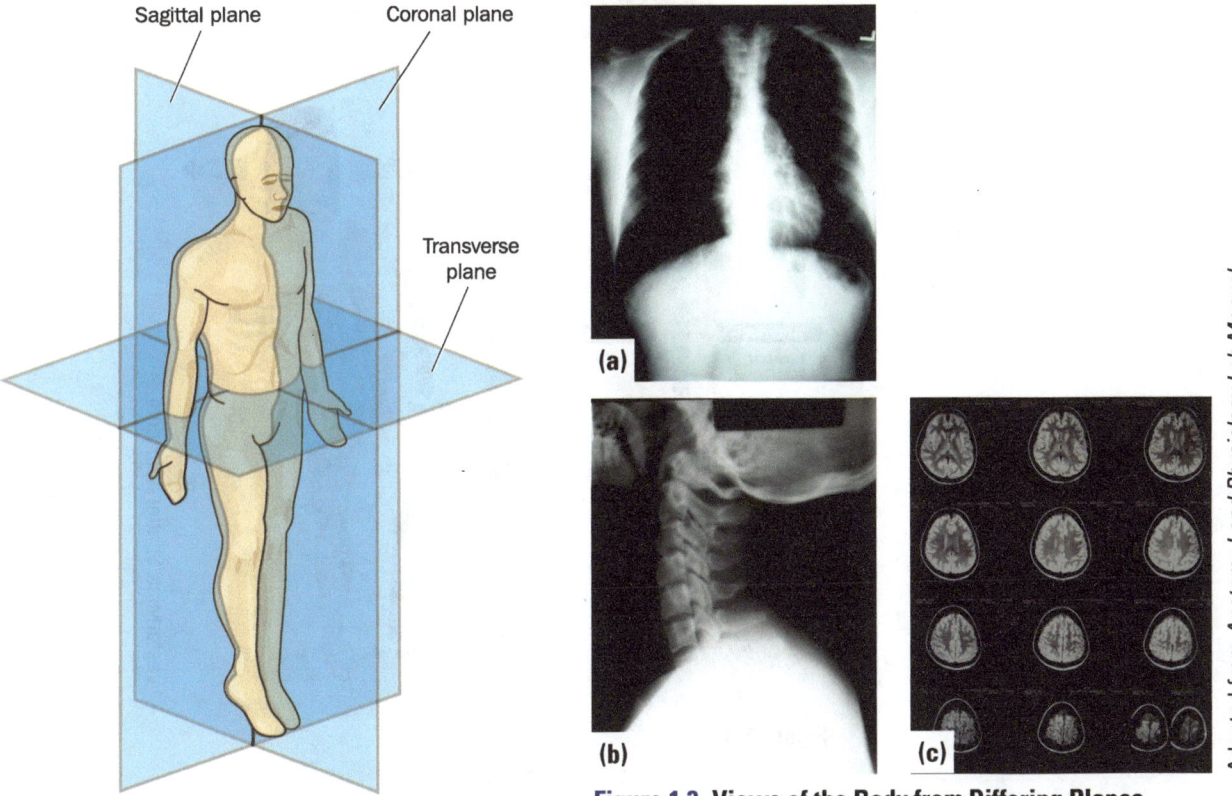

Figure 1.2

Figure 1.3 Views of the Body from Differing Planes.
(a) X-ray of the thorax in the frontal plane. (b) X-ray in the sagittal plane. (c) CAT scan of the brain in the transverse plane.

Anatomical Positions

- Anterior (ventral)—toward the front of the body; for example, the abdomen is anterior to the spine
- Posterior (dorsal)—toward the back of the body; for example, the spine is posterior to the abdomen
- Superior (cranial)—toward the head; for example, the head is superior to the shoulder
- Inferior (caudal)—toward the feet; for example, the feet are inferior to the knees
- Medial—toward the midline; for example, the neck is medial to the shoulder
- Proximal—toward the body or point of origin; for example, the shoulder is proximal to the neck

- Distal—away from the body or point of origin; for example, the hand is distal from the neck
- Supine—lying on one's back; for example, patient lying on hospital bed
- Prone—lying on one's abdomen; for example, patient receiving massage therapy on back
- Fowler's position—sitting position; for example, patient sitting while having nebulizer treatment
- Semi-Fowler's position—sitting but leaning slightly back
- Trendelenburg—head lower than the rest of the body; for example, draining secretions of lower lungs
- Lateral recumbent—lying down on one's right or left side; for example, recovery position in an emergency to prevent aspiration, or to drain secretions from the affected lung placed in upper position

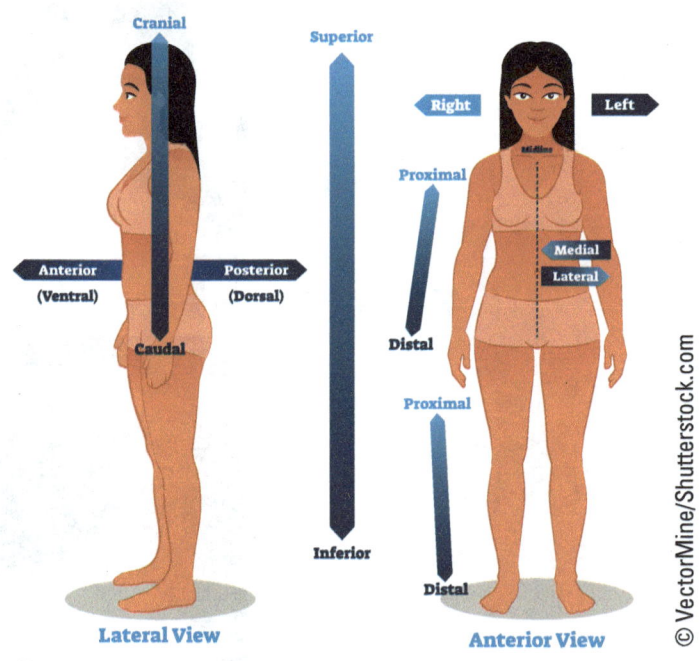

Figure 1.4

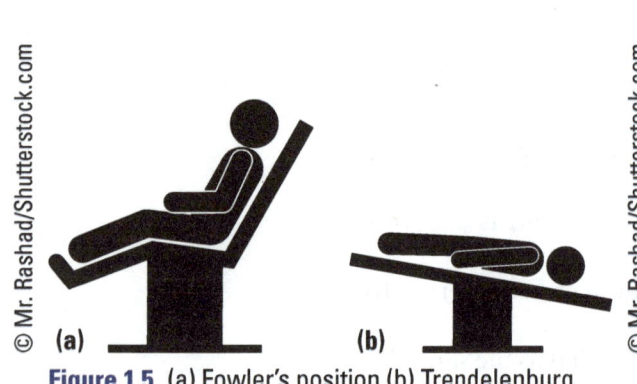

Figure 1.5 (a) Fowler's position (b) Trendelenburg position

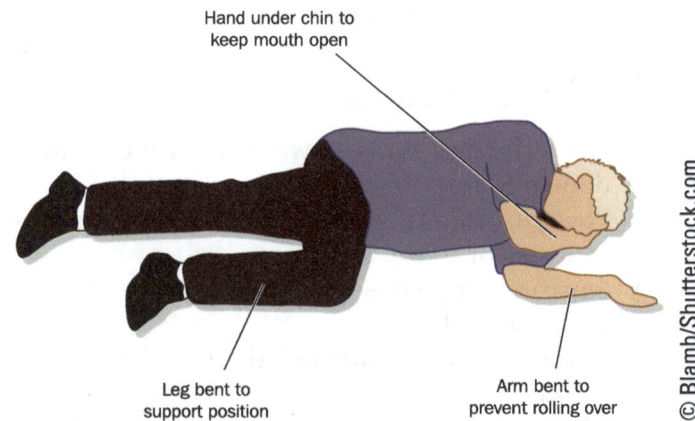

Figure 1.6 Lateral recumbent

Clinical Notes—Anatomical Position

Commonly used terms for patient positioning includes Fowler's, semi-Fowler's, supine, and right or left lateral.

The diaphragm is the major muscle of breathing. It is attached to the inferior end of the rib cage and separates the thoracic and abdominal cavities.

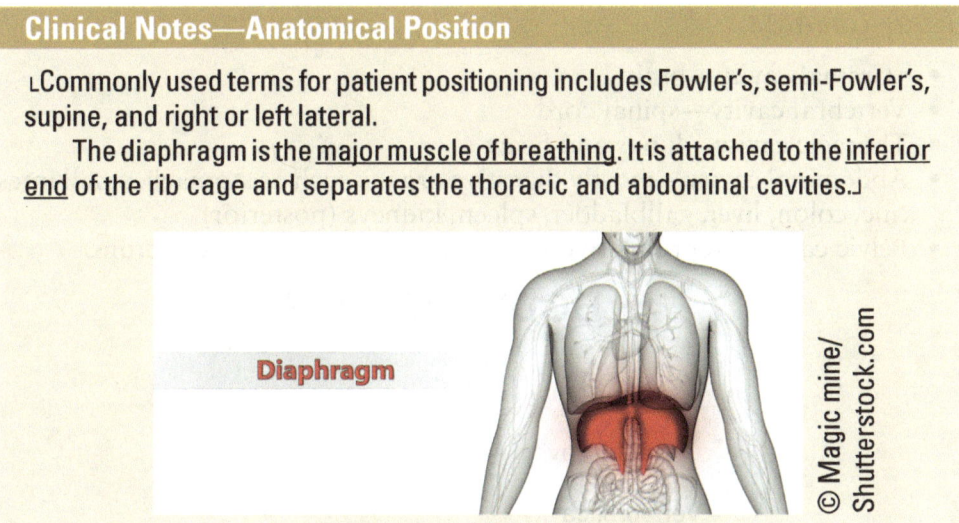

Abdominal Regions

- Right upper quadrant (RUQ)—mostly liver, gallbladder, right kidney (posterior)
- Left upper quadrant (LUQ)—stomach, pancreas, spleen, and left kidney (posterior)
- Right lower quadrant (RLQ)—mostly digestive system, appendix
- Left lower quadrant (LLQ)—mostly digestive system, sigmoid colon

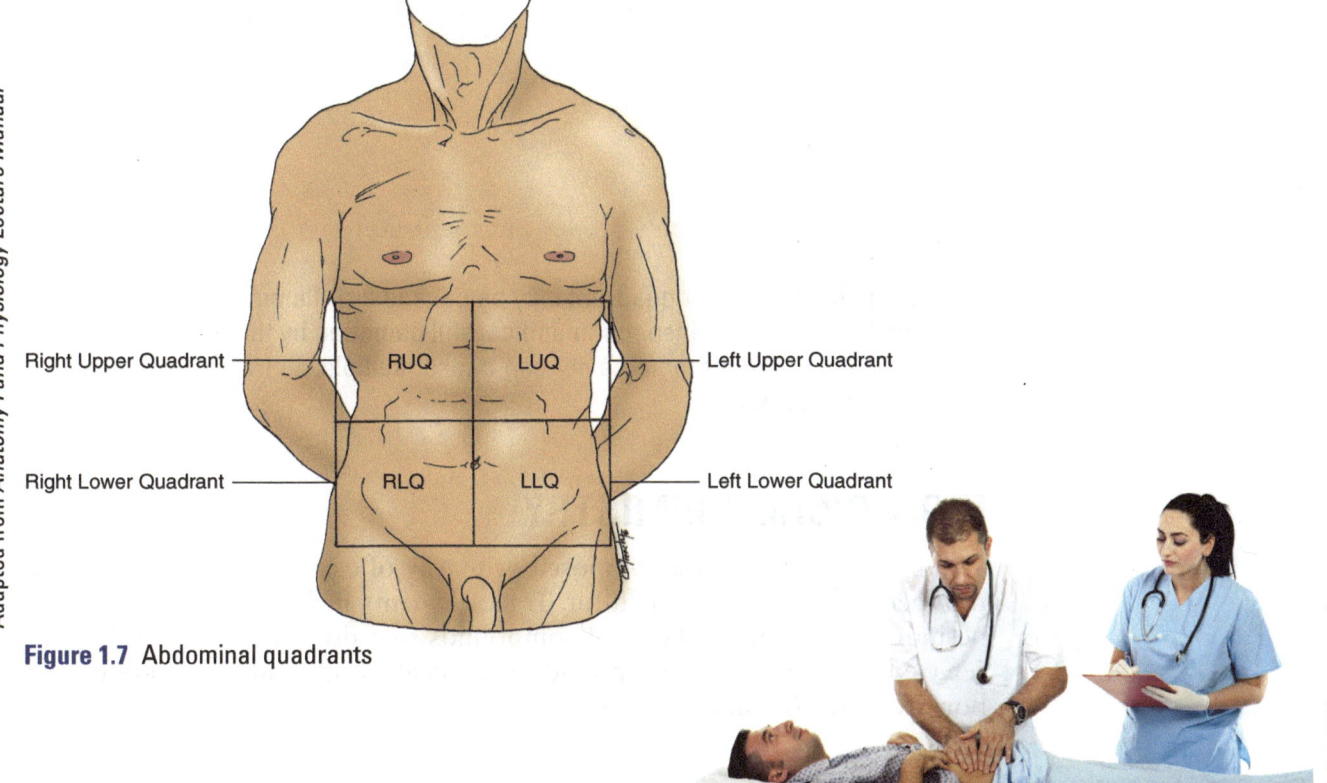

Figure 1.7 Abdominal quadrants

Figure 1.8

Body Cavities

- Cranial cavity—brain
- Vertebral cavity—spinal cord
- Thoracic cavity—lungs and heart
- Abdominal cavity—mostly digestive organs such as stomach, small intestine, colon, liver, gallbladder, spleen, kidneys (posterior)
- Pelvic cavity—reproductive system, urinary bladder, and rectum

BODY CAVITIES

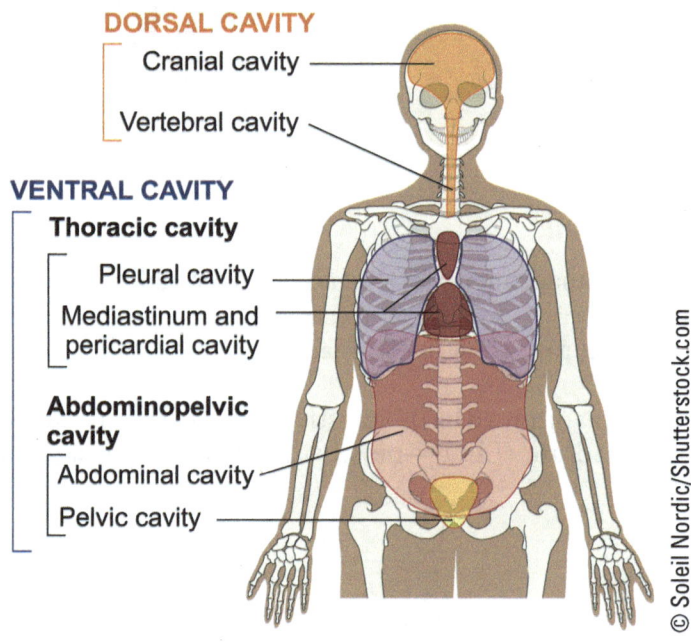

Figure 1.9

Clinical Notes—Abdominal Region and Body Cavity

Pain in the right upper quadrant may be caused by inflammation of the liver; verified by abdominal inspection and careful palpation by the physician.

Trauma to the thoracic cavity can cause multiple rib fractures and lead to pneumothorax.

BIOLOGICAL CHEMISTRY

All living organisms are composed primarily of carbon, hydrogen, oxygen, nitrogen, phosphorus, and sulfur. Trace elements include iron, iodine, magnesium, and calcium. Chemical compounds that do not contain carbon are referred to as inorganic compounds. Chemical compounds that contain carbon are called organic compounds.

Inorganic Compounds

Inorganic compounds can be found in both the living and nonliving. Water and salt are essential components of all living organisms. They simply do not require a living organism to assemble them and are not based on carbon chains.

> **Dehydration and Over-Hydration Can Affect Body Function**
>
> Dehydration is the lack of sufficient water in the body to maintain homeostasis.
>
> **Causes:**
> - Excessive sweating
> - Vomiting/diarrhea
> - Hormone imbalance
> - Kidney malfunction
> - Insufficient intake
>
> **Results:**
> - Blood pressure drops
> - Fainting
> - Exhaustion
> - Brain cells shrink and may cause mental confusion, coma, and death.
>
> Over-hydration is too much water in the body to maintain homeostasis.
>
> **Causes:**
> - Rapid intake of excessive amounts of water
> - Hormone imbalance
> - Kidney malfunction
>
> **Results:**
> - Blood pressure rises
> - Brain cells swell and may cause mental confusion, coma, and death.

Water

Water is an incredibly versatile molecule in the living organism. Because it has charges on the molecule, as mentioned earlier, it not only causes water molecules to stick together, but it can also attract ionic bonds to separate enough to form ions. Those individual ions are then available to activate functions in the body at incredible speed. Every time we move and whenever we think, ions move through our muscle and brain tissue causing them to perform what they do best. Water can act as a *solvent* to dissolve a large number of compounds, particularly in the bloodstream, carrying oxygen, carbon dioxide, salts, sugars, vitamins, and nutrients. Water becomes a major component passing through the digestive system, approximately 9,000 ml per day. It can even be used to break large molecules apart by a process known as *hydrolysis*. It has the ability to carry heat energy so it can be used to transport the heat given off by chemical reactions deep inside the body to the skin for release into the surrounding environment. Water can even be used to surround our brain and spinal cord as a cushion against potential injury when it forms cerebrospinal fluid. With all of the functions water performs, it is easy to comprehend why it accounts for approximately two-thirds of our body weight.

Salt

Salts are the result of ionic bonds formed when atoms give electrons to or accept them from another atom. Most individuals think of table salt, sodium chloride, when the term *salt* is mentioned, but it is only one type of salt. As mentioned earlier, ions are very important for the normal function of both the nervous and muscular systems. They are required to activate our muscles and nerves to cause movement, conduct information about the environment to the brain, and allow brain cells to talk to each other. There are three primary salts that are essential to the normal functioning of the human body:

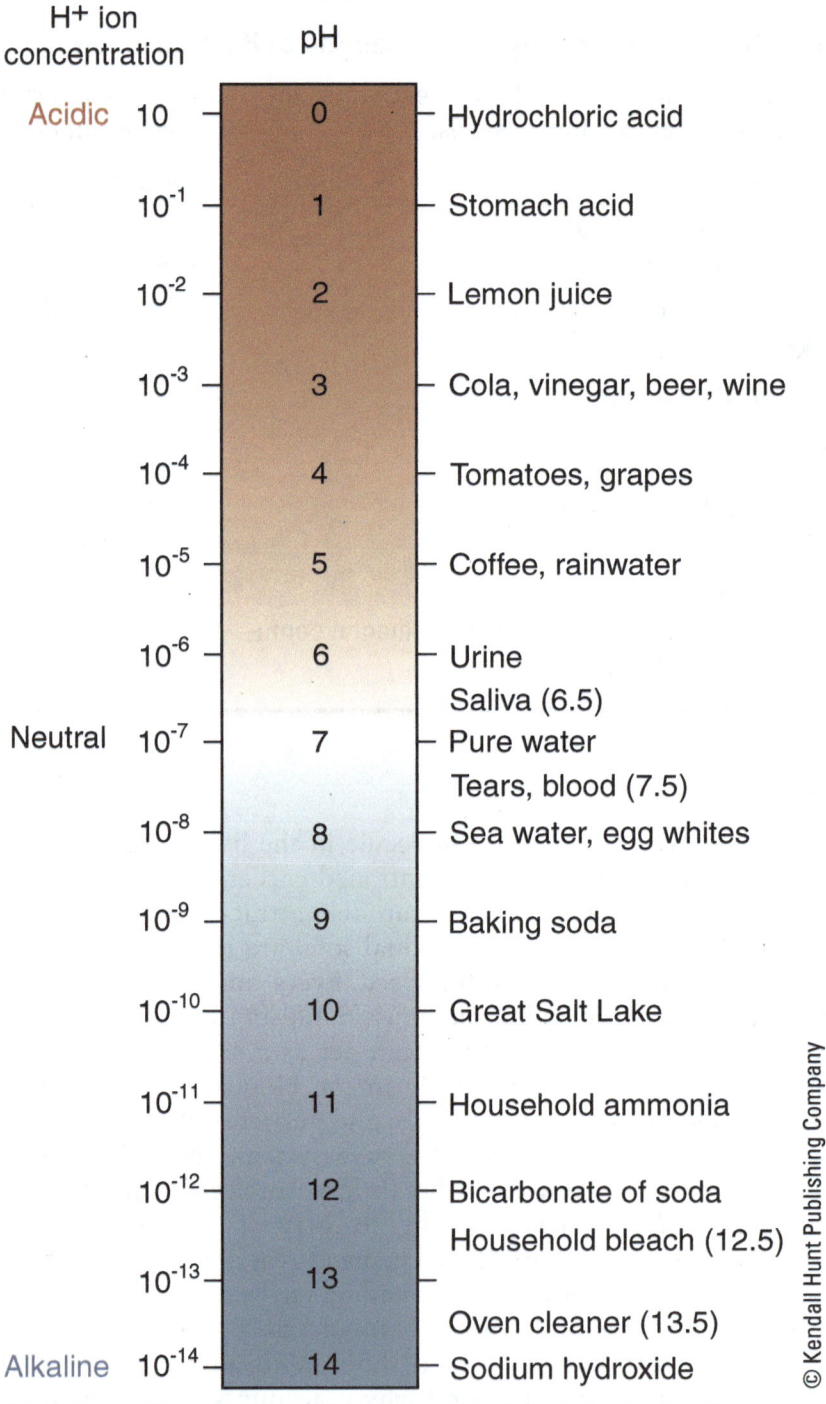

Figure 1.10 pH Scale. The pH scale, which ranges from 0–14, indicates the H concentration of a solution. A pH of 7 is considered neutral. Values less than 7 are acidic and values greater than 7 are basic.

sodium chloride, potassium chloride, and calcium phosphate. Our nerves require sodium and potassium ions to transmit information. Once stimulated, muscle must have calcium ions to activate internal structures to cause contraction. Calcium phosphate is a major component in our bones. Without calcium phosphate in sufficient quantities, our bones become thin and break easily as might occur in osteoporosis.

Acids and Bases

Remember that the hydrogen atom contains one proton and one electron, or one positive charge and one negative. If hydrogen gives its one electron away, it is left with a single unbalanced proton or hydrogen ion. There is no other case in which there will be only protons by themselves. It is helpful for us to keep track of them. Scientists have designed a scale defining the concentration of hydrogen ions in a solution. It is known as the pH scale (power of hydrogen scale) with the H standing for the hydrogen ion (Figure 1.10).

Using this scale, ranging from 0 to 14, 7 is neutral. Solutions with pH numbers less than 7 are defined as acid, and those with numbers greater than 7 are considered to be alkaline or base. The farther away the pH of a solution is from neutral, or 7, the stronger it is. So, acid at a pH of 4 is stronger than one at a pH of 6.

When equal amounts of acids and bases of equal strengths are combined, they will cancel each other out or be neutralized because their pH will average out to be 7, or neutral. The pH of our blood is extremely important and must be regulated very carefully. Normally our blood pH is slightly alkaline, with a pH between 7.35 and 7.45. There are chemicals in our bloodstream known as buffers that can adjust for small changes in pH to keep it within the homeostatic range. If, however, the blood pH becomes too acidic or too alkaline and buffers are unable to absorb the variations, we must add neutralizing base or acid to the bloodstream to maintain the pH at the appropriate level.

> **pH scale** a scale representing the concentration of hydrogen ions in a solution.
>
> **Acid** a solution with a pH less than 7.000.
>
> **Base** a solution with a pH greater than 7.000.
>
> **Buffers** chemicals that can adjust for small changes in pH.

Organic Compounds

The field of chemistry concerned with the study of carbon chains and the hydrogen, oxygen, and nitrogen atoms attached to them is known as *organic chemistry*. The carbon atom has four electrons in its outermost shell that it is able to share by covalent bonding. Carbon atoms share electrons with each other, forming carbon chains. Molecules that are formed by chains of carbon atoms are an essential component of all living organisms. In most cases, these carbon chains are surrounded by hydrogen atoms but occasionally they will also have oxygen or nitrogen atoms attached as well. Other atoms may also be attached to the carbon chains but carbon, hydrogen, oxygen, and nitrogen are the primary elements found in organic chemistry.

When looking at the organic chemistry that occurs in a living organism, the scientist is studying biochemistry. There are three basic biochemical nutrients essential for human life. Those three nutrients are listed on every package or box containing the food we eat. They are: (1) carbohydrates which break down to sugars; (2) lipids, which are fats and oils; and (3) proteins, which are chains of amino acids.

> **Carbohydrates** sugars.
>
> **Lipids** fats and oils.
>
> **Proteins** chains of amino acids.

Carbohydrates

Carbohydrates have been in the news so much recently. How many carbs should we eat? Is there a difference between simple and complex carbs? Should we eat no carb diets?

There are three types of carbohydrates:

1. *Monosaccharides* are individual molecules of sugar. The most important one to us is *glucose*. Our brain runs on glucose and depends on a constant supply of it from our bloodstream.

2. *Disaccharides* are two sugars bonded together. The most common is sucrose, which is also known as table sugar. It is what we put on our food or in our drinks. All disaccharides contain one molecule of glucose in addition to another monosaccharide. Our bodies simply snap them in half and use the glucose. The other sugar can be rearranged into glucose if necessary. These double sugars are a very rapid source of glucose but can raise the amount of sugar in the bloodstream too fast.
3. *Polysaccharides* are complex sugars, which are chains of many sugars bonded together. An example of a polysaccharide is starch. Starchy foods include potatoes, rice, corn, and pasta. They also contain glucose but it is not as easily accessible as molecules containing only two sugars. It takes longer to free up the glucose so the individual's blood sugar level does not rise too rapidly. The end result is more energy being available for an extended period of time. If there is too much glucose in our bloodstream, it is converted into a complex sugar known as *glycogen*. Later, when our sugar level gets too low, as in cases of starvation or when we do not eat for extended periods of time, glycogen can be broken down into glucose molecules, which are released into the bloodstream to maintain blood sugar at homeostatic levels.

Lipids

Lipids are fats and oils. Water is polar and is a major component of our body. Lipids, which are nonpolar, must be handled in a special way because they are insoluble. Fats are attached to a special molecule designed to carry three fats at a time in the bloodstream. This molecule is known as *glycerol*. When three fatty acids are attached to *glycerol*, they form a molecule known as a *triglyceride*. This is how our bodies move and store most fats. Having too many triglycerides may result in these large molecules causing blockages in our blood vessels with the potential for heart disease.

There are other important lipids as well. Phospholipids form our cell membranes. Steroids are hormones—chemical messages moving through our bloodstream communicating between cells at a distance. Low-density lipids (LDLs) are often referred to as bad cholesterol, which also sticks to our blood vessels. High-density lipids (HDLs), also known as good cholesterol, assist in removing LDLs from our bloodstream.

Proteins

Proteins are composed of *amino acids*. Amino acids are used sort of like toy building blocks. A structure can be built out of the blocks, then, when it is no longer needed, disassembled so something else can be constructed. Proteins are assembled by chaining amino acids into a specific sequence. Proteins have many uses. For example, they may be enzymes to speed up chemical reactions. Structural components such as ligaments to hold bones together and tendons to hold muscle to bone are composed of protein. They may be the major structure causing muscle contraction. Many of the components within the cell are proteins. Even the bloodstream contains a number of plasma proteins.

> **Comprehension Check-up**
>
> 1. Complex chains of sugars are known as _____.
> 2. Fats are transported and stored in the body as _____.
> 3. Why would a body-builder want to consume large amounts of protein?.
>
> 1. polysaccharides
> 2. triglycerides
> 3. Protein is composed of chains of amino acids. Those amino acids are also an essential component of muscle. As the body builder works out, he or she will need additional amino acids to build larger muscles.

Nucleic Acids

In order to make substances in our cells, it is necessary to have the instructions kept in a chemical form known as nucleic acids. They are enclosed in the nucleus of each cell. There are two types of nucleic acids: DNA and RNA. The actual instructions to make anything our body was designed to construct are in the chemical form of *deoxyribonucleic acid,* or DNA. DNA is often referred to as a set of recipe books containing instructions for every structure and chemical process that occurs in the body. Every cell possesses, at some time in its life, the complete set of DNA. Each instruction in our DNA to make one substance or structure is known as a *gene*. Just as a cookbook contains one recipe after another, DNA is composed of thousands of sequences of genes chained together. When some substance needs to be made in our cell, it is necessary to copy its gene. The original recipe stays safely within the nucleus, then, the copy is sent out to the cell where those chemicals or structures can be constructed according to the plan in our DNA. The process of copying the original recipe and setting up a system by which new molecules can be assembled is accomplished by another nucleic acid known as *ribonucleic acid,* or RNA. For example, if a cell needs to produce a specific enzyme, the DNA exposes the gene containing the recipe for that enzyme. RNA, on the other hand, makes a copy of that gene and then takes it out of the nucleus where the enzyme can be produced elsewhere within the cell.

> **Nucleic acids** molecules used to form or implement cellular instructions.

> **DNA** a set of nucleic acids forming cellular instructions for very structural and chemical process that occurs in the body passed on from one generation to the next.

> **RNA** nucleic acids used to form cellular products from instructions (genes) within the DNA.

ENERGY

All matter possesses energy. While energy can be neither created nor destroyed, it can be converted from one form to another. The energy can be used to perform some kind of *work* or process, it can be chemically stored, or it may be lost as heat.

Kinetic and Potential Energy

When energy is actively being used for work, it is designated as *kinetic energy*. When that energy is available but is currently not in the process of making some event or process happen, it is referred to as *potential energy*. To illustrate the difference, think of the chemical energy in gasoline as being kinetic energy when it makes the engine in an automobile run. Potential energy would be in the gas in the tank. It is available for use whenever needed, but until it enters the engine, that energy remains available but unused. In our bodies, kinetic energy is needed to activate many processes that are essential to life such as the contraction of muscle and the pumping of ions across cell membranes. On the other hand, it is very helpful to have energy stored and available whenever it is needed. That potential energy is most efficiently stored as fat.

Transfer of Energy

All energy on the surface of the Earth comes from the sun, but that solar energy is not accessible to humans because we lack the ability to capture and use it. Instead, that energy only causes us to become sunburned. Energy cannot be created or destroyed but it can be converted from one form to another. Through photosynthesis, plants can capture the solar energy and store it. Then humans, as animals, eat those plants or eat animals that eat plants and convert that energy into a form they can use. It is a chemical process using all of the types of chemical reactions discussed and requires different enzymes each step of the way. Generally, the *nutrients* taken in are disassembled, the energy is extracted, and then is converted into a form that can be used to perform work. That work involves making muscles contract as well as running many different kinds of pumps moving ions across the cell membrane with that kinetic energy. The process of removing potential energy from nutrients is difficult—75% of that energy gets away from us and becomes heat. It is necessary to constantly remove heat from our bodies because not doing so would cause us to heat up so much it would unravel enzymes and stop essential chemical processes.

When energy is extracted from the nutrients there needs to be a way to transfer that energy to its appropriate location for use. This is done chemically by a *high-energy compound* known as adenosine triphosphate (ATP). Breaking the name down to its component parts makes it fairly easy to follow. The prefix, "tri-" refers to three, indicating that there is an adenosine with three phosphates attached. The phosphates have what are known as high-energy bonds between them. In other words, it takes a lot of energy to make these bonds, but when those bonds are broken the energy is released. To accomplish this process, a high-energy compound containing two phosphates known as *adenosine diphosphate* (ADP) will attach a third phosphate if the energy is available to form adenosine triphosphate. This reversible reaction is:

$$ADP + P + Energy \rightarrow ATP$$

This process is similar to charging a battery. Once it is captured, that potential energy can be released and used however needed. Charged batteries can be put in a flashlight, CD player, MP3, remote-controlled car, clock, or anything else requiring energy into which the battery will fi t. The same is true in our cells. Whenever energy is needed, the third phosphate is broken off ATP, releasing the energy:

$$ATP \rightarrow ADP + P + Energy$$

> **Adenosine Triphosphate (ATP)** the high-energy compound by which energy is transferred within cells.

> **Comprehension Check-up**
> 1. Each individual instruction in our DNA is known as a _____.
> 2. Energy is transferred in the body in the form of _____.
>
> 1. ATP 2. gene

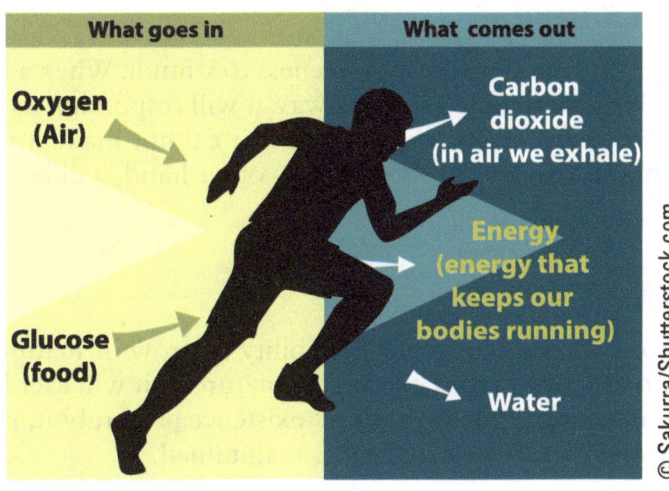

Figure 1.11

> **Clinical Notes—Carbohydrates**
>
> When carbohydrates and lipids are ingested, carbohydrates will be used first as the source of energy by the body. When there is an increased intake of carbohydrates, they are turned into fatty acids and converted to triglycerides (lipids) that can cause high cholesterol levels.

CRITERIA FOR LIFE

Humans are affected by both living and nonliving things in the environment. It will be useful in our study to define the criteria that are required to determine if something is living or not.

Living organisms have the ability to respond both to internal needs and to changes in the environment; this also includes the ability to produce offspring. Bear in mind, the organism or object in question must meet all of the criteria sometime during its existence to be considered living—this may not be as obvious as it appears to be on the surface.

Consider three objects to determine if they are living or nonliving: this textbook, a virus, and a rabbit. A virus is a piece of cellular instructions, DNA or RNA, surrounded by a protein coat or envelope.

> **Criteria for Life**
> › Organization
> › Response to stimuli
> › Growth
> › Reproduction
> › Movement
> › Metabolism and excretion

Organization

Organization refers to the status of the organism or object's internal contents. Are internal structures present that have particular functions in specific areas or are the contents totally random? Are the internal structures of this book arranged in organized patterns or are they random? The intent is to be organized and the content is in distinct groups. The internal content of the virus is also highly organized. If we viewed the rabbit internally, we would rapidly determine that its organs were organized and in specific locations.

Response to Stimuli

The next factor to consider is responsiveness to stimuli. When a living organism is touched or manipulated in some way, it will respond of its own accord. We can touch or even shake this book, but it cannot make a response; the same is true of the virus. The rabbit, on the other hand, will be very quick to respond.

Growth

Another factor necessary for life is the ability to grow or mature. This book cannot grow of its own accord and it is as mature as it will ever be. The virus also remains unchanged throughout its existence. The rabbit, however, will grow rapidly and mature as long as it is maintained.

Reproduction

Reproduction is another important defining factor. Can the object or organism produce offspring just like itself? This book cannot reproduce. The virus can reproduce but not of its own accord—the virus injects its genetic material into a living cell, causing that cell to produce more viruses. The rabbit, on the other hand, is the big winner in this category. It is important to understand that within this category we include the ability to pass on genetic traits to offspring and the potential for changes, or mutations. Although the printing press can produce many books, each book cannot reproduce itself. Genetic changes can occur in viruses. In fact, it is often the rapid mutation of viruses that causes serious outbreaks of diseases around the world every year. While rabbits do change their genetics over time, those changes, without human intervention, would normally occur over hundreds or even thousands of years. Reproductive organs also produce hormones that influence growth, development, and tissue repair.

Movement

Living organisms also have the ability to move. Many can change position or location to accommodate their needs and all have some kind of movement internally. Rabbits move quickly in their environment, but their internal movement is also continually active throughout life. The transport of viruses is the result of some other cause such as air currents, water flow, or bodily fluid. If this book starts moving on its own, get rid of it—fast.

Metabolism and Excretion

All living things need energy to do whatever processes are essential for life. The energy needed to maintain life comes from the sun. Living organisms must find a way to either convert the energy from the sun into a form they can use or consume other organisms that have made that conversion. That process falls into a category of chemical reactions known as *metabolism*. The ingredients in food that are used by organisms to extract energy, form new body components, and assist body functions are called nutrients. Once an organism has obtained the energy it needs from nutrients, the byproducts are removed by the process of *excretion*. This book does possess energy—if the pages are burned, the book will give off heat energy, but it did not, by its own chemical reactions, convert energy into the form it has. A virus is not involved in the conversion of energy either. The rabbit eats plants that have captured solar energy and stored it chemically as glucose (sugar). Glucose is a nutrient from which the rabbit extracts this chemical energy to be able to perform its own functions. By-products from the breakdown of glucose, such as carbon dioxide and water, as well as unusable plant material are excreted from the rabbit's body.

All six of these criteria must be present at some time during the organism or object's existence for it to be considered living. The book may have one of the criteria (organization), but it lacks the others. Viruses, even though they are active in the process of causing diseases, are not living because they do not possess all of the criteria for life. The rabbit possesses all six criteria, allowing us to consider it as a living organism. Humans meet all of these standards, although it is more obvious for some criteria than others.

Requirements to Sustain Life

There are some requirements that humans must have in order to maintain life.

Oxygen

Humans need oxygen to extract the maximum amount of energy from the nutrients they take in. For example, some very active muscles consume large quantities of energy and have such a strong demand for oxygen that they possess special molecules called *myoglobin* to hold extra oxygen to meet the need. Deprive brain cells of oxygen for more than 6 minutes and they begin to die. There are some organisms that do not require oxygen for survival, but to humans it is essential.

Water

Water is found almost everywhere in our body. It is transported in our bloodstream. It bathes the outside of our cells. Approximately 60 to 80% of our body weight is due to water. Running short of liquid (dehydration) can cause our blood to thicken (increase viscosity) and draw some of the water out of cells making it more difficult for them to perform their functions.

Requirements to Sustain Human Life

› Oxygen
› Water
› Nutrients
› Heat
› Atmospheric Pressure

Severe loss of water can have serious results. Over-hydration is also possible if an individual drinks excessive quantities of water in a very short period of time. The resulting dilution of blood causes cells to swell and interferes with the rate of chemical reactions. The consequences can also be very serious.

Nutrients

Nutrients are not just the chemicals from which to extract energy. They are also used to form proteins essential for the construction of muscle and enzymes. Fats are needed to make cell membranes. Nucleic acids form DNA and RNA, which form cellular instructions and are essential components in the transfer of information. Other nutrients such as vitamins and minerals are essential for chemical reactions to occur.

Heat

A constant internal body temperature is maintained to allow all normal bodily functions to occur. If we become too hot or too cold, chemical reactions essential for life are unable to continue at rates necessary to sustain life.

Atmospheric Pressure

The air around us pushes on us much like stepping into the ocean causes the water to press on us from all sides. This is known as *atmospheric pressure*. It is the difference in pressures between the air outside of us and the air inside our lungs, created by changing chest volume that allows us to breathe and to exchange oxygen and carbon dioxide with the air.

Without supplying all of these requirements humans would not be able to survive.

> **Comprehension Check-up**
> 1. List the six criteria essential for an organism to be considered living.
> 2. A lack of which requirement to maintain life would cause blood to thicken?
>
> 1. Organization, Responsiveness to Stimuli, Growth, Reproduction, Movement, and Metabolism/Excretion
> 2. Water

HOMEOSTASIS

In the human body, or in any living organism for that matter, there is a need to maintain certain parameters such as temperature, water balance, hormone levels, blood pressure, and oxygen and carbon dioxide concentration within a constant range. If this is not accomplished, the individual becomes sick. The term for regulating various parameters inside a living body within a specific range is called homeostasis. More precisely, homeostasis is defined as the dynamic constancy of the internal environment of the body. In order to keep things constant, each parameter requires a *control center* to be able to cause changes when the level in the body is not where it should be. This center receives input from sensors, or *receptors,* that determine the current

> **Homeostasis** regulating various parameters within the body so that each is maintained within a specific range.

level in the body and compares that reading to a standard, or *set point,* that indicates what it should be. If the control center determines that the current level in the body is out of range, it will activate a system known as an *effector* to produce the appropriate change to bring that parameter back within the homeostatic range.

For example, let's say the temperature set point of my body is 98.6°F. The control center for temperature is found in an area of my brain known as the *hypothalamus*. Temperature sensors inform the hypothalamus about my current temperature. If my actual temperature goes over the set point—for example, 99.5°F, my hypothalamus will activate the cooling process (effector) and I will sweat. If, on the other hand, my internal temperature becomes too cold, my hypothalamus will cause me to shiver—a process that generates heat and warms my body. By constantly comparing my actual body temperature with the set point, my hypothalamus maintains my internal body temperature at a constant level to regulate homeostasis.

Negative Feedback

The process by which the body maintains parameters within a specific range is known as negative feedback. It is considered negative feedback because the response of the effector results in the controller reversing its direction of control. For example, if the level of thyroid hormones in the blood, detected by receptors in the hypothalamus, began to reach the lower levels of its normal range (set point), the controller, the hypothalamus, would stimulate the thyroid gland (effector) to increase its hormone production resulting in an increase in thyroid hormones in the blood. As the level of this hormone increases it reaches the upper limit of the desired level. The hypothalamus then decreases its stimulation of the thyroid gland causing thyroid hormone levels in the bloodstream to decrease, which will again cause a reversal of the controller as the levels reach the lower limit again. And so the process continues causing the levels to weave back and forth within the homeostatic range.

Positive Feedback

There are times that require exceptional changes in the body, such as childbirth. For this situation, positive feedback is used. *Positive feedback* is the special case in which the response actually increases the need. It enhances the initial stimulus (same direction). For instance, a woman is nine months pregnant, the gestation period is complete and it is time for labor to begin. Positive feedback is required. The hypothalamus (control center) causes the pituitary gland (effector) to release a small amount of the hormone oxytocin, which causes a slight uterine contraction. This contraction stimulates the release of more oxytocin causing an even stronger contraction. The cycle continues with each contraction stimulating more oxytocin until birth occurs. There are additional steps, but the general process is due to positive feedback.

Intrinsic and Extrinsic Regulation

When the control process occurs within the system it is regulating, we call this *intrinsic regulation (auto-regulation)*. When the control comes from a source outside the system it is designated as *extrinsic regulation*. For example, the digestive system produces hormones that tell its own digestive organs to perform tasks essential for the breakdown of food. That is intrinsic regulation because it comes from within the system. The brain can also alter our digestive system by speeding it up or slowing it down. Because the brain is outside the digestive system, its control of digestion is extrinsic regulation.

Clinical Notes—Cell Injury

While living organisms have the ability to respond to both internal needs and to changes in the environment, the human cells have the ability to meet physiologic demands to maintain a state of *homeostasis*, or stability. However, if cells are subjected to harmful agents, inadequate essential nutrients, reduced oxygen supply, stress, and infection, this can lead to cell injury. Continuous cell injury can lead to cell death.

Clinical Scenario—Abdominal Region

During a basketball game, two players collided where one player's elbow hit the other player in the left upper quadrant of his abdomen. The player who was hurt is in extreme pain. What possible internal organs may have been affected?

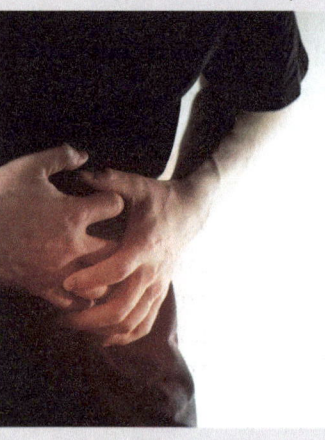

© Sjstudio6/Shutterstock.com

Clinical Scenario—Cell Physiology

To build one's muscles, foods or drinks rich in proteins are needed. What cellular structure or organelle is involved in protein synthesis?

© Syda Productions/Shutterstock.com

Clinical Scenario—COPD and Anatomical Position

A 63-year-old man has been a chronic smoker for over 20 years. He complains of seasonal chronic cough with lots of yellowish sputum, and has difficulty of breathing. On auscultation, his lungs indicate coarse crackles (secretions) and wheezing. His chest x-ray shows:

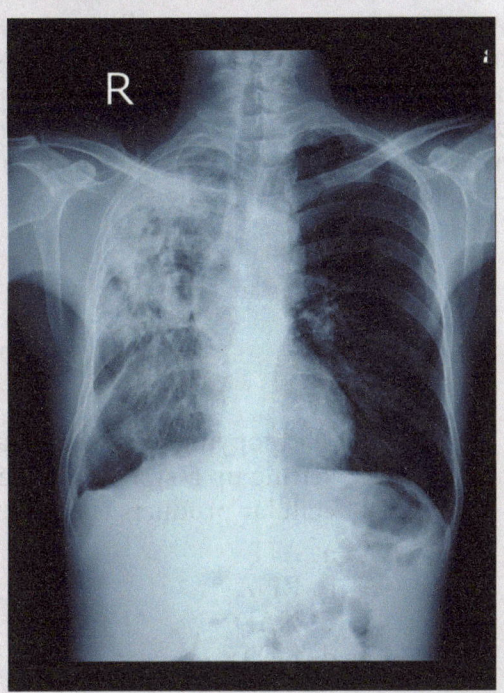

What part or section of the lungs have secretions? What <u>anatomical position</u> can help drain the secretions?

Clinical Scenario—Oxygen Therapy

A patient complains of difficulty of breathing and chest pain. A standard medical order is to administer oxygen to the patient. What is the purpose of giving oxygen to the patient?

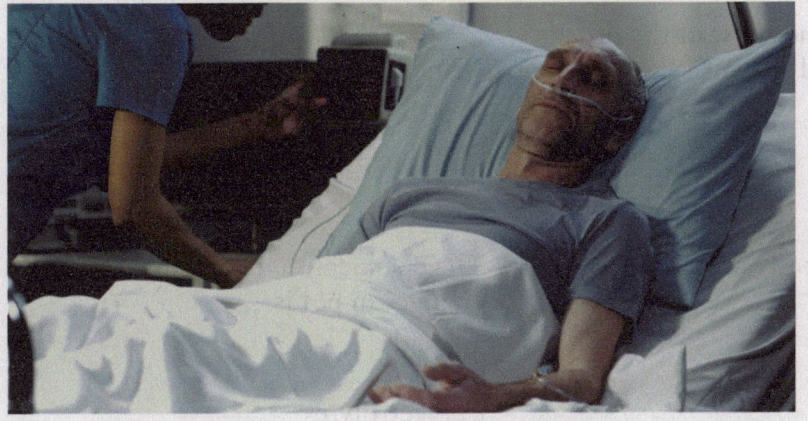

CLINICAL TERMS TO REVIEW

Anterior, posterior, superior
Supine, prone, lateral
Semi-Fowler's
Trendelenburg
Right and left lateral
Right and left upper quadrant (RUQ/LUQ)
Right and left lower quadrant (RLQ/LLQ)
Body cavities—pleural, abdominal, cranial, pelvic, vertebral, and pericardial
Oxygen, oxygen therapy, carbon dioxide
Electrolytes—sodium, potassium, chloride
Carbohydrates
Lipids—triglycerides
Lipids—cholesterol: high-density lipoprotein (HDL), low-density lipoprotein (LDL)
Proteins—amino acids
DNA/RNA—deoxyribonucleic acid/ribonucleic acid
COPD—Chronic Obstructive Pulmonary Disease

Test Yourself

1. The liver and gallbladder are found in the _____ quadrant of the abdomen.
   a. RUQ
   b. LUQ
   c. RLQ
   d. LLQ

2. Viewing sections of the body from above would be looking in the _____ plane.
   a. Sagittal
   b. Frontal (coronal)
   c. Horizontal (transverse)
   d. Diagonal

3. When considering the regions of the body, the neck is located in the
   a. Cephalic region
   b. Pelvic region
   c. Cervical region
   d. Abdominal region

4. What cavity would you find most of the digestive system, kidneys, and spleen?
   a. Thoracic cavity
   b. Abdominopelvic cavity
   c. Cranial/vertebral cavity
   d. Pericardial cavity

5. Fats and oil in the human body are collectively referred to as
   a. Carbohydrates
   b. Lipids
   c. Vitamins
   d. Proteins

6. The copying of the original DNA gene and setting up a system for the assembly of new cellular products is accomplished by
   a. ATP
   b. RNA
   c. NAD
   d. FAD

7. The type of energy given off by ATP to make muscles move is known as _____ energy.
   a. Recurrent
   b. Potential
   c. Dormant
   d. Kinetic

8. Examples of inorganic compounds are the following, except
   a. Water
   b. Salts
   c. Acid and bases
   d. proteins

9. Urine has a pH of
   a. 5.0
   b. 6.0
   c. 7.0
   d. 8.0

10. The chemical that can adjust for small changes in pH is
    a. Acids
    b. Bases
    c. Electrolytes
    d. Buffers

Cells, Tissues, and Organ Systems

Chapter 2

LEARNING OBJECTIVES

Upon completion of this chapter, you will be able to:

1. Describe the structure and function of the human cell.
2. Describe the structure and function of the plasma membrane, and the different ways substances move across it.
3. Explain the different processes and phases of cell division and describe why it is essential for human life.
4. Identify and describe the four categories of body tissues and provide characteristics of each.
5. Describe tissue repair.

CHAPTER OUTLINE

Introduction to Cells, Tissues, and Organ Systems

Levels of Organization of the Human Body

- Cellular Level
- The Plasma Membrane
- Cell Division
- Tissue Level

The Organ and Body Systems

INTRODUCTION TO CELLS, TISSUES, AND ORGAN SYSTEMS

This chapter introduces you to the microscopic structures of the human body, which include the cells and tissues. The organ systems will provide you with a general view of all the body systems discussed in this book.

Studying the cells is an important step to understanding how the entire human body works. When cells function normally, the tissues and organs function normally, too. When the cells become injured or damaged, they can affect the tissues and organs, partly or entirely. The different organ systems have defense mechanisms to protect from any agents that can cause injury to a part or the entire organ. Examples of these agents are physical and chemical agents such as trauma, extreme heat, irritants and toxic chemicals; and microorganisms such as bacteria and viruses.

Clinical Notes—Bacterial Infection

When the body has been exposed to pathogenic bacteria (harmful bacterial that can cause a disease), the cellular response to the bacterial invasion is inflammation of the area. The inflammation is a protective defense reaction that allows fluid, blood cells, and nutrients to accumulate in the site of injury. The affected area appears swollen and reddish. There is also weakness or loss of function in the affected area. When the inflammation becomes systemic (spreads out), there will be fever and an increase of white blood cells. Microorganisms can easily invade and cause infection in the body by penetrating the skin or touching contaminated objects, by inhaling bacteria from a cough or sneeze or by ingesting contaminated food.

The clinical example above shows the importance of knowing how the cells and tissues react when injured. When there is a disruption in the normal metabolic state of the cells, tissues, or organs, the body has to respond with specific defense mechanisms.

Clinical Notes—Tissue Repair

When cells or tissues become injured or damaged caused by trauma, irritants, or microorganisms, they undergo the process of healing or repairing by initiating inflammation. The inflammation delivers blood cells (white blood cells), more blood or fluid to the area, and nutrients. The healing or repairing process includes <u>regeneration</u> where new cells replace the injured ones; <u>fibrosis</u> where a scar tissue (collagen fibers) develops to cover the wound. Regeneration occurs when cells divide by mitosis (discussed later in this chapter).

WOUND HEALING

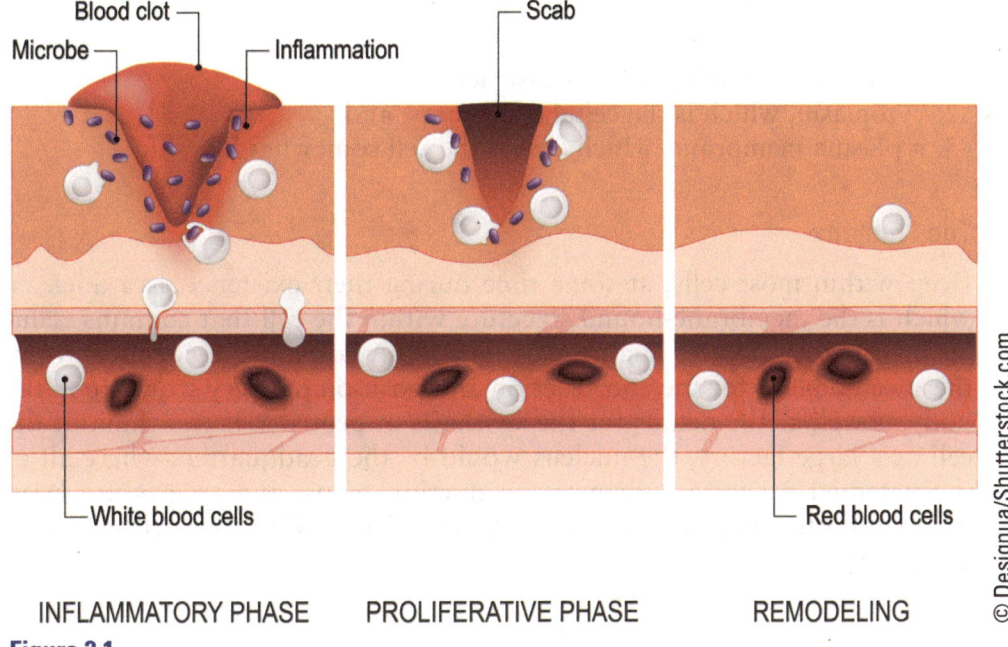

Figure 2.1

LEVELS OF ORGANIZATION OF THE HUMAN BODY

⌞The different levels of complexity of the human body include:

- Biochemical level—refers to the different biochemical reactions that involve catalysts, heat (activation) energy, enzymes, inorganic compounds, and organic compounds.
- Cellular level—refers to the cells, the basic structural unit of any organism, that contain its own minute organs called organelles performing metabolic functions.
- Tissue level—refer to groups of cells that perform specific functions. There are four basic types of tissues—epithelial tissue, muscle tissue, connective tissue, and nervous tissue.
- Organ level—refers to two or more types of tissues that together perform common functions, such as the heart, lungs, stomach, liver, skin, and kidneys.
- Organ system level—refers to the group of organs that comprise a specific system such as the integumentary, muscular, skeletal, circulatory, respiratory, renal, and other systems.
- Organism level—refers to the human organism as a whole, the person or patient.

⌞The biochemical level is presented in Chapter 1.⌟

Cellular Level

All living organisms are composed of cells, the basic unit of life, but not all cells have the same function. The contents and complexity of cells vary depending on the internal structures needed to perform each cell's particular task. All cells contain (Figure 2.2):

1. a nucleus containing cellular instructions;
2. cytoplasm, which is the cellular contents; and
3. a plasma membrane, which covers the cell somewhat like skin.

The Nucleus

> **Nucleus** the membrane-bound structure within the cell that contains cellular instructions (DNA).

Deep within most cells, at some time during their existence, is a nucleus, which is the membrane-bound structure within the cell that contains cellular instructions (DNA) (Figure 2.3). The DNA contains, in chemical form, the specific plans for the construction and position of every structure in the body as well as the details for the synthesis of proteins. If you think of the cell as a large factory, the nucleus would be the headquarters where all of the patented processes, chemical production plans, and recipes are kept. Also within the nucleus is RNA for the transfer of DNA instructions to the cytoplasm.

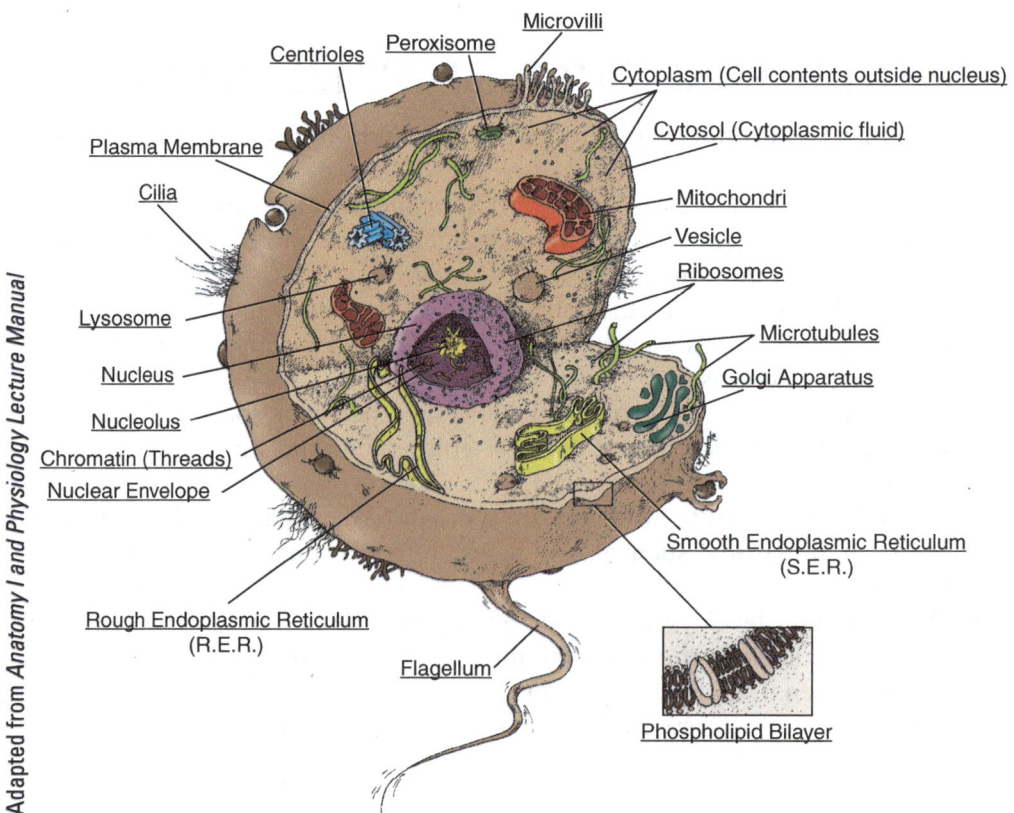

Figure 2.2 The Cell. The contents and complexity of cells vary depending on the internal structures needed to perform each cell's particular task. This prototype of a cell illustrates the most common structures found in most cells, not necessarily human cells.

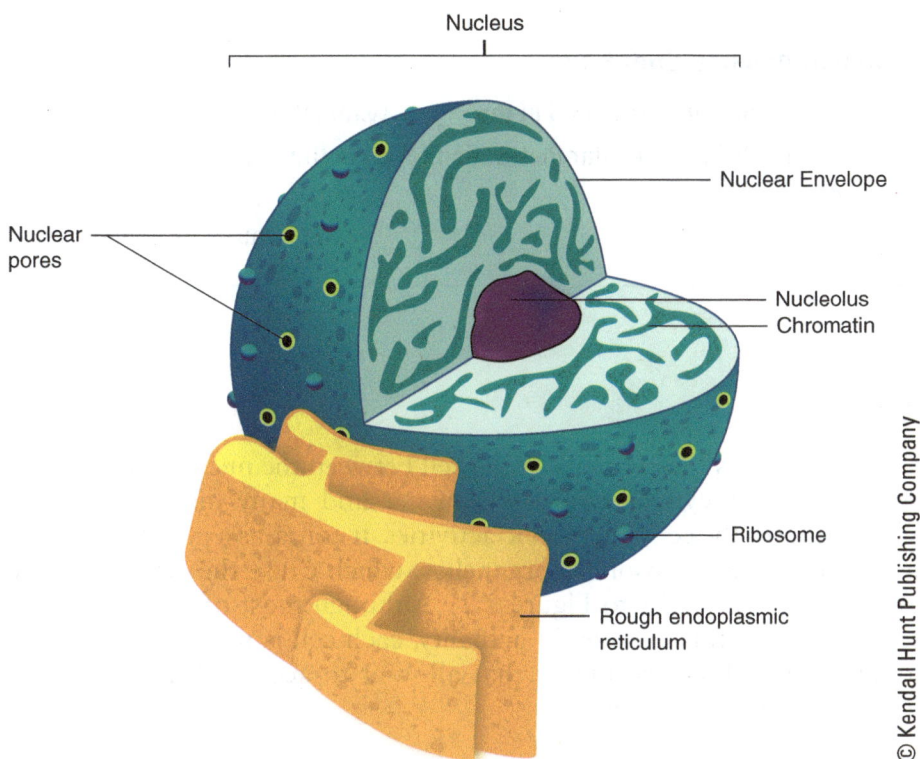

Figure 2.3 The Nucleus. This structure contains instructions for the cell in the form of deoxyribonucleic acid, or DNA.

Nuclear Envelope

Because the DNA is critical for the life of the cell, it is enclosed in a membrane known as the *nuclear envelope* that isolates and protects it from the rest of the cellular contents. This membrane is the same as the one enclosing the cell except that it has pores, or holes, somewhat like a noodle strainer or colander. These pores are large enough to allow easy access to the information inside yet small enough to prevent the DNA from leaking out.

Nucleolus

When we look at the cell under a microscope, the nucleus usually appears to be bluish-purple in color with a couple of dark-staining spots known as *nucleoli*. Normally, substances made by the cell are not produced in the nucleus. However, ribosomes, which are composed of RNA and are used to assist in the production of proteins, are produced in the nucleoli.

Chromatin

The majority of the nucleus contains a dark-staining substance known as chromatin, which is the threadlike network of DNA spread out for easy access to its genes. A gene is a single instruction within the DNA that designates the composition of a chemical to be produced by the body or the location and shape of an organ or structure. Just like a cookbook is filled with recipes one right after another, the genes are connected to each other in the chromatin. When a cell divides, the chromatin condenses into structures designated as *chromosomes*.

> **Ribosomes** structures composed of RNA that assist in the assembling of proteins.

> **Chromatin** during the working phase of a cell, the DNA spreads out for easy access to genes.

> **Gene** a single instruction or plan within the DNA.

> **Comprehension Check-up:**
> 1. Ribosomes are produced in the _____ within the nucleus.
> 2. Each individual cellular instruction found in the DNA is known as a _____.
>
> 1. nucleolus 2. gene

The Cytoplasm

Composition

The cytoplasm is the contents of the cell between the nucleus and the plasma membrane. The cytosol consists of water and many dissolved substances which support the cell's metabolic activities. It contains a gel-like fluid known as *cytosol*. In the cytosol are organelles, which cause the cell to perform its specific function (refer to Figure 2.2). Although most of the organelles will be discussed in this chapter, in actuality, each cell is unique in function and contains only those organelles that allow it to accomplish the processes it contributes to the organism.

Cytoplasm the contents within the cell.

Organelles

Organelles are membrane-enclosed structures within the cell that perform a specific function. The organelle is composed of a membrane containing, in most cases, *enzymes* that catalyze specific chemical reactions within the cell. The organelle's membrane isolates the enzymes inside from the rest of the cell, protecting both the contents of the organelle as well as the cytoplasm. Organelles include:

- Mitochondria
- Endoplasmic reticulum
- Golgi apparatus
- Lysosomes
- Peroxisomes

Organelles membrane-enclosed particles within the cell that perform a specific function.

Mitochondria

Mitochondria are organelles that produce ATP through aerobic metabolism and are referred to as the "powerhouse of the cell." They possess two membranes—the outer membrane gives the organelle a bean shape; the inner membrane is folded. Enzymes for the production of ATP are attached to the inner membrane. The mitochondria contain a substance known as *matrix*. Chemical reactions for the aerobic (oxygen dependent) extraction of energy from nutrients occur in the matrix. Then, the chemical energy is transported to the area between the two membranes where it is transformed into ATP. The more active a cell is, the more mitochondria it contains. A very large quantity of ATP is used by muscle. Up to 40% of the bulk of our heart is composed of mitochondria.

Mitochondria organelles that produce ATP through aerobic metabolism. They are referred to as the *powerhouse of the cell*.

Endoplasmic Reticulum

When a cell produces a product, it uses the endoplasmic reticulum organelle to aid in the synthesis and/or modification of proteins and other substances.

Endoplasmic reticulum organelles that produce cellular products.

There are two forms: rough endoplasmic reticulum (RER) and smooth endoplasmic reticulum (SER).

Rough Endoplasmic Reticulum

Rough endoplasmic reticulum (RER) is involved with the production of proteins. It receives its name from the ribosomes attached to the outside. Ribosomes are the sites where proteins are assembled. Remember that proteins are chains of specific amino acids that need to be folded into a specific shape in order to function. For example, enzymes, which are composed of protein, cannot catalyze reactions unless they are folded. As the protein is made, it is fed into the endoplasmic reticulum where it is made into an exact shape. By attaching the ribosomes to the outside of the rough endoplasmic reticulum, the new protein has rapid access to the folding mechanism. For example, cells in the pancreas, which produce digestive enzymes, contain large amounts of rough endoplasmic reticulum.

Smooth Endoplasmic Reticulum

Smooth endoplasmic reticulum (SER) does not have ribosomes attached to the outer surface and therefore produces nonprotein substances such as lipids. For example, the testes, which produce the steroid hormone, testosterone, contain large amounts of smooth endoplasmic reticulum. The absence of the ribosomes causes the surface of the SER to appear smooth on micrographs. In some cases, the smooth endoplasmic reticulum is used for storage. This occurs primarily in skeletal muscle where the endoplasmic reticulum stores calcium ions.

Golgi Apparatus

Once a substance has been produced by the endoplasmic reticulum, it is not simply released into the cytoplasm. Cellular products need to be placed in containers just like a factory would place its items in boxes, jars, bottles, or cans. This is the task of the Golgi apparatus, an organelle that condenses cellular products and encloses them in a membrane similar to that enclosing the cell. This isolates the product from the rest of the cytoplasm. Sometimes those products are toxic or caustic and would cause great damage within the cytoplasm if not contained.

> **Golgi apparatus** organelles that concentrate cellular products and enclose them in a membrane like that enclosing the cell.

Lysosomes

The prefix "lys-" means to break apart. Lysosomes are organelles that contain *enzymes* to digest or destroy substances within the cell. Some cells such as phagocytes, eat other dead cells, foreign organisms, and particles like dust or pollen that enter our bodies. When these cells take that material into the cell, lysosomes fuse with the undesired substance, then release their enzymes to destroy those substances. This allows the cell to then use the resulting byproducts. When the cell dies, the lysosomes break, causing the cell to self-destruct.

> **Lysosomes** organelles containing enzymes to digest or destroy substances within the cell.

Peroxisomes

Peroxisomes are organelles containing enzymes that convert toxic chemicals into peroxide or that insert oxygen into a molecule to change the compound

> **Peroxisomes** organelles containing enzymes that convert toxins to peroxide or insert oxygen into a molecule to change it to a less harmful substance.

into a less-harmful substance. There are other enzymes in the peroxisome that convert toxic peroxide into harmless oxygen and water.

For example, occasionally a person consumes or produces substances that are toxic to the body. Even some of the foods we eat, such as lima beans or cassava, may contain toxic substances that must be detoxified. Certain drugs may also contain harmful substances or may be converted by our body into toxic compounds that must be rendered harmless. Cells in the liver, active in the detoxification of potentially harmful substances in the blood, contain peroxisomes.

Centrioles

Before a cell divides, it makes a copy of its DNA and forms the original and copy into X-shaped nuclear structures known as *chromosomes*. As the cell divides, spindle fibers appear to pull the original set of DNA toward one cell and the copy toward the other. Spindle fibers are essential for the separation of chromosomes during cell division. The cell, however, spends relatively little time dividing and most of its time working so the spindle fibers are not needed a majority of the time. Centrioles are organelles that store spindle fibers when they are not in use.

> **Centrioles** organelles that store spindle fibers, essential for the separation of DNA during cell division.

Cytoskeleton

Just like the human skeleton provides shape, structure, protection, and attachment for various organs in the body, so does the cytoskeleton structure provide the same functions to the cell. There are several components to the cytoskeleton; however, in this section only two will be considered: microtubules and microfilaments.

> **Cytoskeleton** structures within a cell that provide shape, structure, protection, and attachment.

Microtubules

Microtubules are the largest structure of the cytoskeleton. They can assist in maintaining a general cellular shape and can be spindle fibers for movement of chromosomes during cell division. Some cells possess microtubules contained within stiff, hairlike projections known as *microvilli*, which increase surface contact with substances outside the cell. Other hairlike projections containing microtubules, known as *cilia*, create a constant motion to move material within an organ. Male sperm have a tail, known as a *flagellum*, composed of microtubules that whip back and forth to cause the cell to move as it delivers the father's share of DNA to the awaiting egg.

Microfilaments

Microfilaments are the smallest members of the cytoskeleton. They are the components of muscle that cause it to shorten. When you contract muscles in your upper arm to bend your elbow, thousands of muscle fibers that make up that muscle will shorten. Each muscle fiber contains billions of microfilaments that slide together to cause the muscle fiber to decrease in length.

The Plasma Membrane

The plasma membrane (Figure 2.4) is the outer covering of the cell, or the "skin." It is able to keep substances either in or out of the cell just as our skin provides the same function for our body. Plasma is a substance or material

> **Plasma membrane** the outer covering of the cell; the cell's "skin."

> **Comprehension Check-up:**
>
> 1. Proteins are produced and folded by the _____.
> 2. The organelles that produce large amounts of ATP through aerobic metabolism and are known as "powerhouses of the cell" are _____.
> 3. Cells that take in foreign organisms contain organelles filled with enzymes to destroy or digest that organism. These organelles are known as _____.
>
> 1. Rough endoplasmic reticulum 2. Mitochondria 3. Lysosomes

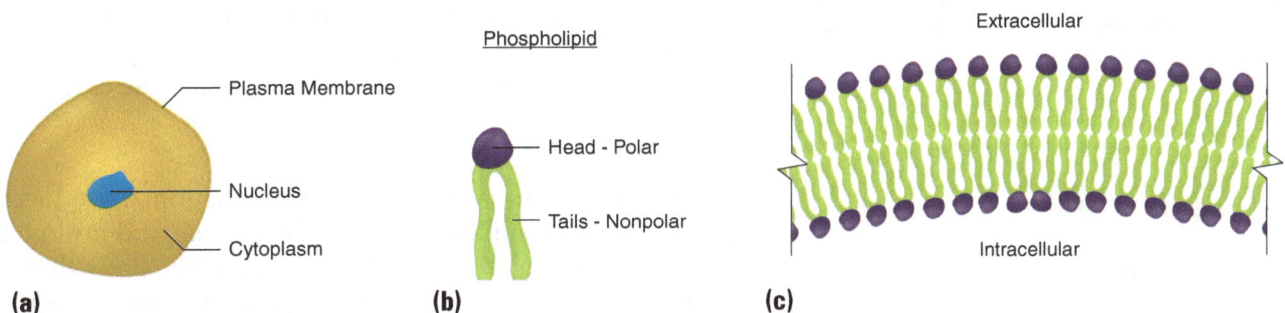

Figure 2.4 The Plasma Membrane. (a) The plasma membrane enclosing the cell; (b) a phospholipid—the major component of plasma membranes; (c) the double phospholipid membrane.

that is in constant change, as is the cell membrane. It is possible for the cell to produce special proteins inside and insert them through the plasma membrane as needed to increase transport of ions across the membrane. The plasma membrane is selectively permeable. The membrane is constantly changing its shape as well as its composition.

Phospholipids

The plasma membrane is made up primarily of phospholipids (Figure 2.4b). These molecules have a head that is polar—that is, it has charges on it. Dangling from the head are two lipid tails, which are nonpolar. Phospholipids provide the link between polar and nonpolar substances that allows cells to form and make life possible. The outside of the cell is composed of a layer of phospholipids with their polar heads sticking together like magnets and with their tails facing inward. There is another layer of phospholipids lining the inside of the cell. The second layer of phospholipid heads faces the inside of the cell; the tails face the other layer of phospholipid tails (Figure 2.4c). This double layer of lipids prevents anything polar or charged, such as ions, from passing through the membrane and keeps the outer fluid and inner contents separated.

Channels

There are times when it is necessary to allow charged molecules or ions to pass through the cell membrane. In order to accomplish this task, the cell makes tube-like *channels* out of protein, providing access for ions to move between

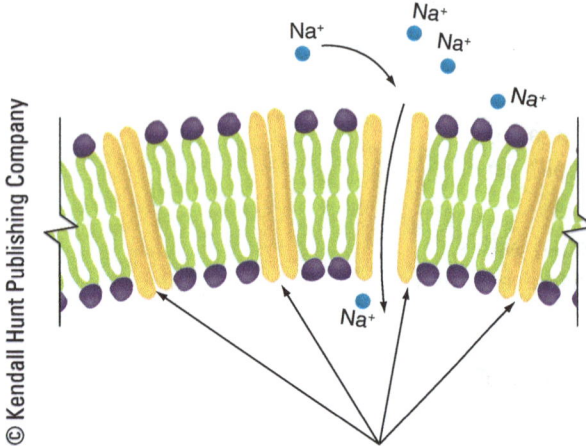

Figure 2.5 Ion Channels. These channels allow direct access of ions to either side of the cell membrane.

the intracellular and extracellular fluids as appropriate (Figure 2.5). Because the channels are across both layers of phospholipids, they can allow direct access of ions to either side of the cell membrane. Channels are very specific about the ions they allow through. There are, for example, sodium channels that only allow sodium ions and water to pass through. Calcium channels only allow access to calcium ions and water. Sodium ions cannot pass through a calcium channel and vice versa. Channels are not open all of the time. Instead, they have gates on them that open temporarily to allow movement of their specific ions. The gate may be on the inside or the outside of the channel depending on where the highest concentration of ions normally occurs. Channels can allow movement in either direction through the membrane. Cells actually use the movement of ions to relay chemical signals such as nerve impulses to the cell.

Receptor Sites

When a cell is ready to perform its function, it produces another protein known as a *receptor site* and inserts it through the plasma membrane to alert the body that it is ready. The section of the receptor site on the outside of the cell has a very specific shape to which a chemical message with the appropriate shape can attach (Figure 2.6). The receptor site is like a lock into which only one specific key, or chemical message, can fi t. When the body is ready for the cell to respond, it produces a chemical message shaped to fi t on the receptor site. When the receptor site is occupied, an internal change occurs within the cell, causing it to carry out a specific function. An example of a function would be when growth hormone binds to its receptor and causes the miotic rate of the cell to increase. There may be many chemical messages circulating throughout the body at any given moment, but only those cells possessing the appropriate receptor site will respond.

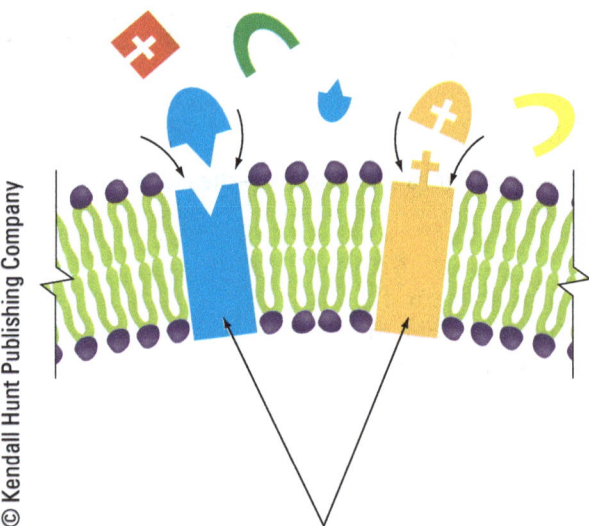

Figure 2.6 Receptor Sites. These sites allow chemical messages with the appropriate shape to attach to the cell.

Transport Across the Plasma Membrane

There are two general methods by which substances can move across the plasma membrane when the opportunity becomes available: passive and

---

Comprehension Check-up:

1. The plasma membrane is primarily constructed of _____.
2. A change in the activity of a cell can occur if a chemical message attaches to a _____.

1. phospholipids 2. receptor site

active transport. Active transport requires energy while passive transport does not. Both active and passive processes play critical roles in the normal functioning of the body.

Passive Transport

Passive transport results when substances move from an area of higher concentration to one of lower concentration without the use of ATP energy causing the substance to spread to areas containing less. There are two types of passive transport that are very similar: diffusion and osmosis.

Diffusion

Diffusion is the passive movement of a substance from an area of high concentration into an area of lower concentration. An example of diffusion would be to place a drop of food coloring into a glass of water (Figure 2.7). The drop of coloring entering the water contains a high concentration of dye. It moves outward into areas of lower concentration and the color spreads. Diffusion continues, without stirring, until the entire glass is filled with dyed liquid.

In our bodies, for example, there are many potassium ions on the inside of a cell and relatively few on the outside. When the gate on a potassium channel opens, potassium ions can flow through the open channel from its highest concentration inside the cell toward the lower level outside. It is important to understand that diffusion through the membrane can go in either direction but it will always move passively from areas of high concentration to those of lower concentration (Figure 2.8).

> **Diffusion** passive movement of a substance from an area of higher concentration into an area of lower concentration.

Osmosis

Osmosis is the diffusion of water. When a drop of food coloring is placed in a glass of water, the dye diffuses into the water, as was discussed above. But if diffusion is considered from the perspective of water, outside the drop of dye is a high concentration of water. In the center of the food coloring is a low concentration of water. Water diffuses from its higher concentration into its low concentration in the center of the dye. It may also be viewed as water being drawn into an area concentrated by some other substance in an

> **Osmosis** the diffusion of water. It may also be viewed as water being drawn into an area concentrated by some other substance in an attempt to dilute that other substance.

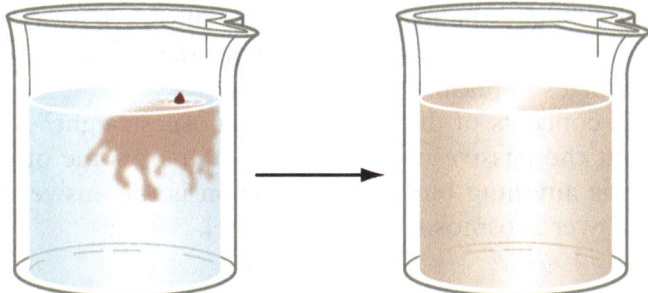

Figure 2.7 Diffusion. Molecules in a solution are in constant motion. They collide into other particles, causing the particles in the solution to move from its area of higher concentration into an area of lower concentration until, eventually, the distribution of that substance is even throughout.

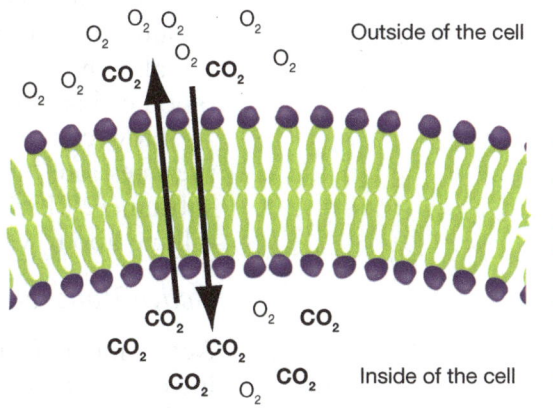

Figure 2.8 Diffusion. Substances flow through a plasma membrane from areas of higher concentration into those of lower concentration.

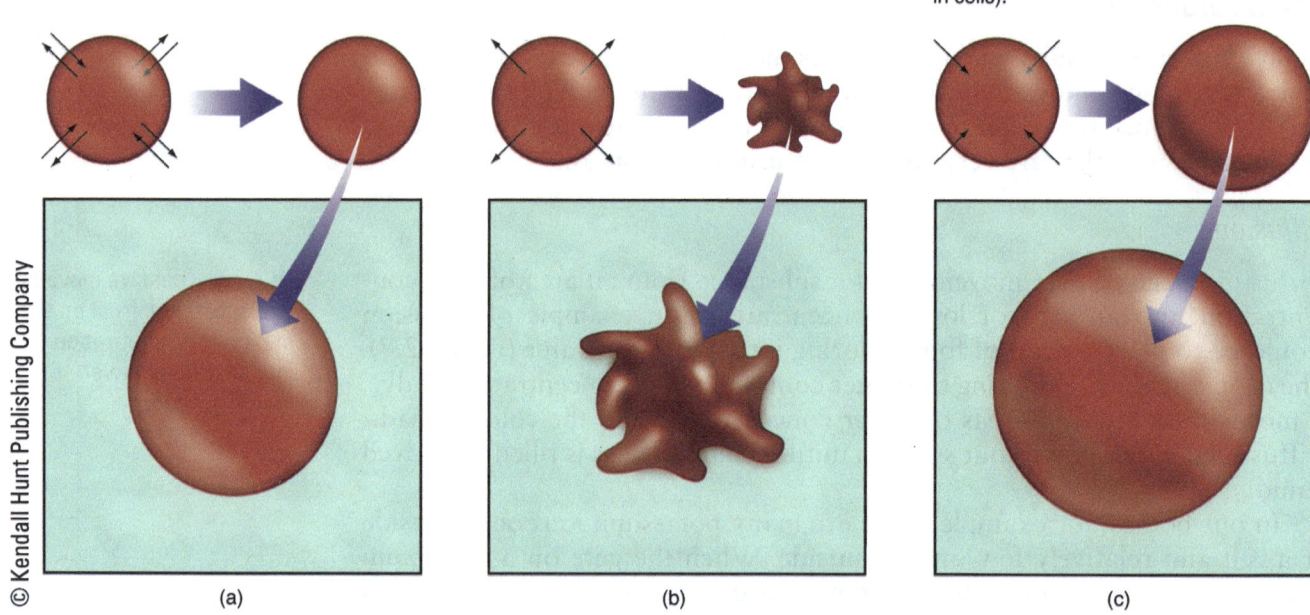

Figure 2.9 Osmosis and Red Blood Cells. (a) When the concentration of the liquid outside a red blood cell is the same as inside, water is drawn in at the same rate it is drawn out and the red blood cell size does not change. (b) If the concentration of the liquid outside a red blood cell is greater than inside (hypertonic), water is drawn out and the red blood cell shrinks. (c) If the concentration of the liquid outside a red blood cell is less than inside (hypotonic), water is drawn into the red blood cell causing it to swell.

attempt to dilute that other substance (Figure 2.9). Think of a flat sponge soaking up water by osmosis.

Our bodies contain a preponderance of water with dissolved substances in it. This water is constantly moving through plasma membranes from an area of higher concentration to an area of lower concentration in order to maintain homeostasis. Under active transport, first-line active transport requires energy in the form of cellular ATP while passive transport moves molecules by means of their own kinetic molecular energy.

Since most cells in the body are surrounded by fluid, if the concentration of substances in that fluid changes, the osmosis of water into or out of the cell also changes. This is illustrated in Figure 2.9 when considering how a red blood cell is affected by changes in plasma concentration.

How can you keep the concepts of diffusion and osmosis straight? Ask the question; "Is this about the passive movement of water or some other substance?" If the answer is anything but water, diffusion is the answer. If talking about water, the answer is osmosis.

Active Transport

Active movement requires energy. If a substance is moved from an area of lower concentration into an area of higher concentration, it needs to be pushed by means of active transport. Pushing the car uphill is active transport because of the energy required.

There are two types of active transport that are very important in cellular function: ion pumps and large-particle transport.

Ion Pumps

Sometimes it is necessary to move a low concentration of a specific ion on one side of the plasma membrane into a higher concentration on the other side. Our plasma membranes contain proteins known as *ion pumps,* which require energy in the form of ATP to push these substances through the cell membrane into a higher concentration. An example of a commonly found ion pump is the sodium-potassium pump which pumps sodium ions out of the cell and potassium ions into the cell (Figure 2.10). Ion pumps are exceedingly important in the plasma membrane to maintain the concentrations of various substances within the cell at their homeostatic level.

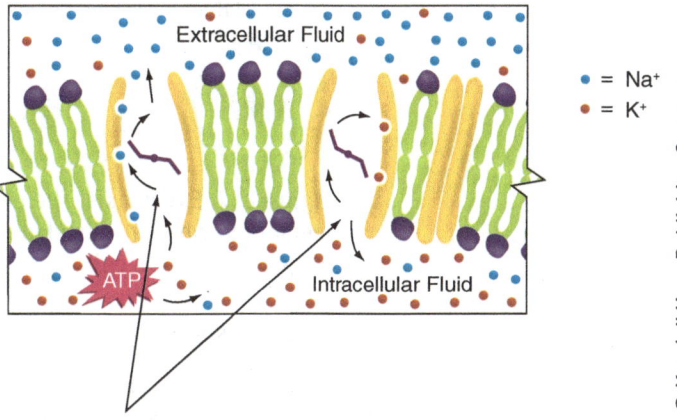

Figure 2.10 Ion Pumps. These pumps move a low concentration of a specific ion on one side of the plasma membrane into a higher concentration on the other side. In order for ions to be pumped through the channel against a concentration gradient, ATP releases energy from ADP.

Large-Particle Transport

Large-particle transport is the process by which substances too large to pass through channels can move through the plasma membrane. While this requires active transport, the processes involved are not yet well understood. There are two forms of large-particle transport: endocytosis and exocytosis.

Endocytosis

The inward movement of a large particle is known as *endocytosis*. The cell surrounds and engulfs a large particle with its plasma membrane (Figure 2.11). Two very important factors occur in the process:

- At no time does the plasma membrane open. Holes in the plasma membrane would allow excessive water to be drawn into the cell by osmosis, causing the membrane to rupture. The production of holes in foreign cells is a process used by our body's defense system.
- The particle taken in is surrounded by the same phospholipid layers as the plasma membrane. Once inside, it is isolated from the contents of the cell. The particle taken in may be hazardous to the cell or it may produce toxins or other substances that are harmful. By coating those large particles with a double phospholipid membrane, those potentially dangerous substances do not have access to the rest of the cell.

There is a distinction made between the inward movement of a liquid or a solid. Inward movement of a liquid is known as *pinocytosis;* inward movement of solids such as dirt, dust, dead cells, and foreign organisms is *phagocytosis*.

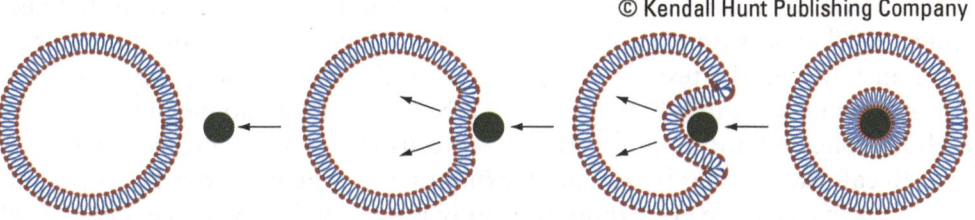

Figure 2.11 Endocytosis. The cell surrounds and engulfs a large foreign particle with its plasma membrane.

© Kendall Hunt Publishing Company

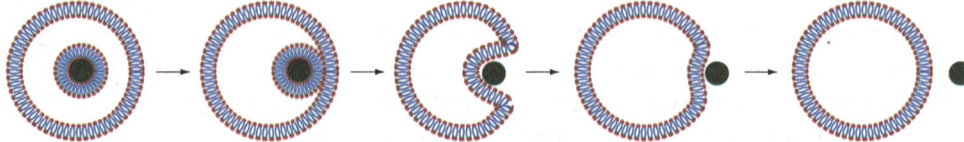

Figure 2.12 Exocytosis. The cell then moves the enclosed product to the plasma membrane where its envelope is incorporated into the cell's membrane as the substance is ejected from the cell.

Exocytosis

The outward movement of a large particle is known as *exocytosis* (Figure 2.12). The endoplasmic reticulum makes a product and sends it to the Golgi apparatus to be coated with plasma membrane material. The cell then moves the enclosed product to the plasma membrane where its envelope is incorporated into the cell's membrane as the substance is ejected from the cell. Again, the plasma membrane never opens. How fast does either of these processes occur? It happens every time a person moves. It is the method by which rapid thoughts are transferred. Cells talk to each other or stimulate muscle contraction by secreting chemical messages by exocytosis.

> **Comprehension Check-up:**
> 1. The passive movement of a substance from an area of higher concentration to an area of lower concentration is known as _____.
> 2. When energy is used to push sodium ions from a low concentration inside the cell into a high concentration outside the cell, what method is used to do so?
> 3. The outward movement of a large particle is called _____.
>
> 1. diffusion
> 2. Ion pumps which are forms of active transport
> 3. exocytosis

Cell Division

When it is time for cells to divide, the single cell referred to as the *parent cell* splits to create new cells. There are two types of cell division: mitosis and meiosis.

Mitosis

Cells spend most of their life working during what is known as *interphase*, the interval of growth and other normal metabolic cellular activities between phases. When it is time for a cell to divide, however, all but reproductive cells (sperm found in the testes and ova found in the ovaries) divide by a process known as mitosis, in which all of the DNA (46 chromosomes) in the parent cell is replicated and each newly formed cells receives 46 chromosomes.

Recall from earlier in the chapter that there are genes in our DNA that are instructions to make everything our body is designed to produce. During cell mitosis, those genes are formed into distinct groups known as *chromosomes*.

> **Mitosis** the process of cell division accomplished by all but reproductive cells during which all of the DNA (46 chromosomes) in the parent cell is replicated and each newly formed cell receives 46 chromosomes.

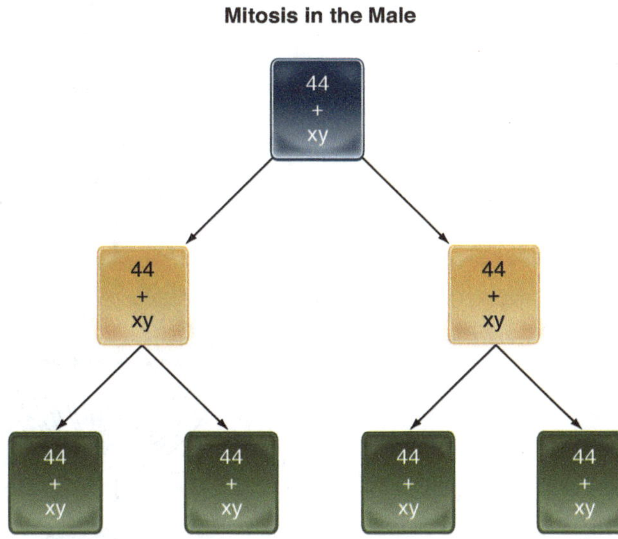

Mitosis in the Male

Figure 2.13 Mitosis in the Male. All of the DNA (46 chromosomes) in the parent cell is replicated and each newly formed cell receives 44 autosomes plus 2 sex chromosomes × 4, totaling 46 chromosomes.

It is like arranging all of the recipes in the kitchen into cookbooks. There are 23 pairs of chromosomes in all, with one of each pair coming from each parent. Before the cell divides, it performs *DNA replication* of all 46 chromosomes and attaches the original and the copy together. The object of mitosis is to separate the copy from its original and form the single cell into two identical cells, each containing the same 46 chromosomes (Figure 2.13).

To keep track of the process of mitosis, scientists have defined four phases to signify what the cell is doing at that time (Figure 2.14):

1. Prophase
2. Metaphase
3. Anaphase
4. Telophase

The cell does not spend much time in any phase. The term is simply a method for describing where the cell is in the process, similar to saying a child is an infant, and then a toddler, preschooler, elementary student, middle schooler, high school student, and a graduate. It is a way of identifying a stage of life. The same is true of the stages of cell division.

Prophase

During the first phase, *prophase*, the cell is forming the DNA that had been in the form of chromatin into distinct chromosomes. During this phase, the chromosomes duplicate. It will not be using the DNA during this process, so the nuclear membrane is not needed and is disassembled and disappears.

Metaphase

During *metaphase*, the second stage of mitosis, the chromosomes are moved into a straight line across the center of the cell. Spindle fibers, which were stored in the centrioles, now appear and attach on both sides of the line of chromosomes.

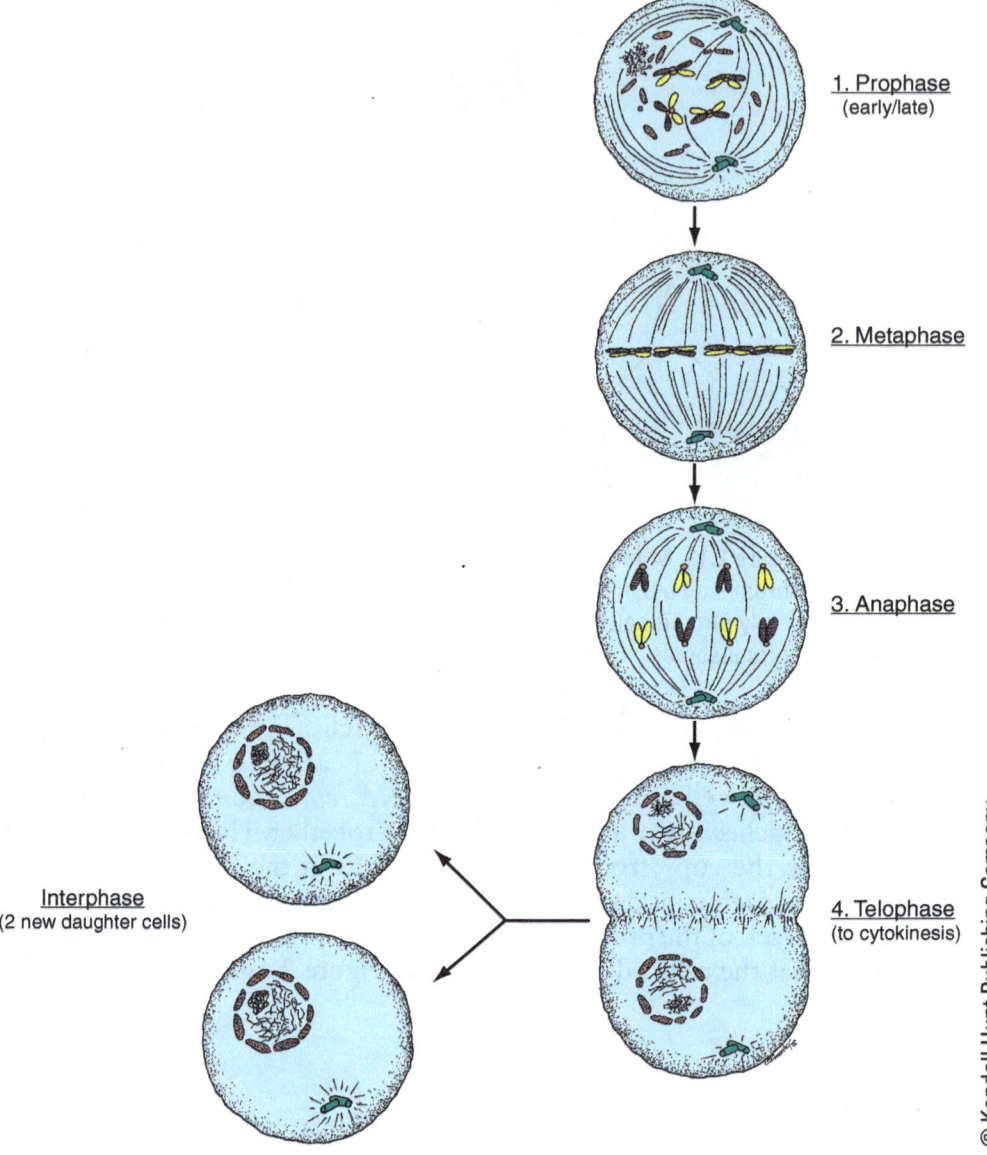

Figure 2.14 Mitosis and Interphase.

Anaphase

It then appears that the spindle fibers tug on the chromosomes separating the original and the copy, pulling them in opposite directions. As these identical sets of chromosomes begin to move apart, it signals the start of the third phase of mitosis known as *anaphase*.

Telophase

As each set of chromosomes continues to move apart, the plasma membrane eventually begins to pinch in between the two sets as the actual formation of separate cells is about to be accomplished. This is the final stage of mitosis known as *telophase*. The process of separating into two identical cells is known as *cytokinesis*.

Meiosis

Reproductive cells, which are sperm and eggs, need half the original number of chromosomes because the child receives 23 chromosomes from each parent. These cells divide by a process known as meiosis, in which each newly formed cell contains only half the original number of chromosomes (23). It is similar to mitosis in that it has the same four phases, but they do not have the same results, and reproductive cells cycle through two rounds of cell division in meiosis (Figure 2.15). The result of the first division in meiosis, meiosis I, is that the original and the copy of the chromosomes do not separate. Instead, half of the sets of chromosomes with their copies go to one cell and half to the other, resulting in 23 copied sets of chromosomes in each cell (Figure 2.15). Also, during the first stage of meiosis, genes are traded between like chromosomes by a process known as *crossing over*. A gene from the father's chromosome trades places with the same gene from the mother. As a result, the chromosome passed on to the child will be slightly different and the two sides of the family will become genetically blended into a unique individual with a mix of both sets of parental traits. During the second division of meiosis, meiosis II, the original and the copy separate as in mitosis, resulting in

> **Meiosis** cell division performed only by reproductive cells in which each newly formed cell contains only half the original number of chromosomes (23).

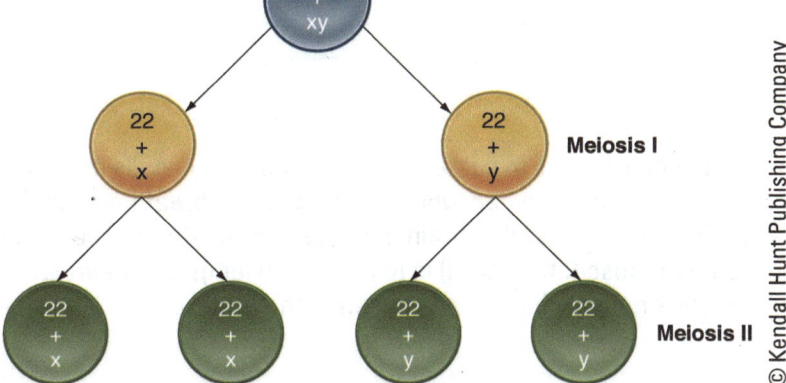

Figure 2.15 Meiosis in the Male. Each newly formed cell contains only half the original number of chromosomes (23).

Stem Cells

Many organs of the body have cells, known as stem cells, with the ability to mature into whatever specific type of cell is needed by that organ system. Blood stem cells can become whatever kind of blood cell is needed most at the time. Heart stem cells can apparently replace damaged heart cells when directed to do so. An interesting factor about brain stem cells is that they are able to become supporting non-nerve brain cells that are needed at the time but are apparently inhibited from becoming nerve cells in the brain. Roughly half of the volume of the nervous system is supporting cells designed to protect and support the brain nerve cells. Ongoing studies are underway to discover how to stimulate stem cells to form brain nerve cells. At the same time, it is important to determine the side effects of stimulating stem cells to become new nerve cells.

four cells from the original one but with each containing half the original number of chromosomes. Each is slightly different due to the crossing over that occurred during the first division.

Crossing over explains why we do not look exactly like our brothers and sisters unless we are identical twins. It also creates variation in humans so we are not clones of each other, making life considerably more interesting. Inbreeding decreases the differences in genes and increases the potential for defects to occur. Males produce sperm and females ova or eggs. Sperm and ova are collectively known as *gametes*. When a sperm is able to insert its chromosomes into an ovum to make a complete set of 46 chromosomes (23 pairs), the process is called *fertilization*.

> **Comprehension Check-up:**
> 1. What phase of mitosis is represented by a straight line of chromosomes across the center of the cell?
> 2. List three differences between mitosis and meiosis.
>
> 1. Metaphase
> 2. (1) Mitosis occurs in one round while meiosis requires two rounds. (2) Mitosis results in the daughter cells having 23 chromosomes. (3) During meiosis I, crossing over of genes occurs between chromosomes. This does not occur during mitosis.

> **Clinical Notes—Oxygen and Brain Cells**
>
> One of the requirements to sustain life is oxygen. Brain cells begin to die after 4-6 minutes without oxygen when a patient stops breathing and the heart stops beating. This can lead to brain damage or brain death. The goal of cardiopulmonary resuscitation (CPR) is to keep both lungs and heart working, in order to move or supply blood and oxygen to the brain.

Tissue Level

There are four general types of tissues in the human body: epithelial tissue, connective tissue, muscle tissue, and nervous tissue.

Epithelial Tissue

Epithelium or epithelial tissue is a protective layer of cells that covers the body and each organ. It is found on the outer layer of the skin and internal linings of some hollow structures such as the respiratory passages, blood vessels, heart, stomach, and urinary bladder. The functions of the epithelium include protection of the underlying tissue, absorption of nutrients from digestion, and secretion of substances into the glands such as mucous glands. It also acts as a protective barrier on the skin to prevent loss of water from the body. The epithelium has no blood vessels. It has a basement membrane composed of collagen, reticular fibers, and protein molecules. There are several types of epithelial tissues:

- Simple squamous epithelium—composed of a single layer of thin, flat, scale-like cells found in the alveoli, linings of blood vessels, pleura, pericardium, kidneys, and lymph vessels.

- Simple cuboidal epithelium—composed of a single layer cube-like cells found in kidney tubules and glands. They carry sweat oil to the skin surface and hair, and transport enzymes and secretions to the digestive tract.
- Simple columnar epithelium—composed of a single layer of tall, thin cells found in the linings of the stomach and small intestines. Its function is absorption of digested foods.
- Stratified epithelium—composed of more than one layer of cells with four different types:
    1. stratified squamous epithelium—found in outer layer of skin, lining of the mouth, throat, upper esophagus, vagina, and anus
    2. stratified cuboidal epithelium—found in sweat glands, urethra, ovary, and salivary glands
    3. stratified columnar epithelium—found in male urethra, larynx, and mammary glands. They are rare
    4. transitional epithelium—many layers of cells that appear to change shape into squamous or cuboidal when stretched; found in urinary bladder

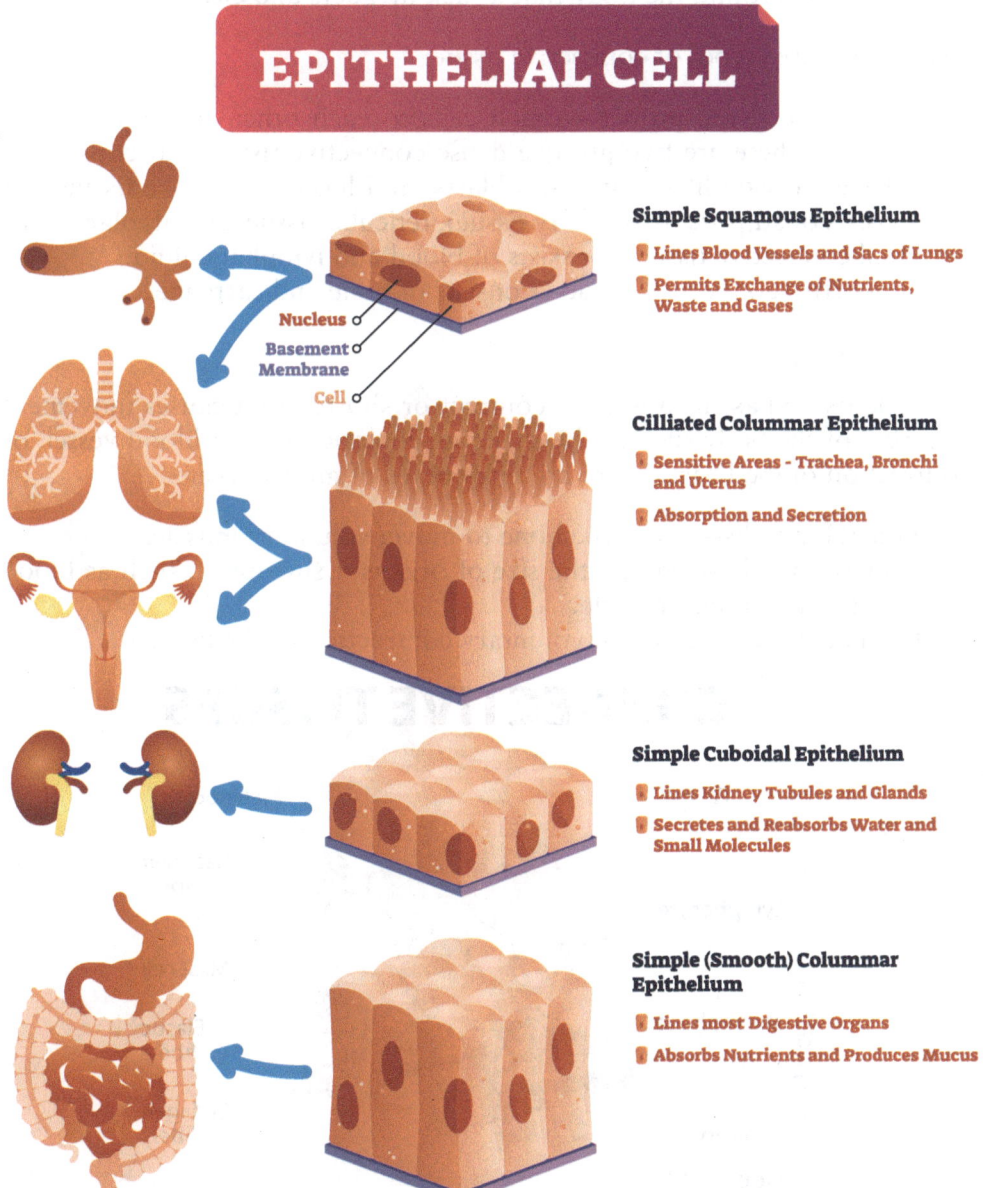

Figure 2.16

Connective Tissue

Connective tissue consists of fibers and cells that serve as the framework of most organs throughout the body. It provides protection, insulation, and energy reserve; defends or repairs the body and transport substances within the body. There are three basic components of connective tissue:

1. Extracellular matrix—surrounds and holds the cell and tissue together, and consists of fibers, proteins, carbohydrates, and water with dissolved nutrients.
2. Connective tissue fibers—composed of collagen fibers, elastic fibers, and reticular fibers; collagen fibers connect bones to each other as ligaments; elastic fibers stretch and recoil such as in large arteries, ligaments, and tendons; reticular fibers are found in lymph nodes and spleen.
3. Connective tissue cells—macrophages, fibroblasts, adipocytes, wandering cells; macrophages perform phagocytosis and are the first line of defense against foreign organisms; wandering cells are white blood cells that travel around the body that attack invaders and clean debris.

There are three types of connective tissue:

1. Connective tissue proper—attach cells to each other and hold organs in place. There are two groups: dense connective tissue that consists of collagen, elastic fibers, and fibroblasts; and loose connective tissue that consists of adipose tissue (fatty tissue), reticular tissue, and areolar tissue.
2. Fluid connective tissue—consists of blood and lymphatic fluid.
3. Supportive connective tissue—consists of bone and cartilage.

Muscle Tissue

Muscle tissue has the ability to contract or shorten that causes movement of a part of the body such as the skeleton, constriction of a blood vessel, or contraction of the heart. There are three types of muscle tissue:

1. Skeletal muscle—causes the bone to move; voluntary muscle.
2. Smooth muscle—changes the size of organs it surrounds, such as blood vessels; involuntary muscle.
3. Cardiac muscle—causes involuntary contraction of the heart.

CONNECTIVE TISSUES

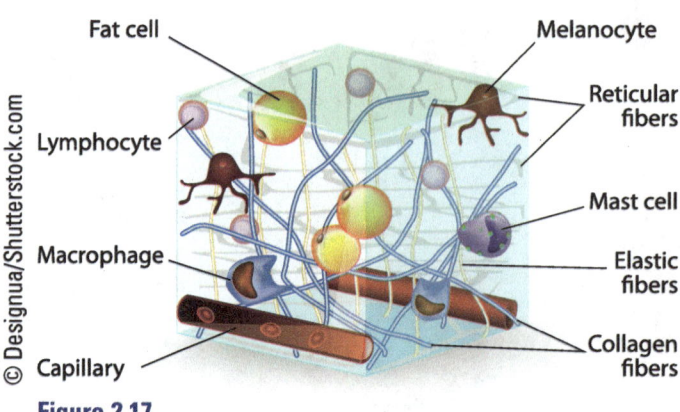

Figure 2.17

Types of Muscle Tissue

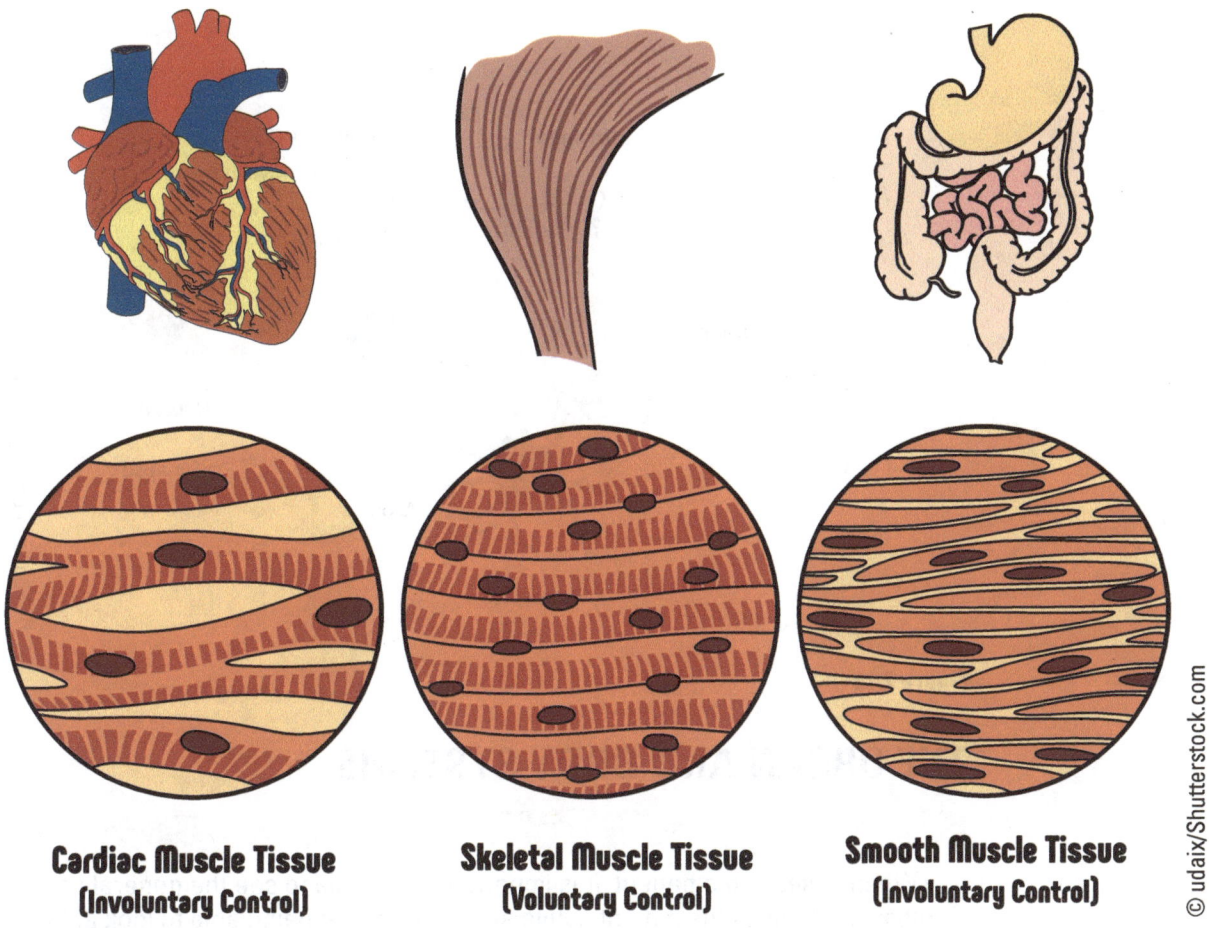

Figure 2.18

Nervous Tissue

Nervous tissue is composed of nerve cells called neurons and glial cells, which are found in the brain, spinal cord, and nerves. Neurons possess irritability or excitability. When stimulated, they create an electrical impulse that travels through the nerve cell and transmits to another neuron by conduction. The skeletal muscle can only move when stimulated by a neuron. The glial cells provide protection and defense of neurons and also increase their ability to conduct electrical impulses. There are three kinds of neurons:

1. Motor neurons—innervate muscle to cause contraction resulting in movement or stabilization of the body.
2. Sensory neurons—transmit information received from the senses (touch, temperature, vision, hearing, taste, etc.) to the spinal cord and brain for analysis and response.
3. Interneurons (association neurons)—integrate the information received from the sensory neurons and coordinate the information to the motor neurons. Most neurons are interneurons.

NEURONS AND NEUROGLIAL CELLS

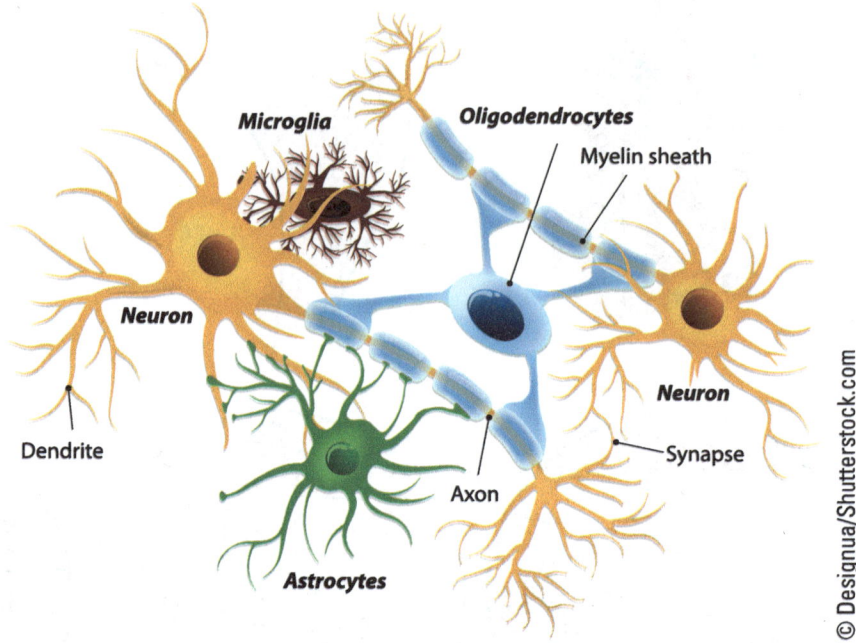

Figure 2.19

THE ORGAN AND BODY SYSTEMS

> **Clinical Notes—General Assessment of a Patient**
>
> When assessing a patient, it is important to be able to see the general condition or status of the patient, while at the same time being able to look at the smaller details. This is how you will assess patients in the clinical setting—viewing and observing the patient as a whole just as you enter the patient's room. The moment you meet and communicate with your patient you may be able to observe some clues or hints that the patient is experiencing or telling you. It is also important that you review the patient's chart or medical record so you will be prepared before you see the patient.

The different body systems for clinical assessment include:

- The integumentary system—skin and underlying connective tissues
- The skeletal system—bones and joints
- The muscular system—skeletal muscle attached to bones
- The nervous system—brain, spinal cord, and nerves
- The endocrine system—hormones and glands
- The cardiovascular system—blood, blood vessels, and heart
- The respiratory system—upper and lower respiratory structures, and lungs
- The lymphatic and immune system—lymph, lymphatic organs, and lymphocytes
- The digestive and nutrition system—oral cavity, pharynx, esophagus, stomach, small and large intestines, liver, pancreas
- The renal or urinary system—kidneys, ureters, and urinary bladder
- The reproductive system—male and female reproductive structures

Clinical Scenario—Pre-Op Patient Assessment (Assessing a Patient before Surgery)

A patient has been scheduled for surgery. A general patient assessment is needed to ensure that the patient is ready and would not have any complications during and after the surgery. A routine patient assessment and preparation would include the following:

General Assessment of the Patient

- Integumentary—to check for skin color, local skin infection, and skin diseases
- Cardiovascular system—to check for vital signs that include heart rate, respiratory rate, body temperature, and blood pressure; heart rhythm and oxygen saturation
- Respiratory system—to check the airways, breathing, and abnormal breath sounds; to check lung function, such as lung volumes and expiratory flows
- Gastrointestinal system—to check for abdominal pain, discomfort, distention, bowel habits, appetite, intolerance for food, nausea, and vomiting
- Musculoskeletal system—to check for skeletal malformations such as kyphoscoliosis, weakness in limbs, or other restrictive abnormalities that could affect the patient's ability to breath after surgery
- Nervous system—to check for alertness, orientation to time and place
- Extremities—to check for problems in the upper and lower extremities

Laboratory Investigations

- Complete Blood Count (CBC)—to check for hemoglobin level to ensure that anemia is not present, to check the white blood cells for any bacterial or viral infection, and to check for any cellular abnormalities
- Electrolytes—to check for electrolyte imbalance that includes sodium (Na), potassium (K), chloride (Cl), and calcium (Ca); abnormal levels of these electrolytes can cause cardiac arrhythmias during surgery that can lead to cardiac arrest
- Pulmonary function testing—to check for lung function abnormalities such as obstructive, restrictive disorder, vital capacity for adequate cough so patient can expectorate secretions to prevent complications such as pneumonia after surgery
- Blood gases—to check for adequate oxygenation, and normal level of carbon dioxide
- Urinalysis—to check for abnormalities in the urine sample, such as red blood cells, white blood cells, presence of protein and glucose, and other cellular structures
- Clotting and platelet function—to check for possibility of bleeding or hemorrhage during and after surgery
- Nutrition—to check for any underlying malnutrition that may affect the patient's ability to heal after surgery

(*continued*)

Blood Typing and Blood Products

- Blood typing—to check for the patient's blood type—whether it's A, B, AB, or O
- Blood products—to prepare blood products such as whole blood for possible blood transfusion in case there would be excessive blood loss during surgery. A specific procedure is performed called cross-matching to check for compatibility of the blood donor and recipient

After performing patient assessment, the health care team can make the decision to proceed with the surgery. This depends on the outcome or results of the physical examination, patient history, and lab investigations.

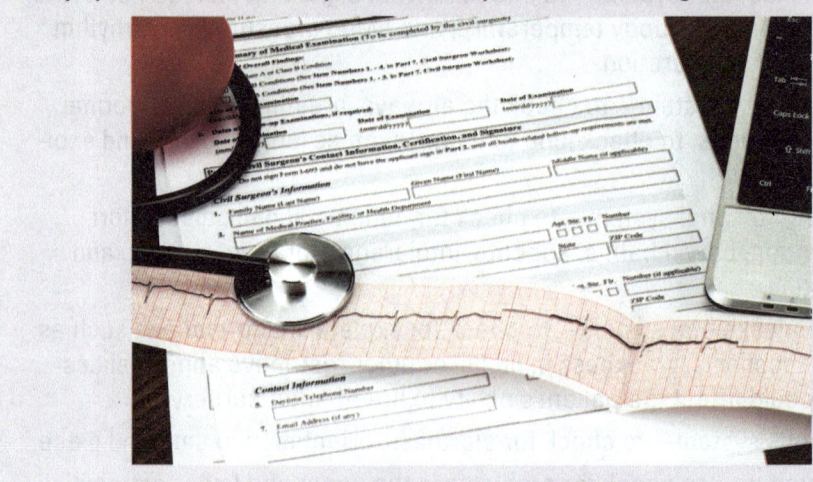

CLINICAL TERMS TO REVIEW

Tissue repair, regeneration, fibrosis
Biochemical level, cellular level, and tissue level
CPR—cardiopulmonary resuscitation, oxygen supply, brain damage, brain death
Epithelium, connective tissue
Skeletal muscle, smooth muscle, cardiac muscle
Electrical impulse, motor neurons, sensory neurons
Patient assessment
CBC—complete blood count, bacterial infection, viral infection
Cardiac arrhythmias
Blood gases, pulmonary function testing
Blood typing, blood products, blood transfusion, cross-matching
Urinalysis
Clotting and platelet function
Receptor sites, mitosis, meiosis
Nucleus, mitochondria, plasma membrane, lysosomes, nucleolus

Test Yourself

1. The primary structure that forms the plasma membrane is
   a. Phospholipid
   b. Triglyceride
   c. Sodium ion
   d. Microfilament

2. The organelles involved in the detoxification of substances in the cell are known as
   a. Lysosomes
   b. Peroxisomes
   c. Endoplasmic reticulum
   d. Mitochondria

3. The phase of cell division in which the chromosomes start to pull apart and move in opposite directions is known as
   a. Metaphase
   b. Prophase
   c. Telophase
   d. Anaphase

4. Movement through the plasma membrane that does not require ATP energy is called
   a. Reactive transport
   b. Inactive transport
   c. Passive transport
   d. Active transport

5. The largest amount of ATP is produced in the cell organelles known as
   a. Lysosomes
   b. Peroxisomes
   c. Endoplasmic reticulum
   d. Mitochondria

6. Cellular instructions are found in the
   a. Cytoplasm
   b. Nucleus
   c. Endoplasmic reticulum
   d. Plasma membrane

7. Movement of a substance from an area of high concentration to an area of lower concentration is known as
   a. Reconstitution
   b. Osmosis
   c. Endocytosis
   d. Diffusion

8. When the cell is not in the process of dividing, it is considered to be in
   a. Interphase
   b. Dormancy
   c. Resting state
   d. Recuperation

9. The type of tissue that covers all organs and the linings of hollow organs is called
   a. Fibrous tissue
   b. Connective tissue
   c. Nervous tissue
   d. Epithelium

10. The term for more than one layer of cells is
    a. Glandular
    b. Simple
    c. Stratified
    d. Transitional

11. Fluid connective tissue consists of
    a. Urine and sweat
    b. Mucus
    c. Blood and lymph
    d. Digestive fluids and enzymes

12. Muscle that moves the bone is called
    a. Cardiac muscle
    b. Skeletal muscle
    c. Rough muscle
    d. Smooth muscle

13. A nerve cell transmits its information to another cell by a process known as
    a. Synergy
    b. Refraction
    c. conduction
    d. induction

14. The muscle that has the ability to cause itself to contract and is found in the heart is
    a. Cardiac muscle
    b. Skeletal muscle
    c. Rough muscle
    d. Smooth muscle

15. The organ system that consists of hormones and glands is
    a. Lymphatic system
    b. Endocrine system
    c. Digestive system
    d. Renal system

The Integumentary System

Chapter 3

LEARNING OBJECTIVES

Upon completion of this chapter, you will be able to:

1. Describe the components and function of the integumentary system and explain how it helps to maintain homeostasis.
2. List and describe the anatomical characteristics and functions of each layer of the skin.
3. List and describe the anatomical characteristics of the skin's accessory structures.

CHAPTER OUTLINE

Introduction
- Components
- Functions

The Skin
- Epidermis
- Epidermal Ridges
- Skin Color
- Dermis
- Hypodermis (Subcutaneous Tissue)

Accessory Structures
- Hair
- Exocrine Glands
- Nails

INTRODUCTION

This chapter introduces you to the largest organ of the human body. The integumentary system, or simply the skin, covers about 22 square feet or 2 square meters, and weighs approximately 16% of the total body weight. Its primary function is to protect the human body as a whole.

When you meet or see your patient for the first time, you need to take note of the patient's skin color. This is in addition to making observations on how the patient responds to you, and how the patient describes his or her problem, which is what we call the "chief complaint" in medical language.

To describe a patient's skin color, the terms used include: "pink" for normal color (light-skinned patients), "pale" for anemia or shock, "cyanotic" (bluish-gray) for lack of oxygen, "flushed" (red) for hypertension or fever, and "jaundice" (yellow) for excess bile pigment or hepatitis.

An important aspect of studying the integumentary system, or the skin, is the different layers of the skin in relation to burn injuries caused by fire, hot liquids, corrosive acids, gases, electricity, or exposure to radiation.

When considering all of the substances in the environment that come in contact with our body, it is amazing that almost nothing passes through our

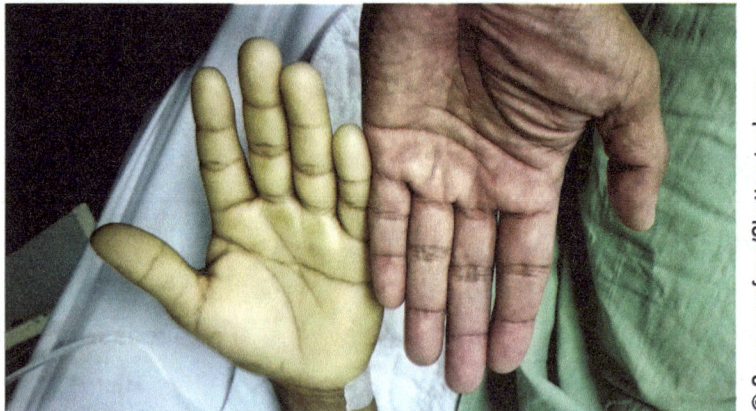

Figure 3.1

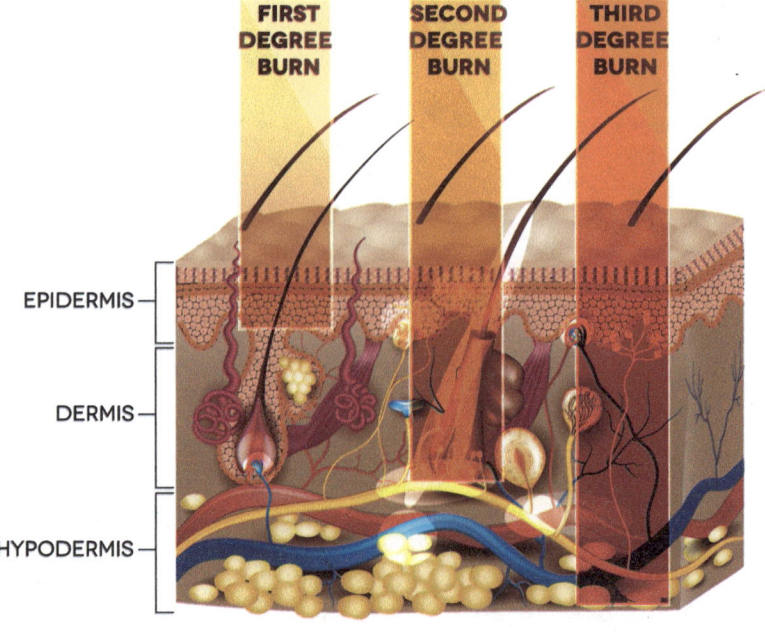

Figure 3.2

skin to cause us harm. Our skin is sensitive enough to monitor the environment yet remains a strong barrier against invasion of external organisms. It produces fluids to cool the body and oils to soften the skin and hair. Even though the skin is the largest organ of the human body, it is sensitive to touch, temperature, pain, and pressure throughout its entire surface. Major damage to the integumentary system results in serious trouble, not only from external invasion but also from fluid loss from within. However, if damaged, it can repair itself.

Components

The integumentary system or cutaneous membrane, is composed of the skin and its accessory structures (Figure 3.3). It is comprised of two distinct layers—the epidermis, or outermost layer, and the dermis, or deeper layer. Loose connective tissue known as the *subcutaneous tissue,* or hypodermis, attaches the cutaneous membrane to underlying tissue.

Within the skin are accessory structures such as hair follicles, glands, and sensors that detect both the external environment of the body and the condition of the integumentary system.

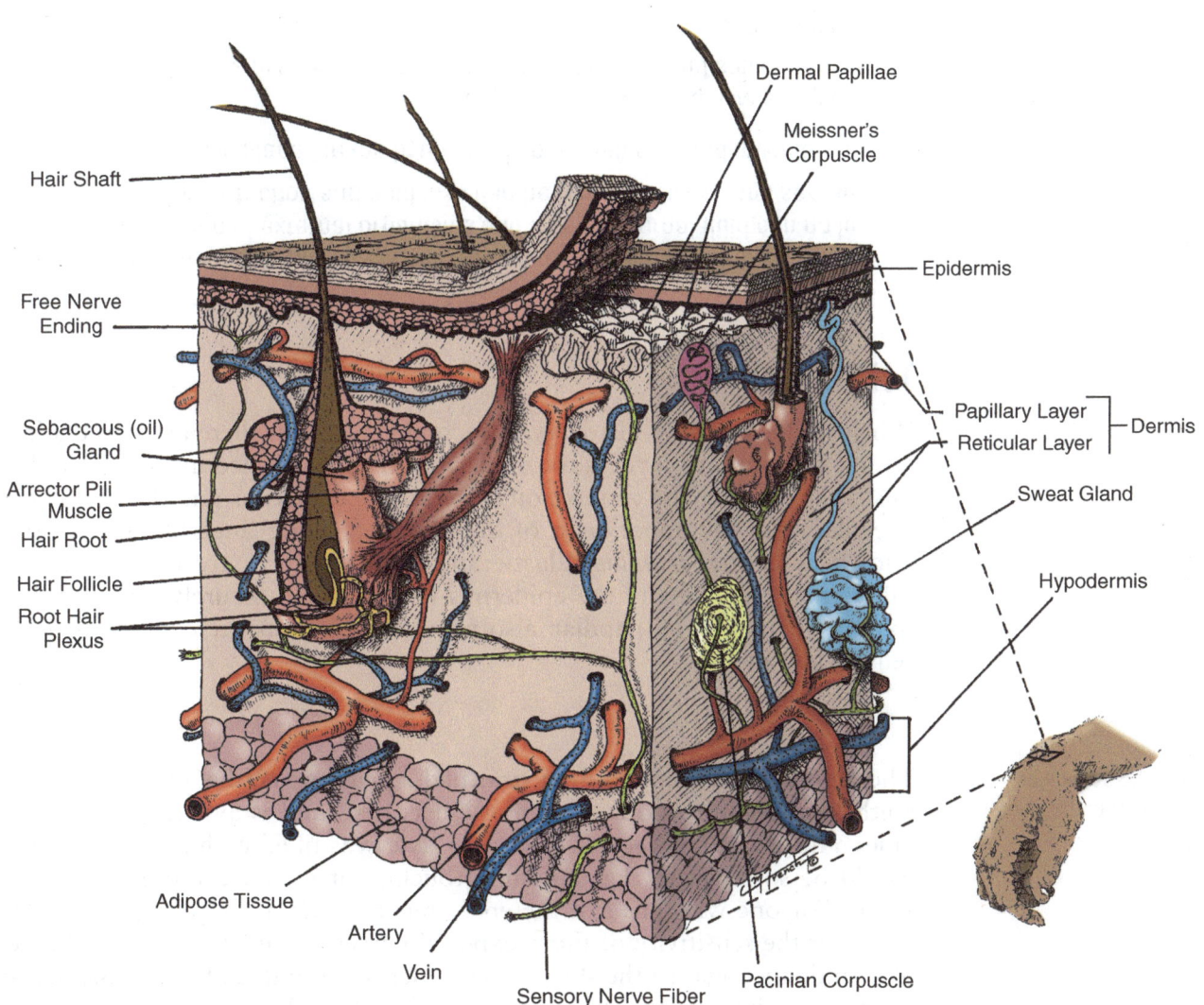

Figure 3.3 The Integumentary System.

Adapted from *Anatomy I and Physiology Lecture Manual*

Functions

The integumentary system provides many functions for the body:

- It acts as a barrier between the outside environment and the internal structures.
- It is supplied with bacterial growth-inhibiting (bacteriacidal) secretions.
- Capillaries in the skin transfer heat transported by the bloodstream from the center of the body to the surface for release into the atmosphere.
- It can excrete water, salt, and some waste products.
- It is packed with sensory receptors to detect touch, pain, temperature and pressure.
- The skin produces vitamin D, which is essential for the absorption of calcium.
- The cutaneous membrane also prevents loss of internal fluid from the body.
- The subcutaneous tissue holds the skin onto underlying muscle. It is also a major location for the storage of high-energy fat in adipose tissue.

> **Comprehension Check-up:**
> 1. Why would a burn that damages only the cutaneous membrane be so dangerous?
> 2. Sometimes physicians administer a subcutaneous injection to a patient. Where will the medicine be injected?
>
> 1. If the cutaneous membrane is damaged, the barrier that normally prevents invasion of external organisms and loss of internal fluid can no longer function. Infection and fluid loss can become extensive and serious.
> 2. The injection is placed in the hypodermis between skin and muscle.

THE SKIN

When we look at skin histology, it becomes clear that there are two distinct layers: The outmost layer is the epidermis; firmly attached inside the epidermis is the dermis. The connection between the epidermis and dermis is very irregular with the upper layer of the dermis forming fingerlike projections known as *dermal papillae*. Elastic and collagen fibers are found between the dermal papillae and the epidermis to hold them securely together. The folding of the dermal papillae also provides more surface contact area for sensory receptors.

Epidermis

Epidermis the avascular outermost layer of the skin.

The epidermis is composed of stratified squamous epithelium. Recall that epithelium contains no nerves (non-innervated) and is avascular. This provides a distinct advantage. If this outer layer was filled with nerves, our skin would be so sensitive that anything touching it would cause extreme irritation. Anyone who has lost the epidermis through a scrape is exceedingly aware of the sensitivity of those exposed nerves. Even bandages would feel like sandpaper against the skin. If the epidermis contained blood vessels, just rubbing a hand over the skin could cause bleeding. Also, invading organisms would have easier access to the bloodstream, so infection would be a constant problem. The epidermal tissue can be divided into five layers, or *strata*, each having a unique function (Figure 3.4). The epithelium is built on a basement membrane.

Stratum Basale

The deepest layer of the epidermis just above the basement membrane is known as the *stratum basale,* or the *stratum germinativum* (Figure 3.5). It is here that mitosis occurs. The cells are pushed outward as additional cells in the stratum basale divide.

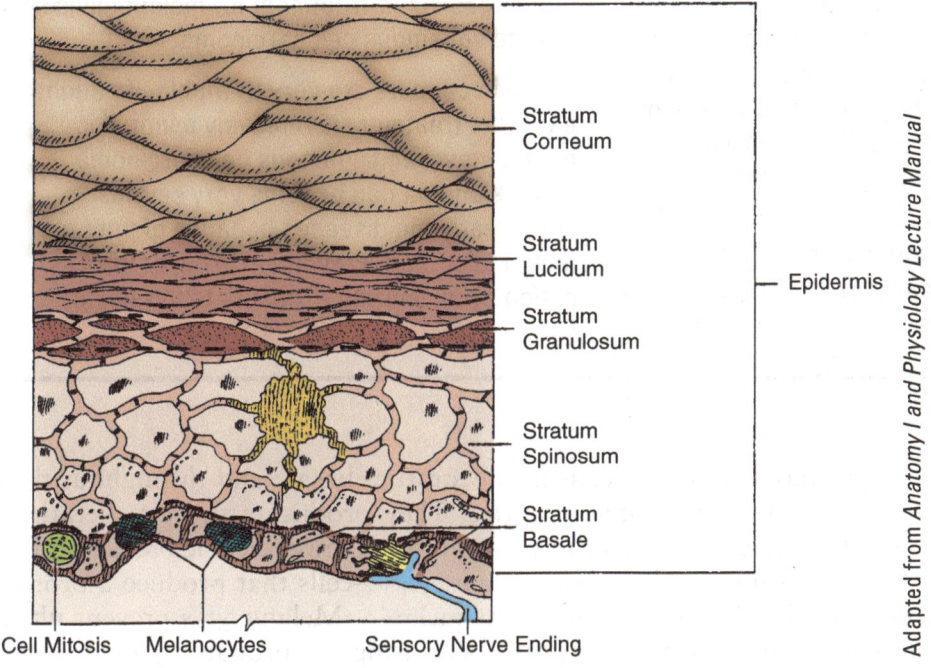

Figure 3.4 The Epidermis. The uppermost layer of the skin is composed of stratified squamous epithelium.

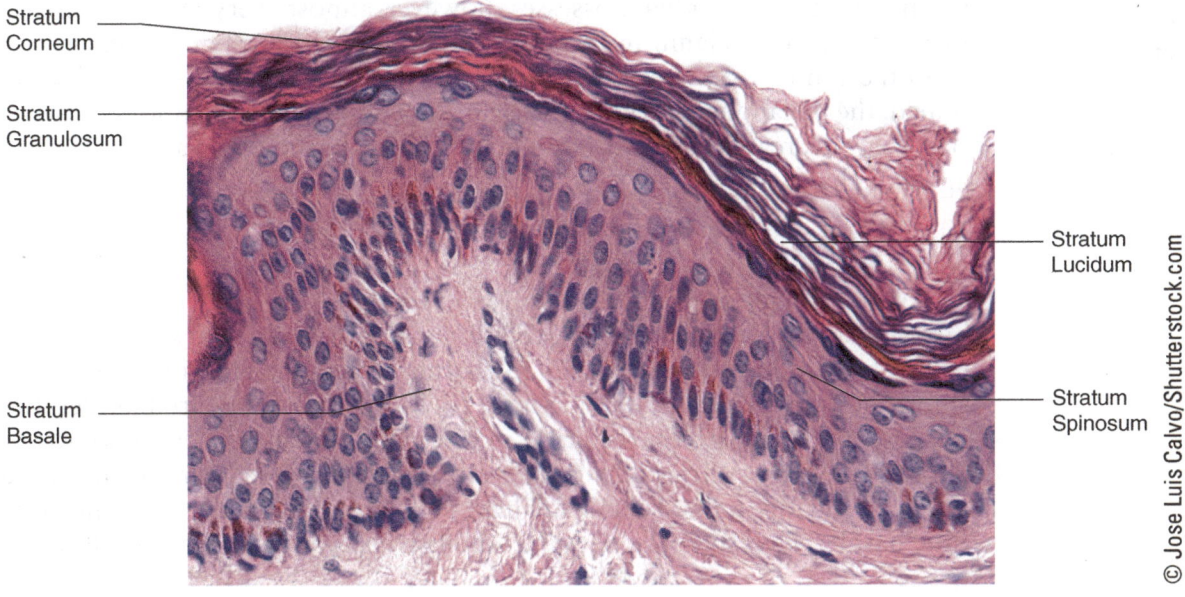

Figure 3.5 The Epidermis.

> **Skin Cancer**
>
> By knowing the layers of the epidermis, it is easier to comprehend the origin of the three basic types of skin cancer. They are:
>
> - Basal cell carcinoma, which accounts for at least 75% of all skin cancers, originates in the stratum basal of the epidermis. These cancerous tumors may be sores that ooze and do not heal, yellow or white areas that resemble scar tissue, or pearl-like grayish nodules. They grow slowly and rarely spread. There are several methods for removal of the cancerous tissue including surgical excision, laser, freezing, radiation, or creams that either destroy cancer cells or stimulate the immune system to defend against them.
> - Squamous cell carcinoma, which originates in the stratum corneum, forms a scaly, reddish, dome-shaped tumor that grows slowly but may spread to other organs. The treatment involves removing the tumor and, depending on the probability of its spreading, may require further treatment such as chemotherapy.
> - Melanoma arises from melanocytes in the stratum basale. They form black or brown irregularly-shaped patches with irregular borders typically larger than 6 millimeters in diameter. They appear as a mole that grows or changes color. Treatment for melanoma is similar to squamous cell carcinoma.
>
> Skin cancers are very treatable. If caught early, the treatment is relatively minor. If ignored, however, squamous cell carcinoma and melanoma can become very serious life-threatening diseases because of their potential to spread. Prevention is by far the best option by minimizing exposure to ultraviolet light.

The most abundant cells in the stratum basale and throughout the epidermis are known as *keratinocytes*. They produce a protein, keratin, that helps to bind cells together to provide for an impermeable barrier.

Also within this layer are melanocytes—cells that produce a protective, brownish-black pigment known as melanin. Melanocytes prevent ultraviolet light from altering DNA, thus decreasing the probability of skin cancer. When a person is exposed to ultraviolet light, melanocytes increase their production of melanin. It is the content of melanin in the skin that is responsible for skin color. Genetics play a major role in determining the basic amount of melanin that an individual possesses; however, almost everyone will increase their amount of melanin in response to exposure to ultraviolet light either from the sun or from a tanning bed. If regular periodic exposure is discontinued, the melanocytes decrease their production of melanin and the darker skin color decreases as the melanin-containing cells move toward the surface and slough off.

Stratum Spinosum

The second layer is known as the *stratum spinosum* (Figure 3.5). When histologists first began studying the skin, the preservative used to prepare epidermal tissue caused the cells to shrink. When the cells in the stratum spinosum shrank, their connections to each other appeared under the microscope to be spines, resulting in the name. Today, different solutions are used to preserve tissue and the shrinking does not occur so spines are not seen but the name remains. These cells have migrated from the stratum basale. They provide a dense defensive layer to protect underlying tissue.

Melanocytes cells that produce protective brownish black pigment known as *melanin*.

Melanin brownish-black particles that protect cells against ultraviolet light.

> **Burns**
>
> Burns, caused by heat, electricity, radiation, or chemicals result in damage, in most cases, to the integumentary system. Burns have been classified into three degrees of severity.
>
> - A first-degree burn involves damage to the epidermis only. These burns cause redness and pain. A common example of a first-degree burn is sunburn.
> - A second-degree burn damages the dermis as well as the epidermis. This type of burn results not only in redness but also blisters and significant pain. Severe sunburns may also result in blistering. As the damaged tissue is replaced, the outer layers peel from the healthy tissue beneath it.
> - A third-degree burn reaches below the epidermis and dermis destroying the skin's protective barrier. Because nerve endings are commonly destroyed within the third-degree burn, there is a loss of sensation. The damaged skin may be black because of burned skin or white because of a loss of blood flow resulting from damaged blood vessels. Third-degree burns can be a very serious injury because foreign invaders now have access to the body. There may also be loss of internal fluid through a severe burn. Depending on the extent of damage, the third-degree burn may be covered and protected by antibiotic cream while healthy tissue or scar tissue replaces the destroyed tissue, or, if it is extensive, skin grafts may be required to provide a framework upon which tissue replacement can occur.
>
> It is possible for internal burns to occur as well as a result of chemicals ingested or radiation, however, the vast majority of burns occur on the outer surface of the body.

Stratum Granulosum

As the keratinocytes are pushed farther from the stratum basale, they actively produce a fibrous protein known as keratin. The manufactured keratin is packaged within the cells and gives the appearance that these cells are filled with granules. As a result, this distinct layer becomes known as the *stratum granulosum* (Figure 3.5). The fibrous nature of keratin makes it useful for holding cells together, giving the skin additional strength against tearing as these cells, relatively far from their nutrient source, begin to die.

> **Keratin** a fibrous protein produced in the epithelium that assists in holding dying cells together in the outer layers of the epidermis.

Stratum Lucidum

Heading toward the surface, the cells begin to flatten and turn translucent, or clear, as they fill with keratin—in other words, they become *keratinized*. This layer is known as the *stratum lucidum* (clear layer) (Figure 3.5). Because the stratum lucidum contains large amounts of strong keratin fibers, it becomes thickest where it is needed most—the palms and the soles of the feet. In other areas it is a thin translucent band.

Stratum Corneum

The outermost layer of the epidermis, known as the *stratum corneum*, is composed of dead cells held together by keratin (Figure 3.5). Under the microscope, these cells appear to be shredded. The cytoplasm, including organelles and nucleus, is gone at this point. This provides a protective layer of dead cells. As cells continue to be pushed outward, they are shed from the stratum corneum and flake off or are washed away during bathing. On the average, cells remain in the stratum corneum about two weeks.

> **Comprehension Check-up:**
>
> 1. Why would an albino—a person who does not have active melanocytes—be at high risk for sunburn and skin cancer?
> 2. Which stratum of the epidermis thickens to form calluses on our hands and feet?
>
> 1. Melanin protects skin cells against high energy ultraviolet light. Cells may be damaged or DNA altered potentially causing skin cancer. An individual lacking melanin lacks protection against ultraviolet light.
> 2. The stratum lucidum thickens to form calluses especially on the palms.

Epidermal Ridges

There are contoured epidermal ridges on our fingertips that provide a rough surface with which objects can be grasped. These ridges, supported by the underlying dermal papillae in the dermis (described later), form unique patterns known as *fingerprints* on our finger tips (Figure 3.6). Each has epidural ridges distinct from anyone else, making it possible to use fingerprints for specific identification. Even otherwise identical twins have different fingerprints.

Skin Color

As was discussed earlier, the genetically determined content of melanin in the skin plays a significant role in skin color. People of African descent possess greater levels of melanin that do those of European heritage. It is not, however, the only factor contributing to the color of the skin, especially for those not producing as much melanin as dark-skinned individuals. Carotene, a yellowish-orange pigment found in some vegetables like carrots or certain squashes, is also found in the skin. Again, genetic differences in carotene production may affect skin color, particularly in individuals of Oriental descent. Even though the epidermis does not contain blood vessels, there are many capillaries just under the basement membrane, which gives skin with lower levels of melanin a pink color. In fact, skin color can change substantially as a result of blood flow. If blood flow decreases, the skin becomes pale. If there is a decrease in blood flow and the available blood is low in oxygen, the skin

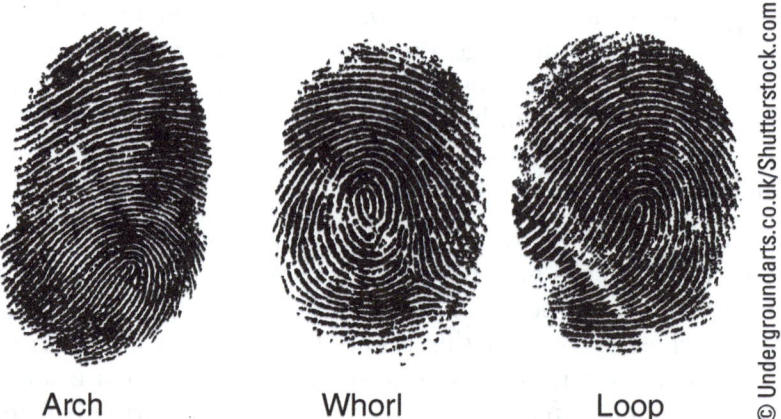

Figure 3.6 Patterns of Epidermal Ridges. These ridges form unique patterns on our fingertips known as *fingerprints*.

appears bluish (cyanotic) or gray in color. On the other hand, increased blood flow through the dermis can cause the skin to take on a red appearance, as might occur with sunburn as the skin tries to heal from the damage caused to it or to remove the heat from the area. Cherry red color may occur with carbon monoxide poisoning as dermal blood vessels dilate. For these reasons, skin color can be a very useful diagnostic tool for medical professionals.

When red blood cells die, a yellowish-brown chemical known as *bilirubin* results from the breakdown of these dead cells. Bilirubin is placed into the bloodstream to be removed by the liver and secreted into our digestive system for excretion. It is bilirubin that causes feces to be brown in color. If an individual has difficulty removing bilirubin from the bloodstream fast enough, it may accumulate in the skin, causing it to take on a yellowish-brown tint known as *jaundice*. This condition is commonly seen as a side effect in someone with liver disease although there are other causes as well.

It is easiest to see color differences in individuals possessing the least amount of melanin, such as Caucasians. When medical practitioners work with a darker-skinned patient, it is still possible for them to check areas of the body containing less melanin, such as the fingers or nail beds or eyelids, for signs of skin color abnormalities.

> **Comprehension Check-up:**
> 1. Do any two people have identical fingerprints?
> 2. If I come across an accident and find a Caucasian victim whose left arm is gray in color but the rest of his body is pink, what should be my major concern?
>
> 1. Everyone has differing fingerprints, including identical twins.
> 2. The gray color of the limb indicates decreased blood flow or low blood oxygen. It might indicate that some object is blocking circulation or there is severe blood loss to the limb.

Dermis

While the epidermis is non-innervated and avascular, and essentially acts as a barrier to isolate the contents of the body from the world outside, the dermis is the deeper layer that is exceedingly active in providing nutrients to the epidermis through blood vessels, sensing the environment by nerve sensory receptors, and glands that produce fluids to cool the body and oils to soften the skin. Defensive cells that reside in the dermis are ready to go into action against any possible invasion.

Dermis deeper layer of the skin containing sensory receptors, blood vessels, hair follicles, and secretory glands.

Layers of the Dermis

The dermis contains two sections: the outermost layer is known as the *papillary layer* while the inner one is referred to as the *reticular layer*.

Papillary Layer

The papillary layer is formed from nipple-shaped projections from the upper surface of the dermis. These projections, known as *dermal papillae*, are composed of areolar tissue. They fit into the epidermis, creating a strong

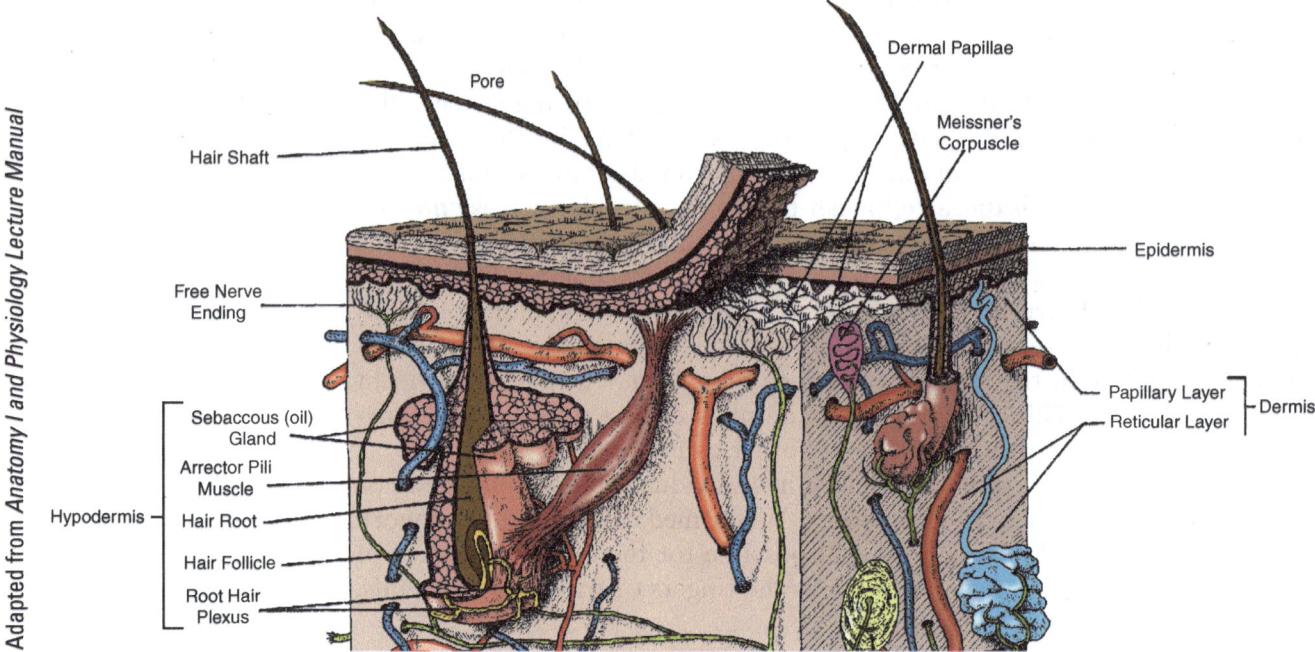

Figure 3.7 Layers of the Dermis. The thin papillary layer is composed of papillae—nipple-shaped projections that increase surface contact with the epithelium. This allows more sensory receptors to detect environmental contact with the skin than would be possible if the skin was flat. The deep reticular layer contains accessory structures of the skin such as hair follicles, glands, and deep sensory receptors.

attachment between dermis and epidermis, and contribute to the formation of fingerprints. They increase surface contact area, allowing a greater number of sensory receptors to monitor the epidermis.

Reticular Layer

Deep to the papillary layer is dense irregular connective tissue comprising the bulk of the dermis (Figure 3.7). It is in this region that the accessory structures of the skin are located. These structures monitor the environment and provide for defense, secretions, and the ability to decrease body temperature.

Structures in the Dermis

Nestled in the dense irregular connective tissue comprising the dermis are hair follicles and several structures associated with hair such as arrector pili muscles, oil glands, and nerve endings wrapped around the hair bulb (Figure 3.7). Sweat glands are also abundant. Sensory receptors populate the dermis more densely in some areas than others, such as the fingers and lips, to provide the brain with more precise information about location and identification of the object sensed. Capillaries supply nutrients and remove waste not only for the epidermis but for the dermal structures mentioned as well.

Skin and Sensation

Just below the epidermis are sensory receptors that detect objects, liquids, or even drafts of air that contact the skin. The most common receptors sense pain, touch, temperature, or pressure. A single sensor can detect only one type of sensation, so a pain sensor cannot register touch or temperature. There are separate sensors for hot and cold as well. In any given area of the skin, separate receptors for each sensation are clustered together. The closer

> **Tattoos—Permanent Body Art**
>
> Tattoos can be attractive, artistic, clever—a means of identification or even rebellion. They are also permanent. The cause for the longevity is because colored ink is inserted through needles below the basement membrane of the epidermis. If the ink was only placed in the epidermis, the color would fade in a matter of weeks as the cells containing the design were pushed to the surface and sloughed off. In order for the ink to remain, it must be inserted into the dermis. It is important that the artist is careful not to infect the client because the injections penetrate the protective barrier of the epidermis. Once the ink is in place, it remains throughout the life of the recipient. If the desire to remove the tattoo arises, it is necessary to again penetrate into the dermis to remove the additional color. Lasers are used as the most efficient method for tattoo removal.

the receptors, the more sensitive the skin. For example, the sensors in our lips and fingers are packed very closely together. If an individual gets a splinter in his finger, he has no trouble determining exactly where the injury occurred. The receptors on the back, however, are spread out. It is sometimes difficult to locate an itch on the back in order to know exactly where to scratch.

Skin and Extensibility and Elasticity

Between the skin and the underlying muscle is the subcutaneous tissue, which consists of collagen and elastic fibers holding the cutaneous tissue to muscle. The collagen fibers provide a strong attachment whereas the elastic fibers allow the skin to stretch and then be pulled back to its original position. This is not the case in the feet and palms, however, where the cutaneous tissue is tightly bound to the underlying structures so there is no slippage when we grasp an object or when we walk.

Hypodermis (Subcutaneous Tissue)

As stated previously, the hypodermis contains collagen and elastic fibers to hold the cutaneous tissue to underlying muscle and other structures. This layer is composed of loose areolar connective tissue (Figure 3.8). Embedded within the hypodermis are bundles of sensory neurons that carry information from the skin toward the spinal cord and eventually to the brain for interpretation. Larger blood vessels are found in the subcutaneous tissue as well. Veins commonly seen in the back of the hands and arms are running through the hypodermis.

Accumulations of adipose tissue are also found in the subcutaneous layer. The sex hormones estrogen and testosterone affect the distribution of fat deposits differently in males and in females. Estrogen, which is in higher levels in females, causes increased adipose tissue in the breasts and thighs. Testosterone, predominant in males, tends to cause increased fat on the anterior of the trunk and neck. Body-builders work hard at reducing the amount of subcutaneous fat to an absolute minimum to increase definition of their muscle mass. The amount of adipose tissue found in the hypodermis, along with its distribution, has been cause for concern. Obese individuals can be seen in two general shapes that are often compared to fruit, those having a pear shape—that is, the bulk of their fat deposits are on the hips—tend to fare considerably better health-wise than those who are apple-shaped—those

> **Hypodermis (subcutaneous tissue)** loose connective tissue between the cutaneous membrane and underlying muscle or other underlying structures.

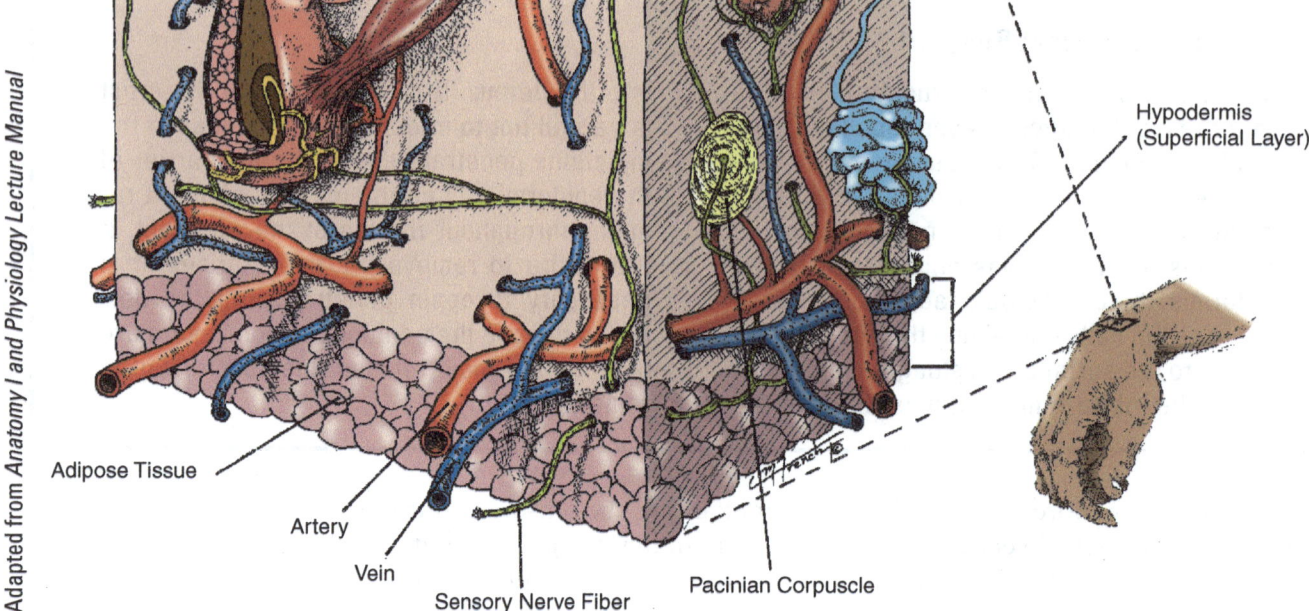

Figure 3.8 The Hypodermis. This layer contains collagen and elastic fibers to hold the cutaneous tissue to underlying structures.

Do We Really Have Separate Sensory Receptors for Hot and Cold?

You can prove to yourself that you have separate hot and cold sensors by a simple demonstration. Take three cups and fill each ¾-full with water. Place hot water in one cup, cold in another, and lukewarm water in the third. Put two fingers on one hand in the cup containing hot water and two fingers on the other hand in the cup of cold water. Remain in place for one minute, then put the same two fingers of both hands in the lukewarm water. If temperature sensors detect both hot and cold, then the lukewarm water will feel the same temperature in both hands. If there are separate sensors for hot and cold, the fingers that were in the hot water will feel cool and the fingers that were in the cold water will feel warm in the lukewarm water.

whose fat storage is primarily around the trunk. When there is an excessive amount of fat on the trunk, it is usually accompanied by an increased concentration of internal fat deposits around the abdominal organs. Apple-shaped individuals tend to have a high incidence of adult onset (Type II) diabetes mellitus, uncontrolled blood glucose. The poor blood sugar control causes increased circulation of lipids, potentially leading to stroke or heart attack. As a result, subcutaneous fat content can be a diagnostic tool in determining the level of risk for certain serious diseases.

Comprehension Check-up:

1. Which layer of the dermis contains hair follicles and sensory receptors?
2. Larger blood vessels and accumulations of adipose tissue are found in the _____.

1. Reticular layer   2. hypodermis (subcutaneous tissue)

Clinical Skills—Handwashing and Hand Sanitizers

The spread of bacteria and viruses is very common in healthcare facilities where patients, doctors, and other healthcare providers and staff can get infected. The safety of everyone from infectious diseases is a priority. The single most important infection control is handwashing.

There are two methods of hand cleaning:

1. Wash your hands with soap and water vigorously for 15-20 seconds after each patient contact. Focus on the fingertips and underneath, extend washing up to the wrist level. Dry your hands using paper towel. Turn off faucet with paper towel, and discard it.

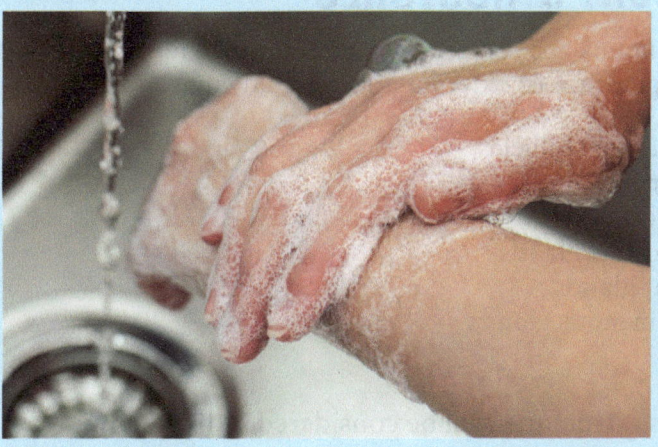

2. Use alcohol-based hand sanitizers. Alcohol content must be at least 60%.

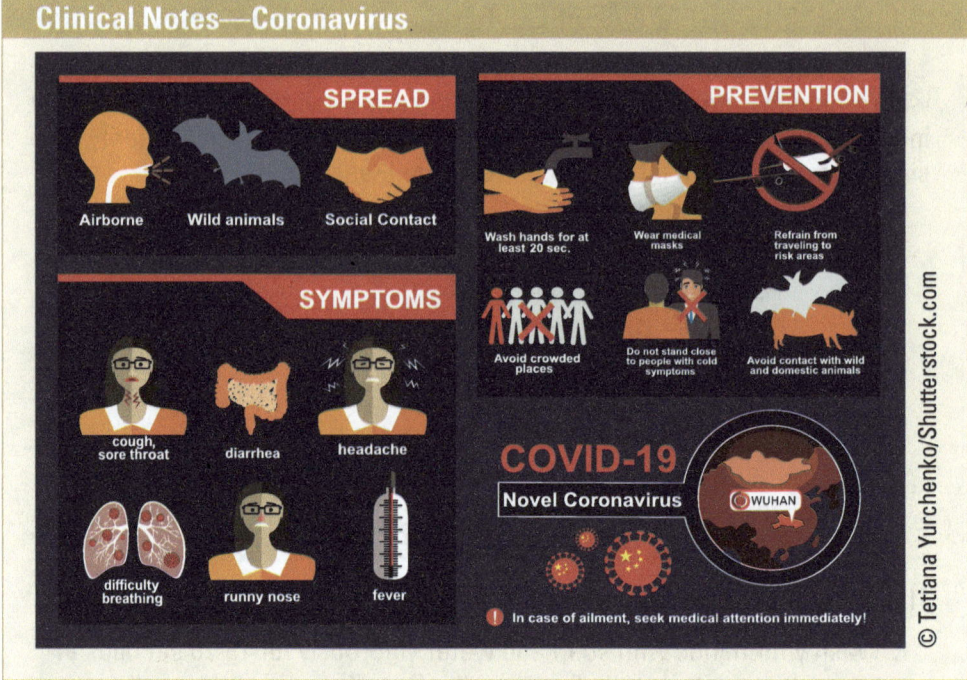

ACCESSORY STRUCTURES

The skin contains **accessory structures** that provide additional functions beyond the separation of external and internal environments:

- Hair and hair follicles
- Arrector pili muscles
- Exocrine glands
  1. Sebaceous glands
  2. Sweat glands
  3. Cerumenous glands
- Nails

Each of these structures will be considered individually.

> **Accessory structures in the integumentary system** structures in the skin that provide additional functions beyond the separation of external and internal environments.

Hair

Hair provides a slight level of insulation, but more importantly, nerves wrapped around its base can sense bending of the hair. This allows us to be aware of close objects even before they come in direct contact with our skin. Hair is actually an outgrowth of the epidermis. Hair follicles produce hair as they add cells to the proximal end, causing the hair shaft to lengthen. The shape of the hair shaft has an effect on the type of hair an individual possesses. If the shaft is round, the hair is straight. Oval-shaped shafts result in wavy hair and hair flat on two sides is very curly.

Structure of the Hair Follicle

The stratum basale of the epidermis forms a deep, bulb-shaped hair follicle that produces hair. As mitosis occurs within the follicle, the cells are added to the hair, causing it to grow outwardly. The center of the hair is a soft core

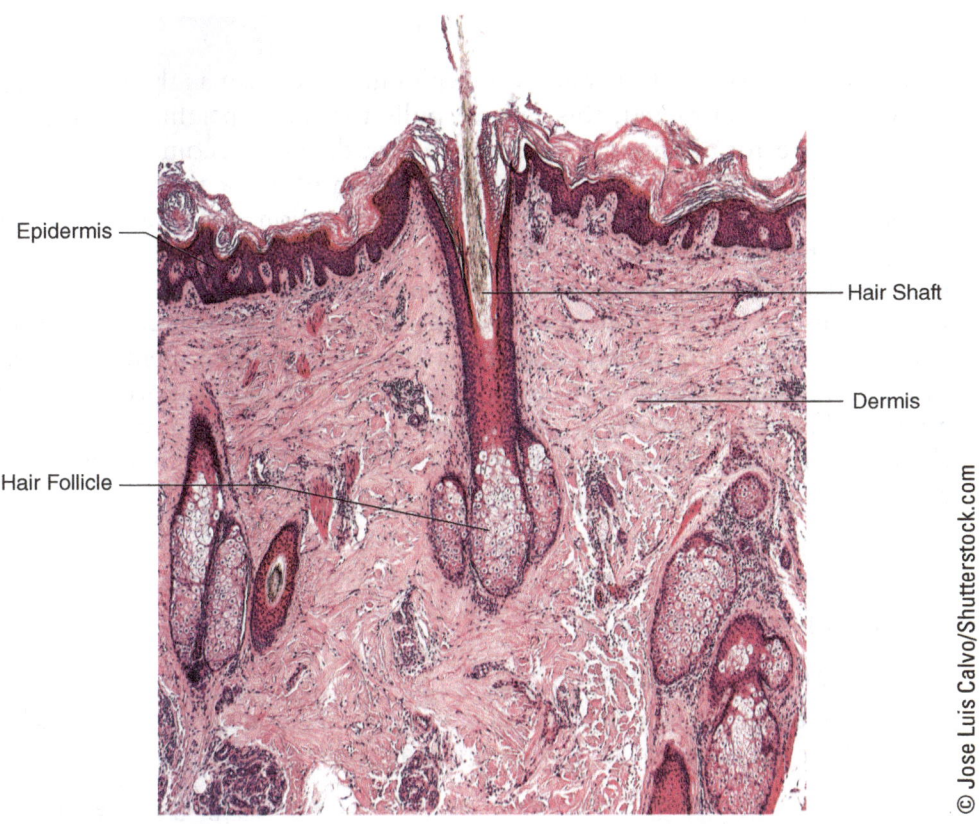

Figure 3.9 Hair Follicle.

surrounded by an outer layer of cells. The section of the hair below the scalp is called the *root*. The shaft is actually the part of the hair that is above the scalp (Figure 3.9).

Hair Color

Inside the follicle are melanocytes, which add melanin to the hair as it is produced. A small amount of melanin results in blond hair; black hair is caused by a high production of melanin. Hair color is determined genetically. As a person ages, the melanocytes in the hair follicle die and stop producing melanin. Reduced amounts of melanin in the hair result in a gray color. White hair results from the absence of melanin.

Growth of Hair

Hair growth and length are obviously not the same at all locations on the body. A major factor is with the length of time a hair follicle produces a single hair. For example, each hair on your head will be produced by its follicle for two to five years. After that time, that particular hair is terminated and falls out and a new hair shaft will then begin forming. Fortunately for us, the hair on most of the rest of our body is terminated every few months. Hair growth can be affected by nutrition, drugs, radiation, or hormone levels. High levels of testosterone, the hormone responsible for sex drive, is also responsible for facial, underarm, and pubic hair in men. Excessive levels in men tend to lead to baldness. Women also produce a smaller amount of a testosterone-like hormone that also causes underarm and pubic hair, but facial hair production is not usually increased substantially.

Arrector Pili Muscle

Attached to the shaft of the hair is smooth muscle known as the *arrector pili muscle*. When contracted, this muscle pulls the hair upright. On humans, when the arrector pili muscle pulls on the hair, the skin becomes bunched up to form a "goose bump." This process is a form of work that causes an increased need for more ATP—recall that the production of ATP gives off heat, so contraction of the arrector pili muscle, as well as shivering when we are cold, increases metabolism to increase temperature. Contraction of the arrector pili muscles can be easily seen in other animals. Watch a cat when a dog walks by—the fur on the cat suddenly stands up, making it seem to increase in size. Its tail will also double or triple in diameter. More importantly, when an animal is cold, the arrector pili muscles make the fur stand up, thickening the fur coat and trapping air. This serves as an insulating blanket and helps retain heat produced by the additional metabolism.

> **Comprehension Check-up:**
> 1. Hair color is caused by _____.
> 2. The reason the hair on our head is so long compared to the hair on rest of our body is because it grows for what length of time?
>
> 1. melanin 2. 2–5 years

Exocrine Glands

Exocrine glands secrete their products through a duct or tube. In the skin, the epidermis forms into cavelike exocrine glands containing cells that secrete through a duct onto the surface. Those glands produce a specific material essential to the normal functioning of the body to maintain homeostasis.

Sebaceous Glands

Associated with the hair follicle is a sebaceous gland that produces an oily, waxy protective substance known as *sebum*. The secretions from this gland help keep the hair and surrounding skin soft as well as inhibit bacterial growth. It can also slow water loss from the tissue underlying the skin. Their production is also affected by sex hormones, especially during puberty when the output of the sebaceous glands increases significantly in many individuals.

> **Sebaceous gland** glands associated with hair follicles that produce an oily, waxy, protective substance know as *sebum*.

Sweat Glands

The primary function of the sweat glands is to provide fluid on the surface of the skin that will evaporate and cool the body. There are two general categories of sweat glands: apocrine glands and merocrine glands. *Apocrine glands* are found around the armpits and groin and are associated with hair in those areas. Their purpose is not well understood but they appear to be analogous to scent glands in other animals. They are also active during times of extreme stress or pain. They produce a cloudy solution containing fatty material in which bacteria can thrive. Those bacteria can produce an unpleasant odor most humans do not like. As a result, antiperspirants that inhibit secretion from these glands or scents that cover the smell produced by the bacterial growth are very popular. Apocrine glands are not active until puberty, which is why little children do not get body odor like adults. *Merocrine glands* are found almost everywhere else. They produce sweat, which is mostly water with some ions and waste products. As sweat evaporates on the surface, it draws heat from the skin, cooling the blood running through it. Internal heat from deep inside the body can be transported by the bloodstream to the skin for release into the environment. The process of expelling additional heat is accelerated by sweating when metabolism has increased.

Cerumenous Glands

Our ear canals contain a waxy substances produced by modified sweat glands known as *cerumenous glands*. This ear wax prevents drying, cracking, and infection and provides protection against organisms crawling into this opening. Protective hairs also prevent invasion. The cerumen, or wax, flows slowly toward the outside. If that flow is inhibited, a person may have enough of a buildup to affect their ability to hear.

Nails

Nails protect the ends of our fingers and toes (Figure 3.10). The main body of the nail is known as the *nail plate*. It is attached to the underlying finger or toe by the *nail bed*. The nail actually is produced by the *nail root* under the skin. It then grows outwardly from the root. As the nail lengthens, the distal end of the nail becomes unattached and is known as the *free edge*. The layer of cells contacting the outer surface of the nail plate commonly known as the *cuticle* is anatomically referred to as the *eponychium*. There are times when disease processes elsewhere in the body can actually affect the structure of the nail.

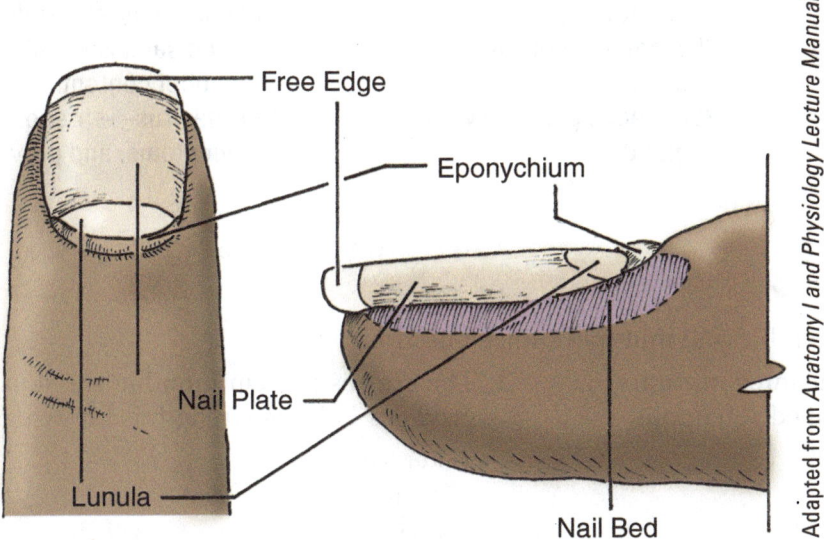

Figure 3.10 The Nail. This structure protects the ends of our fingers and toes.

Comprehension Check-up:

1. The skin and hair are kept soft due to secretions from _____ glands.
2. Our bodies are cooled by fluid secreted by _____ glands.

1. sebaceous 2. merocrine

Clinical Scenario—Patient with Thermal Burns on the Skin

What possible complications can happen to a patient with thermal burns on the skin?

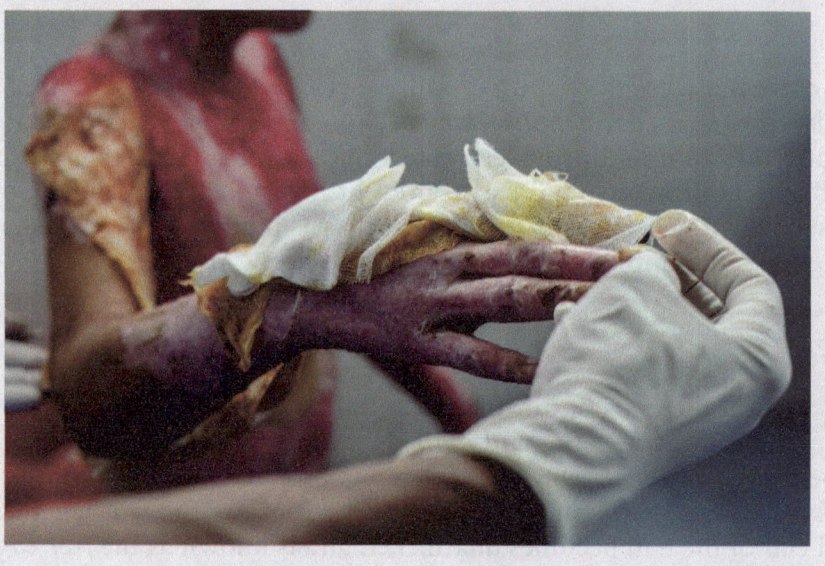

CLINICAL TERMS TO REVIEW

First-degree burns, second-degree burns, third-degree burns, thermal burns
Epidermis, dermis, subcutaneous tissue

Melanocytes, melanin
Skin cancer, cell carcinoma, melanoma
Sebaceous glands, sweat glands

Infection control, handwashing, hand sanitizers with 60% alcohol content
Coronavirus—spread, symptoms, and prevention

Test Yourself

Choose the best answer to the following multiple choice questions:

1. The outermost layer of the integumentary system composed of stratified squamous epithelium is known as the
   a. hypodermis.
   b. dermis.
   c. epidermis.
   d. hyperdermis.

2. The fibrous protein found in the skin that provides holding strength is known as
   a. keratin.
   b. sebum.
   c. melanin.
   d. collagen.

3. Nipple-shaped projections of the dermis that attach to the epidermis and increase surface contact for increased sensory reception form the
   a. papillary layer.
   b. stratum granulosum.
   c. reticular layer.
   d. stratum corneum.

4. The outermost layer of the epidermis that consists of dead cells is known as the
   a. papillary layer.
   b. stratum granulosum.
   c. reticular layer.
   d. stratum corneum.

5. The skin is held to underlying muscle by
   a. ligaments and tendons.
   b. reticular fibers.
   c. hyaline and elastic cartilage.
   d. collagen and elastic fibers.

6. Gray-colored skin is often an indication of
   a. carbon monoxide poisoning.
   b. decreased flow of low-oxygen blood.
   c. infection.
   d. liver problems.

7. The smooth muscle that raises the hair and also forms "goose bumps" is known as the
   a. arrector pili muscle.
   b. erector spinae muscle.
   c. periformis muscle.
   d. serratus anterior muscle.

8. The section of a hair above the scalp or epidermis is known as the
   a. blade.
   b. shaft.
   c. root.
   d. petiole.

9. Cerumenous glands are typically found in the
   a. toes.
   b. nasal passages.
   c. armpits and groin.
   d. ears.

10. The unattached section of the nail is known as the
    a. nail bed.
    b. nail plate.
    c. free edge.
    d. eponychium.

Chapter 4

The Skeletal System

LEARNING OBJECTIVES

Upon completion of this chapter, you will be able to:

1. Describe the anatomy of bone.
2. Identify the classifications of bones and provide examples of each.
3. Describe the components and function of the skeletal system.
4. Identify and provide examples of each classification of joints and explain how they facilitate movement in the body.
5. Describe how the bone grows, remodels and repairs.

CHAPTER OUTLINE

Introduction

Characteristics of the Bone
- Overview of Function
- Anatomy of Bone

The Skeleton
- The Axial Skeleton
- The Appendicular Skeleton

Joints (Articulations)
- Joint Classifications
- Joint Movements

Bone Growth, Remodeling and Repair

INTRODUCTION

The skeletal system provides the general framework of the human body and serves as a protection to many internal organs and structures. There are 206 bones in an adult human body. The skeletal system facilitates the movement of the body and the extremities.

It is impossible to build a house or building without a framework. In the same way, the human body needs a skeletal framework to support it and protect the internal organs. The brain is enclosed by the skull. The lungs and heart are enclosed by the thoracic cage. The body is kept standing straight by the vertebral column, pelvic girdle (hip bone), and lower extremities. The skeletal system also responds to changes in workload and activity levels.

One of the most common bone injuries is fracture of the hips. Individuals who are 60 years old and older are vulnerable to hip fractures. There is a higher rate of osteoporosis (loss of bone tissue) among older individuals. There is loss of calcium that causes the bones to be thinner and weaker when a person gets older. According to the Health Encyclopedia of the University of Rochester Medical Center, there are 1.5 million Americans who suffer from hip fractures due to osteoporosis. Surgery is needed for hip replacement. Patients undergo rehabilitation by doing physical therapy exercises.

Another common bone injury is hip dislocation among young adult individuals; over 90% of those who suffered a hip dislocation are males, as reported by the National Center for Biotechnology Information (December 2019).

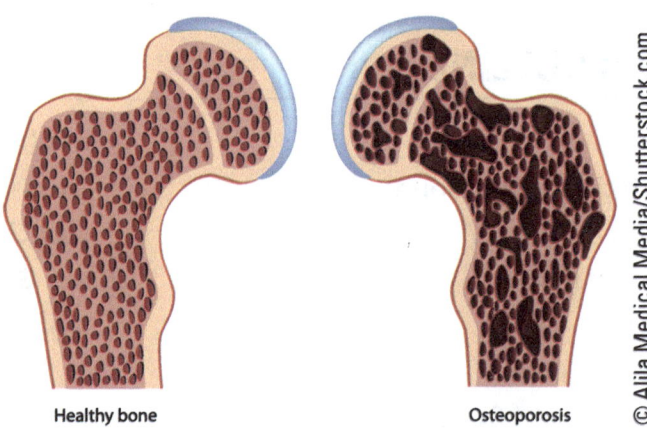

Figure 4.1

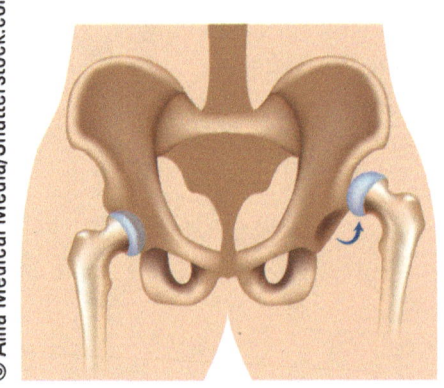

Figure 4.2

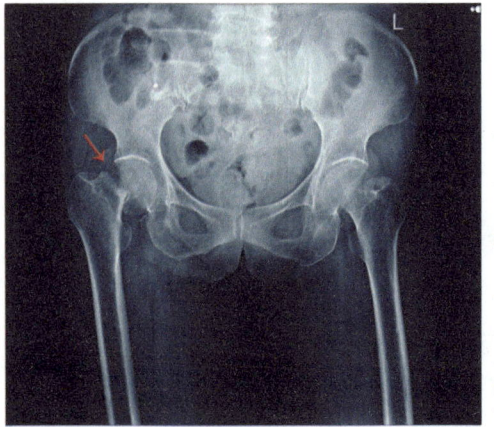

Figure 4.3

CHARACTERISTICS OF THE BONE

Overview of Function

The skeletal system provides five functions for the human body:

1. Body movement—Muscles attached to bone allow them to move, giving humans the ability to perform most of the activities they accomplish throughout the day. The position of joints, as well as the placement of muscle attachments, allows the bones to act as levers so that larger or stronger movements can be accomplished.
2. Protection—Organs critical to our survival are surrounded by or next to bone. Our brain is completely enclosed in the cranium for maximum protection. The heart and lungs are surrounded by the rib cage. Even some of the abdominal organs are protected by ribs.
3. Support—Organs, membranes, and other supporting structures may be attached to the skeletal system, allowing them to be suspended in place.
4. Hemopoiesis—The process of forming blood cells is known as *hemopoiesis*. Deep within bone is blood-forming tissue known as red marrow. It is more plentiful in children, then decreases in quantity with age.
5. Storage of inorganic salts—To maintain a constant homeostatic level of some inorganic ions in the bloodstream, it is necessary to have a reservoir for the storage of the excess. Bone functions as a location for storage of some ions that can be released back into the bloodstream as needed.

Anatomy of Bone

Bones can be categorized by shapes (long, short, flat, or irregular), as well as by internal structure (spongy or compact).

Shapes

Bones can be described in four general shapes (Figure 4.4):

1. *Long bones* are so named because their general shape is longer than they are wide or column-shaped. These bones are found in the extremities.
2. *Short bones* are more round or cube-shaped. They are found in the wrist and ankle. It is interesting to note that the short bones of the ankle are actually larger than the long bones in the fingers and toes. It is not the size that matters, but the shape.
3. *Flat bones* are rather easy to detect because they are flat on two sides. Ribs and the sternum fall into this category. Although it is not perfectly flat, some of the bones of the cranium, which encloses the brain, are also considered flat bones.
4. *Irregular bones* do not fall into any of the other categories. Some examples of irregular bones are vertebrae and some skull bones.

Structure of a Long Bone

We begin our discussion of the skeletal system with the study of the long bone (Figure 4.5). The ends of the long bone are called the epiphyses. Internally the ends are composed of spongy (cancellous) bone, which when dried resembles a dried sponge. The area of the epiphysis that comes in contact with other bones is covered with hyaline cartilage, referred to as *articular*

> **Epiphyses** the ends of long bones. They are composed of spongy bone.

> **Spongy (cancellous) bone** in dried bone, it resembles a dried sponge.

Chapter 4 The Skeletal System

TYPES OF BONES

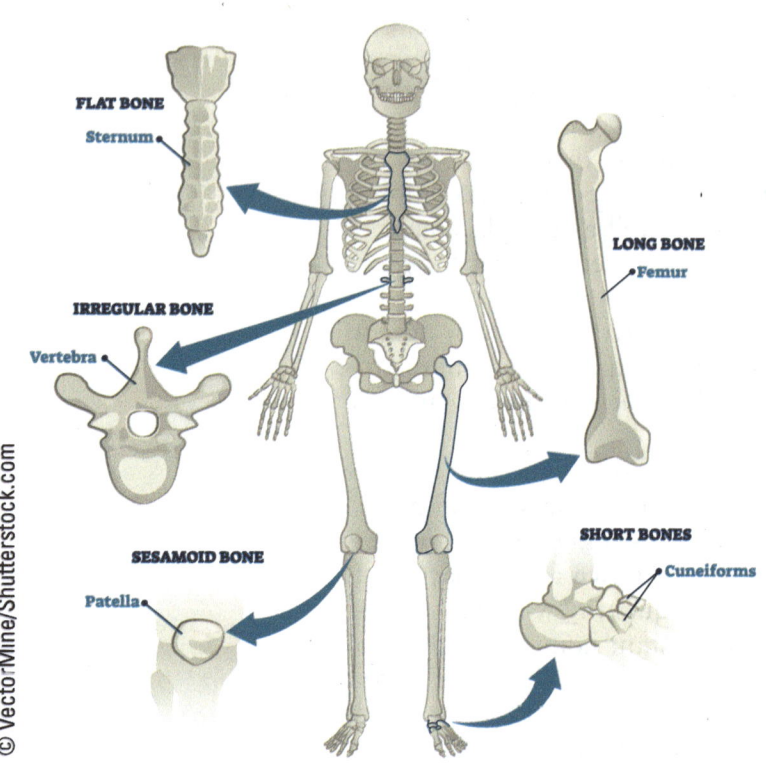

Figure 4.4

Figure 4.5 Long Bone Anatomy. Long bone consists of a tubular section composed of compact bone (known as the diaphysis) with a spongy bone epiphysis on each end.

Adapted from Anatomy I and Physiology Lecture Manual

> **Diaphysis** the tubular section of a long bone composed of dense bone. Inside the diaphysis is the medullary cavity.
>
> **Compact (dense) bone** thick, densely packed layers of bone.
>
> **Red marrow** blood cell forming tissue found in bone.
>
> **Yellow marrow** adipose tissue found in the medullary cavity of long bones.

(hyaline) cartilage, to cushion movements between bones. There is also a tubular section in the long bone known as the diaphysis. The walls of the diaphysis are several times thicker than those of the epiphysis and are composed of compact (dense) bone, which contains dense layers of bone. Within the diaphysis is the medullary cavity, which is lined by a membrane called the *endosteum*. In children the medullary cavity is filled with red marrow, which is blood-forming tissue. As the individual ages, the red marrow is replaced by adipose tissue referred to as yellow marrow. Not all red marrow in every bone is replaced by yellow marrow, but in the long bones of adults, it is converted into adipose tissue. Red marrow may still be found in the other types of bones, such as the sternum or pelvis. Covering the outer surface of the diaphysis and part of the epiphysis is a tough thick membrane filled with nerves and blood vessels known as the *periosteum*.

> **Comprehension Check-up:**
> 1. Short bones are found in the _____ and _____.
> 2. The end of a long bone is referred to as the _____.
>
> 1. wrist and ankle 2. epiphysis

Structure of Other Bones

Flat bones are composed of spongy bone that is sandwiched between two layers of compact bone (Figure 4.6). Short and irregular bones are similar to the epiphysis of a long bone in that there is an outer layer of compact bone with a spongy bone center. The spaces in spongy bone usually contain red marrow.

INTERNAL STRUCTURE OF A BONE

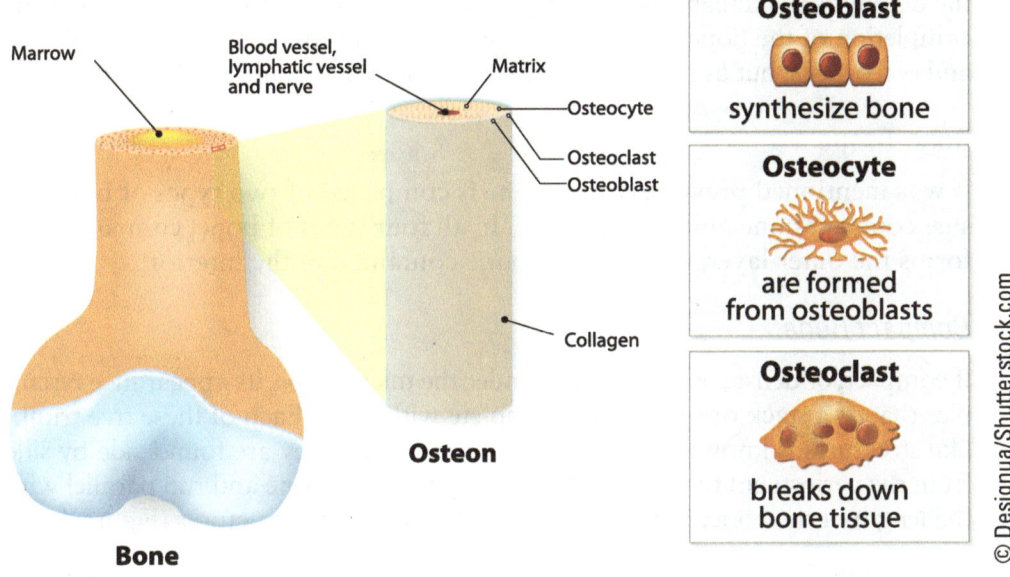

Figure 4.6

Histology

Before discussing bone anatomy, it will be helpful to define cells and structure that are involved with the production or breakdown of bone.

Bone Matrix

Bone cells are fixed cells surrounded by a dense, hard substance known as *bone matrix*. This material between bone cells is composed primarily of mineral deposits and collagen. The matrix becomes hardened by calcium and phosphate deposits. After death, the bone cells die, but the calcium phosphate matrix is unchanged, resulting in skeletal remains that can be found even hundreds of years later.

Types of Bone Cells

Bone cells are surrounded by matrix, so it is necessary to deal with them differently than how we deal with other types of cells in the body. The anatomical term for bone is *osteo-*. Three types of cells are associated with bone.

1. An osteocyte is a bone cell surrounded by matrix.
2. Cells that form bone matrix are known as osteoblasts. Recall that a "blast" is an immature cell or a cell that makes something. An osteoblast is going to move where it is needed and form bone matrix around itself. Once the osteoblast is entrapped within its matrix, it will differentiate into an osteocyte.
3. Some cells specialize in breaking down bone; these are called osteoclasts. When osteocytes die, both the organic material of the cell and the hard surrounding matrix need to be removed. The osteoclasts break down the matrix so that the dead osteocytes can be consumed by phagocytes. If the blood content of calcium and phosphate needs to be increased, osteoclasts can break down the matrix when instructed to do so by hormones and add those ions to the bloodstream.

Osteocyte bone cells surrounded by a hard bone matrix.

Osteoblasts bone-forming cells that produce bone matrix around themselves and then differentiate into osteocytes.

Osteoclast a cell that is able to break down the bone matrix around osteocytes.

As long bones develop in the child, osteoblasts continue to build bone around the outside, causing the long bone to grow thicker. This is referred to as *ap- positional growth*. At the same time, osteoclasts hollow out the inside of the diaphysis to increase the size of the medullary cavity. It would seem that the completion of the bone-building process would end the function of osteoblasts and osteoclasts, but as is discussed later, their activities continue throughout life.

Bone Tissue

It was mentioned previously that bone is composed of two types of bone tissue: compact bone and spongy bone. In all four types of bone, compact bone forms the outer layer, with spongy bone contained in the interior.

Compact Bone

If compact, or dense, bone is viewed under the microscope, its appearance resembles that of a stack of tree trunks, complete with rings. Each of these tree trunk-like structures is known as an osteon, or bone unit. They are found side by side from the periosteum to the medullary cavity in a long bone and run parallel with the length of the bone. Osteons consist of the following structures (Figure 4.7):

- A *central canal* is also known as the Haversian canal. In the center of each osteon is a hollow core containing blood vessels. Osteocytes are living tissue, so they need a supply of nutrients and oxygen as well as a means to dispose of waste. Blood vessels in the central canal provide for the metabolic needs of osteocytes.
- Around the central canal are concentric rings of bone known as *lamellae*. Think of a candle that has been dipped in wax five or six times. If a cut is made across the candle, the wick in the center is similar to the blood vessels in the central canal. Rings of wax surround the wick just like the lamellae encircle the canal.

> **Osteon** the structure of which bone is constructed. It consists of a central canal containing blood vessels surrounded by lamellae, rings of bone. Within the lamellae are spaces, lacunae, which contain bone cells, osteocytes.

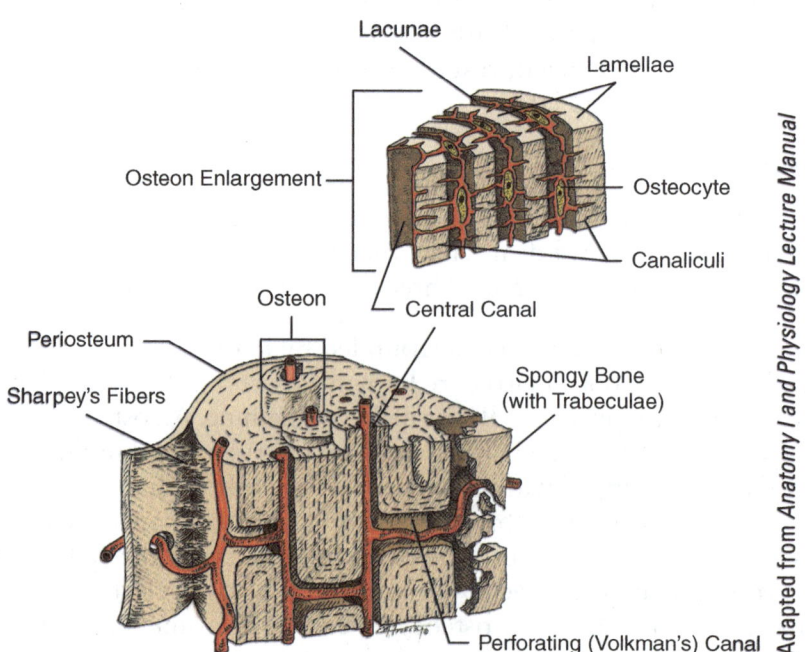

Figure 4.7 Osteons, or Bone Units. Osteons are found side by side from the periosteum to the medullary cavity.

- On each ring are spaces where osteocytes live. These spaces are known as *lacunae*.
- A problem for a cell surrounded by a concrete-like wall is that it cannot migrate to obtain nutrients. Nutrients are in the blood vessels in the central canal, but the osteocytes may be some distance away and surrounded by bone matrix. Hairlike cracks known as *canaliculi* that run from the central canal to the lacunae or between lacunae develop through which tissue fluid can travel. Blood is not flowing through the canaliculi. The blood remains in the blood vessels in the central canal. Fluid from the capillaries in the canal can flow slowly through the canaliculi to nourish the osteocyte and to haul away waste.
- Because the osteons run parallel to the length of the diaphysis, it is necessary to have *perforating canals* that run perpendicular to the osteon's central canal that can provide a pathway for blood vessels to reach each central canal.

Spongy (Cancellous) Bone

As mentioned earlier in this chapter, the epiphysis of a dried long bone contains what looks like a dried sponge known as *spongy (cancellous) bone* (Figure 4.8). The structures within spongy bone that give it this appearance are known as trabeculae. This latticework essentially spreads the weightbearing capabilities of the long bone over the entire epiphysis.

> **Trabeculae** latticework of bone found in spongy bone; is formed primarily to support weight.

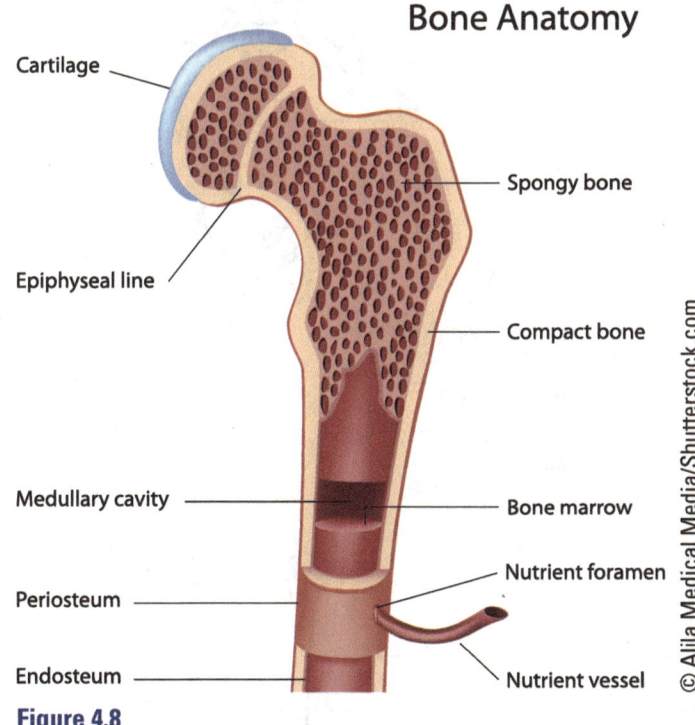

Figure 4.8

Comprehension Check-up:
1. The dense, hard substance around a bone cell is known as _____.
2. The latticework found in spongy bone is known as _____.

1. matrix 2. trabeculae

THE SKELETON

> **Axial skeleton** the bones found in the head and trunk.

> **Appendicular skeleton** the bones found in the upper and lower extremities.

Study of the skeleton provides more than just knowledge of bones. It becomes a reference point that can be used to identify regions of the body. The bones that make up the skeleton can be divided into two categories: the axial skeleton and the appendicular skeleton (Figure 4.9a). The axial skeleton is composed of the bones found in the head and trunk. The appendicular skeleton consists of the bones that make up the extremities or appendages.

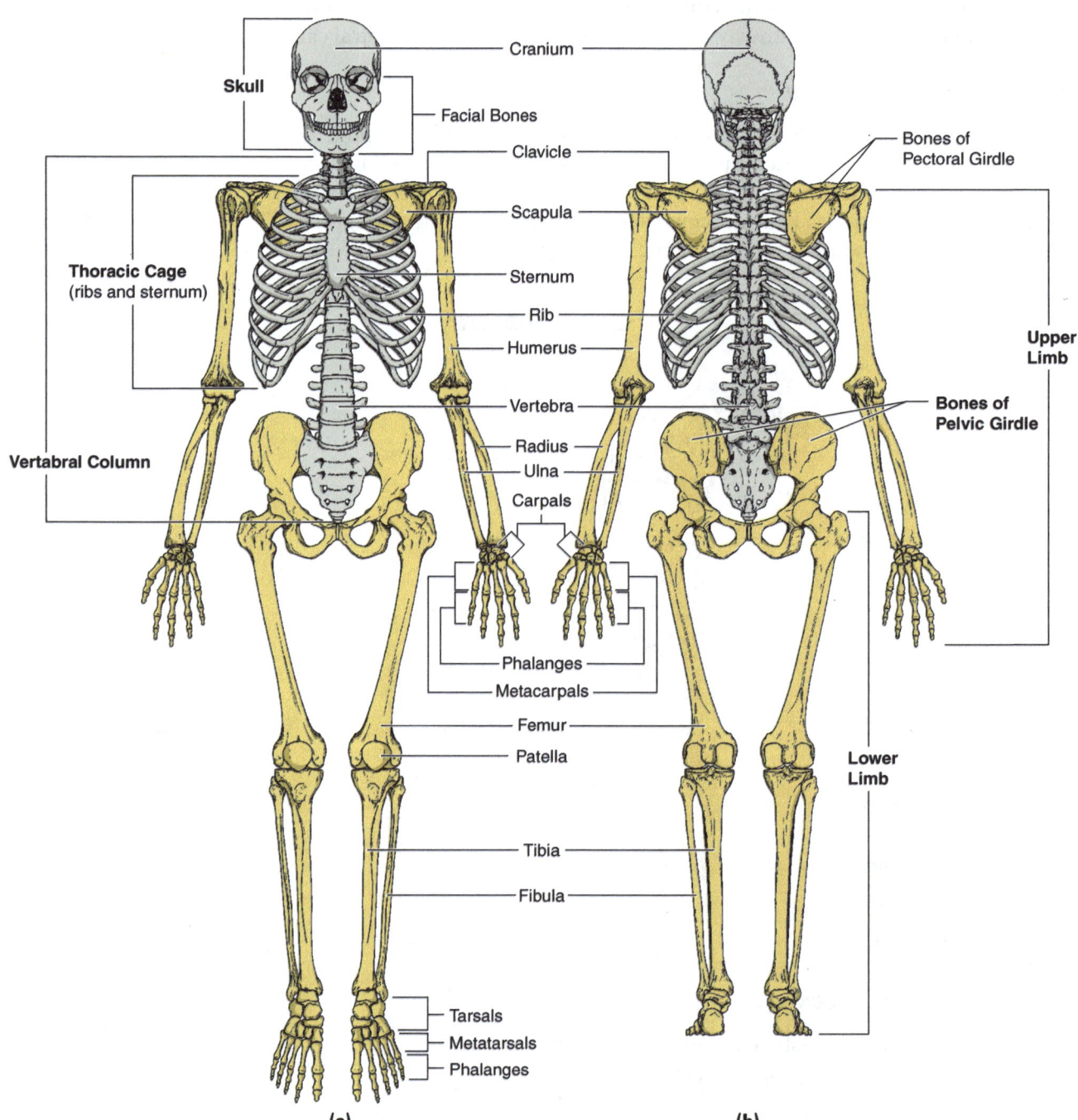

FIGURE 4.9 Skeleton. (a) Anterior view of the skeleton. (b) Posterior view of the skeleton.

The Axial Skeleton

The axial skeleton includes the skull, vertebrae, and bones of the thorax. Each area will be considered separately.

The Skull

The skull can be divided into two general sections: cranium and facial bones. In most cases, there are complementary bones on the left and right sides.

> **Skull** the bones comprising the head.

The Cranium

The cranium consists of bones surrounding and protecting the brain. It is composed of the following bones (See Figures 4.10–4.13):

- The *frontal bone* forms the forehead. It also includes the upper portion of the eye sockets. There is a single frontal bone.
- The *parietal bones* form the top of the skull. They are two relatively large plates of bone that cover a majority of the brain.
- The *temporal bones,* as the name implies, are found at the temples. These paired bones are found on the lateral skull and also contain the ear canal.
- The *occipital bone* is found in the back of the head and forms the base of the skull. In the occipital bone is a large hole through which the spinal cord enters the head.
- The *sphenoid bone* is shaped like a bat or butterfly. It lies just anterior to the temporal bone and forms the anterior inferior cranium. It is a single bone that runs to both sides of the head. From a lateral view of the skull, it would appear that there are left and right sphenoid bones but it is actually one continuous bone.
- The *ethmoid bone* is shaped somewhat like a walnut. It is found in the anterior medial area of the cranium. On the upper surface of the ethmoid bone is a depression that contains many holes, known as the cribriform plate. The olfactory nerves, which sense smell, pass from the nasal cavity through the holes of the cribriform plate into the brain.

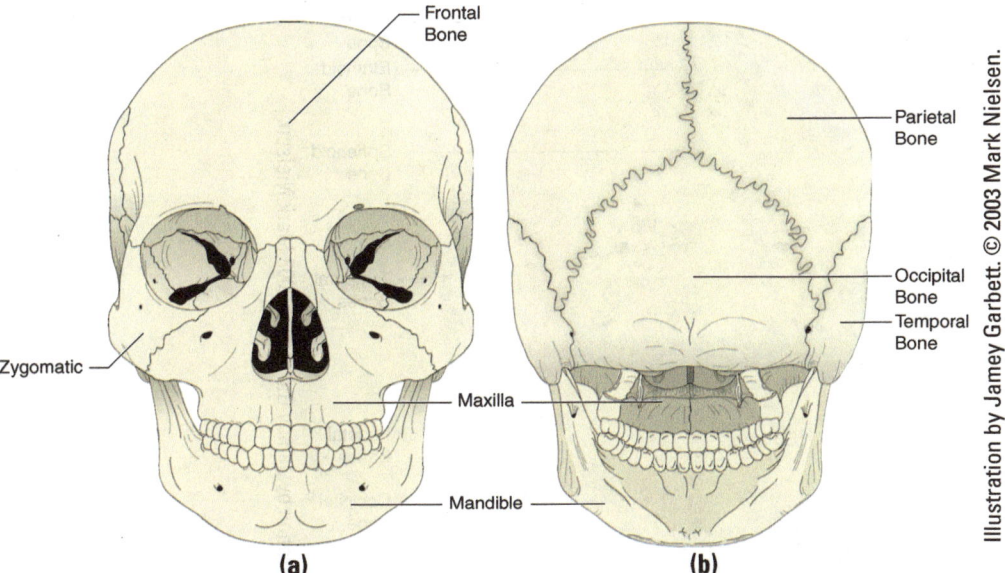

Figure 4.10 Anterior and Posterior Views of the Skull. (a) Anterior, (b) Posterior.

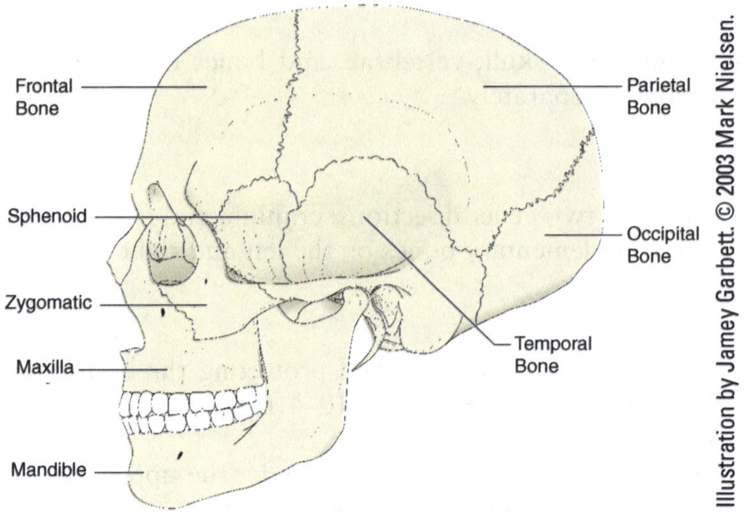

Figure 4.11 Lateral View of the Skull.

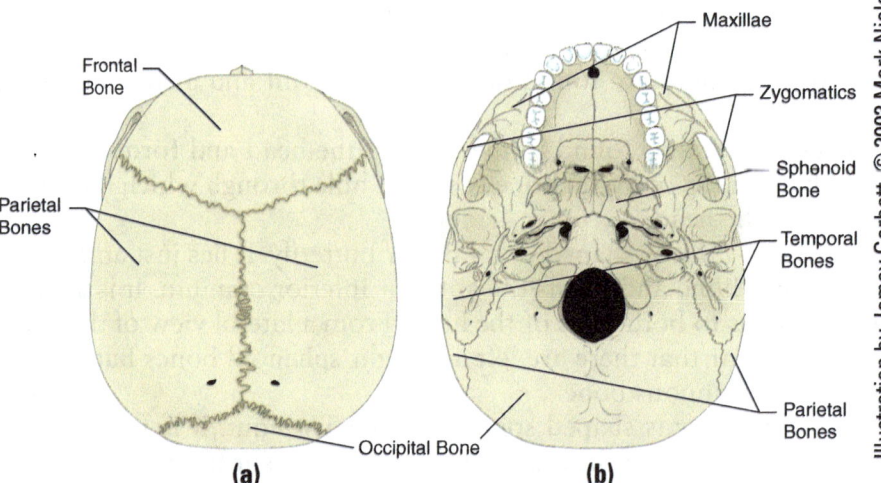

Figure 4.12 Superior and Inferior Views of the Skull. (a) Superior. (b) Inferior.

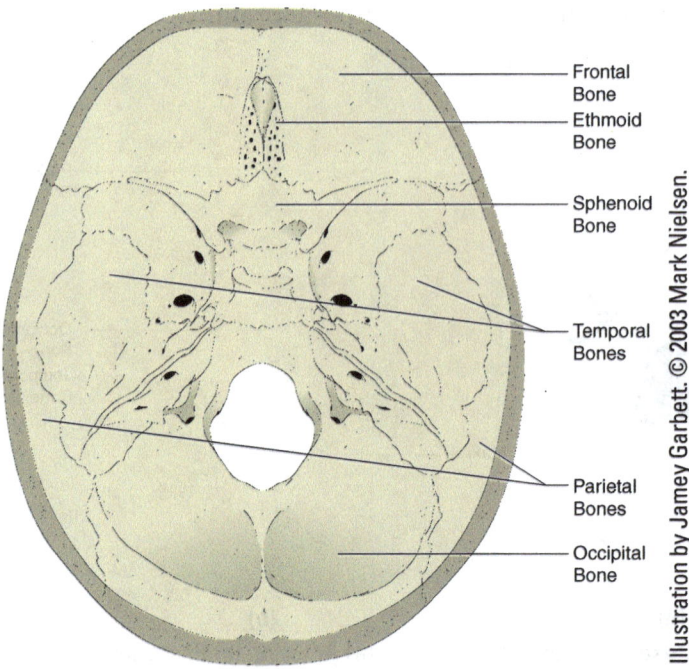

Figure 4.13 Superior View of the Internal Cranium.

Facial Bones

The major bones that comprise the face are the following:

- The *maxilla* is commonly referred to as the *upper jaw*. It contains the upper set of teeth. The two maxillae also form the roof of the mouth.
- *Zygomatic bones* are the anterior portion of the cheekbones. The arch forming the cheekbone is actually a combination of extensions or processes from the zygomatic and temporal bones.
- The *mandible* is commonly known as the *lower jaw*. It contains the lower set of teeth. The posterior superior portion of the mandible terminates as mandibular condyles, which fit into the temporal bone, forming the temporomandibular joint, and allows the mandible to move as the individual chews.

Other Associated Bones or Markings

It is helpful when studying the skull to locate a few associated areas of interest. The *hyoid bone* is a U-shaped bone found in the neck (Figure 4.14). It is not attached to any other bone but rather is held in place by muscle. It becomes an anchor for muscle movements that involve swallowing and tongue movement.

The *paranasal sinuses* are spaces within some skull bones (Figure 4.15). Paranasal sinuses resonate the voice to give it a fuller sound and also decrease the weight of the bones in which they are found. There are four paranasal sinuses:

1. Frontal
2. Sphenoid
3. Maxillary
4. Ethmoid

The sinuses are lined with mucous membranes to protect against invasion. If the membranes become inflamed because of infection, swelling occurs and the protective mucus is unable to drain. The sinuses fill with fluid, altering the individual's voice. Sinus infections can be painful as the pressure within these cavities pushes against sensitive sinus tissue.

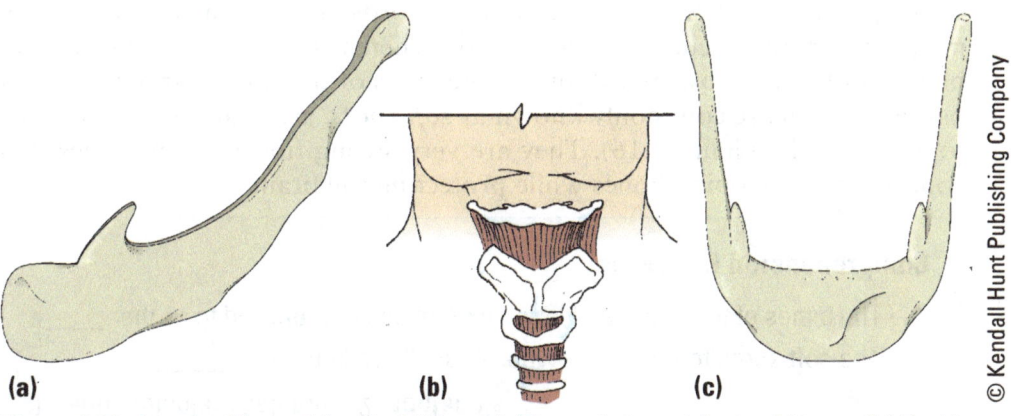

Figure 4.14 Anterior View of the Hyoid Bone. (a) Lateral view of the hyoid bone. (b) Hyoid bone in the neck. (c) Anterior view of the hyoid bone.

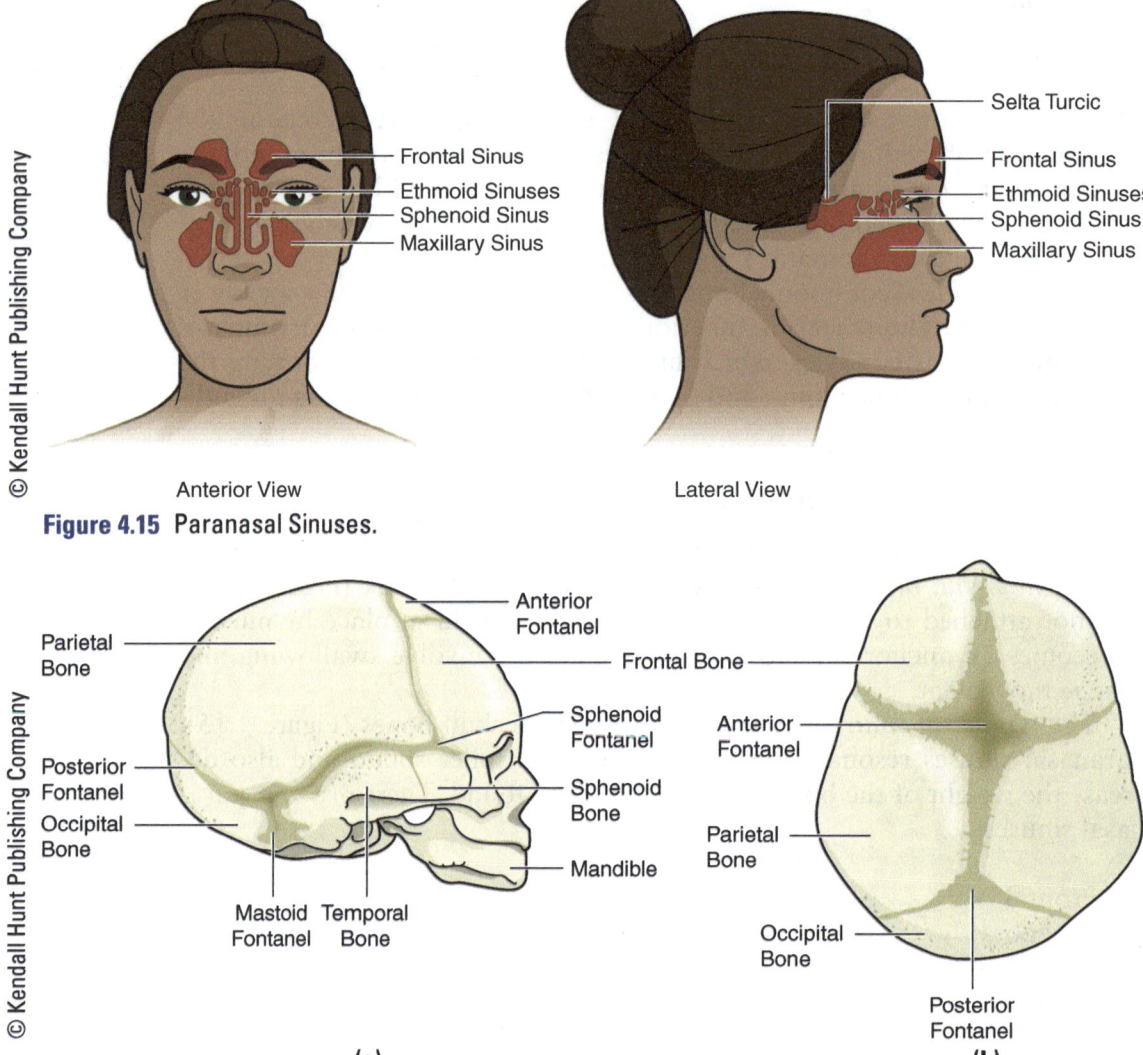

Figure 4.15 Paranasal Sinuses.

Figure 4.16 Fetal Skull. (a) Lateral view. (b) Superior view.

The Fetal Skull

As mentioned earlier in this chapter, the skull in the newborn is not yet completely ossified to allow the brain to continue to grow after birth. It also allows flexibility of the skull during birth. The baby's skull is about one-fourth the entire length of the newborn's body. In the adult, the skull is only about one-eighth the total body length. In other words, most of the brain development has already occurred prior to birth, whereas the rest of the body completes most of its growth after birth. The areas of the skull that have not yet ossified, which are commonly known as *soft spots,* are anatomically referred to as fontanels (Figure 4.16). They are very strong fibrous membranes that connect the developing bones while protecting the brain.

Comprehension Check-up:

1. The bones of the upper and lower extremities are referred to as the _____.
2. The soft spots found in a newborn's skull are known as _____.

1. appendicular skeleton   2. fontanels

The Vertebral Column

The **vertebral column** is a flexible set of stacked interlocking vertebrae that provide attachment for the head on the superior end, for the ribs in the middle thoracic section, and for the pelvic bones connected to the sacrum (Figure 4.17). There is a space within the vertebral column that contains the spinal cord.

Each vertebra has the same basic structures found at every level of the spine. The following are structures of the vertebra (Figure 4.18):

- The *body* of the vertebra is the area shaped like a thick disc that is flat on its superior and inferior surfaces. The bodies are the anterior portion of the vertebral column. They are stacked on top of each other all the way down the spine.
- The *spinous process* is a projection toward the back. These can be felt when running the fingers down someone's back over the spine.
- *Transverse processes* project laterally. All processes serve as attachments for muscles.
- The *vertebral foramen* is a large hole found in each vertebra through which the spinal cord passes.
- The *intervertebral disc* is a gel-filled pad found between the bodies of each vertebra. They are shock absorbers designed to prevent damage to the vertebrae. If they become weak, they may bulge or rupture, causing pain or numbness as they push on nerves exiting the spine.

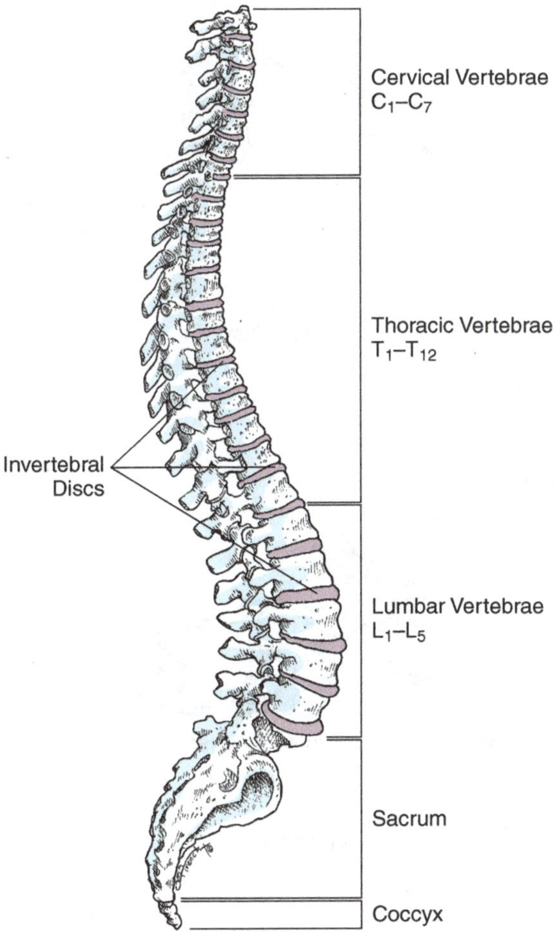

> **Vertebral column** a flexible set of stacked interlocking vertebrae.

Figure 4.17 A Vertebral Column.

Cervical Vertebrae C_1–C_7
Thoracic Vertebrae T_1–T_{12}
Invertebral Discs
Lumbar Vertebrae L_1–L_5
Sacrum
Coccyx

illustration by Jamey Garbett. © 2003 Mark Nielsen.

Adapted from *Anatomy I and Physiology Lecture Manual*

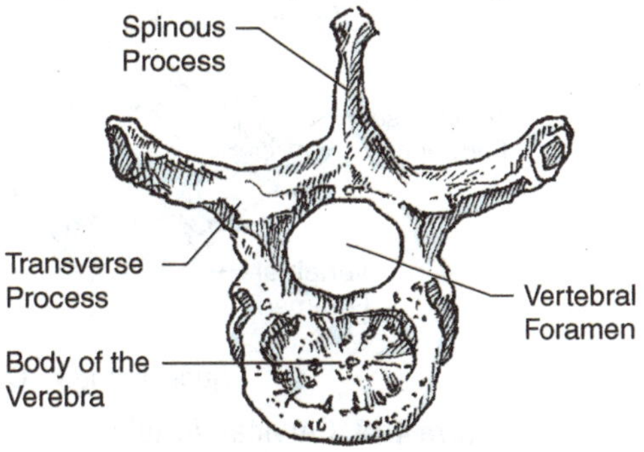

Figure 4.18 A Vertebra—Superior View.

Spinous Process, Transverse Process, Body of the Verebra, Vertebral Foramen

The vertebral column can be divided into sections involving the neck, chest, and lower back.

- *Cervical vertebrae* are found in the neck. Humans have seven cervical vertebrae. Cervical vertebrae are easy to identify even when mixed randomly with other bones of the spine. The body of the cervical vertebra is small, but the identifying feature is that on both sides of the body are holes known as transverse foramina (Figure 4.19). Arteries pass through these holes. Only cervical vertebrae have transverse foramina.
- *Thoracic vertebrae* are located in the posterior chest. There are 12. Thoracic vertebrae typically have a spinous process that points downward (Figure 4.20).
- *Lumbar vertebrae* are in the lower back. There are five. Their bodies are quite large, and their spinous process is blunt, as though it was snipped off (Figure 4.21). Viewed from the back, lumbar vertebrae resemble a moose head.
- The *sacrum* is part of the pelvis. It is actually five bones that fuse together as the individual ages (Figure 4.22).
- The *coccyx* is commonly known as the tailbone and consists of three or four bones fused together.

Adapted from *Anatomy I and Physiology Lecture Manual*

Figure 4.19 Cervical Vertebra.

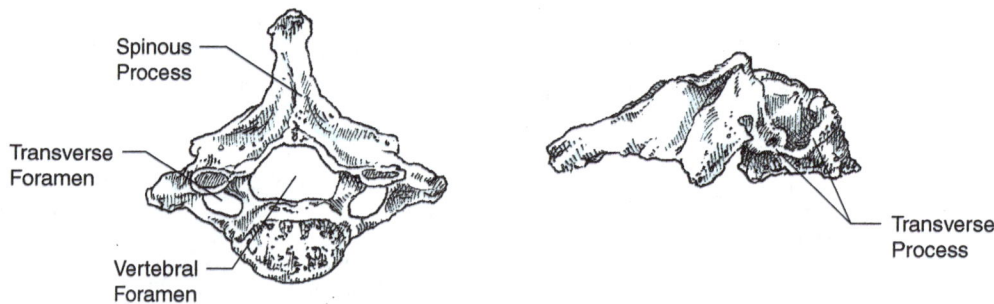

Adapted from *Anatomy I and Physiology Lecture Manual*

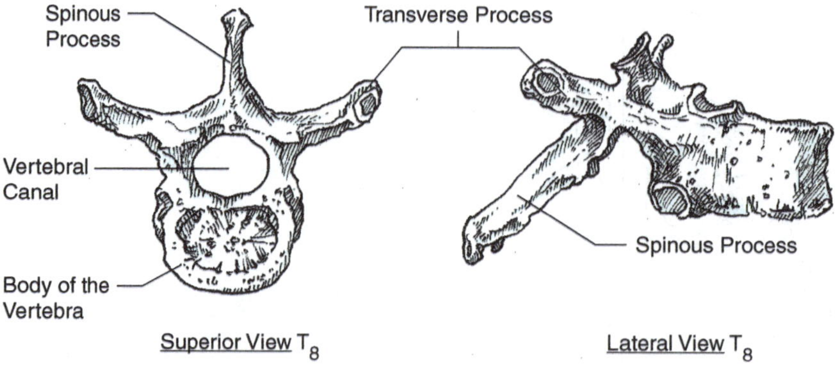

Figure 4.20 Thoracic Vertebra.

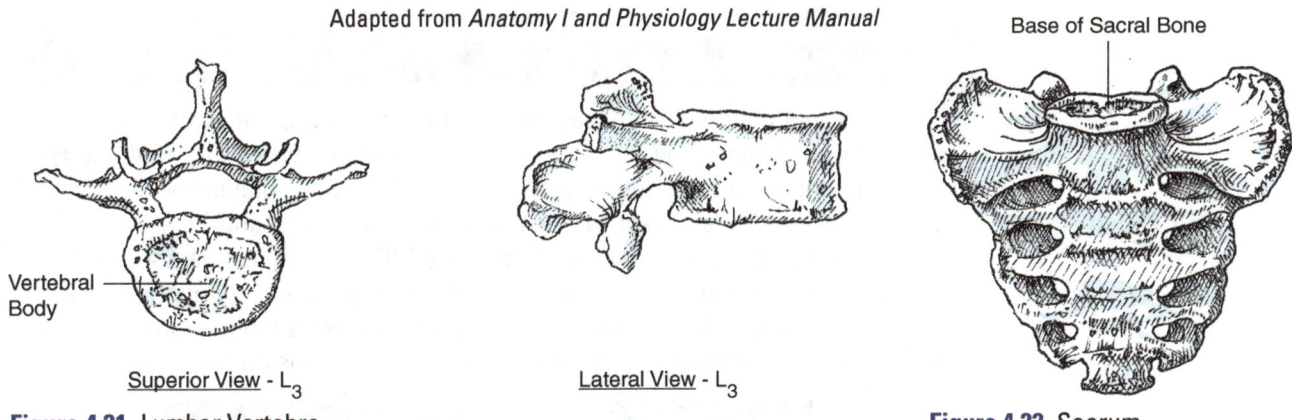

Figure 4.21 Lumbar Vertebra.

Figure 4.22 Sacrum.

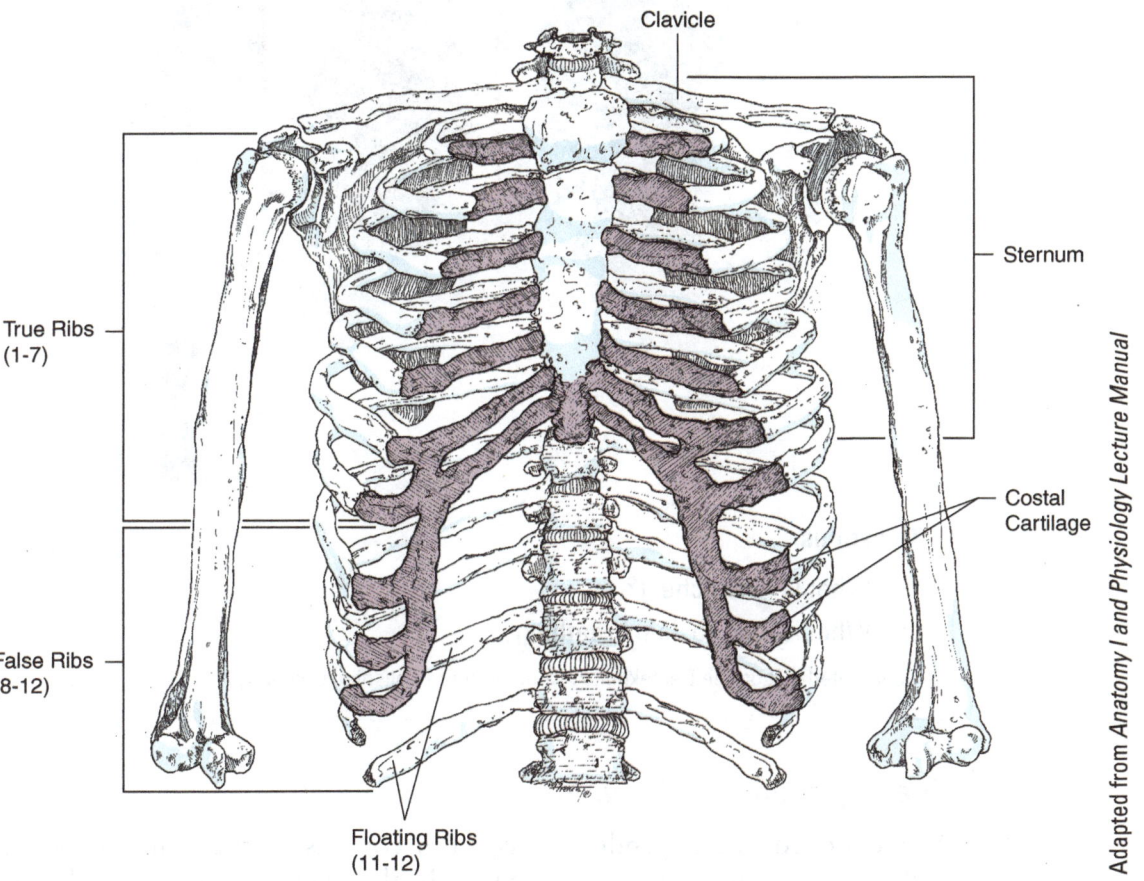

Figure 4.23 Rib Cage.

The Thoracic Cage

Surrounding and protecting the organs of the chest is the **thoracic cage** (rib cage). In the front is the *sternum*, which is commonly known as the breastbone. Attached between the sternum and vertebrae are 12 pairs of *ribs* (Figure 4.23). The upper seven pairs of ribs that attach directly to the sternum are considered true ribs. The lower five pairs are known as *false ribs*. Their cartilage attaches to the cartilage of other ribs rather than to the sternum, or they do not attach on the anterior chest at all. The two most inferior false ribs that are unattached on the anterior end are called floating ribs.

> **Thoracic cage** twelve pair of ribs attaching directly or indirectly to the sternum in the anterior thorax which provide protection to visceral organs within the thoracic cavity.

> **Clinical Scenario—Initial Assessment of a Patient with Chest Trauma (Fractured Ribs)**
>
> A 21-year-old male suffered trauma to his chest from a car accident. There was a forceful impact on his chest because he wasn't wearing a seatbelt. He was brought to the emergency room by the EMTs. He complains now of severe pain on his left lower chest. He has shortness of breath, and looks pale. His vital signs are: Heart Rate (HR) or Pulse Rate (PR) is 115/min; Respiratory Rate (RR) is 30/min; Blood Pressure (BP) is 90/60 mmHg. On auscultation, the physician notices that there is decreased breath sounds on the left lung. A chest x-ray has been ordered that shows several fractured ribs on the left chest.
>
>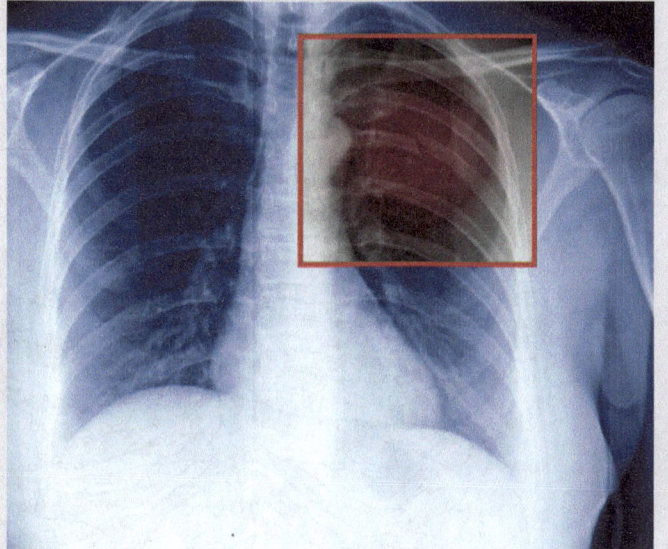
>
> **Questions:**
> 1. What is a flail chest?
> 2. What is paradoxical breathing?
>
> Contributed by Eugene Demekhin. © Kendall Hunt Publishing Company

The Appendicular Skeleton

As mentioned, the appendicular skeleton consists of the bones associated with the upper and lower extremities. It also includes the pectoral girdle consisting of the clavicle and scapula through which the upper extremity attaches to the trunk. The lower extremity attaches to the trunk through the pelvic girdle. The arrangement of both the upper and lower extremities is the same in that there is a large single upper bone followed by two smaller bones in the forearm and calf. At the distal end are eight short bones making up the wrist; in the ankle there are seven. Both the palm and instep of the foot contain five bones. The fingers or toes, referred to as *digits*, on the end of each extremity consist of 14 bones.

The Pectoral Girdle and Upper Extremities

The upper extremities are not welded to the trunk by nonmoving attachments. Instead, most of the upper appendage is held in place by muscle.

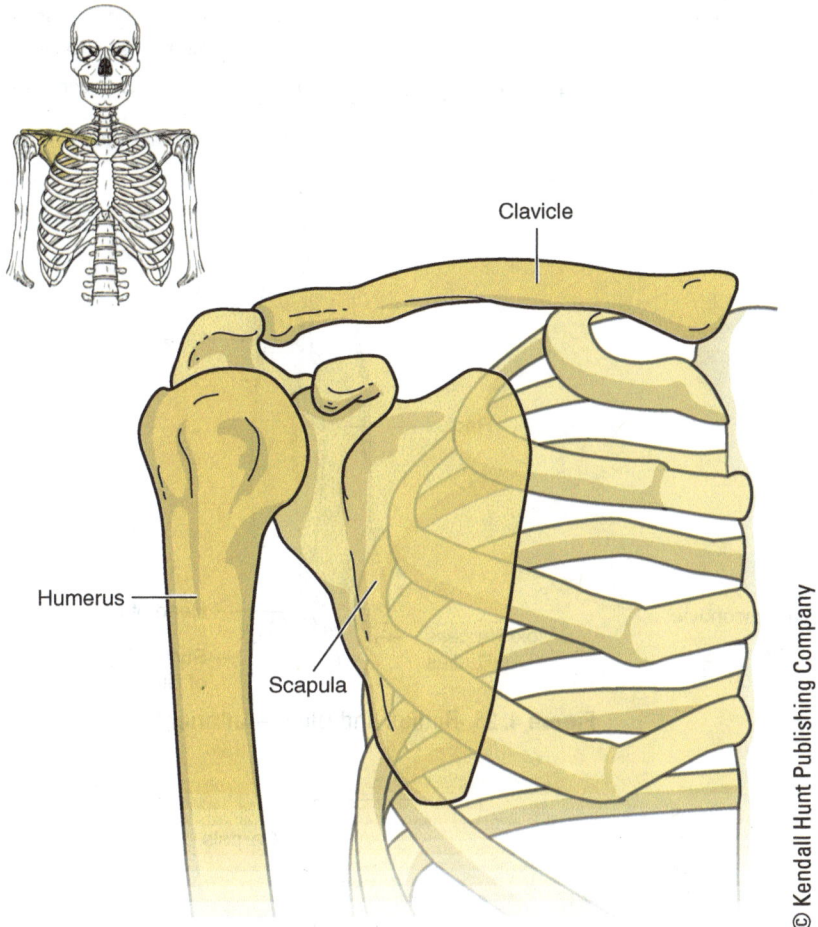

Figure 4.24 Anterior View of the Pectoral Girdle.

This forms what is known as the shoulder, or *pectoral girdle*. The bones involved are:

Pectoral Girdle (Figure 4.24)

- The *scapula*, commonly called the *shoulder blade*, is found in the upper back. It contains the shoulder socket into which the upper extremity is attached. The scapula is attached to the trunk and held in place by muscle.
- The *clavicle*, commonly known as the *collarbone*, attaches the pectoral girdle with the sternum. It is located in the upper anterior chest between the sternum and the shoulder. It is the most commonly broken bone in the human body.

Upper Extremity

- The *humerus* is the single bone in the upper arm. It has a ball on the proximal end to allow movement in all directions (Figure 4.25).
- The *radius* is the lateral bone in the forearm on the same side as the thumb (Figure 4.26). When looking at the radius by itself, note that the proximal end has a round disc, like the radius of a circle.

Chapter 4 The Skeletal System

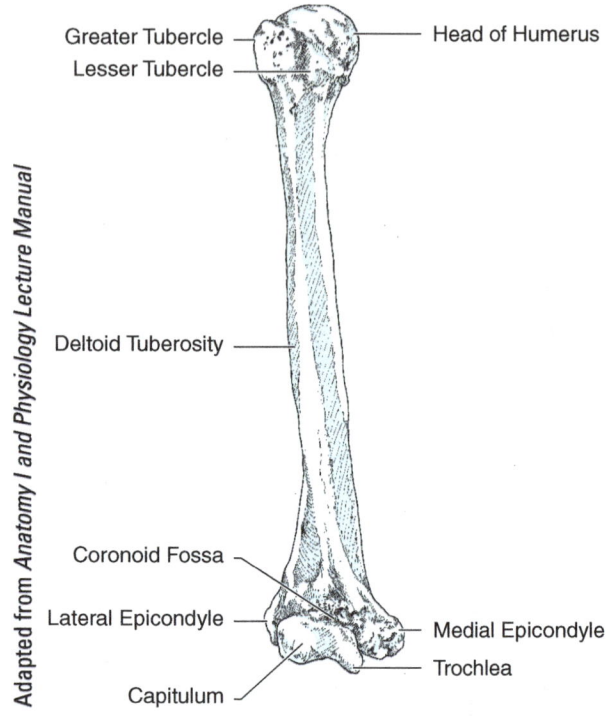

Figure 4.25 Humerus—Anterior View.

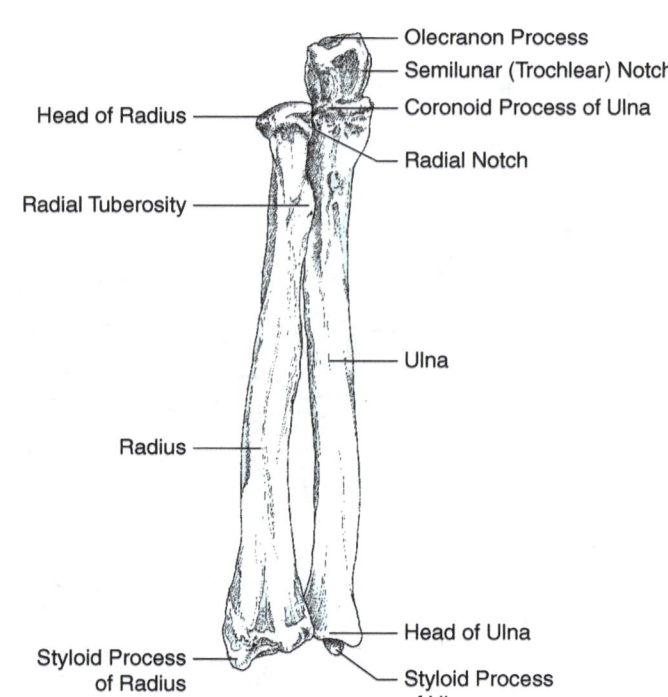

Figure 4.26 Radius and Ulna—Anterior View.

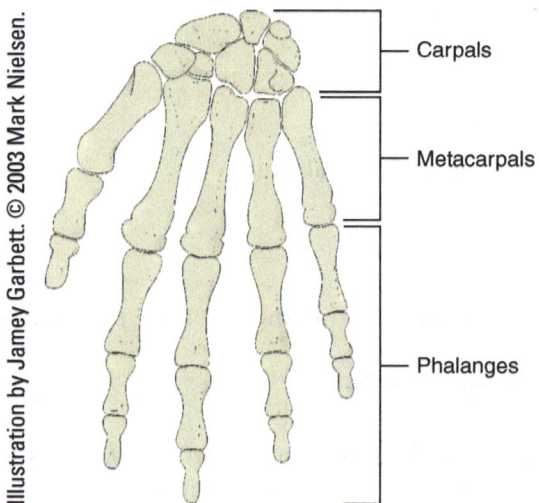

Figure 4.27 The Hand—Anterior View.

- The *ulna* is on the medial side of the forearm and forms the elbow (Figure 4.26). Looking at the ulna separated from the rest of the skeleton, you will notice that on the proximal end is a notch that forms a "U," as in *ulna*.
- The eight short bones that form the wrist are known as *carpals* (Figure 4.27).
- The five bones of the palm distal to the carpals are known as *metacarpals* (Figure 4.27).
- *Phalanges* are the bones of the fingers (Figure 4.27). A single finger bone is known as a *phalanx*. There are two phalanges in the thumb and three in each of the fingers (digits).

The Pelvic Girdle and Lower Extremities

The pelvic girdle is the location for the attachments of the lower extremities to the body. There are three bones that, together, form the os coxa, commonly known as the *pelvic bone,* which articulates (forms a joint) with the sacrum to form the pelvic girdle.

The *pelvic girdle* is made up of three paired bones that fuse together with age (Figure 4.28):

- The *ilium* is the large, flat upper bone of the pelvis. It projects more laterally in the female than in the male.
- The *ischium* is the inferior posterior bone of the pelvic girdle. Located on the ischium is the *ischial tuberosity,* upon which we sit.
- The *pubis* is a relatively small bone found in the anterior pelvis. During childbirth, the joint between the two pubic bones stretches to allow the baby to pass through more easily.

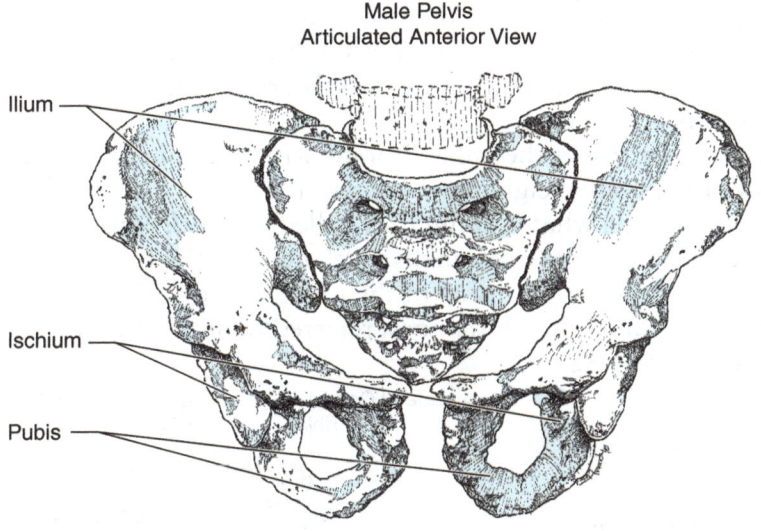

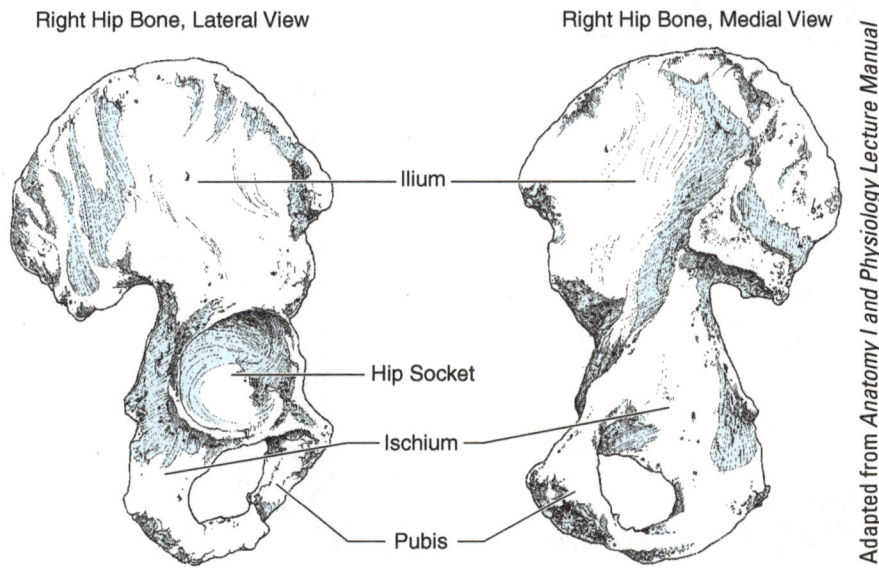

Figure 4.28 Pelvic Girdle.

Lower Extremity

- The *femur* is the longest bone in the body (Figure 4.29). It is located in the thigh. Like the humerus, it has a ball on the proximal end, known as the *head,* which articulates with the pelvis to form a deep socket with greater stability but less mobility than the shoulder.
- The *tibia* is the larger bone on the medial side of the leg (Figure 4.30). The ridge that runs down the front of the tibia is commonly referred to as the *shin*. When running, the pull of the anterior muscles of the lower extremity will sometimes cause small cracks in the tibia (shin splints).

When looking at the tibia by itself, it is helpful to note that the top forms a "T." Some individuals refer to it as the *tubby tibia* when trying to recall the relationships of the bones of the calf.

- The *fibula* is the smaller bone on the lateral side of the leg (Figure 4.30). It has often been referred to as the *little fib*.
- The *patella* is commonly known as the kneecap. It is a bone embedded in the patellar tendon that runs over the knee (Figure 4.31).
- At the distal end of the leg are the *tarsals,* which are the bones of the ankle (Figure 4.32). Although all of the tarsals are named, there are two of particular importance to us:
  - The *talus* is at the top of the ankle. It joins with the tibia and fibula to receive all of the body weight on the top of the foot.
  - The *calcaneus* forms what is commonly called the *heel*. Muscles in the posterior calf attach to the calcaneus to plantarflex the foot.
- The five *metatarsals* form the instep of the foot (Figure 4.32).
- As with the fingers of the hand, the toes, referred to as *digits,* are known as *phalanges* (Figure 4.32).

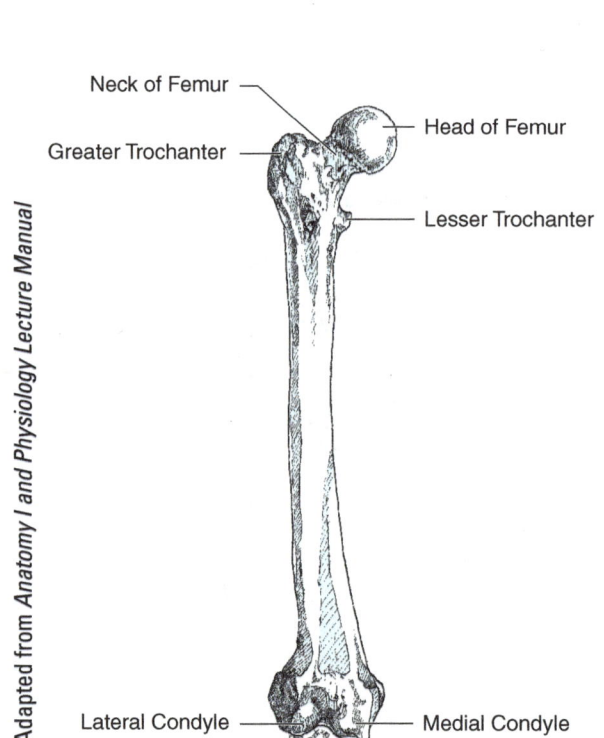

Figure 4.29 Femur—Anterior View.

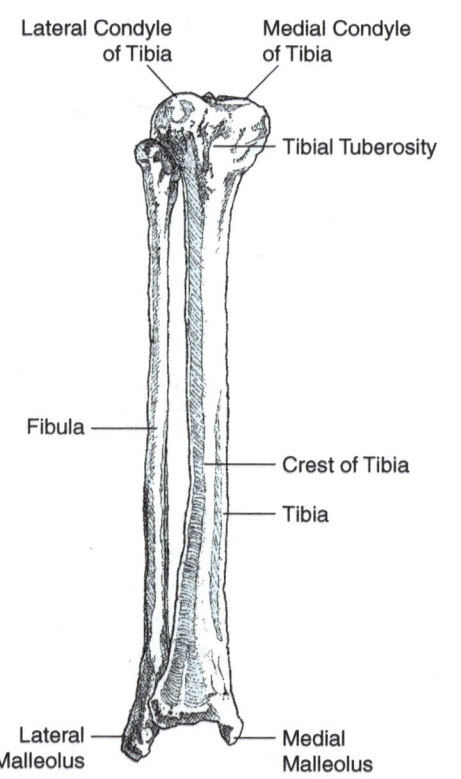

Figure 4.30 Tibia and Fibula—Anterior View.

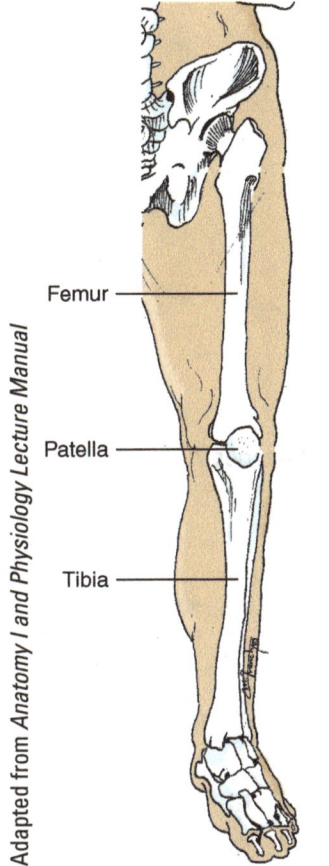

Figure 4.31 Patella—Anterior View.

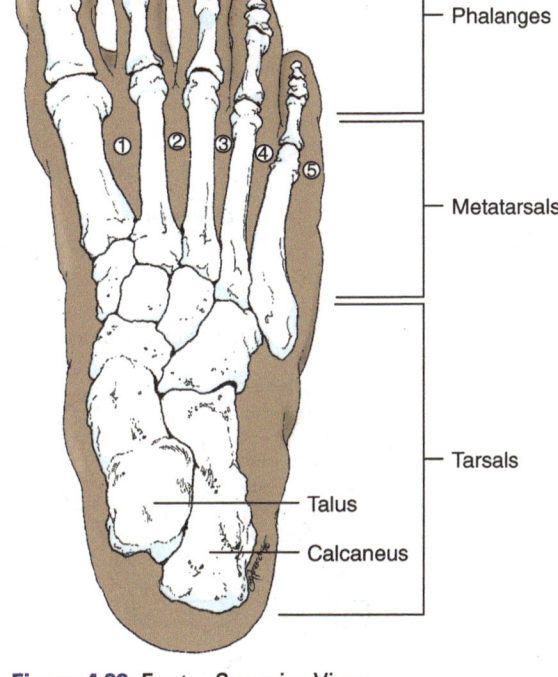

Figure 4.32 Foot—Superior View.

> **Comprehension Check-up:**
>
> 1. The anatomical name for the shoulder blade is the _____.
> 2. The three bones fused together in the adult where the lower extremity attaches to the body is known as the _____.
>
> 1. scapula   2. pelvic girdle

Major Structural Differences between Sexes in the Pelvic Girdle

The pelvis is able to provide more clues about the sex of skeletal remains to the anthropologist, archeologist, or forensic pathologist than any other area of the skeletal system. Overall the pelvic girdle is wider in females than in males. The larger pelvic cavity provides more space for the uterus to grow as the fetus develops. The pelvic inlet (sometimes referred to as the "true pelvis" and commonly known as the birth canal), is the large hole within the pelvis in females. It is larger than in the male (Figures 4.33 and 4.34). In the anterior pelvis the pubic bones form an inverted "V"' known as the *subpubic angle,* inferior to these bones. In males the subpubic angle is at a much steeper angle than in females (Figures 4.35 and 4.36). There are other sex-determined variances in the pelvis that can provide the trained scientist with critical information about the gender of a discovered skeleton.

Chapter 4 The Skeletal System

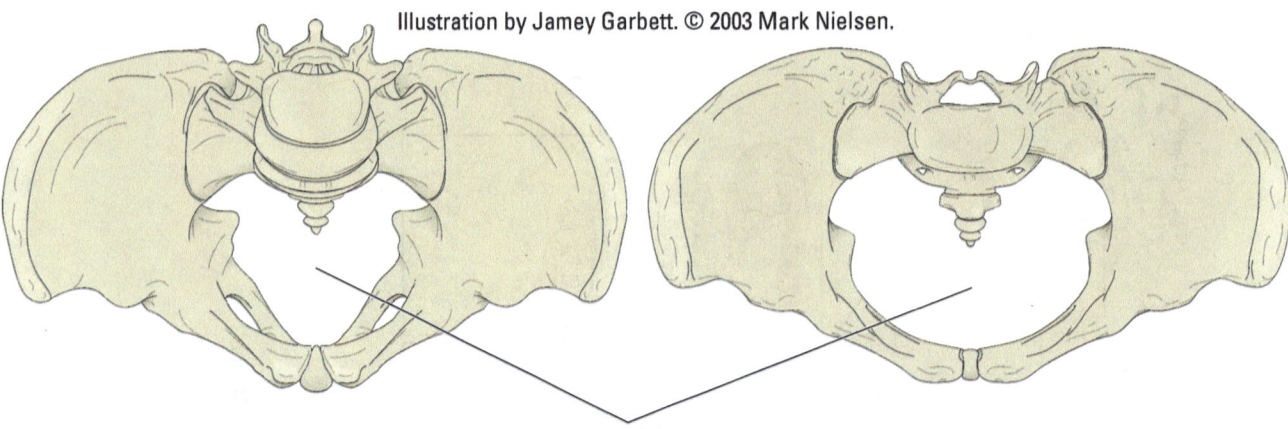

Figure 4.33 Female Pelvis—Superior View. **Figure 4.34** Male Pelvis—Superior View.

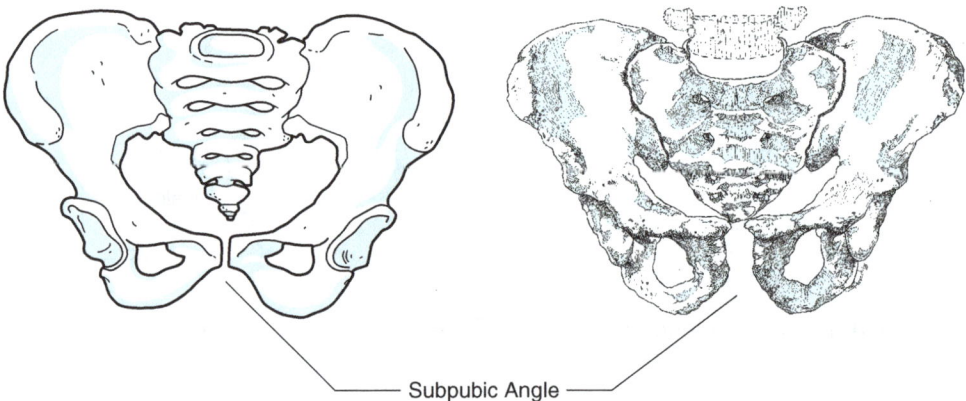

Figure 4.35 Female Pelvis—Anterior View. **Figure 4.36** Male Pelvis—Anterior View.

Clinical Notes—Knee Injury (Knee Cap or Patella)

The most common injuries of the knee include knee cap fractures, dislocation of the knee cap, and when the knee is bent at 90° forcing the knee to absorb the pressure of body weight. Sports such as basketball, soccer, football, gymnastics, and wrestling commonly cause knee injuries. The most common signs are pain and swelling. Treatment of knee injury depends on the type and severity of injury. Treatments include RICE therapy (rest, ice, compression, and elevation), physical therapy, or surgery when it is severe. Athletes need to perform stretching and strengthening exercises on their quadriceps to minimize risk of knee injury.

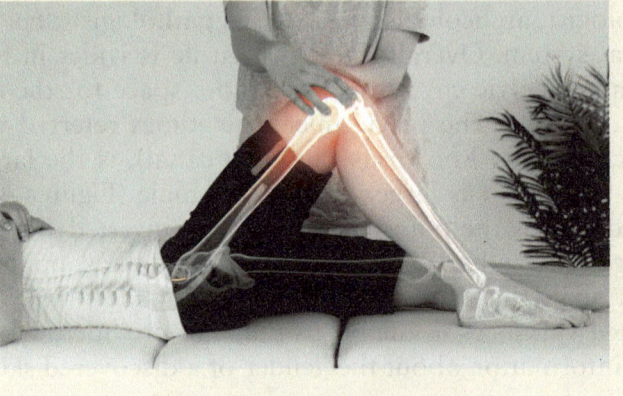

JOINTS (ARTICULATIONS)

Two or more bones in contact with each other form joints, or articulations. Bones are held together by ligaments. Along with the shape of a bone's structures, it is the placement of the ligaments that have a major effect on the movement of a joint. For example, one of the phalanges is shaped to slide against the next, but the medial and lateral placement of ligaments allows them to move in only one direction. The proximal ends of the humerus and femur are shaped like a ball allowing those bones to pivot in their retrospective sockets in all three planes.

> **Articulations** two or more bones in contact with each other.
>
> **Ligaments** strong connective tissue, primarily collagen fibers, which hold bone to bone.

Joint Classifications

The classification of joints may be based on the structural features between bones, or it may be identified by the extent of movement each will allow. The following provides a list based on structural features.

Fibrous Joints

Fibrous joints are composed of dense connective tissue; these result in essentially nonmoving joints. They are found between most bones of the skull (Figure 4.37). There are other fibrous joints as well, including the joint between our teeth and bone and also the distal radioulnar and tibiofibular joints.

> **Fibrous joints** the contact of bones forming a nonmoving joint.

Cartilagenous Joints

Cartilagenous joints, as the name implies, are connected by cartilage and are somewhat movable. They are needed where limited movement is essential. Two examples will describe their use. The first is found between the ribs and sternum. This rib-shaped cartilage allows the ribs to raise and lower without cracking or breaking the connection (Figure 4.38). The sternum is

> **Cartilagenous joints** joints composed of cartilage between bones forming a somewhat movable joint.

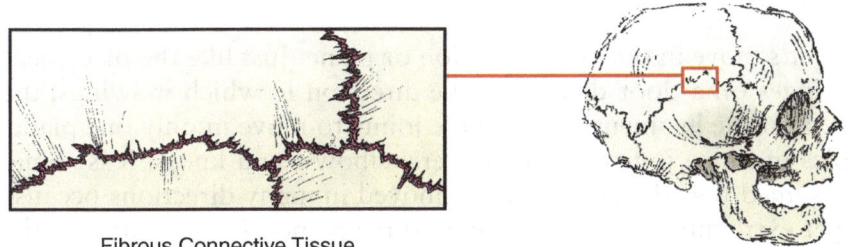

Figure 4.37 Fibrous Joint Attaches Nonmoving Bones.

Figure 4.38 Cartilage between ribs and sternum allows a slight amount of movement when breathing.

Adapted from *Anatomy I and Physiology Lecture Manual*

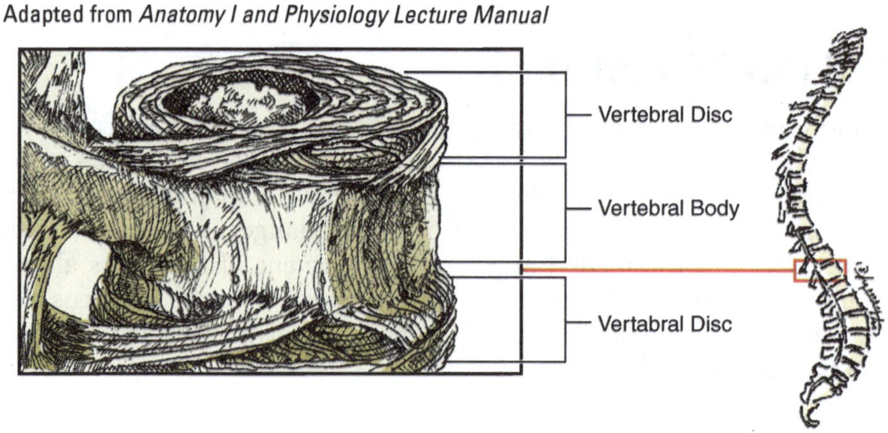

Figure 4.39 Intervertebral discs not only act as pads between the bodies of each vertebrae, but they also allow a small amount of rotation of each vertebrae.

also able to move in the anterior direction as these cartilaginous joints twist when we breathe deeply. Another example is the intervertebral disc. As mentioned before, the discs are gel-filled pads between the bodies of the vertebrae (Figure 4.39). These discs are composed of fibrocartilage, which provides a very tough outer membrane enclosing a gel-like material. Intervertebral discs account for about one-third the total length of the spine.

Synovial Joints

Synovial joints are composed of structures that allow them to be freely movable. They are found primarily in the extremities (Figure 4.40). The position of the ligaments around these joints and the shape of the bones where they come into contact play major roles in determining the direction of movement. They are also well lubricated by synovial fluid, a thick fluid roughly the consistency of egg whites that minimizes irritation between the ends of the articulating bones.

The following are examples of synovial joints (Figure 4.41):

- *Hinge joints* move in only one direction or plane. Just like the placement of the hinges on a door determine the direction in which it swings, the placement of the ligaments hinge these joints to move in only one plane. Examples of hinge joints are the fingers, elbows, and knees. It is sometimes assumed that the elbow can be moved in many directions because the upper extremity moves freely in all three planes. In actuality it is the shoulder that has the capacity to move in many planes. If the shoulder is not allowed to move, the elbow can swing in only one plane.
- *Pivot joints* also move in only one plane. In this case, one bone pivots on another. Consider, for example, the first two cervical vertebrae. The first cervical vertebra pivots on the second to allow the individual to shake his or her head "no."
- *Gliding (planar) joints* form as two relatively flat bone surfaces glide against each other. An example of a planar joint is found between carpal bones as they glide against each other in the wrist. Some planar joints move on one plane although others may glide in two planes.
- *Condyloid joints* move in two planes. They are found in the knuckles (metacarpophalangeal joint), where the fingers can be bent toward the palm or they can be spread out side to side.

> **Synovial joints** freely movable joints.

[Peroneal is the Grk. = to Fibular]

Adapted from *Anatomy I and Physiology Lecture Manual*

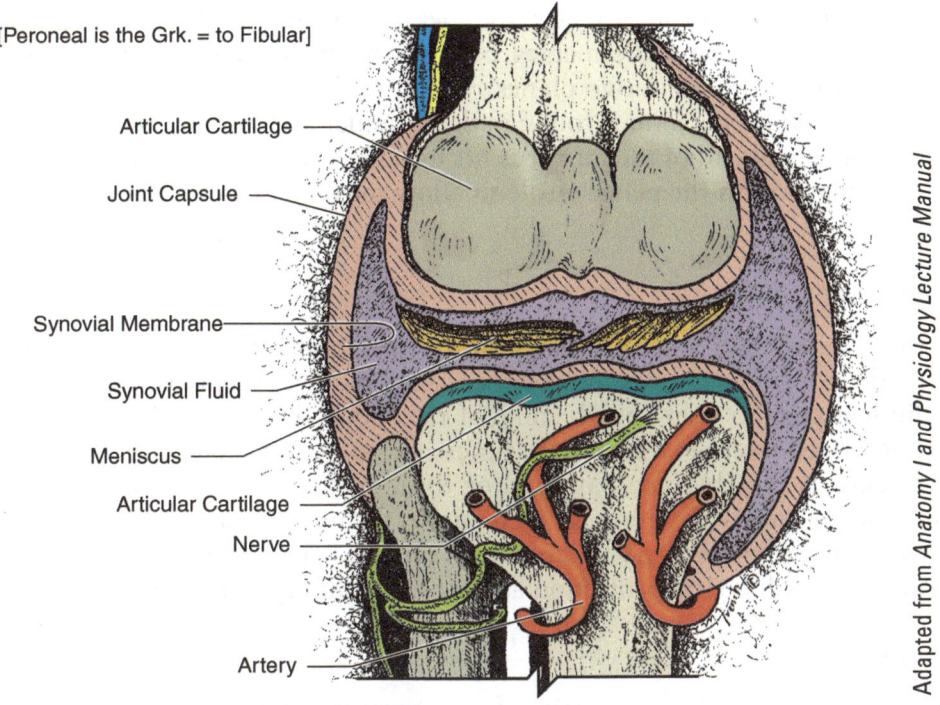

Figure 4.40 **Synovial Joint.** Anterior view of a synovial joint found in the knee.

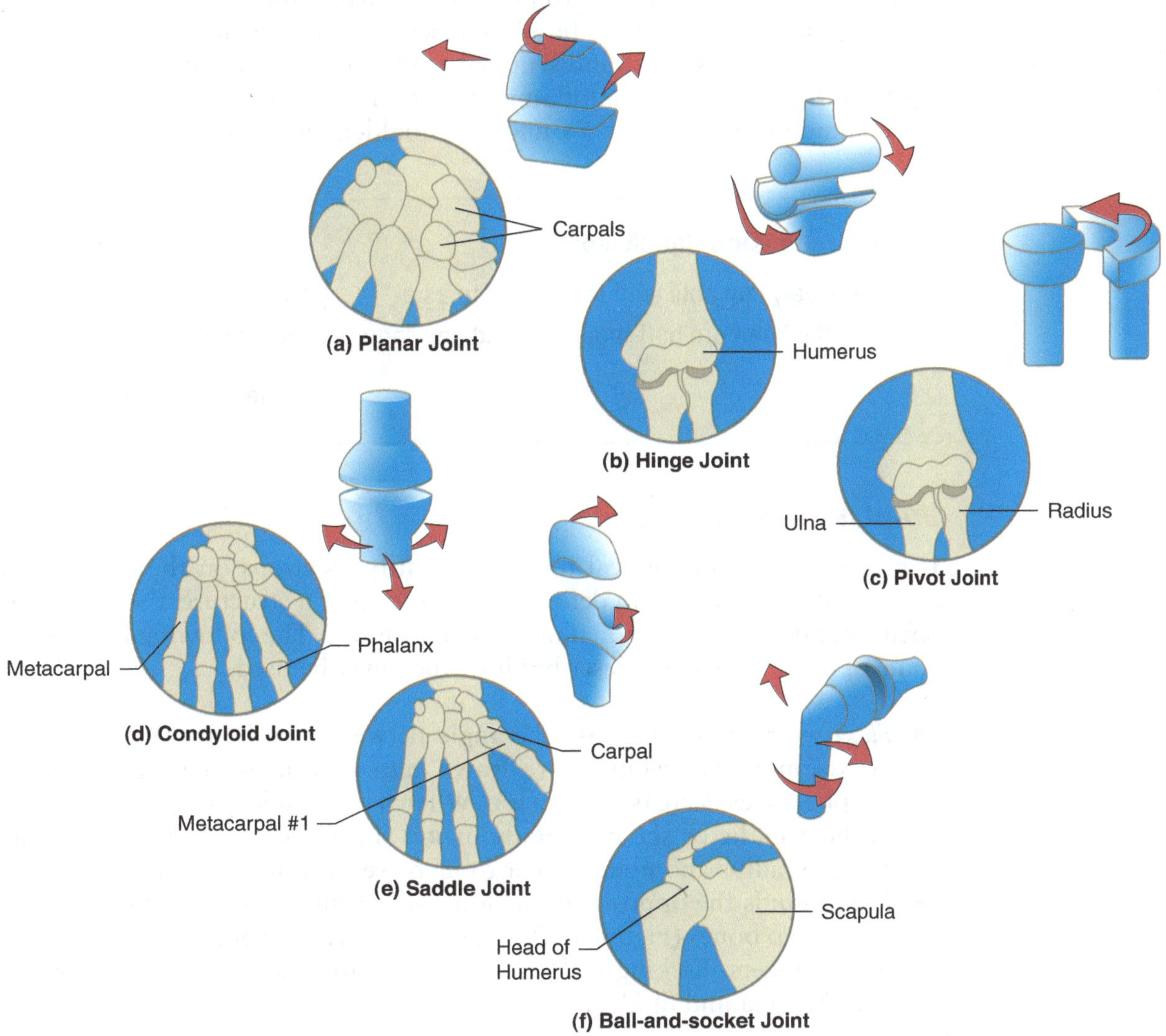

Figure 4.41 Types of Joints.

- A *saddle joint* is found where the metacarpal for the thumb joins with one of the carpals. The metacarpal "sits in the saddle" to allow movement of the thumb in two planes.
- A *ball-and-socket joint* is found in the shoulder and hip, where the ball on the proximal end of the humerus fits into the pectoral girdle or the femur articulates with the pelvic girdle to allow movement in all three planes.

Meniscus

> **Menisci** shock-absorbing fibrocartilage pads found in the knee joint.

In the knee are half-moon-shaped fibrocartilage pads known as menisci. Because there is so much stress on the knee as people run and jump, the knee has additional pads to help absorb some of the shock. They are able to provide only a limited level of stress relief. If pressure on the knee becomes excessive, the meniscus may become damaged or torn. It is a common athletic injury. Joints are packed with nerves, so an injury to the menisci can be quite painful. The options are either to try to repair torn cartilage or to remove it. Its removal means there will be less protection against future injury to the femur or tibia.

Bursa

> **Bursa** fluid-filled pad providing protection between potential sources of irritation.

In some locations, the movement of structures against each other can be a potential source of irritation. To protect the surrounding tissue, a fluid-filled pad known as a bursa may be placed between these sources of irritation. They are found mostly around synovial joints, and they lessen friction resulting from body movements. Inflammation of a bursa is known as *bursitis*. It is a common source of pain around the shoulder, knee, and hip.

> **Comprehension Check-up:**
> 1. A freely movable joint is referred to as a _____ joint.
> 2. In the knee are half-moon-shaped cartilage pads known as _____.
>
> 1. synovial 2. menisci

Joint Movements

Just as there are anatomical terms to identify locations on the body or to assist medical personnel in defining the position of a structure, there are also terms that describe directional movement of joints. The following list defines some of those terms and describes how they may be used.

- *Flexion* means to decrease the angle between two bones (Figure 4.42). For example, as the fingers are drawn in to make a fist, the angle between the phalanges decreases in flexion. When a bodybuilder flexes the muscles in the arm, he or she commonly makes the angle between the humerus and the radius and ulna smaller to increase the bulk of the muscles.
- *Extension* is the opposite of flexion and means to increase the angle between two bones (Figure 4.42). Opening the hand from a fist is extension of the fingers. Kicking increases the posterior angle between the femur and the tibia and fibula.

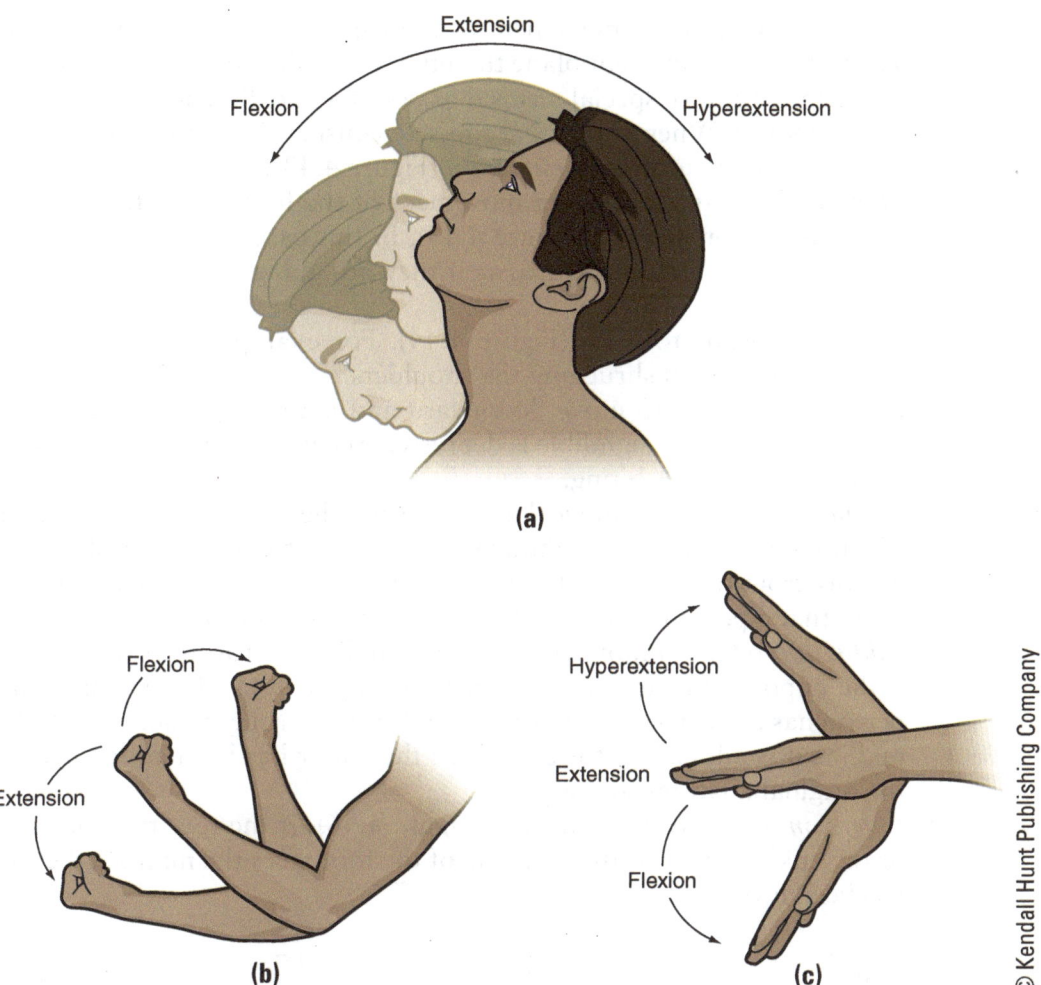

Figure 4.42 Examples of Flexion, Extension, and Hyperextension.

- *Hyperextension* occurs when a joint is extended beyond the normal anatomical position (Figure 4.43). For example, if in the anatomical position the wrist is flexed, the hand bends forward, decreasing the angle between the hand and the forearm. Extension of the hand places it back in its original position. If, however, the hand is bent in the posterior direction, it is now extended farther than the anatomical position and is said to be hyperextended. The same action occurs when a person bends his or her head back to look up at the ceiling. This is hyperextension of the neck. Some joints are designed to allow hyperextension; others are not. Hyperextension of a joint not intended to move that far can cause considerable discomfort. For example, many ball players are aware of the pain of hyperextended fingers. Very often, when the abnormal hyperextension of a joint occurs, the result is injury to the ligaments holding the bones in place.

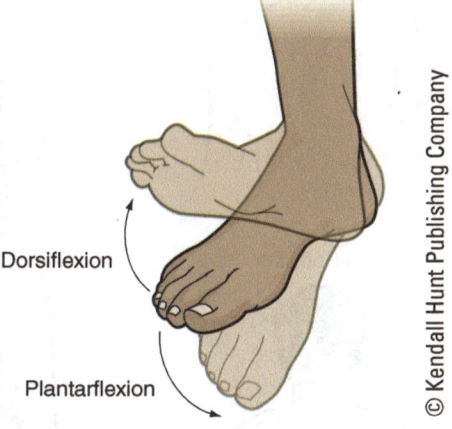

Figure 4.43 Dorsiflexion/Plantarflexion.

- *Dorsiflexion* refers to movement of the feet in the anatomical position. The feet are in a different plane than the rest of the body. For this reason, it is helpful to have special terms that describe the direction in which the foot is moved. When the toes are raised, causing the individual to stand on the heels, the term is *dorsiflexion* (Figure 4.43).
- *Plantarflexion* again involves movement of the foot but in the opposite direction of dorsiflexion (Figure 4.43). In this case the foot is flexed so that the toes point downward, as if they are going to be planted in the ground.
- *Elevation* means to raise (Figure 4.44). For example, elevation of the shoulders results in shrugging the shoulders.
- *Depression* means to move downward and is the opposite of elevation (Figure 4.44). If the mandible is depressed, the mouth opens. Elevation of the mandible causes biting.
- *Abduction* means to move laterally from the midline or to the side (Figure 4.45). If a person abducts the upper extremity, he or she raises it away from the body to the side. One way to remember this term is to recall that a person who is abducted is kidnapped or taken away.
- *Adduction* refers to moving toward the midline or toward each other. It is the opposite movement of abduction (Figure 4.45). For example, if a person has abducted the right femur, when it is brought back toward the midline it is adducted. It is as though the femur has been added back to the original anatomical position.
- *Inversion* means to turn inward (Figure 4.46). If the toes turn medially or the ankle is bent so that the sole of the foot faces the midline, the foot has been inverted.

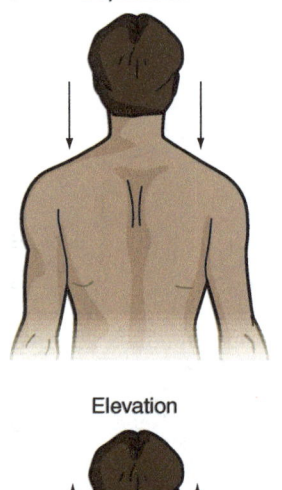

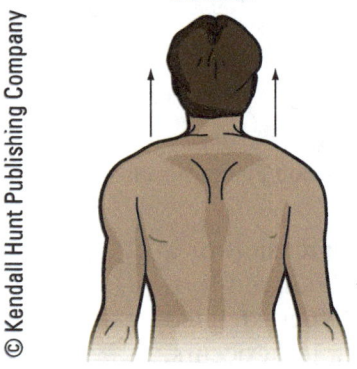

Figure 4.44 Depression/Elevation.

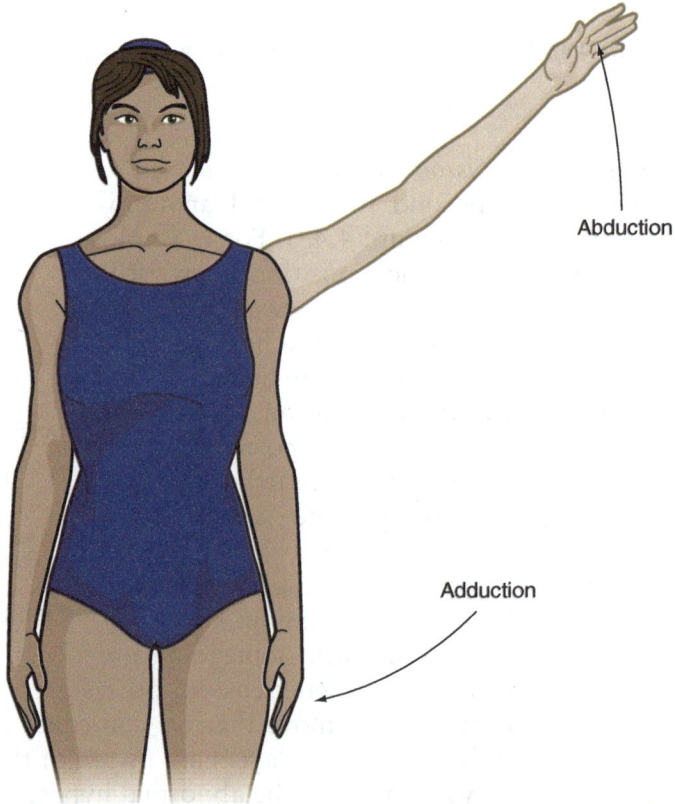

Figure 4.45 Abduction/Adduction.

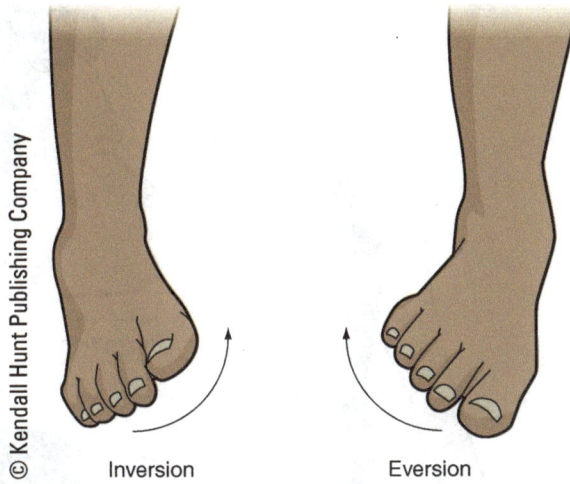

Figure 4.46 Inversion/Eversion.

- *Eversion* is the opposite of inversion and means to turn outward (Figure 4.46). Using the previous example, eversion of the foot would cause the sole of the foot to face the lateral side.
- *Supination* usually refers to the position of the hand. In the anatomical position, the palms face forward. Rotating the palm so that it faces toward the front of the body is known as *supination* (Figure 4.47).
- *Pronation* is the opposite of supination. Rotating the palm toward the back when in the anatomical position is known as *pronation* (Figure 4.47).
- *Protraction* refers to moving in the anterior direction, away from the body. If a person protracts the mandible, he or she pushes the lower jaw outward (Figure 4.48).
- *Retraction* is the opposite of protraction and means the body part is returned to its original anatomical position. The protraction of the mandible given in the previous example will be followed by retraction as it is brought back to its starting position (Figure 4.48).
- *Rotation* means to twist. For example, when a person rotates the head, he or she turns it from side to side (Figure 4.49).
- *Circumduction* is moving in a circle, as might occur in a person's arm if drawing a large circle on a whiteboard or wall. The distal end of the limb moves in a circle (Figure 4.49).

Not only are the terms for movement useful in describing the events involving bones, they will also be useful when describing the actions of muscles that cause those movements.

Comprehension Check-up:

1. The movement term for pointing the toes downward is _____.
2. Moving in the anterior direction away from the body is known as _____.

1. plantarflexion 2. protraction

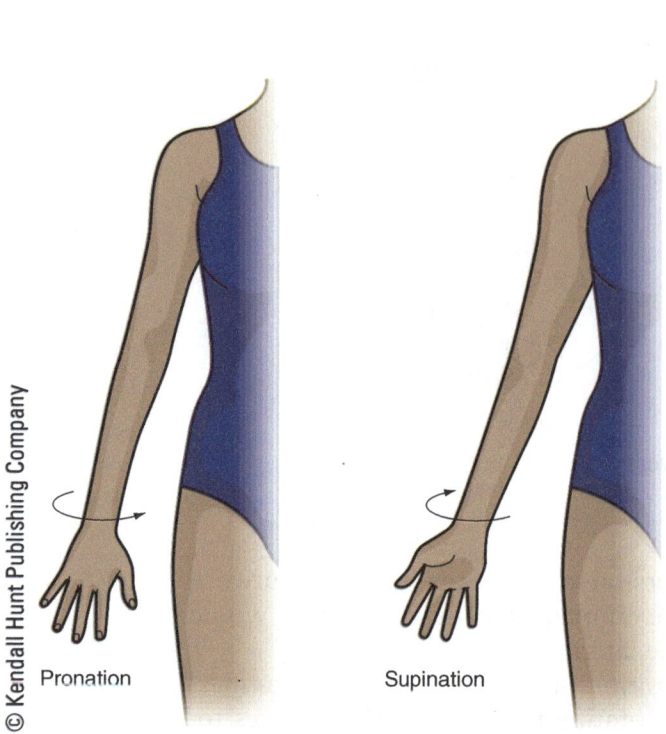

Figure 4.47 Pronation/Supination.

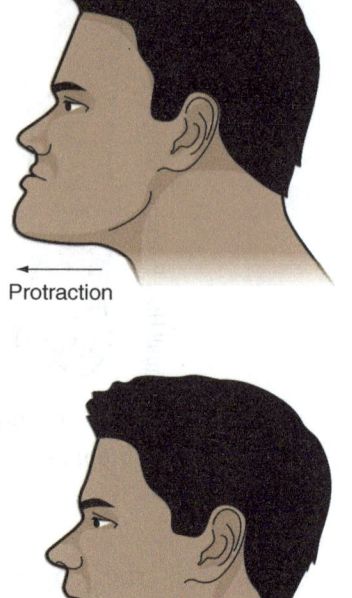

Figure 4.48 Protraction/Retraction.

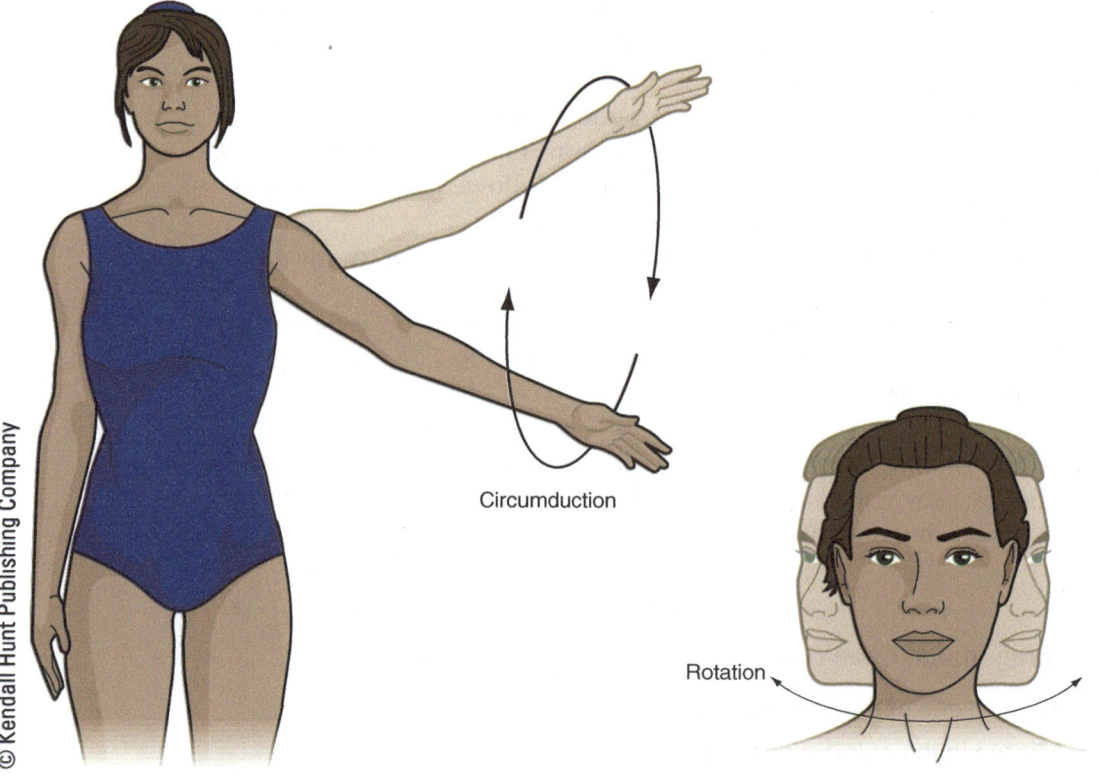

Figure 4.49 Rotation and Circumduction.

BONE GROWTH, REMODELING, AND REPAIR

Ossification is the process of bone growth and formation. There are two types of ossification: intramembranous and endochondral. In intramembranous ossification, the bone grows from the connective tissue such as the mesenchyme, and not from the cartilage. Intramembranous ossification occurs during the formation of the skull, clavicle, mandible, and maxilla. In endochondral ossification, the bone grows from the cartilage in growth plates. Growth plate (epiphyseal plate) is the site of growing tissue near the end of a long bone called epiphysis.

In intramembranous ossification, osteoblasts form the bones by depositing bony matrix in the collagen fibers in the membrane. Osteoblasts are large cells found on the external surface of the bone. They are responsible for synthesis and control of calcium and mineral deposits during the initial bone formation and later during the bone remodeling. These cells secrete the extracellular matrix that binds with calcium and hardens the matrix. This is called the osteoid, which is the non-mineralized portion of the bone matrix that hardens to form into a bone tissue. The osteoid becomes highly vascularized which forms the spongy bone. Spongy bone is also known as cancellous or trabecular bone. The outer layer of compact bone is formed first with spongy bone filling the interior. The spongy bone is reshaped throughout the process to allow space for red marrow within the center of the bone. The red

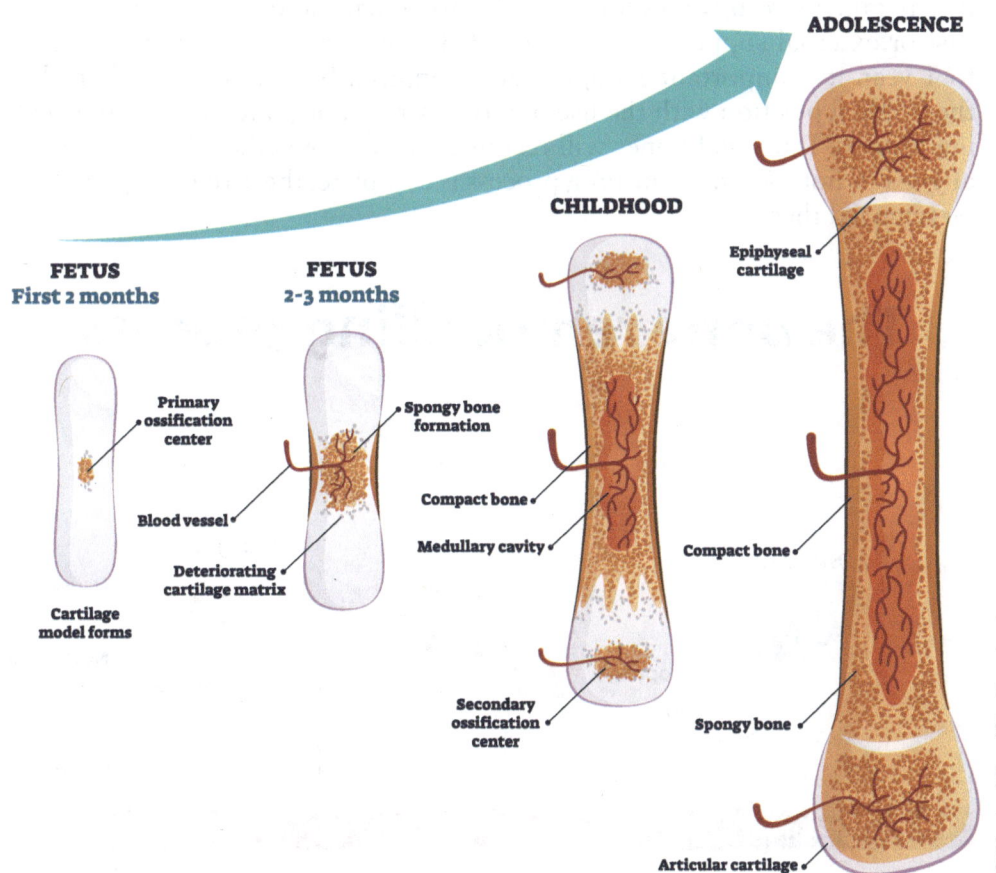

Figure 4.50

bone marrow produces the red blood cells and platelets and about 70% of lymphocytes of the human body.

In endochondral ossification, the hyaline cartilage is eventually replaced with bony tissue. Most bones of the human body are formed in this way. Chondrocytes proliferate in the growth plate that increases the length of the bone. Chondrocytes are cells found in cartilage connective tissue. They produce the cartilage matrix, which primarily contains collagen. An increase in bone length also increases the height of an individual. However, at around 18 and 20 years of age, skeletal maturity occurs when all cartilage growth stops; epiphyseal or growth plate closes; all cartilage is replaced by bone to form the epiphyseal line; and the epiphysis and diaphysis fuse together.

Bone remodeling is the process of removal of old, mineralized bone by osteoclasts (resorption), and formation of new bone by osteoblasts. Osteoclasts are large, multinucleated cells that breaks down bone tissue. Osteoblasts are cells that developed from the periosteum found in the outer layer of the bone. They deposit new bone and, therefore, play an important role in bone remodeling. Bone remodeling happens throughout life after birth. Growth hormone, insulin-like growth factor (IGF), and glucocorticoids stimulate bone remodeling and bone cell differentiation.

Wolf's law of bone formation states that when a bone is subjected to stress or force, the bone tissue would adapt to the stress by remodeling and becoming stronger. When a bone is injured, its blood supply is damaged also. This results in hematoma which forms a blood clot. A callus, consisting of a new cartilage and spongy bone, replaces the blood clot, in order to hold the injured bone tissue together, or to stabilize a bone fracture during repair. It may take up to 6 weeks for the callus to be hard and strong enough for a cast or external support to be removed. The healing or repair may take up to 1 year. It is important though that the injured bone tissues be aligned in the original position with the use of a rod, screws, or plates, during the healing process. Damaged bones will reconnect or heal whether their position is aligned or not. When the healing process is complete, the screws or plates can be removed then.

The bone remodelling process

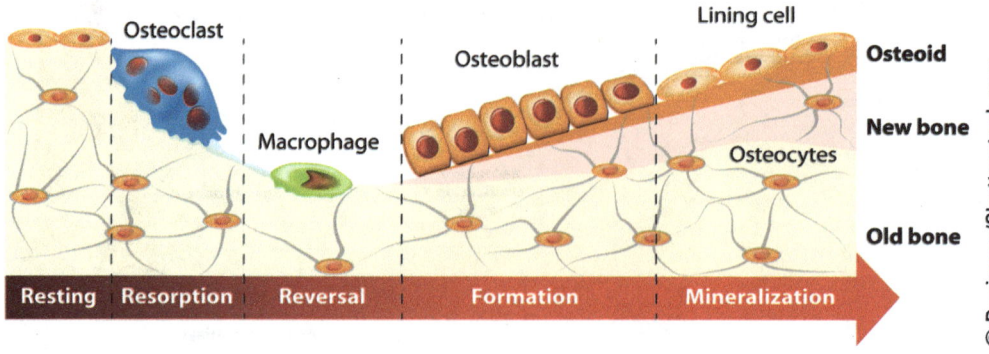

Figure 4.51

Stages Of Healing Bone Fracture

![Stages of bone fracture healing showing four stages: Week 1 Stage 1 Hematoma formation with hematoma and rupture of blood vessels; Week 2-3 Stage 2 Fibrocartilaginous Callus formation with internal callus (cartilage, fibrous tissue), external callus, spongy bone trabeculae, and new blood vessels; 1-4 Months Stage 3 Callus ossification; 4-12 Months Stage 4 Bone remodeling with consolidated fracture and bony callus of spongy bone.]

© Timonina/Shutterstock.com

Figure 4.52

Clinical Scenario—Gait Rehabilitation Program

A 50-year-old male with post-stroke left hemiparesis participated in a 6-week gait rehabilitation program. This patient has severe weakness on the left side of his body (left hemiparesis) that was caused by injury to the right side of his brain leaving him with a stroke (cerebrovascular accident or CVA). CVA occurs when blood flow to the brain stops causing the brain cells to die. About 80% of stroke survivors are caused by hemiparesis. The 6-week gait training consisted of treadmill walking for 3 sessions per week. The gait speed was measured before and after the training and showed an increase to 0.58 m per s from 0.38 m per s.

© Peakstock/Shutterstock.com

Contributed by Dave Smith. © Kendall Hunt Publishing Company

> **Clinical Skills—Simple Gait Evaluation**
>
> A simple way to evaluate gait from a healthy patient would be to perform the following steps where the patient would not have discomfort or losing balance:
>
> 1. Walk heel to toe
> 2. Walk on the toes
> 3. Walk on the heels
> 4. Hop in place
> 5. Do a shallow or small knee bend
> 6. Rise from a sitting position without assistance

CLINICAL TERMS TO REVIEW

Osteoporosis, hip dislocation
Red marrow, yellow marrow, long bones, flat bones, compact bone, spongy bone
Skull, cranium—occipital bone, temporal bone, parietal bone
Facial bone—maxilla, mandible, zygomatic bone
Hyoid bone
Vertebral column—vertebral foramen, intervertebral disc, cervical vertebrae, thoracic vertebrae, lumbar vertebrae, sacrum vertebrae, coccyx
Thoracic cage, scapula, clavicle
Humerus, radius, ulna, phalanges
Femur, tibia, fibula, patella
Pelvic bone, sacrum, ilium
Joints—synovial, fibrous, cartilaginous
Chest trauma, fractured ribs, flail chest, paradoxical breathing
Flexion, extension, dorsiflexion, plantarflexion, elevation
Abduction. Adduction, inversion, depression, supination, pronation
Retraction, rotation, circumduction
Ossification
CVA—cerebrovascular accident
Gait training, rehabilitation

Test Yourself

Choose the best answer to the following multiple choice questions:

1. Bones in the extremities, except for the wrist and ankle, are what type of bone?
   a. irregular
   b. short
   c. flat
   d. long

2. Long bones contain an internal space where, in children, blood-forming tissue known as _____ is found.
   a. yellow marrow
   b. red marrow
   c. periosteum
   d. bone matrix

3. When looking at compact bone histology, you will see rings of bone known as
   a. lacunae.
   b. canaliculi.
   c. lamellae.
   d. perforating canals.

4. Before bone forms, it first begins with a model made of
   a. muscle tissue.
   b. adipose tissue.
   c. pseudostratified squamous epithelium.
   d. hyaline cartilage.

5. The type of bone found in the epiphysis of long bones and between the two plates of flat bone is
   a. spongy bone.
   b. areolar bone.
   c. compact bone.
   d. bone marrow.

6. The axial skeleton includes the
   a. head and extremities.
   b. trunk only.
   c. pectoral and pelvic girdles and extremities.
   d. head and trunk.

7. All of the following are bones of the cranium EXCEPT the
   a. occipital bone.
   b. maxilla.
   c. temporal bones.
   d. sphenoid bones.

8. A fluid-filled pad found mainly around synovial joints that decreases the irritation caused by movement of structures in the area is known as a
   a. fibroid.
   b. bursa.
   c. meniscus.
   d. keloid.

9. The joints between the fingers and at the elbow and knee are known as _____ joints.
   a. planar
   b. condyloid
   c. hinge
   d. pivot

10. The term for moving an extremity laterally (away from the body) is
    a. abduct.
    b. plantarflex.
    c. adduct.
    d. dorsiflex.

Chapter 5

The Muscular System

LEARNING OBJECTIVES

Upon completion of this chapter, you will be able to:

1. Describe the components and functions of the muscular system.
2. Describe the types of muscle tissue and provide examples of each.
3. Describe how muscles work to create movement.
4. Describe how different forms of exercise affect muscle.
5. Demonstrate how to perform a clinical skill such as the manual muscle testing (MMT).

CHAPTER OUTLINE

Introduction
Overview of Function
Types of Muscle Tissues
- Skeletal Muscle
- Smooth Muscle
- Cardiac Muscle

Skeletal Muscle Structure
- Gross Anatomical Structures
- Microscopic Structure
- Myofibrils

Neuromuscular Junctions
- Motor Neuron
- Motor End Plate
- Motor Unit

Location and Action of Muscles
- Sternocleidomastoid
- Deltoid
- Trapezius
- Pectoralis Major
- Latissimus Dorsi
- Biceps Brachii
- Triceps Brachii
- Rectus Abdominis
- External/Internal Abdominal Obliques
- Transverse Abdominis
- Quadriceps Femoris
- Hamstrings
- Gastrocnemius
- Tibialis Anterior

Muscles and Movement
- Types of Contractions

Energy Needed for Contraction
- Anaerobic Metabolism
- Aerobic Metabolism

Effect of Exercise on Muscle
- Strength Training
- Aerobic Training
- Bodybuilding

Muscles Working Together

INTRODUCTION

The muscular system is one of the most interesting topics to study, not only for its clinical relevance and applications to health sciences but also for its importance in the sports and fitness industry.

There are about 700 muscles in the human body which comprise about 40% of the total body weight. The muscles utilize ATP (adenosine triphosphate) to convert it to energy that can be used by the body, which was discussed in Chapter 1.

Cardiac muscles make the heart contract in order to circulate blood around the body. The main muscle of ventilation, the diaphragm, makes us breath. Smooth muscles in the stomach help digest or break down foods so nutrients can be absorbed and used by the body's physiological and metabolic needs.

In the sports and fitness industry, individuals and athletes undergo training or do workouts to build their muscles for better performance, endurance, and overall fitness. Through exercise, muscles can be developed and strengthened to help protect the bones from injury, maintain balance and coordination of the body, and improve blood flow to the entire body. In short, muscles are vital to facilitating movement of the body.

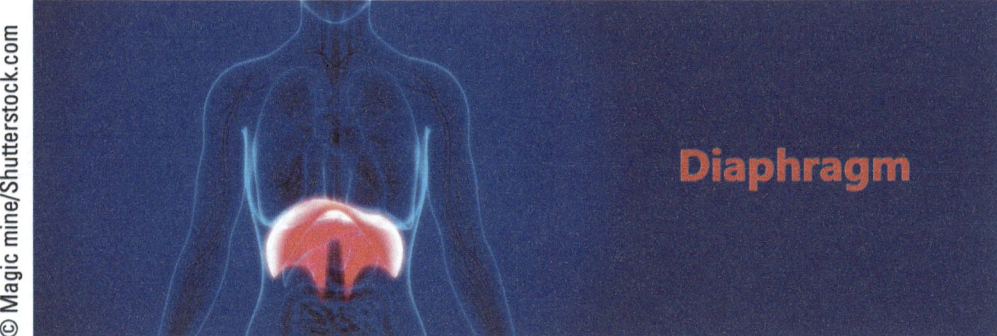

Figure 5.1

Figure 5.2

OVERVIEW OF FUNCTION

All types of muscle are designed to contract. It is the configuration or attachment of muscle that causes differences in function. Generally their actions can be defined as four functions.

1. Body movement—Muscle attached to bone can form levers to allow the skeleton to move. Movement also occurs in the walls of organs as muscle constricts to change internal size or shape. As a result, substances can be propelled through the organ, or the movement through the organ system can be altered. It is the squeezing of the heart that pushes blood through arteries to transport substances throughout the body.
2. Maintenance of posture—Although it is obvious that we must contract some muscles to help us stand erect or to prevent us from sliding out of a chair, there are some subtle contractions affecting posture that may be easily overlooked. When you pick up an object, you have to counterbalance the weight of the object to keep from falling over. Your brain stimulates muscles to maintain your posture. If you lean to one side or bend forward, not only will some muscles contract to cause the bending, but other muscles will become involved to maintain balance.
3. Production of body heat—The contraction of muscle requires a great deal of energy in the form of adenosine triphosphate (ATP). The production of ATP generates a considerable amount of heat. Much of that heat is used to maintain a constant body temperature. In most activities, excess heat is transported through the bloodstream to the skin for release into the atmosphere. If, on the other hand, a person becomes cold, he or she may shiver. This momentary contraction of muscle generates heat to help increase body temperature.
4. Stabilization of joints—When muscles pull on bones they can exert enough force to create stress on joints. Muscle tone (that is, tension in the muscles) helps stabilize the joints involved in the movement and help maintain posture. If a person's muscle tone weakens, primarily due to lack of physical exercise, there is a greater possibility that joint injury may occur. Weak muscle tone combined with poor posture often results in back trouble or increased risk of injuries when engaging in occasional physical exertion, such as sports.

TYPES OF MUSCLE TISSUES

There are three types of muscle, each with its own structure and specific function.

Skeletal Muscle

As the name implies, skeletal muscle is attached to bone to allow movement. In utero (as the fetus develops in the uterus) developing muscle cells fuse together in threadlike structures known as muscle fibers. When viewed under the microscope, skeletal muscle appears to have bands, or *striations*, running through each muscle fiber (Figure 5.3). Skeletal muscle does not have the ability to contract on its own; it must wait for an impulse from a nerve to stimulate contraction. For this reason, skeletal muscle is considered voluntary—our muscles wait for a consciously controlled coordinated plan

> **Skeletal muscle** attached between bones to cause movement. Under the microscope, skeletal muscle contains striations throughout each muscle fiber. Skeletal muscle contraction requires conscious stimulation and its movement can be fast.

> **Muscle fibers** threadlike strands of muscle formed in utero as muscle cells fused together. Each muscle fiber contracts as a unit.

Types of Muscle

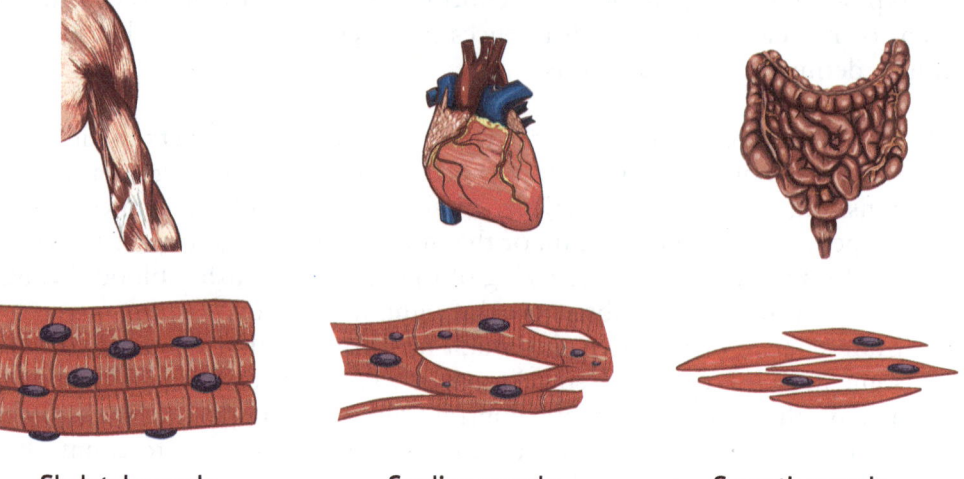

Skeletal muscle   Cardiac muscle   Smooth muscle

Figure 5.3

for movement. Some skeletal muscles (e.g., respiratory muscles and back muscles for balance maintenance) can also be voluntarily controlled. Sometimes the plan may be very complex in nature, such as swinging a bat to hit a baseball. Because of the reliance on neural signals for skeletal muscle to contract, nerve damage, which can prevent the impulse to contract from being received by the muscle fiber, may result in paralysis.

Smooth Muscle

Smooth muscle has no striations (bands) when viewed under the microscope (Figure 5.3). It is found in the walls of hollow organs such as the stomach, intestines, and blood vessels. The purpose of this muscle type is to change the size and shape of an organ in order to move its contents. For example, smooth muscle in the wall of the small intestine contracts rhythmically to propel food through the digestive system. Smooth muscle also spirals around the walls of most blood vessels to alter the diameter of the vessel. They are constantly changing diameter to alter the amount of blood flow to various areas of the body.

Smooth muscle may cause itself to contract, or it may be stimulated by unconscious control from our brain. We do not have to think about when to contract smooth muscle in our digestive system or blood vessels. Smooth muscle is considered involuntary because it does not require conscious stimulation from the brain for contraction to result.

Cardiac Muscle

Cardiac muscle is found exclusively in the heart. Its function is to pump blood throughout the cardiovascular system as a means of transporting substances throughout the body. Microscopically cardiac muscle is also striated, similar in appearance to skeletal muscle, but it has a distinguishing feature known as intercalated discs (Figure 5.3). Intercalated discs are connections between muscle cells that transmit the impulse to contract from one muscle cell to the next, enabling the heart to contract in a coordinated manner. Cardiac muscle is autostimulating; that is, it has the ability to cause itself to

---

> **Smooth muscle** found in the walls of hollow organs and blood vessels to change the size or shape of an organ in order to move its contents, or, in the case of a blood vessel, alter the flow of blood. Smooth muscle does not possess striations as are found in skeletal muscle. Its contractions are slow and may be due to stimulation by the autonomic nervous system or initiated by the smooth muscle itself.

> **Cardiac muscle** found exclusively in the heart. Its contraction causes the pumping of blood throughout the body. Cardiac muscle, when viewed under the microscope, contains striations, but also has a distinguishing feature known as intercalated discs. Cardiac muscle is autostimulating.

contract. The heart does not need the brain to tell it to contract, although the brain does have the ability to alter the speed of the heart contraction based on the body's needs. To prove this point, some researchers removed the heart of a chicken and placed it in a chamber that supplied nutrients and gases essential for the heart. It continued to beat outside the chicken for the next 10 years. They believed they had proved their point.

> **Comprehension Check-up:**
> 1. Maintaining a slight amount of contraction to stabilize joints and hold posture is known as _____.
> 2. The type of muscle that has striations and requires conscious stimulation by the brain is known as _____ muscle.
>
> 1. muscle tone   2. skeletal

SKELETAL MUSCLE STRUCTURE

Although we have been discussing three types of muscle, the muscular system involves skeletal muscle only, which is where we will now turn our attention. To understand the function of skeletal muscle it is most beneficial to study both the gross anatomical structures as well as the microscopic anatomy.

Gross Anatomical Structures

In most cases each individual skeletal muscle is attached to at least two bones by bands of connective tissue known as **tendons**. When a muscle contracts, it moves one bone more than the other. For example, muscles may flex the forearm (Figure 5.4). When that happens, the radius and

> **Tendons** bands of connective tissue that attach skeletal muscle to bone.

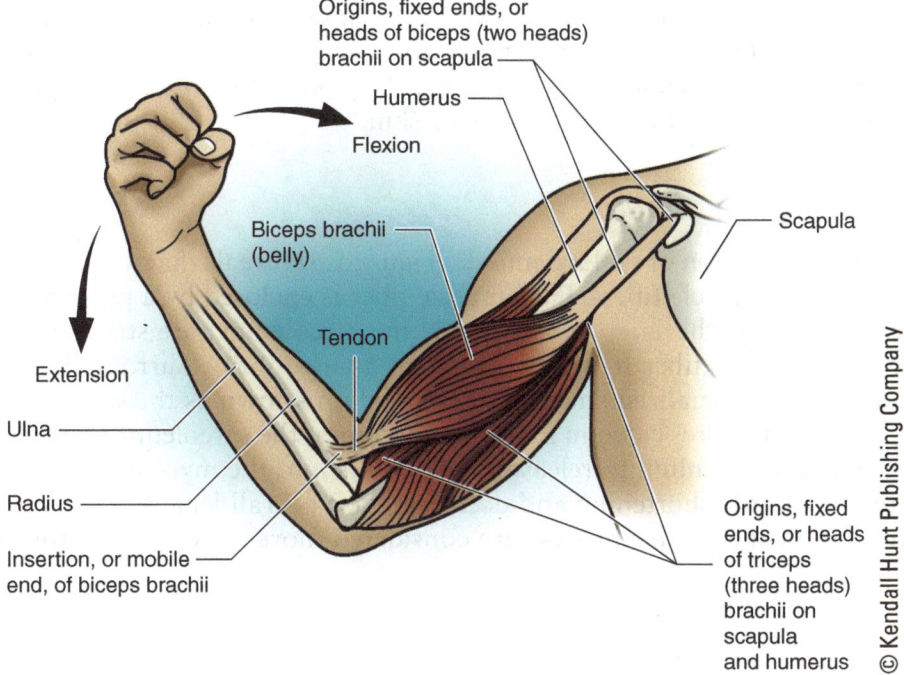

Figure 5.4 Muscle Attachments with Origin and Insertion.

What's in a Name?

Often the name of a muscle is descriptive. For example, the deltoid muscle is named for its triangular shape—the shape of the capital Greek letter delta, Δ The abdominal oblique muscles signify direction because they are at an oblique angle in the abdomen. Some muscles originate in more than one place, designated as having more than one head, such as the biceps brachii (two heads), triceps brachii (three heads), or even quadriceps femoris (four heads). A muscle named by its origin and insertion is the sternocleidomastoid. If the name is broken down to its component parts, it is not nearly as intimidating and its location is clear. *Sterno-* refers to the sternum. *Cleido-* means clavicle. The *mastoid process* is a bony projection on the temporal bone inferior and posterior to the ear. The name tells us where to find it. Its origin (stationary point) is the sternum and clavicle, and it runs to the temporal bone just below the ear (moving portion). Because insertion moves toward the origin, when a person contracts these muscles, his or her head will flex forward. This muscle can be seen in the neck as a person turns the head.

> **Origin** the end of the muscle attached to the bone moving the least.

> **Insertion** the end of the muscle attached to the bone moving the most.

> **Sarcolemma** the plasma membrane enclosing a muscle fiber.

> **Myofibrils** the structures within a muscle fiber that causes muscle to shorten.

> **Sarcoplasmic reticulum** specialized endoplasmic reticulum within the muscle fiber that stores calcium ions.

ulna may move considerably but the humerus moves very little. In this case the humerus is considered a nonmoving bone while the radius and ulna are designated as moving bones. The end of the muscle attached to the nonmoving bone is considered the origin, such as the humerus in the previous example. The end of the muscle attached to the moving bones, the radius and ulna, is called the insertion. In most cases, because of how the fibers are oriented, when a muscle contracts, the insertion moves toward the origin. Thus, when you flex your arm, the ends of the radius and ulna move toward the humerus. Sometimes muscles are named by their origin and insertion. When this occurs, the origin is given first, followed by the insertion.

Muscle is also covered by *fascia*, which separates it from other muscles and provides an attachment for skin if the muscle is superficial. Fascia is composed of fibrous connective tissue to provide isolation between muscle groups and skin. Tendons are dense regular connective tissue and provide strong attachment of muscle to bone or skin.

Microscopic Structure

Closer investigation of skeletal muscle (Figure 5.5) reveals that it is composed of bundles of muscle fibers. Each muscle fiber is enclosed by a plasma membrane known as the sarcolemma. Inside the sarcolemma are structures that cause the muscle fiber to shorten, known as myofibrils. Surrounding each myofibril is a specialized endoplasmic reticulum known as the sarcoplasmic reticulum that stores calcium ions. Stimulation of the sarcolemma causes the sarcoplasmic reticulum to release calcium ions into the myofibrils, resulting in the myofibrils shortening and decreasing the overall length of the muscle fiber. The details of this process are considered more completely in the next section.

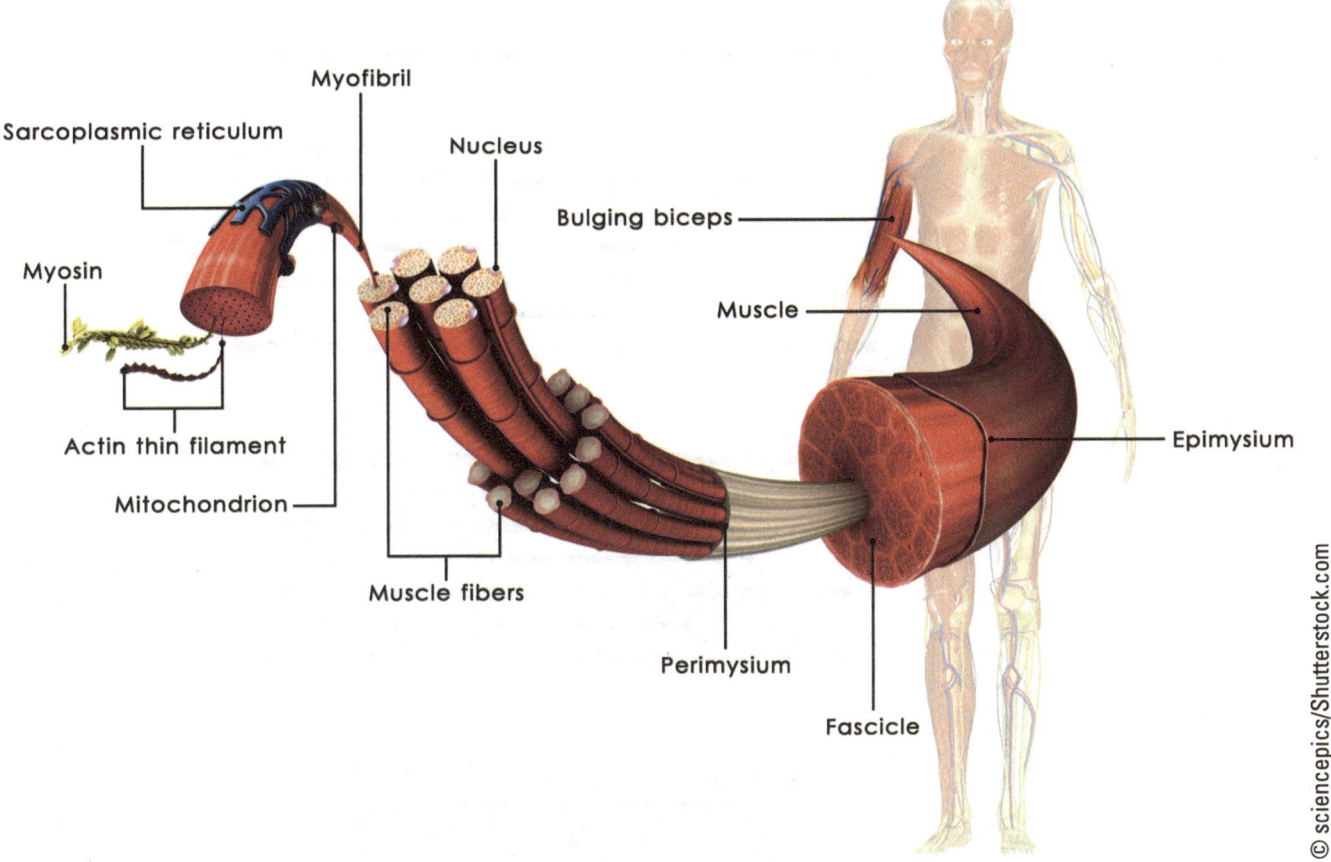

Figure 5.5

Myofibrils

Inside the myofibril are millions of sarcomeres—the functional unit within the myofibril that causes contraction or shortening. *Functional units* are the components within an organ that causes it to accomplish its purpose. The sarcomere consists of two major protein structures:

- Thin filaments: The structure within the sarcomere that slides inward to shorten the overall length of the sarcomere.
- Thick filaments: The structure within the sarcomere that slides the thin filaments inward to decrease sarcomere length.

The arrangement is that a thick filament is surrounded on either end by six thin filaments. When the sarcomere is relaxed, the thin filaments on either end of the thick filament are separated. Contraction occurs at the end of a series of chemical changes in the sarcomere that result in the thick filaments. These changes cause the thin filaments to slide inward toward each other, decreasing the overall length of the sarcomere (Figure 5.6).

> **Sarcomere** the functional unit within the myofibril that causes shortening.

> **Thin filament** the structure within the sarcomere that slides inward to shorten the overall length of the sarcomere.

> **Thick filament** the structure within the sarcomere that causes the thin filaments to slide inward to decrease sarcomere length.

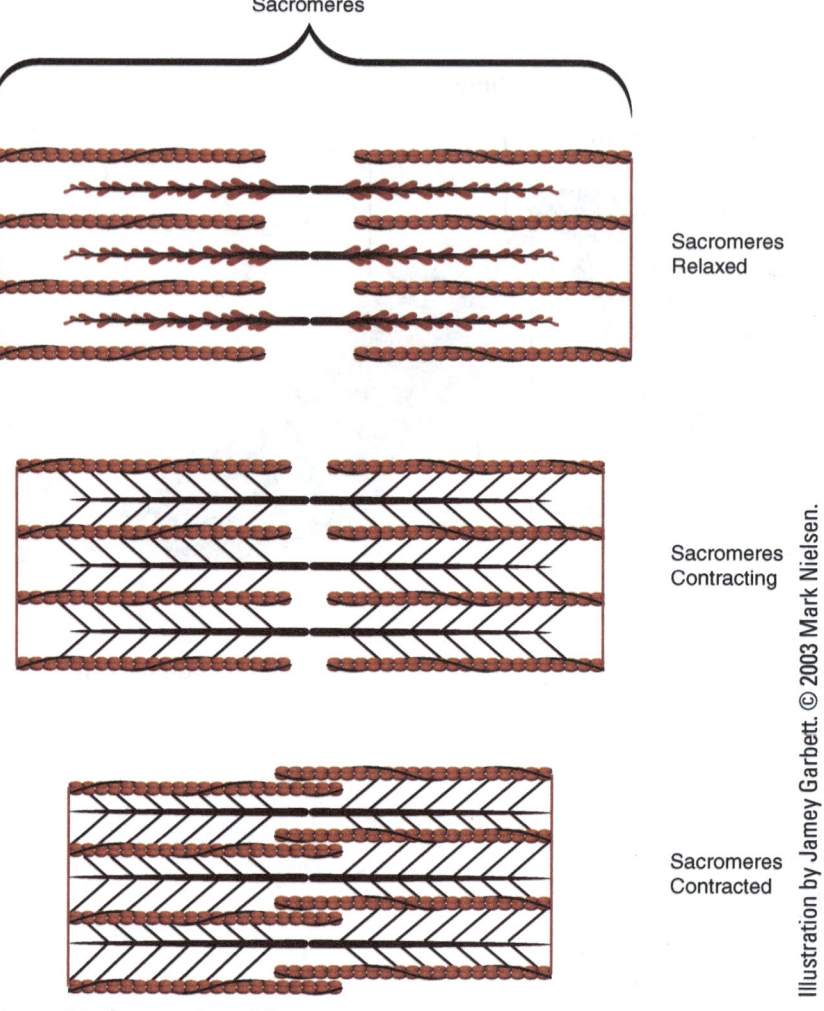

Figure 5.6 Contraction of Sarcomeres.

Sarcomeres are side by side in the myofibril and share common thin filaments (Figure 5.7a). The more sarcomeres there are side by side, the more strength the muscle fiber has. As humans grow, they do not add more muscle fibers. They can, however, add more sarcomeres within the fiber. This is a major goal for a bodybuilder. Adding more resistance to movement by lifting heavy weights increases the number of sarcomeres and, as a result, increases the size and strength of the muscles.

Muscle fibers are also end to end (Figure 5.7b). The more sarcomeres there are in the length of the muscle fiber, the more shortening that occurs. Imagine that a group of individuals went outside, stood in a straight line, held their arms outstretched to the side, and held hands with the people to their sides. Think what would happen if everyone pulled their arms in but still held hands. The people at the end of the line would move a great deal because they would receive the sum of all the shortening that occurred. The longer the line is, the greater the movement at the ends. In our bodies, the longer a muscle is, the greater amount of shortening it will accomplish. For this reason, muscles that operate our fingers are primarily in our forearm and not in our hand.

If those muscles were in our hand, we would barely be able to bend our fingers because they simply would not be long enough. Fortunately, the bulk of the muscles operating our fingers are in our forearm, providing much more length and giving us the ability to clench our hands into a fist.

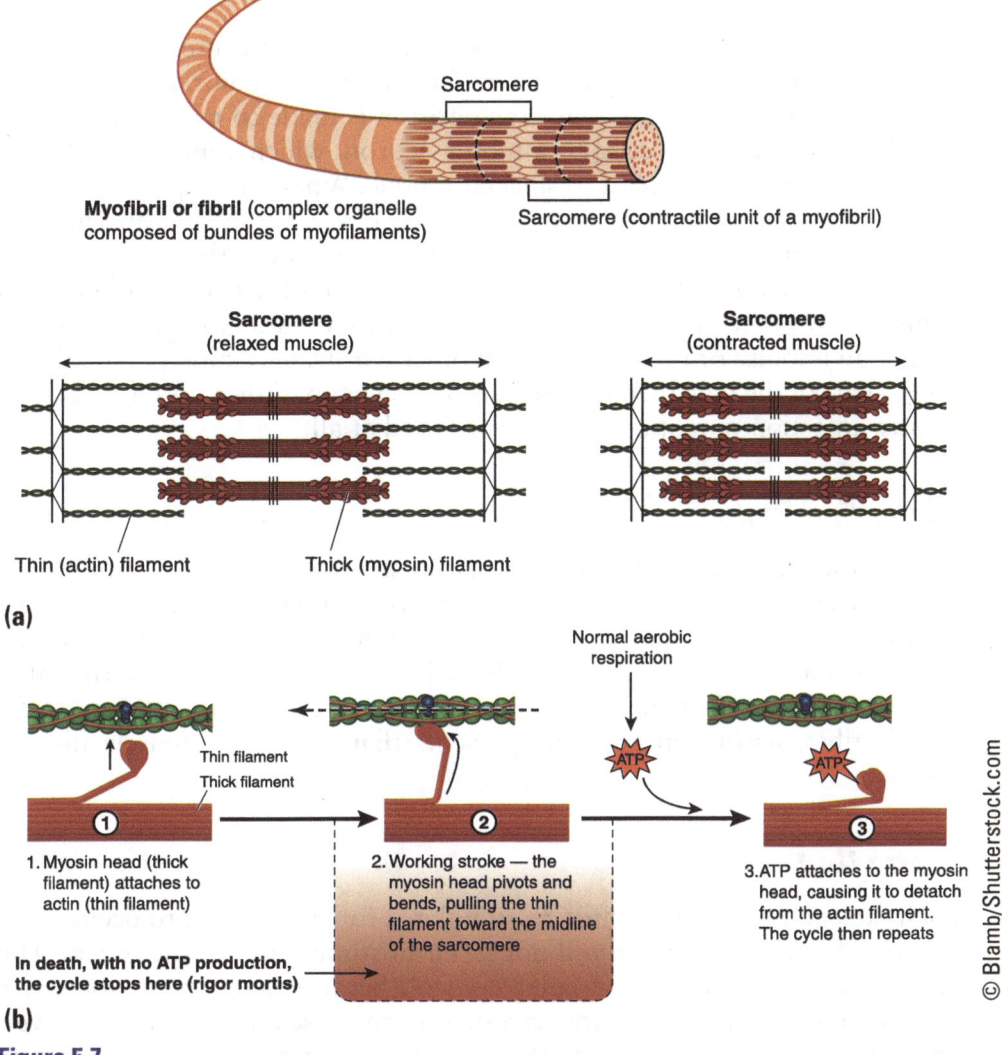

Figure 5.7

> **Comprehension Check-up:**
> 1. The end of the muscle that is attached to the nonmoving bone is known as the _____.
> 2. Describe how muscle contracts.
>
> 1. origin
> 2. A series of chemical reactions result in the thick filaments sliding the thin filaments toward each other.

NEUROMUSCULAR JUNCTIONS

Recall that skeletal muscle does not contract on its own but instead must be stimulated to contract by a nerve. The **neuromuscular junction** is where the nerve comes in close contact with the muscle fibers it innervates. The components of a neuromuscular junction are described in the following sections.

> **Neuromuscular junction** the close contact of a stimulating nerve with the muscle fiber it stimulates.

> **Motor neuron** a neuron that transmits the stimulus to cause muscle to contract.

> **Threshold stimulus** the level of stimulus at which an action potential is created, causing muscle contraction.

> **All or none principle** the response of a neuron to stimulus. Either all of the neuron will respond or none of it will.

> **Action potential** the electrical potential on a muscle or nerve causing a response. In a neuron it is a nerve impulse. In a muscle it is the stimulus to cause contraction.

> **Neurotransmitter** a chemical released by a synaptic knob of one neuron that may cause the creation of an action potential in the next nerve or muscle fiber.

> **Motor unit** a stimulating nerve and the muscle fibers it stimulates.

Motor Neuron

A neuron is a nerve cell. A *motor neuron* is a nerve that causes movement by stimulating muscle fibers to contract. The impulse to contract comes from the central nervous system. Once the decision is made to move, the impulse travels down the spinal cord to a motor neuron, which conducts that stimulation from the spinal cord to skeletal muscle. When the impulses reaching a specific motor neuron reach the *threshold stimulus* (the level at which the motor neuron responds by creating an action potential), the impulse is sent to its associated muscle fibers. The response of a motor neuron to incoming stimuli follows the *all or none principle*, which means that the entire neuron creates an impulse or none of it will. In other words, the action potential does not go part of the way through the neuron then stop. Either the entire neuron creates an action potential or it is not created at all.

Motor End Plate

The *motor end plate* is the region of the sarcolemma that is adjacent to the presynaptic terminal of the motor neuron. The impulse from the nerve, called an *action potential*, is passed on to the motor end plate when a chemical, called a *neurotransmitter*, is released from the presynaptic terminal of the motor neuron. The neurotransmitter transfers the action potential to the muscle fiber, setting up a chain of events that causes the muscle fiber to shorten.

Motor Unit

Each skeletal muscle fiber must be stimulated for contraction to occur. Stimulation of one muscle fiber will not result in stimulation of the others. This enables a fine degree of muscular control, because only those fibers that are required for movement are activated. Each muscle fiber needs to have a stimulating nerve, so it is most efficient when one nerve activates more than one muscle fiber. The stimulating motor neuron and the muscle fibers it stimulates are known as a *motor unit* (Figure 5.8).

These motor units are functionally quite clever. Picture someone juggling an apple, a bowling ball, and an egg. The juggler does not want to grab the egg with the same strength as used for the bowling ball. Likewise, it certainly takes more strength to juggle an apple than an egg. How does this individual create different strengths when handling objects of different weights? It is the number of motor units involved that make the difference. If grabbing the egg, only a few motor units are involved. The apple requires a few more, and the bowling ball involves many more motor units. If an individual decides to pick up an object, before doing so, his or her eyes send information to the brain about the object's size. Past experience allows the person to determine the approximate weight of the object. The brain performs calculations to determine the number of motor units that are needed. Action potentials (nerve impulses) are sent down the spinal cord to the appropriate motor units. As the object is being lifted, adjustments can be made to stimulate more motor units or allow some to relax.

At the same time, as muscles are contracting to pick up an object, muscle tone must simultaneously be maintained to maintain posture, counterbalance movements, and stabilize joints. Motor units involved with muscle tone are activated whenever an object is lifted. Once the complexity of a

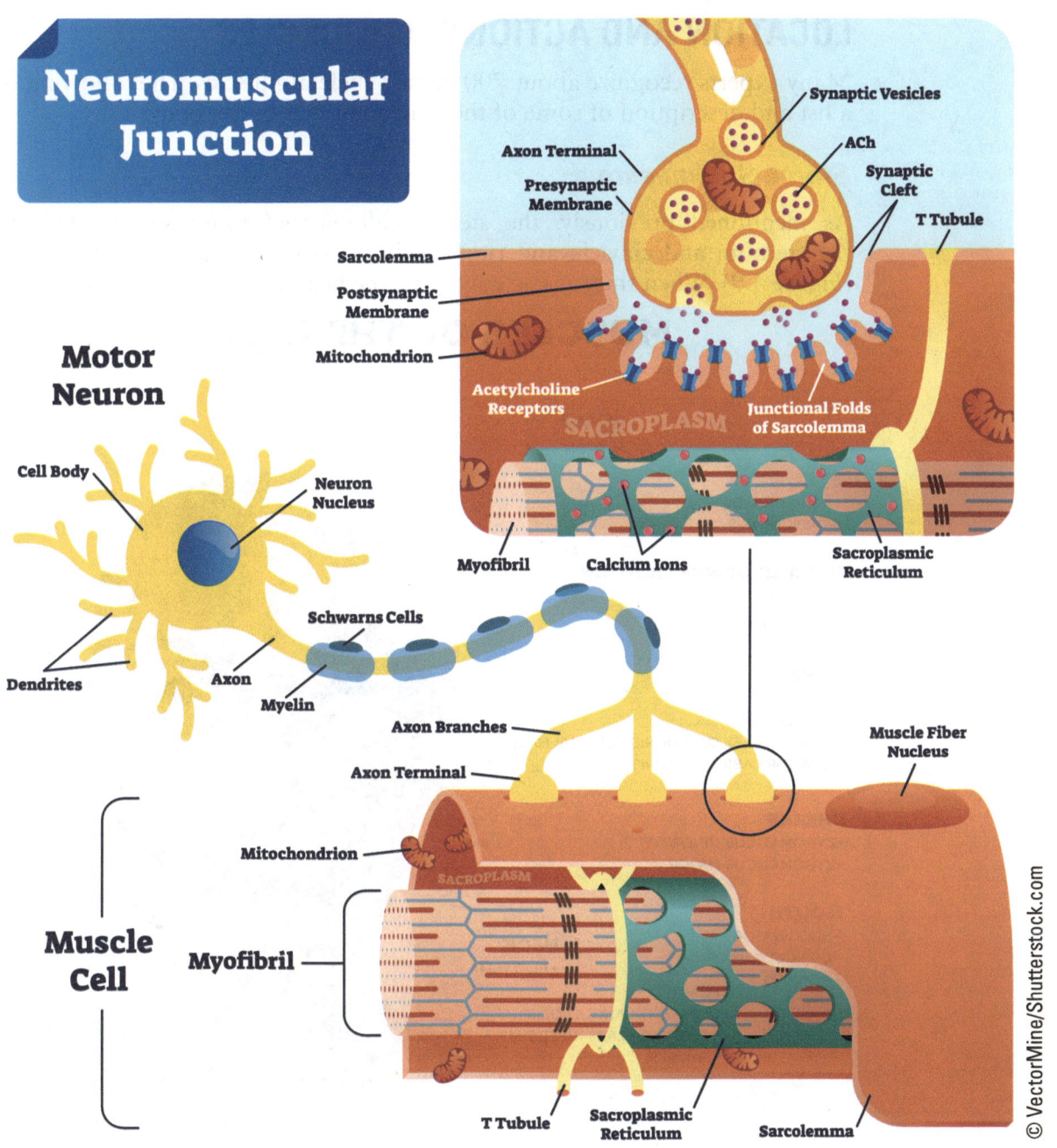

Figure 5.8

single movement is grasped, the understanding of our ability to produce rapid complicated activities, such as playing sports or gymnastics, practicing martial arts, or performing on musical instruments, becomes more fully appreciated.

Comprehension Check-up:
1. A nerve that stimulates a muscle to contract is known as a _____.
2. A nerve and the muscle fibers it stimulates form a _____.

1. motor neuron   2. motor unit

LOCATION AND ACTIONS OF MUSCLES

Many experts recognize about 700 named skeletal muscles. The following is a list and description of some of the major muscles of the body.

Sternocleidomastoid

As mentioned previously, the sternocleidomastoid muscle originates on the sternum and clavicle and runs to the mastoid process behind the ear (Figure 5.9). When this muscle contracts, the head flexes and rotates.

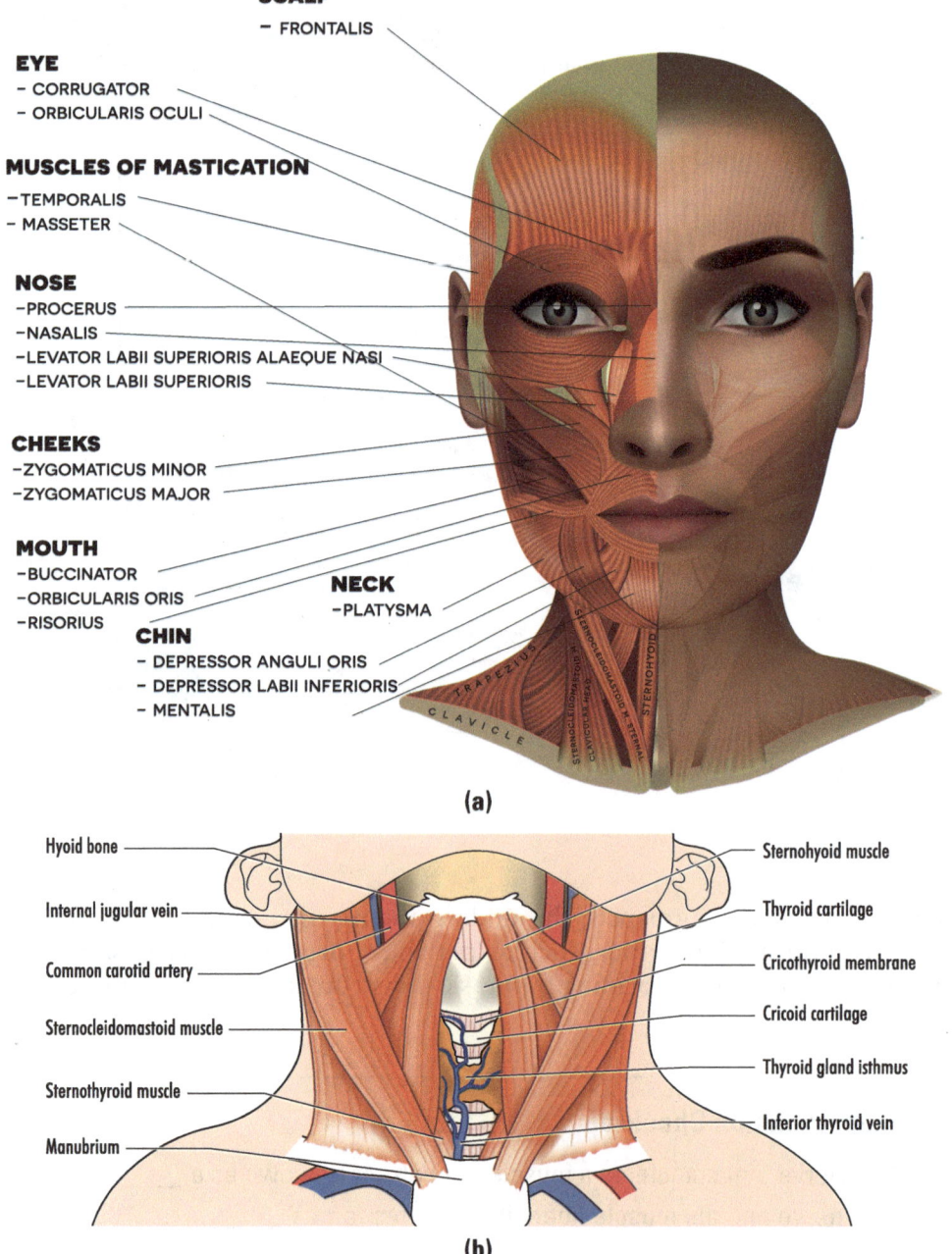

Figure 5.9

Deltoid

The deltoid muscle is so named because it is shaped like a triangle (the capital Greek letter delta looks like a triangle—Δ). It lies over the shoulder, originating on the clavicle and scapula and inserting on the lateral side of the humerus (Figures 5.10). It abducts the humerus.

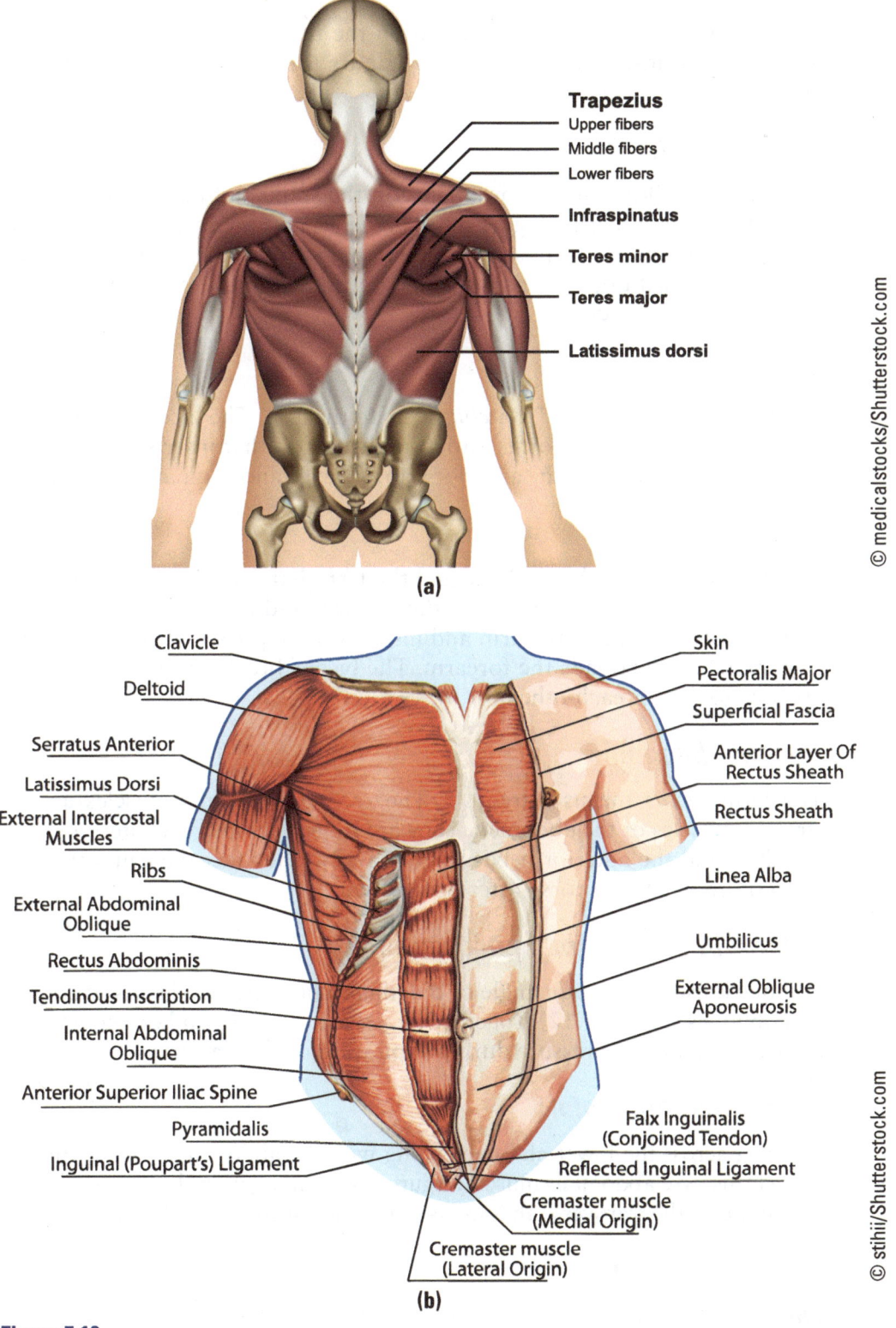

Figure 5.10

Trapezius

The trapezius muscle is found in the upper back. It originates on the spine and inserts on the scapula (Figure 5.10). It elevates the scapula, as in a shrug, and extends the head.

Pectoralis Major

The pectoralis major muscle is found on the upper anterior chest, where it originates and inserts on the humerus (Figure 5.10) to flex, adduct, and medially rotate the arm.

Latissimus Dorsi

The latissimus dorsi muscle originates on the spine and inserts on the anterior side of the humerus (Figure 5.10). When it contracts, it extends and adducts the humerus in the posterior direction. If a person leans back and pulls himself or herself up, the latissimus dorsi muscle brings both humeri toward each other in the back.

Biceps Brachii

Biceps means "two heads." *Brachii* refers to the arm. The biceps brachii originates in two places but forms into a single tendon that inserts on the radius (Figure 5.11). Found on the anterior arm, it flexes and supinates the forearm.

Triceps Brachii

Triceps means "three heads." Because the term *brachii* is also used, it makes sense that this muscle originates in three places and is found in the arm. It is on the posterior side of the arm and inserts on the proximal end of the ulna (Figure 5.11). It extends the forearm. The biceps brachii and triceps brachii are antagonists of each other.

Rectus Abdominis

The term *rectus* means "straight." The rectus abdominis muscles are the straight muscles of the abdomen. They are found between the inferior end of the chest and the anterior pelvis (Figure 5.12). They flex the trunk to allow us to sit up or lift our torso after lying down.

External/Internal Abdominal Obliques

Oblique muscles are at an angle or diagonal. There are two layers of abdominal obliques (Figure 5.12). Together they twist the torso and can constrict to increase abdominal pressure during expiration or defecation.

Transverse Abdominis

Running across the lower abdomen along with the abdominal oblique muscles is the transverse abdominis muscle (Figure 5.12). Along with the diagonally running muscles, the transverse abdominis muscle also increases abdominal pressure.

Quadriceps Femoris

Quadriceps means "having four heads." In other words, this muscle originates in four locations. It is found on the anterior thigh. All four heads

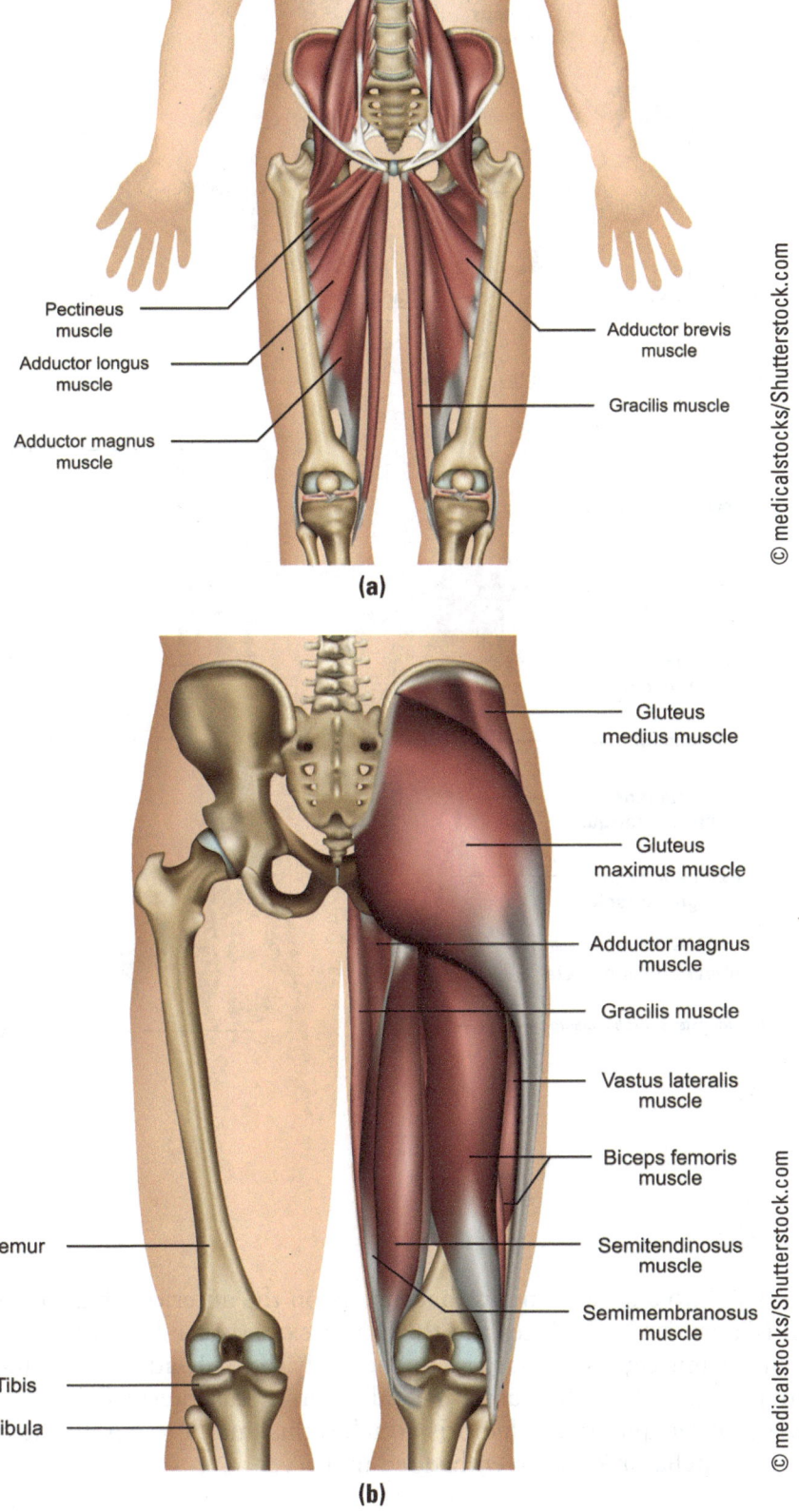

Figure 5.11

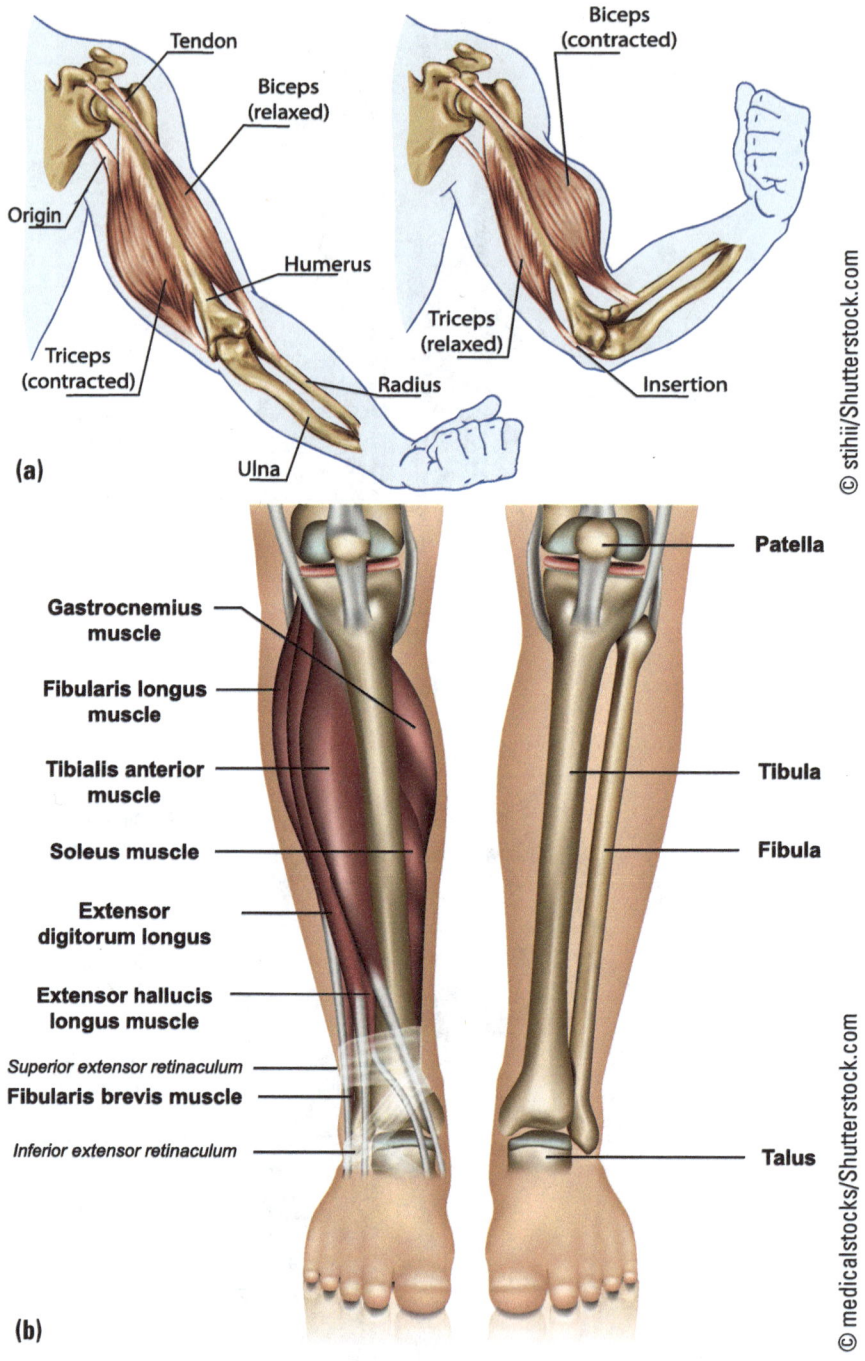

Figure 5.12

combine into a single tendon that inserts on the anterior tibia to extend the calf (Figure 5.13). It flexes the thigh and extends the leg as in a kick. This group of muscles is the largest in the body. Each head has a name: rectus femoris, vastus lateralis, vastus medialis, and vastus intermedius. The patellar tendon of the quadriceps femoris attaches this muscle group to the tibia and has the patella or kneecap embedded in it.

Hamstrings

On the posterior thigh is a group of three muscles commonly known as the *hamstrings* (Figure 5.14). They extend the thigh and flex the leg. The three muscles are the biceps femoris, semitendinosis, and semimembranosus.

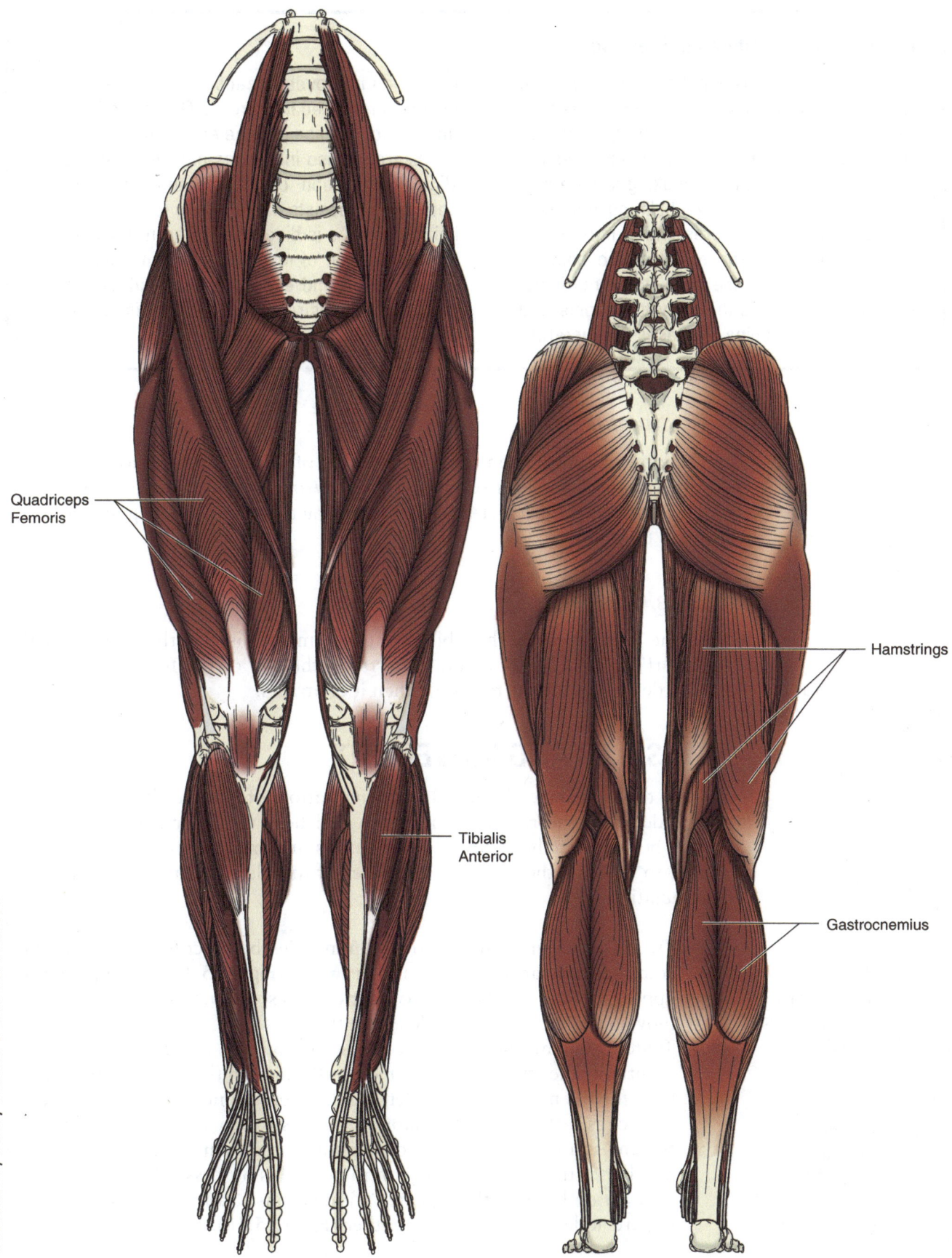

Figure 5.13 Muscles of the Lower Extremity— Anterior View.

Figure 5.14 Muscles of the Lower Extremity— Posterior View.

> **There Is No Substitute for Exercise**
>
> There is an enormous benefit of exercise to the muscular system. Increased activity increases blood supply to muscle, creating improved muscle efficiency. Exercise also increases mitochondria within muscle, making it more efficient at producing ATP. Weight lifting increases the number of sarcomeres within muscle, increasing strength. The result of the benefits listed is an ability to use more stored nutrients. This, in turn, causes the body to become and remain trim as stamina increases. Temperature regulation improves because the body adapts to move increased heat from ATP production to the skin for release into the environment.
>
> Exercise also increases muscle tone, that slight contraction of muscle that maintains posture and assists in stabilizing joints to minimize injury. As is discussed later in this book, the contraction of muscle assists in returning venous blood and excess tissue fluid back to the heart for recirculation, thereby improving circulation.

Gastrocnemius

The large calf muscle on the posterior side of the tibia almost looks like two stomachs, which gives rise to the name *gastrocnemius*. Its tendon attaches to the calcaneus (Figure 5.14). When it contracts, it flexes the leg and plantarflexes the foot.

Tibialis Anterior

As the name implies, the tibialis anterior muscle is slightly in front of the tibia (Figure 5.13). Its tendon inserts on the top of the foot, causing it to dorsiflex (to point the toes up) when it contracts.

MUSCLES AND MOVEMENT

Now that we have discussed the stimulation of muscle, it is possible to consider how the rate of stimulation and how the lever action of muscle with bone affect the types of contractions and movement. Three terms are used to describe the different responses of muscle fibers to varying rates of stimulation:

- Twitch—The single stimulation of a muscle fiber to contract, followed by complete relaxation, is known as a twitch (Figure 5.15). It simply would appear as a muscle jerk. Although twitches sometimes occur, most of our movements are more complex than that.
- Twitch summation—If the speed stimulation of a muscle fiber increases, eventually the stimulation occurs rapidly enough that there is insufficient time for the muscle to completely relax before another stimulus occurs. This causes the muscle to shorten more so than what occurs during a twitch. As a result, the total contraction becomes an accumulation of the number of times it is rapidly contracted. This process is known as twitch summation because the repeated stimulations causing increased shortening are added (or summed) together (Figure 5.16).

One can see the difference between a twitch and twitch summation by using the analogy of a children's game. Take, for example, some children

Twitch a single stimulation of a muscle fiber to contract, followed by complete relaxation.

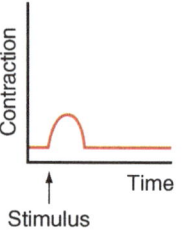

Figure 5.15 Twitch. A single stimulus is followed by complete relaxation of the muscle.

Twitch summation occurs when the speed of the stimulation of skeletal muscle is rapid enough that there is not enough time for complete relaxation before additional contraction begins. The end result is to cause additional shortening of the muscle.

playing a game in which one child says to take three giant steps forward, then go back to the beginning. This is the equivalent of a twitch. Now, using the same scenario, the child is asked to take three giant steps forward then return to the beginning. In this case, however, before the child gets back to the starting point he or she is asked to take three more giant steps forward. The child then again starts returning to the beginning; after taking only one or two steps, the child is asked yet again to take three more giant steps forward. This process is repeated a number of times. Each time the child took only three giant steps forward, but he or she is not starting out from the beginning any more. As a result, the child is much farther down the road. The total length of the child's travel was the sum of all of the movements made in the forward direction. The same process occurs with muscle. If the muscle is stimulated again before it is completely relaxed, then the muscle becomes shorter than a single twitch would accomplish. In fact, the longer this process occurs, the shorter the muscle will become. In this way the body can control the extent of the muscle contractions by the speed with which they are stimulated.

A majority of the movements we make involve twitch summation. As we grasp an object, summation causes our fingers to flex around the item. When we walk or run, twitch summation shortens the muscles in our posterior thigh to propel us forward. The degree to which muscle contraction occurs depends on the speed by which twitch summation is created.

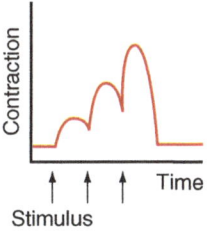

Figure 5.16 Twitch Summation. Repeated stimulation before the muscle has time to completely relax causes additional shortening of the muscle.

- Tetanus—Tetanus is used to describe the response of skeletal muscle to rates of stimulation that are so rapid there is little or no relaxation period, causing the muscle to reach extensive or complete contraction (Figure 5.17). It may occur when an individual grasps an object very tightly. This may also occur if someone grabs an energized electric wire. The electricity activates the muscle 60 times per second, which is fast enough to cause tetanus. The problem is, if muscles flexing the fingers are stimulated to the point of tetanus, the person will involuntarily hold on to the wire and cannot let go.

> Tetanus describes the response of skeletal muscle to rates of stimulation so rapid there is little or no relaxation period. As a result, the muscle reaches extensive or complete contraction.

Clinical Notes—Muscle Strain

Muscle strain is the terminology used to describe damage to a muscle. Muscle strains can range from simple inflammation of a muscle to complete tearing and are graded on a scale of 1-3.

First-degree (or Grade 1) strains involve the partial tearing or damage of a small number of muscle fibers and result in minimal swelling and localized tenderness as well as minimal strength limitations.

Second-degree (or Grade 2) strains involve complete tearing of some but not all muscle fibers and result in increased amounts of swelling and pain as well as moderate strength limitations.

Third-degree (or Grade 3) strains involve the complete tearing of most or all muscle fibers and result in significant swelling, pain, and complete loss in muscle strength.

Treatment for first- and second-degree strains may be conservative and include physical therapy as well as inflammation- and pain-decreasing modalities. Treatment for third-degree strains requires a surgical repair of the damaged tissue.

(continued)

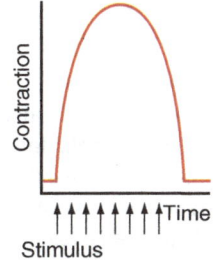

Figure 5.17 Tetanus. Stimulation occurs so rapidly that there is no relaxation, causing the muscle to reach maximum contraction.

Grading Muscle Strain

	Pain	Weakness	Loss of Function
First Degree (Grade 1)	X		
Second Degree (Grade 2)	X	X	
Third Degree (Grade 3)	X	X	X

Grade 1   Grade 2   Grade 3

Modified from © Barks/Shutterstock.com

Contributed by Dave Smith. © Kendall Hunt Publishing Company

Types of Contractions

There are two general types of contractions. Both contain the prefix *iso-*, which means "unchanging." Typically muscle movements involve both types of contractions, although isotonic contractions involve a full range of motion, whereas isometric contractions are primarily intended for stabilization.

Isotonic Contractions

With isotonic contractions the muscle tension or strength (tonic) remains unchanged. What changes is the length of the muscle. For example, if you lift a 50-pound weight, your forearms flex. Throughout the movement, the amount of strength needed to lift that weight remains constant at 50 pounds. To make the movement, however, the muscles moving your forearm shorten, so their length changes.

Isometric Contractions

An isometric contraction results when the length of the muscle does not change while the strength created in the muscle changes. If you stood in a doorway and pushed on the sides of the door frame, your strength would increase as you pushed harder and harder, but the length of your muscles remains unchanged.

> **Isotonic contractions** result when skeletal muscle tension or strength remains unchanged. Muscle length changes.

> **Isometric contraction** results when the length of the skeletal muscle does not change because the resistance against which the muscle pushes does not vary.

Comprehension Check-up:

1. The term for stimulation of a single muscle fiber followed by complete relaxation is _____.
2. The pivot point on a lever is known as a _____.

1. a twitch   2. fulcrum

ENERGY NEEDED FOR CONTRACTION

An enormous amount of energy is required to perform a muscle contraction. Recall that the energy required by the human body is primarily transferred from nutrients in the form of ATP. ATP carries that energy to the muscle fiber to assist in causing the contraction. The process of converting energy in nutrients into ATP is referred to as *metabolism*, also known as *cellular respiration*. To avoid confusion between *cellular respiration* and *breathing*, we will use the term *metabolism* for cellular respiration. There are two methods for obtaining ATP: (1) *anaerobic metabolism*, which involves no oxygen, and (2) *aerobic metabolism*, which requires oxygen to perform its chemical reactions.

Anaerobic Metabolism

The production of ATP without oxygen is more rapid than aerobic metabolism because fewer chemical reactions are involved. Anaerobic metabolism begins with the breakdown of glucose into two molecules of *pyruvic acid* (pyruvate). Several steps are involved in this series of chemical reactions, but in the process of converting glucose into pyruvic acid, two ATPs are produced. If oxygen is available, the next step involves aerobic metabolism. However, if oxygen is still not available, the pyruvate must be converted into *lactic acid* (lactate) (Figure 5.18). This process occurs often in muscle. The lactic acid is transported to the liver, which will perform a series of chemical reactions to convert the lactic acid into glucose when oxygen later becomes available. The soreness of muscles experienced after exercise is not due to the "burning of lactic acid" as previously thought, but due to the tearing of muscle fibers during fast contraction of the muscles.

Aerobic Metabolism

When oxygen is available, the production of ATP can increase greatly. This process is slower than anaerobic metabolism because numerous chemical reactions are involved. In aerobic metabolism, pyruvic acid can be fed into a series of chemical reactions that are able to convert that chemical energy into ATP (Figure 5.19). With oxygen, aerobic metabolism is able to make an additional 34 ATPs per glucose beyond what is produced with anaerobic metabolism. Not only can we extract additional ATP from pyruvic acid, but we can also break down fat and protein as additional sources of energy. Because of the conversion of lactic acid back to pyruvic acid as oxygen becomes available, aerobic ATP production may continue for hours after an activity has ended to replenish depleted supplies.

Adapted from *Anatomy I and Physiology Lecture Manual*

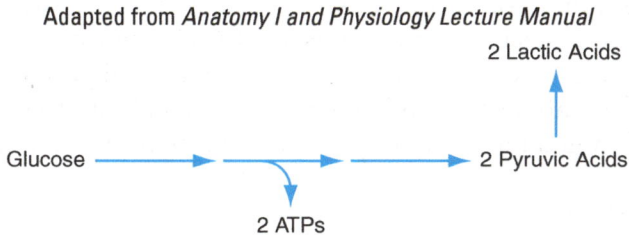

Figure 5.18 Anaerobic Metabolism of Glucose.

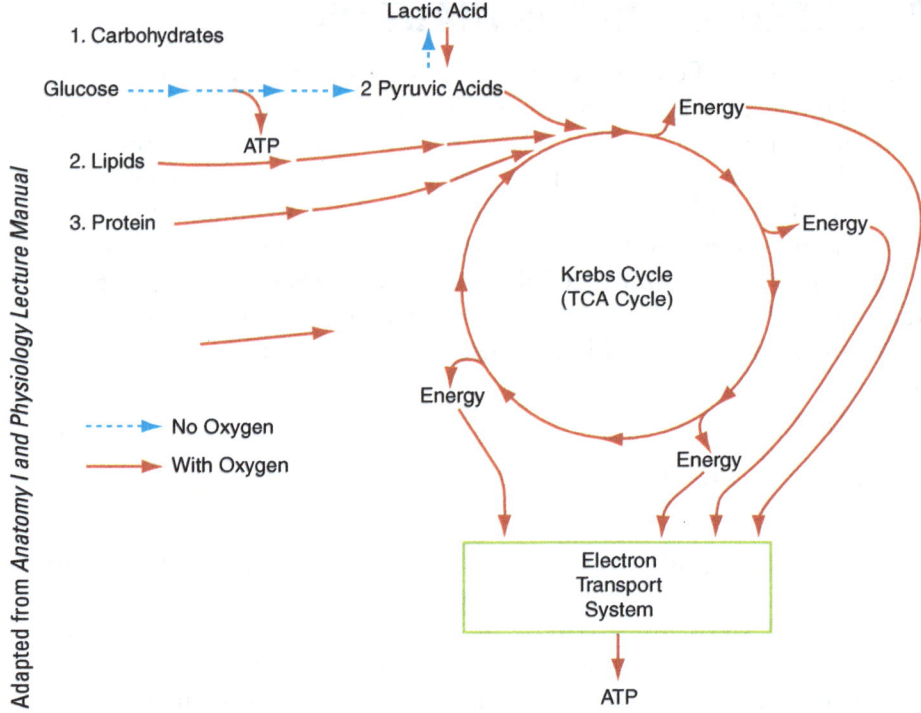

Figure 5.19 Aerobic Metabolism.

EFFECT OF EXERCISE ON MUSCLE

Let's face it, there is no substitute for exercise. Often there is confusion about the type and amount of exercise needed. Is there a relationship between the amount of muscle an individual has and that person's ability to stay trim? If a woman lifts weights, will she develop bulging muscles like a man?

There are three general types of exercise on muscle. They are:

- strength training;
- aerobic training; and
- body building.

After they have been discussed, it should be clearer how they can work together.

Strength Training

When an individual lifts weights on a regular schedule, the body adapts to accommodate the need for additional ATP production. Blood flow increases to muscle and becomes more efficient at optimizing conditions for the muscle contraction. Strength training primarily involves muscles that are able to produce quick, strong contractions and use anaerobic metabolism for their energy production. In order for those muscles to increase strength, they produce more sarcomeres within their muscle fibers. As muscle density increases, metabolic rate after exercise continues for an extended period. The extended higher rate of metabolism, coupled with increased ability to perform aerobic metabolism, is a great asset in losing unwanted bulk or remaining trim.

Aerobic Training

The use of aerobic metabolism for the production of ATP allows muscle to work for longer periods than would be possible by anaerobic metabolism alone. Remember that aerobic metabolism is the method by which we also convert fat into ATP A molecule called *myoglobin* stores oxygen in muscle. Muscle that contains myoglobin is more capable of producing ATP through aerobic metabolism than is muscle that does not possess myoglobin. The more aerobic training we do, the more myoglobin and mitochondria we produce in our skeletal muscle; in addition, we increase capillary density, which means we become more efficient at breaking down fat.

In humans, some muscle works primarily by anaerobic metabolism, that is, without oxygen for breaking down glucose for its source of ATP Other muscles get their ATP from aerobic metabolism. Humans' muscles are actually a combination of both types. To illustrate how this process works, let's discuss chickens. There are two types of meat in chickens: white meat and dark meat. Myoglobin gives dark meat its deeper color. White meat contains less myoglobin. Chickens have white meat in their breast and dark meat in their legs. They can fly but not very far. They spend almost all of their time walking around. Why? The dark meat in chicken legs is able to efficiently produce large amounts of ATP, meaning their leg muscles are able to work for extended periods without becoming fatigued. Because their breast muscles have less myoglobin, the majority of their source of ATP comes from anaerobic metabolism. This means chickens can fly, but not very far. They become fatigued quickly because they run short of ATP Although humans have a mixture of both types of muscle in most places, there is more myoglobin-containing muscle in the lower extremities and less in the upper appendages. As a result, we can work much longer with our legs than with our arms. It is not unusual for people to spend the entire day on their feet. Although strength training involves using all muscles, aerobic training is most efficient when using the lower extremities. Our upper body is much better at performing strong, rapid movements that do not require extended periods of constant contraction.

Bodybuilding

If an individual uses anaerobic muscles for excessive strength training, the body not only makes the availability of ATP more efficient by storing extra glucose in the muscle, it also makes the muscle stronger by adding more sarcomeres inside the muscle fiber. As a result, the muscle fibers increase in diameter. This process is affected to a large extent by genetics and the amount of testosterone in the body, which is found in much higher quantities in men than in women. Usually men interested in building muscle bulk can do so to a greater extent than can women. In addition, the muscle fibers become stressed, so much they tear slightly. The body responds by mending the muscle with scar tissue. This also increases the bulk of muscle.

> **Comprehension Check-up:**
> 1. If glucose is broken down into pyruvic acid and oxygen is not available, the pyruvic acid can be temporarily converted into _____.
> 2. Some muscle that obtains its ATP through aerobic metabolism contains a molecule designed to hold additional oxygen. That molecule is known as _____.
>
> 1. lactic acid   2. myoglobin

> **Anabolic Steroids—Beware!**
>
> Desire for a competitive edge can so easily drive individuals to cross the line into territories that are physically harmful and even illegal. Such is the case with anabolic steroids. These testosterone-like drugs were developed in the 1950s as a supplement for people suffering from muscle-wasting diseases. Because testosterone stimulates an increase in muscle and bone mass, it didn't take long before their use by bodybuilders began to increase. Unfortunately their advantages are far outweighed by their dangerous side effects. It has been found that anabolic steroids increase the risk of heart disease and strokes. They may damage the kidneys and increase the risk for liver tumors as well. High levels of testosterone have been linked to aggressive behavior. Anabolic steroids have led to extremely violent acts referred to as *roid rage*. The concern has become so great that they have been banned in many athletic events.
>
> Women are also susceptible to the same desire for the competitive edge. Steroid use among females also gives them extra strength and speed—at a price. They may experience menstrual irregularities and even atrophy of the uterus and mammary glands. They may develop facial hair.
>
> Many of the anabolic steroids must be injected. There is an increased concern of acquired immunodeficiency syndrome (AIDS) or hepatitis if users become careless. The major concern is that the questionable short-term advantage opens the door to long-term serious health risks.

MUSCLES WORKING TOGETHER

In most cases muscles do not work independently but rather are coordinated to contract together to produce movement. The structure and placement of the muscles in the body enable the precise, coordinated movement to occur. The muscles' roles in a movement can be classified from three different perspectives.

- Prime movers (agonists)—The movement caused by skeletal muscles is intentional. When an action occurs, one muscle is the major cause of that movement. This muscle is considered the prime mover (agonist). For example, if a person flexes and rotates the forearm, one muscle—the biceps brachii—is the prime mover. Other muscles may be involved in the movement, but only one is the major contributor.
- Antagonists—Antagonists are muscles on the opposite side of a joint that relax while the muscles causing movement contract. For example, the biceps brachii was mentioned in the previous section as the prime mover causing flexion of the forearm. Its antagonist, the triceps brachii, is on the posterior side of the upper arm and extends the forearm. In order for the biceps brachii to flex, the arm the triceps brachii must relax.
- Synergists—The term *synergy* refers to a situation in which the collective efforts of a group results in a greater product than the sum of the individual accomplishments. For example, suppose a group of five students forms a company that makes baseball caps. Every night, after class, each student goes home and makes caps. Some are better at it than others, but let's say they make an average of 8 caps per night, or a total of 40 baseball caps in all. One day during a class break, these students are talking and one says, "I can make my quota every night, but what I'm really good at is sewing the bill on the hat." Another says, "You know, I can sew the band inside the cap really fast." That causes one of the other students to point out how good he is at putting the button in the top of the hat. The five students decide one evening to try making hats together so that everyone can do the task they do best. By the end of the evening, they have made 75 hats. That is synergy.

Prime mover (agonist) the skeletal muscle that is the major cause for a specific movement.

Antagonists skeletal muscles on the opposite side of a joint that relax while the muscle causing the movement contracts.

Working together they were able to produce more hats (75) than the sum of their individual efforts (40 hats). Muscles can do the same thing. When they are stimulated together, they are able to achieve more strength, increased range of motion, and greater stability of joints than is achieved through the mere sum of their individual abilities. When they are in the process of working together for this purpose they are considered to be synergists. For example, if a person flexes the shoulder, the biceps brachii, pectoralis major (attached to the anterior chest), and anterior deltoid (covering the shoulder) all work together to provide the ability to increase range of motion.

> **Synergists** skeletal muscles that are stimulated together and are able to achieve more strength, range of motion, or stability of joints than the mere sum of the individual abilities.

Comprehension Check-up:

1. Muscles that contract together to achieve greater strength than the sum of their individual abilities are considered at that time to be _____.

2. Muscles that relax as others on the opposite side of the joint are contracting are regarded as _____.

1. Synergists 2. Antagonists

Clinical Skills—Manual Muscle Testing

Manual Muscle Testing (MMT) is used in physical and occupational therapy as a technique to grade the strength of a patient's muscles. This technique assists a therapist in assessing the areas where muscles need to be strengthened or evaluating the possibility of some other musculoskeletal dysfunction. A therapist performs MMT by providing resistance in specific patterns to the main action of a muscle or muscle group. The clinician would then grade the strength of that muscle or muscle group on a scale of 0-5. It is important to note that testing should be done bilaterally for comparison.

As an example, a patient who "holds test position against maximal resistance" is graded as "5" or "Normal" or "100%." A patient who "holds test position against moderate resistance" is graded as "4" or "Good" or "80%." A patient who "holds test position against gravity" is graded as "3" or "Fair" or "50%." These three different methods of grading or scoring are based on Medical Research Council scale for the numerical scores; the Daniels and Worthingham grading system for the "normal" (descriptive terms) scale; and the Kendall and McCreary approach for the percentage (%).

http://highered.mheducation.com/sites/0071474013/student_view0/chapter8/manuaul_muscle_testin

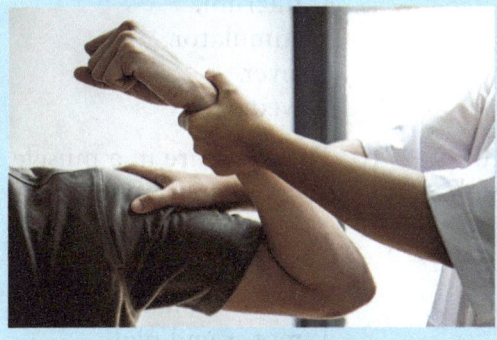

© SuperOhMo/Shutterstock.com

Contributed by Dave Smith. © Kendall Hunt Publishing Company

Chapter 5 The Muscular System

CLINICAL TERMS TO REVIEW

⌐Skeletal, smooth, and cardiac muscles
Sarcomere, myofibrils, sarcolemma
Neuromuscular junction, neurotransmitters, motor unit
Tetanus

Sternocleidomastoid, deltoid, trapezius, latissimus dorsi, pectoralis major
Triceps, biceps
Rectus abdominis, transverse abdominis, external and internal obliques

Quadriceps femoris, hamstrings, gastrocnemius, tibialis anterior
Muscle strain
Manual muscle training⌐

Test Yourself

Choose the best answer to the following multiple choice questions:

1. The type of muscle that has no striations, may contract by its own stimulation or is unconsciously activated by the brain, and is found in the walls of hollow organs and blood vessels is
   a. cardiac muscle.
   b. smooth muscle.
   c. skeletal muscle.
   d. rough muscle.

2. Sarcomeres in the muscle fiber are composed of
   a. thin and thick filaments.
   b. light and dark filaments.
   c. collagen and elastic fibers.
   d. sarcolemma and sarcoplasmic reticulum.

3. Sarcomeres are arranged side by side for increased
   a. longevity.
   b. distance of contraction.
   c. resistance to change.
   d. strength.

4. The chemical released by a nerve at the motor end plate that stimulates a muscle fiber to contract is known as a(n)
   a. hormone.
   b. enzyme.
   c. neurotransmitter.
   d. neurotoxin.

5. When a muscle fiber is stimulated again before it has time to completely relax, it results in additional shortening. This is known as
   a. twitch summation.
   b. fixating.
   c. tetanus.
   d. a reflex.

6. The greatest amount of ATP is produced in humans through
   a. photosynthesis.
   b. anaerobic metabolism.
   c. parthenogenesis.
   d. aerobic metabolism.

7. The type of muscle contraction that results in the length remaining the same but the strength changing is known as a(n) _____ contraction.
   a. hypertensive
   b. isometric
   c. hypotensive
   d. isotonic

8. The impulse from a nerve that stimulates a muscle fiber to contract is known as a(n)
   a. action potential.
   b. isotonic stimulator.
   c. nervous system relay.
   d. neuroactivator.

9. Muscle creates a force acting on the bone. The bone involved is considered to be a
   a. fulcrum.
   b. stimulator.
   c. lever.
   d. wedge.

10. The structure in a muscle fiber that shortens to cause contraction is known as a
    a. sarcoplasmic reticulum.
    b. myofibril.
    c. sarcolemma.
    d. motor end plate.

The Nervous System and the Senses

Chapter 6

LEARNING OBJECTIVES

Upon completion of this chapter, you will be able to:

1. Describe the components and functions of the nervous system and how it helps maintain homeostasis.
2. Identify the four (4) major areas of the brain, and briefly describe how their functions are critical to maintaining normal body functions.
3. Describe how nerve impulses are conducted along nerves and across synapses.
4. Identify the different sensations and briefly describe each.
5. Describe the physiology of the retina and hearing.

CHAPTER OUTLINE

Introduction
Functions of the Nervous System
- Sensory Function
- Integrative Processing
- Motor Stimulation

Nervous System Divisions
- Anatomical Divisions
- Functional Divisions

Components of the Nervous System
- Neurons

Neuroglia
- Astrocytes
- Oligodendrocytes
- Schwann Cells
- Ependymal Cells
- Microglia

Neuron Physiology: Creation of an Action Potential
- Cell Membrane Potential
- Resting Potential
- Depolarization: Sodium Gates
- Repolarization: Potassium Gates
- Transmission of the Action Potential
- Myelinated versus Unmyelinated Axon Conduction
- Physiology of the Synapse

Central Nervous System
- Brain
- Spinal Cord

Peripheral Nervous System
- Cranial Nerves
- Spinal Nerves

The Senses
- Vison
- Hearing and Balance

INTRODUCTION

One of the fascinating body systems to learn is the nervous system. This is the control center of the human body. It also coordinates many activities of the body. It comprises the brain, spinal cord and the nerves. The nervous system, as a control center, plays an important role in stimulating muscle contraction, reacting to the environment for any changes, being sensitive to emotions, and managing learning, memory, retention and reasoning.

As mentioned in Chapter 2, brain cells need blood supply and oxygen. When they are deprived of oxygen, brain cells die that can lead to brain damage or brain death. The brain is part of the central nervous system (CNS). The CNS, when damaged or injured, cannot repair or heal, and does not regenerate.

> **Clinical Notes—Oxygen and Brain Cells**
>
> One of the requirements to sustain life is oxygen. Brain cells begin to die after 4-6 minutes without oxygen when a patient stops breathing and the heart stops beating. This can lead to brain damage or brain death. The goal of cardiopulmonary resuscitation (CPR) is to keep both lungs and heart working, in order to move or supply blood and oxygen to the brain.

An interesting topic to many people, especially the students, is learning "How the Brain Works?" It is the goal of every student to improve one's memory. However, memory alone does not guarantee that it would be for long term. One may easily forget what has been learned recently. Learning, memory and retention do not happen overnight. There are many other factors to consider. Learning and retention need to be "spaced out." It means that studying and reviewing need to take place at different time frames. It takes time, hours and days, for the brain to process what has been learned. The brain cells called neurons and glial cells, need to find "connections" in order to establish a network of brain cells that serve a common purpose, and connect related ideas, concepts or skills. This means that students need to read, review and reinforce what they are learning on a *daily basis*. This can strengthen the connections or networks formed by the brain cells, for long term memory and retention.

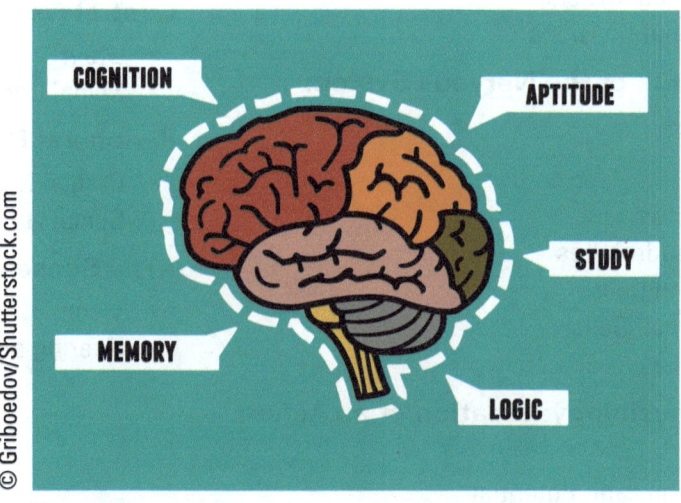

Figure 6.1

FUNCTIONS OF THE NERVOUS SYSTEM

The ability of the nervous system to provide rapid communication between cells allows it to perform three key functions: sensory function, integrative processing, and motor stimulation.

Sensory Function

The nervous system monitors sensory information from both the internal and external environments. This sensory input includes touch, pain, temperature, pressure, and location to keep the brain informed about conditions within as well as outside the body so that it can respond appropriately. Monitoring the surface of our body allows the brain to make informed decisions that safeguard us against potential external danger, changes in our surroundings, and communication with others. The ability to locate all areas of our body goes even beyond the ability to find areas of irritation or stimulation on our skin or knowing how far to reach. It even makes us identify with each part of our body. If we lose the sensation of location, we actually feel that the affected part no longer belongs to us.

Integrative Processing

Integrative processes allow us to interpret sensory information and compare that information with past experience or with a desired homeostatic level designated as the set point. That information can then be used to formulate a decision that results in an appropriate course of action. For example, if I am touched on the back by someone I do not see at that moment, from the degree and speed of contact, my brain interprets these sensations as friend or foe, causing me to respond in a similar manner.

Motor Stimulation

Motor responses refer to the activation of mechanisms that allow the body to make internal changes. Many motor functions involve the conscious stimulation of skeletal muscle as well as unconscious autonomic stimulation of smooth muscle, cardiac muscle, and glands and the altering of internal conditions to maintain homeostasis.

NERVOUS SYSTEM DIVISIONS

There are two general ways in which we can view the nervous system. One is in anatomical terms, and the other is in terms of function. Our primary approach in this book is anatomical.

Anatomical Divisions

There are two anatomical divisions of the nervous system. The first is the central nervous system, where all sensory perception, decision making/integration, and motor control occur. The central nervous system consists of the following:

- Brain
- Spinal cord

> **Central nervous system** the anatomical division of the nervous system that includes the brain and spinal cord.

> **Peripheral nervous system** nerves outside the central nervous system. The anatomical division of the nervous system consisting of cranial nerves and spinal nerves.

The peripheral nervous system consists of the outlying nerves that transmit impulses to skeletal muscle to cause muscle contraction or stimulate glands to secrete specific substances, or nerves that send sensory information about the body's internal and external environment, such as touch, pain, temperature, and pressure to the brain and spinal cord for analysis. The peripheral nervous system consists of the following:

- *Cranial nerves,* which exit through foramina (holes) in the skull
- *Spinal nerves,* which exit between vertebrae

The details of the anatomical divisions of the nervous system are discussed later in this chapter.

Functional Divisions

There are two functional divisions of the nervous system. The somatic division is concerned with conscious decisions and control of skeletal muscle, whereas the autonomic division controls smooth muscle, cardiac muscle, and glands.

Somatic Nervous System

> **Somatic nervous system** the division of the nervous system that is concerned with movement of the body and sensing the environment.

The somatic nervous system is concerned with movement of the body and sensing the environment. The two functions of the somatic nervous system are:

- Skeletal muscle stimulation
- Perception of general senses

Nerves that stimulate skeletal muscle run through our body together with nerves that sense the skin around those muscles. Conscious awareness is a major component in both the decision to move as well as the reception and interpretation of sensory information. Skeletal muscle is considered voluntary because it is consciously controlled by the cerebral cortex.

Autonomic Nervous System

> **Autonomic nervous system** the division of the nervous system that causes unconscious stimulation of smooth muscle and glands and regulation of intensity and speed of heart contractions.

The second system is the autonomic nervous system, in which the control is automatic without the need for conscious decisions. The three major functions of the autonomic nervous system are as follows:

1. Smooth muscle stimulation to alter the diameter of blood vessels and also affect the rate of activity at which some organs function
2. Regulation of the intensity and rate of heart contraction
3. Gland control, which stimulates glands or the prevents glandular secretions to maintain homeostasis throughout the body

> **Sympathetic nervous system** one division of the autonomic nervous system referred to as the *fight-or-flight system* because it prepares the body to defend itself or leave the situation.

The two subdivisions of the autonomic nervous system are the:

- Sympathetic nervous system, commonly known as the "fight-or-flight" system
- Parasympathetic nervous system, also referred to as the "rest-and-digest" system.

> **Parasympathetic nervous system** the other division of the autonomic nervous system referred to as the *rest-and-digest system* because it causes the opposite responses of the sympathetic nervous system to calm the body after being in the fight- or-flight mode.

The somatic nervous system allows us to concentrate on moving through the environment, whereas the autonomic nervous system makes internal adjustments as the need arises without diverting our attention.

> **Comprehension Check-up:**
> 1. The central nervous system consists of the _____ and _____.
> 2. Automatic unconscious stimulation of smooth muscle, cardiac muscle, and glands is accomplished by the _____, a functional division of the nervous system.
>
> 1. brain spinal cord   2. autonomic nervous system

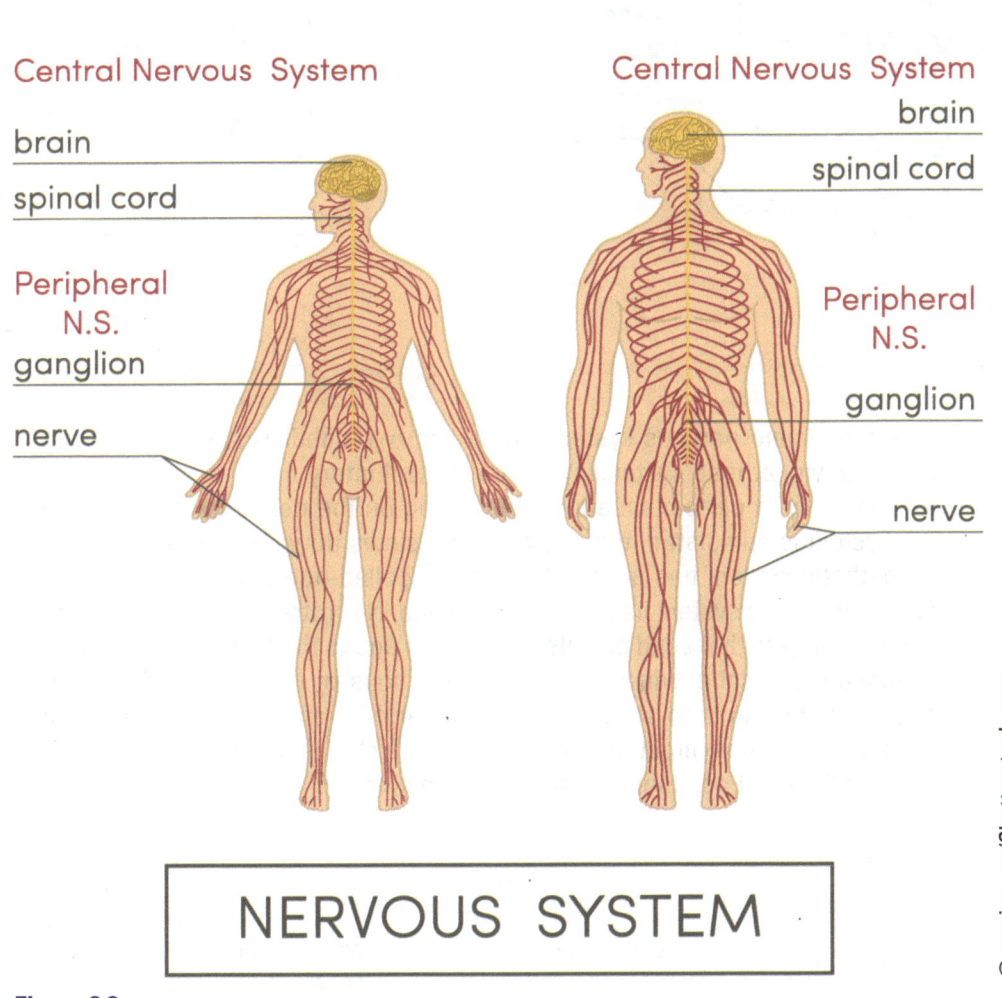

Figure 6.2

Chapter 6 The Nervous System and the Senses

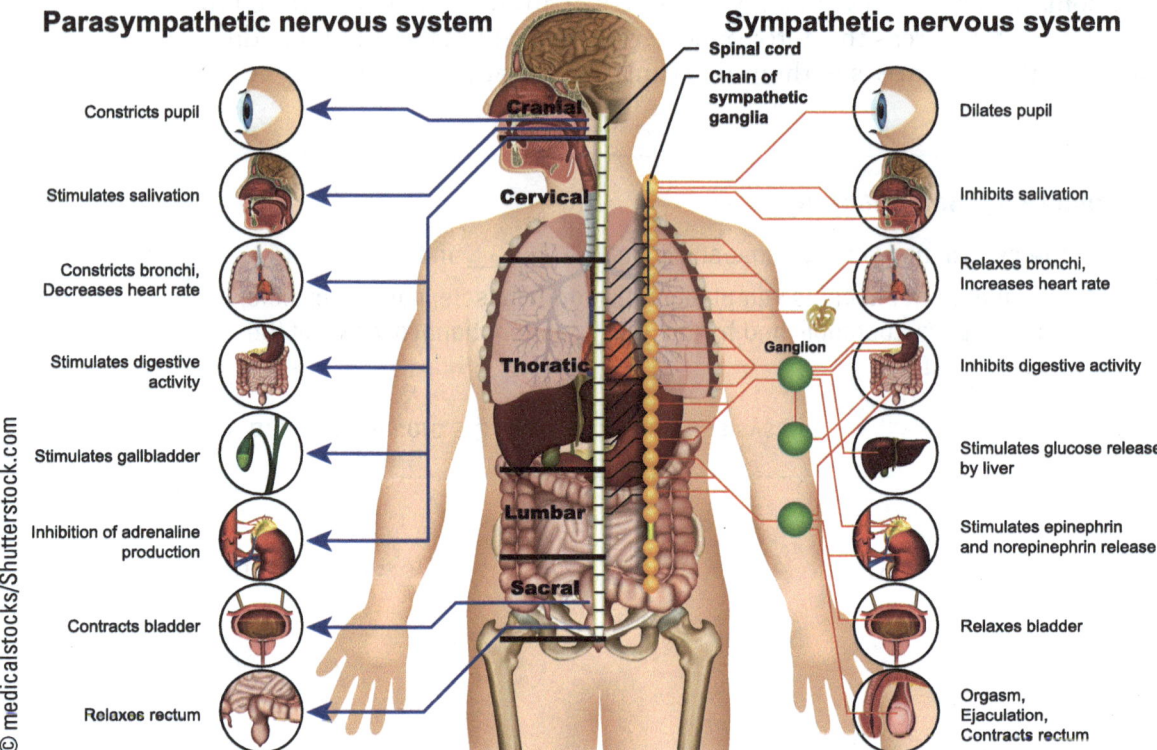

Figure 6.3

Clinical Notes—Sympathetic and Parasympathetic Systems

The autonomic nervous system (ANS) has two subdivisions: sympathetic and parasympathetic systems. Both have involuntary actions, or they react automatically in certain situations depending on the nerve impulses. When one system is stimulated, the other system slows down, or vice-versa. The sympathetic system has a neurotransmitter called norepinephrine, and the parasympathetic system has the acetylcholine. Neurotransmitters are chemical messengers that send signals to the brain and to the body by either firing or activating or inhibiting physiological functions of the body. The sympathetic system reacts to stressful situations and is referred to as the "flight or fight." The parasympathetic system is the "relax and digest." Below is a listing of the most common reactions of both systems:

Organs affected	Sympathetic reaction (flight or fight)	Parasympathetic Reaction (relax and digest)
Heart	increase in heart rate increase in heart contraction	decrease in heart rate decrease in heart contraction
Coronary arteries	dilate	contract
Bronchi	dilate	contract
Eye pupils	dilate	contract
Muscles	contract	relax

Organs affected	Sympathetic reaction (flight or fight)	Parasympathetic Reaction (relax and digest)
Digestion- salivary glands Stomach movement	decrease decrease	increase increase
Urine output	decrease	increase

When a person is faced with a stressful situation such as a burning house, the sympathetic system is activated. When a person is more relax after eating a meal, the parasympathetic system controls and relaxes the person.

When a person has an asthmatic attack where the bronchi are constricted due to bronchospasm, the person has trouble or shortness of breathing (dyspnea). A bronchodilator drug is then given that has a sympathomimetic effect that will dilate or open the bronchi to allow breathing.

COMPONENTS OF THE NERVOUS SYSTEM

When asked about the nervous system, most people know about neurons (nerve cells), but few are aware that there is an abundance of non-nerve cells in the nervous system. These supporting cells are known as *neuroglia*. The action of the nervous system depends on the impulses created and transmitted by neurons, but the neuroglia protect and defend the neurons in addition to providing a means to increase their ability to conduct information.

Neurons

Nerve cells are known as neurons. The following sections describe the anatomy and physiology of the neuron.

Neuron Structure

A neuron consists of a cell body, axon, and terminal end. The cell body contains the nucleus and organelles, and on some neurons, there are extensions of the cell body, known as dendrites, which provide additional locations for attachment of other neurons. (Please note that the dendrites are not part of the cell body.) The cell body narrows to a thin threadlike structure known as the axon; the axon extends from the cell body to its intended receiver, sometimes at a long distance. In some individuals some of the axons run close to a meter in length. Many of the axons have a coating of neuroglia that forms what is known as the *neurilemma (myelin sheath)*. On one end of the axon is a terminal end that forms a junction with either the next neuron, muscle, or gland.

Types of Neurons

Neurons are classified as one of three types based on their morphology (shape) and function. These different morphologies enable them to make connections and conductions suitable for different types of functions.

> **Neurons** the section of the nerve cell that contains the nucleus and organelles. It is also where synapses from other neurons connect.

> **Dendrites** finger-like projections on the neuron cell body that provide additional locations for attachment of other neurons.

> **Axon** the narrow threadlike extension of the cell body that runs to the intended receiver of the nerve impulse.

> **Terminal end** the distal end of a nerve that forms a junction with the next nerve, muscle, or a gland.

> **Sensory neuron** a neuron that transmits sensory information such as touch, pain, temperature, and pressure.

- *Unipolar neurons* begin as a single process (pole) coming out of the cell body that splits to run in two directions. It transmits sensory information such as touch, pain, temperature, and pressure and is often referred to a sensory neuron, or *afferent neuron* (meaning "inward"). They have an axon that runs the entire length of the neuron, with its cell body toward the center. On one end of the axon are sensory receptors that monitor the environment and create an impulse if stimulated. The impulse travels down the axon to the terminal end, where the nerve connects with the next neuron, carrying this information up to the brain for interpretation.
- *Bipolar neurons* have a projection on either side of the cell body. They transmit information for vision and our sense of smell.
- *Multipolar neurons* have multiple processes on their cell body. Surrounding the cell body are dendrites that provide sites for other neurons to attach. They stimulate all three types of muscle and are referred to as motor neurons, or *efferent neuron* (meaning "outward"). The impulse to contract is first received through the sensory receptors of the neuron via the dendrites or cell body, and then is transmitted along the axon to the terminal end attached to the motor end plate of a muscle fiber. Multipolar neurons usually attach to more than one muscle fiber, forming a motor unit.

> **Motor neurons** a neuron that transmits the stimulus to cause muscle to contract.

Synapse

The connection of a neuron with an effector, another neuron, muscle fiber, or gland is known as a synapse. On the terminal end of the neuron are synaptic knobs. These synaptic knobs are close to but not in contact with the postsynapse. The space between the synaptic knob and postsynapse is known as the synaptic cleft.

> **Synapse** the connection of a neuron with another neuron, muscle fiber, or gland.

> **Comprehension Check-up:**
> 1. Axons are coated with myelin-containing cells forming the _____.
> 2. The type of neuron that has multiple processes on its cell body to which other neurons can attach is known as a neuron.
>
> 1. neurilemma or myelin sheath 2. motor

NEUROGLIA

Supporting cells are found in the central nervous system that assist, insulate, and protect the delicate neurons in the brain and spinal cord. Neuroglial cells outnumber the neurons they support accounting for about half of the nervous tissue within the nervous system. There are five types of neuroglia, each having distinct functions (Figure 6.4).

- Astrocytes
- Oligodendrocytes
- Schwann cells
- Ependymal cells
- Microglia

Astrocytes

Astrocytes are found between blood vessels and brain cells (Figure 6.4a). They are so named because they are star-shaped (from *astro,* meaning "star"). They protect the crucial brain neurons. *Astrocytes,* along with specialized capillaries, form what is known as the *blood-brain barrier.* Substances in the bloodstream cannot reach the brain cells without first passing through the astrocytes. These protectors prevent potentially damaging substances from reaching highly critical brain cells. The down side of this arrangement is that if we need to provide the brain with useful chemicals, it is sometimes difficult for them to pass through the blood-brain barrier. Mental difficulties, such as depression or schizophrenia, are symptoms of a chemical imbalance in the brain. Drugs can correct or counteract this imbalance by regulating the transmission of neuron impulses, thereby alleviating the symptoms; however, it often takes weeks for the drug to penetrate the blood-brain barrier. Alcohol, on the other hand, readily passes through the blood-brain barrier and can therefore alter the mental state in a very short period.

Oligodendrocytes

Oligodendrocytes are wrapped around the axons of neurons in the brain and spinal cord (Figure 6.4b). They form the neurilemma, that is, the myelin sheath within the central nervous system. The interior or the brain and the superficial layer of the spinal cord are composed of bundles of myelinated axons forming white matter in the brain and spinal cord. They allow the rapid transmission of enormous numbers of impulses within the central nervous system every second. Neuron cell bodies have no myelin covering and are known as gray matter in the central nervous system.

> **White matter** bundles of myelinated axons in the interior of the brain and superficial layer of the spinal cord carrying nerve impulses within the brain or between the brain and spinal cord.

Schwann Cells

In the peripheral nervous system are myelin-containing cells, known as *Schwann cells,* that form the neurilemma on axons of cranial and spinal nerves (Figure 6.4c). They form the myelin sheath around axons that require rapid transmission of impulses.

> **Gray matter** neuron cell bodies having no myelin covering in the central nervous system.

Ependymal Cells

Ependymal cells produce cerebrospinal fluid (Figure 6.4d). Ependymal cells line the ventricles of the brain and the central canal of the spinal cord. The ependymal cells join with blood vessels to form the choroid plexus which produces cerebrospinal fluid. This clear liquid flows between the cranium and brain and also down the spine between the vertebrae and spinal cord. This fluid serves as liquid protection around the central nervous system, helping to minimize the potential of damage resulting from impact with surrounding bone.

> **Cerebrospinal fluid** a clear protective fluid that flows between the brain and cranium and then between the spinal cord and vertebrae to minimize the potential of damage due to impact with surrounding bone.

Microglia

Microglia are small phagocytes (Figure 6.4e). They wander around in the central nervous system killing and eating unwanted organisms and debris.

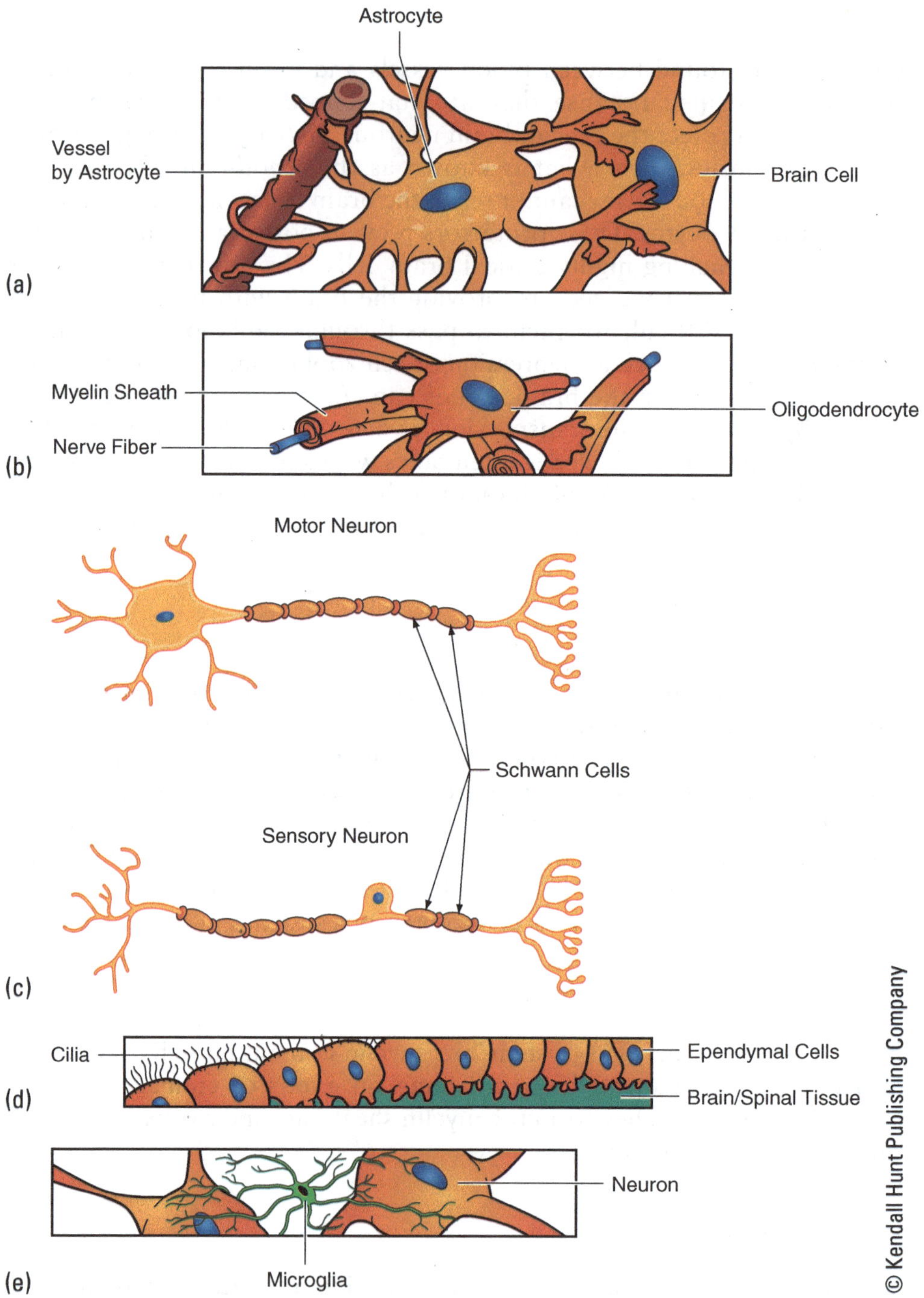

Chapter 6 The Nervous System and the Senses

(a) Astrocyte — Vessel by Astrocyte — Brain Cell

(b) Myelin Sheath — Nerve Fiber — Oligodendrocyte

(c) Motor Neuron — Schwann Cells — Sensory Neuron

(d) Cilia — Ependymal Cells — Brain/Spinal Tissue

(e) Microglia — Neuron

Figure 6.4 Neuroglia. (a) Astrocyte. (b) Oligodendrocyte. (c) Schwann cells. (d) Ependymal cells. (e) Microglia.

© Kendall Hunt Publishing Company

Comprehension Check-up:

1. Star-shaped cells that form part of the blood-brain barrier are called _____.

2. _____ are small phagocytes that protect and defend the central nervous system.

1. astrocytes   2. Microglia

NEURON PHYSIOLOGY: CREATION OF AN ACTION POTENTIAL

Jerking my hand back after grabbing very hot pizza, pitching a baseball, jumping when feeling something crawling across my face in the middle of the night, and breaking out in a sweat are all caused by an action potential. An action potential, also known as a nerve impulse, travels down a nerve to cause some event to occur. It may be the stimulus to cause muscle contraction. It may cause a gland to start or stop its secretions. It may be information about some sensation of which our brain needs to be aware so that it can respond appropriately. The speed with which we create action potentials is incredible. Every movement of skeletal muscle is the result of an action potential sent by the brain or spinal cord. In fact, every thought we have is the result of a series of action potentials, waves of electricity traveling, colliding, and responding across the brain. Understanding how this process occurs provides an enormous amount of insight into the complexity of the human body.

> **Nerve impulse** an action potential traveling through the neuron to its intended receiver.

Cell Membrane Potential

When discussing membrane potential it helps to explain the intent of the term *potential*. This is the "potential to do work" or to cause an event to occur. *Cellular membrane potential* results from positive and negative electrical charges found on ions concentrated on opposite sides of the plasma membrane. Muscle and nerves have an accumulation of ions on both sides of their plasma membranes. It is the differences in concentrations of ions, that is, the electrical charges, on opposite sides of the membrane that provide the drive to cause an action potential.

Resting Potential

On the outside of the cell is a relatively high level of sodium ions. Because the membrane is surrounded by these positive sodium ions, the outside of the cell is positively charged. Attached to the inside of the plasma membrane are negatively charged proteins. The inside of the cell is negatively charged but does also contain positive potassium ions, so the inside of the cell membrane is not as negative as it could be thanks to potassium ions. A charged molecule is considered polar, so too a charged plasma membrane is considered polarized. When muscle or nerve cells are not active, their plasma membranes are polarized with positive charges on the outside and negative charges on the inside. This is known as its resting potential.

> **Resting potential** the electrical state of the plasma membrane of a nerve or muscle cell when at rest. The membrane is positively charged outside and negatively charged inside.

Depolarization: Sodium Gates

The events of the creation of an action potential are identified in the text by numbers within parentheses. We begin with a neuron or muscle fiber in its resting state when a stimulus occurs (1). This stimulus causes gates of sodium channels under the stimulus to open (2). Remember that across the plasma membrane are channels, tube-like structures that have gates on them. When the gate on a sodium channel opens, sodium ions diffuse from areas with the highest concentration (that is, outside the cell) to an area with a lower concentration (that is, inside the cell) (3). As sodium ions diffuse in, they add positive charges to the inside of the membrane and make it less negative. If the gates stay open long enough, sufficient sodium ions diffuse in and cancel the negative charges on the inside of the membrane. At that point, because there are no charges on the inside, we signify that location as *depolarized* (4).

This is generally what is happening with depolarization. As sodium ions enter, they eventually add enough positive charges to the inside of the membrane to cancel the negative charge. In actuality, enough sodium ions diffuse in to cause the inside of the membrane to briefly become positively charged.

Repolarization: Potassium Gates

As soon as the plasma membrane depolarizes, the sodium channels close and potassium channels open (5). Recall that there is a higher concentration of potassium ions inside the cell. When the potassium channels open, potassium ions diffuse out of the cell and take their positive charges with them. As positive charges leave, the inside of the membrane becomes negatively charged inside again; that is, it is *repolarized* (6).

The depolarization of one location on the nerve or muscle fiber causes neighboring sodium channels to open as well. As a result, the neighboring area of the plasma membrane depolarizes. This process continues over the cell in a wave as the entire membrane depolarizes (7). And why is this important? It is the depolarization of the membrane followed by repolarization that is a nerve impulse or the stimulus for a muscle fiber to shorten.

Now that the membrane is back to its original electrical state, being positive outside and negative inside, it is electrically back to its resting potential, but the ions are backward. To keep the concentrations as they should be, we use pumps to push the sodium ions out of the cell and potassium ions back in again (8). Adenosine triphosphate (ATP) is required for this process. Once the sodium and potassium ions are back in their original location, the cell is back to its resting state, where it stays until another stimulus occurs (9).

> **Comprehension Check-up:**
> 1. When a neuron is not stimulated, it has positive charges on the outside of the plasma membrane and negative charges on the inside. The cell is considered to be in its _____.
> 2. The term that describes the canceling of the negative charge inside a cell is _____.
>
> 1. resting rate   2. depolarization

Transmission of the Action Potential

Depolarization of the plasma membrane followed by repolarization is known as an *action potential*. Once the action potential has begun in one area of the neuron, it will travel across the entire nerve. If the action potential begins on the cell body of a multipolar neuron, it will travel over the cell body until it reaches the axon. Many neurons have axons covered by a myelin sheath. Ions cannot penetrate the myelin sheath, and it would, at first, seem that the action potential would stop. However, there are places where ions can

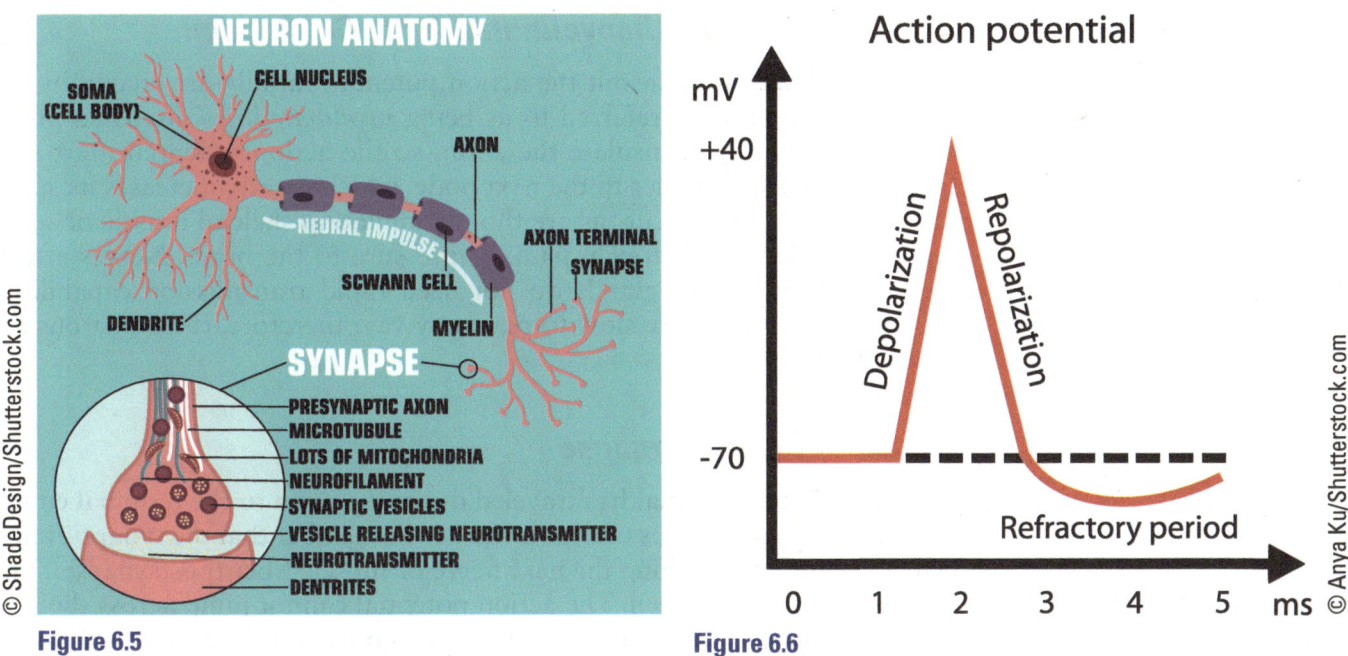

Figure 6.5

Figure 6.6

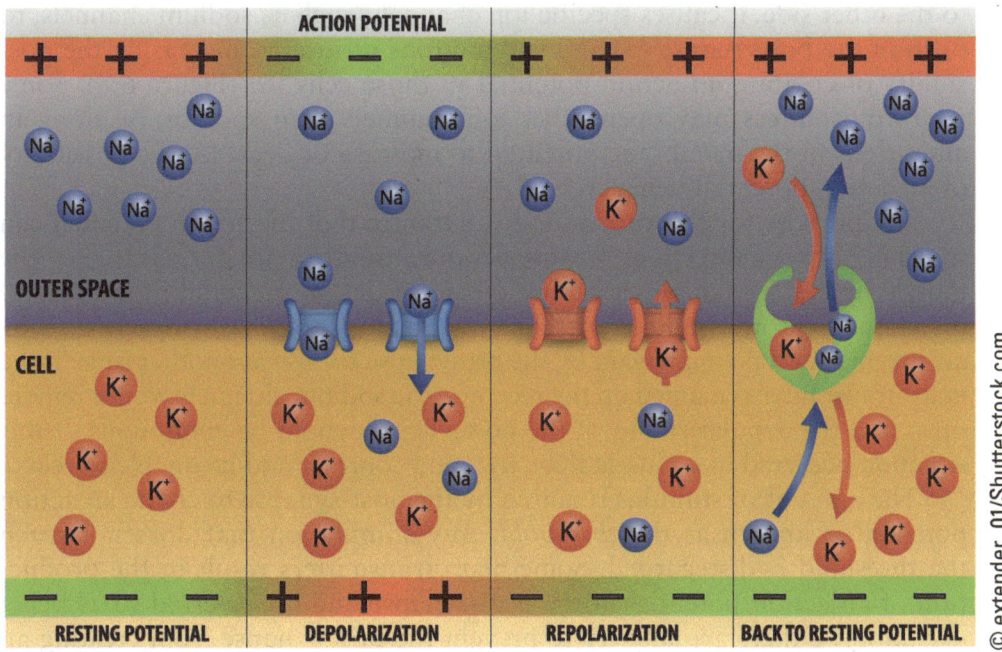

Figure 6.7

exchange. Between the neuroglia forming the neurilemma are spaces where the axon is exposed. These gaps are known as **nodes of Ranvier**. Depolarization on the proximal side of a neuroglial cell provides enough stimulus to open sodium channels in the next node of Ranvier. In essence, the action potential skips over the myelin-containing cells of the neurilemma, jumping from node to node. The advantage is that it is not necessary to depolarize every millimeter of the axon. By skipping from node to node, the action potential travels down the axon at greatly increased speed.

> **Nodes of Ranvier** the gap between myelin-containing cells wrapped around the axon. This is where the axon's plasma membrane is exposed and where sodium and potassium ions exchange through the cell membrane to cause an action potential.

Myelinated versus Unmyelinated Axon Conduction

Nerves that need to transmit the action potential rapidly are coated by the myelin sheath and are referred to as being *myelinated*. As discussed previously, these neuroglia insulate the axon, so the action potential must skip from one node of Ranvier to the next node, which greatly increases its speed of travel. Myelinated neurons are those stimulating skeletal muscle or sending information about touch, pain, or pressure to the brain. Nerves affecting smooth muscle or glands do not need rapid transmission capabilities because these organs are slow to react anyway; therefore, these neurons are *unmyelinated*.

Physiology of the Synapse

Once the action potential has traveled down the axon to the terminal end, it causes the synaptic knobs to depolarize (1). Remember that the synaptic knob is next to but not touching the next nerve or muscle fiber (postsynapse). The synaptic cleft lies between. The action potential cannot jump across the synaptic cleft. Instead, once the synaptic knob depolarizes, it releases a chemical known as a neurotransmitter into the synaptic cleft (2). The neurotransmitter crosses the cleft (3) and attaches to receptors on the postsynapse that are designed to accept the neurotransmitter (4). When this chemical is attached to the other side, it causes specific ion channels, such as sodium channels, to open in the postsynapse (5) and sodium ions diffuse in, increasing the possibility of creating an action potential in those cells (6) (Figure 6.8). Some neurotransmitters may open other ion channels than sodium, resulting in alterations of the membrane potential to increase or decrease the possibility of an action potential being created.

The purpose for this process is to guarantee that the action potential can travel in only one direction. Because neurotransmitters are not released by the muscle fiber, for example, it cannot create an action potential in the reverse direction. Neurotransmitters attached to receptors are rapidly deactivated by an enzyme and taken up again by the presynapse for use at another time. If the neurotransmitter remained on the receptor, the sodium channels would remain open and the depolarization of the postsynapse would be continuous. If this situation occurred in a muscle fiber, it would contract and be unable to relax.

The amount of stimulus required by the post-synapse to create an action potential is known as the *threshold*. Any stimulation that does not reach the threshold is disregarded. Some neurotransmitters result in the production of an action potential in the post-synapse and are referred to as being *excitatory*. Others actually try to prevent the post-synapse from creating an action potential; these are called *inhibitory*.

> **Neurotransmitter** a chemical released by a synaptic knob of one neuron that may cause the creation of an action potential in the next nerve or muscle fiber.

> **Comprehension Check-up:**
> 1. The action potential can skip down the axon from node to node to greatly increase the speed of the impulse if the axon is _____.
> 2. The chemical released by the synaptic knob that opens sodium channels on the postsynapse is known as a _____.
>
> 1. myelinated   2. neurotransmitter

THE SYNAPSE

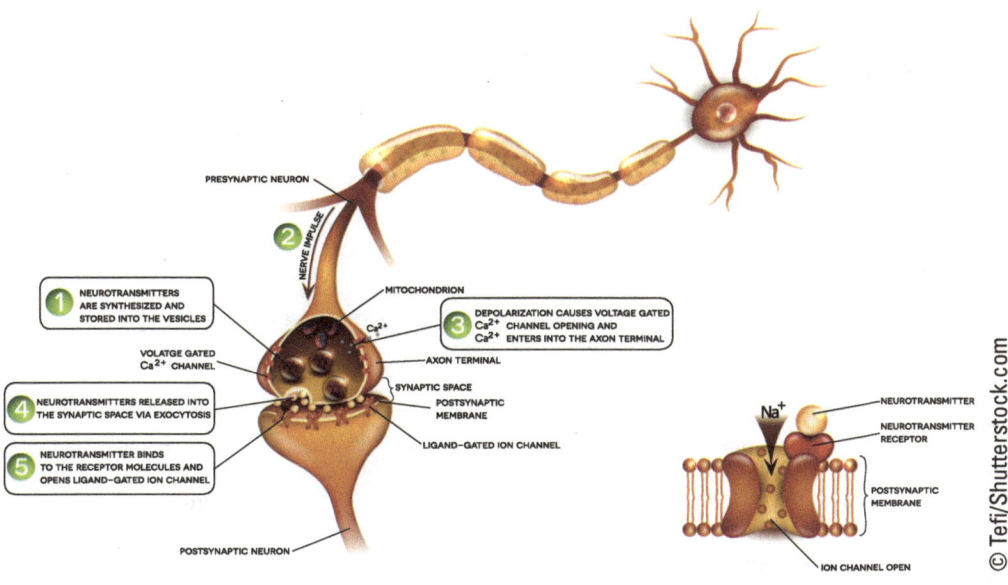

Figure 6.8

CENTRAL NERVOUS SYSTEM

We begin our discussion of the central nervous system by considering the most complex structure in the body: the brain. Each brain cell has roughly 20,000 synapses, with other neurons providing tremendous levels of information exchange.

Brain

The brain can be divided into four major areas: cerebrum, cerebellum, brain stem, and diencephalon (Figure 6.9):
Each of these areas has very different functions critical to maintaining normal body functions.

Cerebrum

The cerebrum is the largest of the four brain areas. It can be divided into a left and right cerebral hemisphere and into four lobes on each side.

Structures of the Cerebrum

When observing the cerebrum it is easy to see that there are ridges or folds on the outside; these are known as *gyri*. The valley between two gyri is known as a *sulcus*. In some cases the divisions between sections of the brain are very deep, resulting in what is known as a *fissure*. The left and right hemispheres of the cerebrum are separated by the longitudinal fissure. A wide band of axons, known as the *corpus callosum*, connects the two hemispheres. The corpus callosum contains more than 200 million axons carrying an estimated 4 billion impulses per second between the two hemispheres to ensure that the left side of the brain knows what the right side is doing and vice versa.

> **Cerebrum** the largest structure in the brain. It is responsible for conscious decision making, stimulation of skeletal muscle, interpretation of sensory information, communication, personality, and behavior.

> **Cerebral hemisphere** the cerebrum can be divided into two halves, the left and right hemispheres.

Ventricles in the Brain

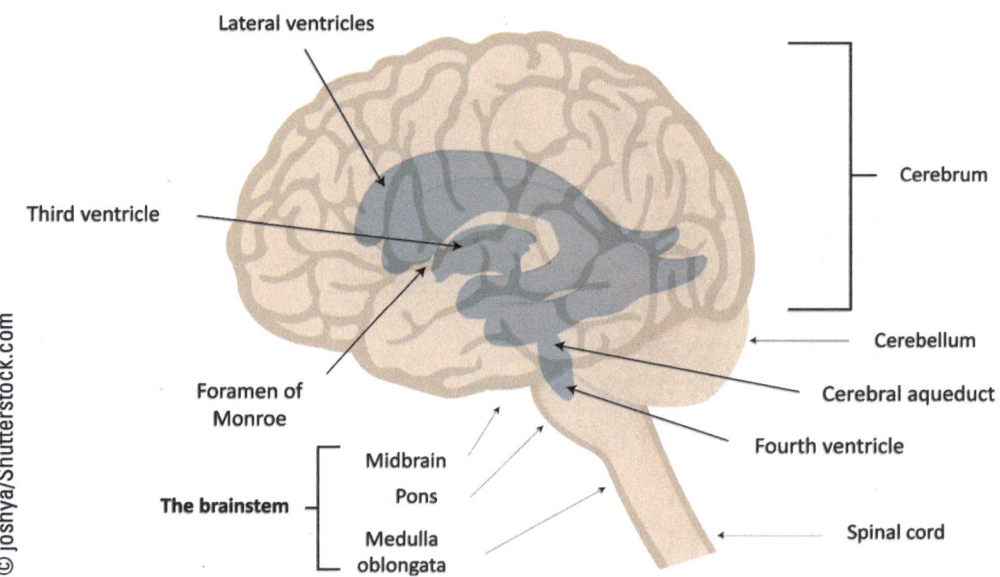

Median section of the brain

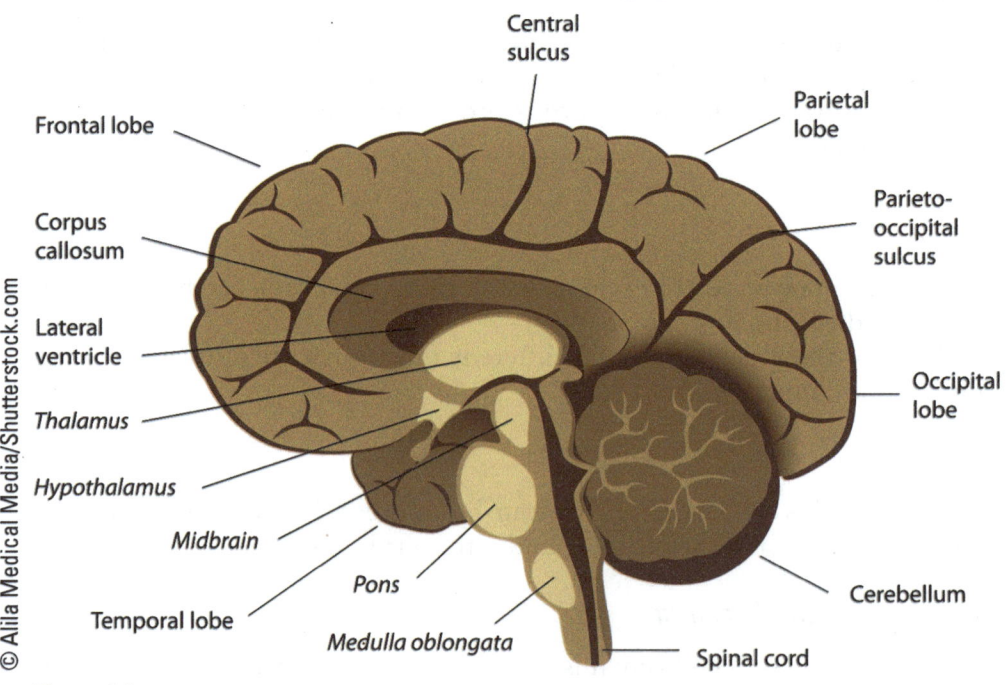

Figure 6.9

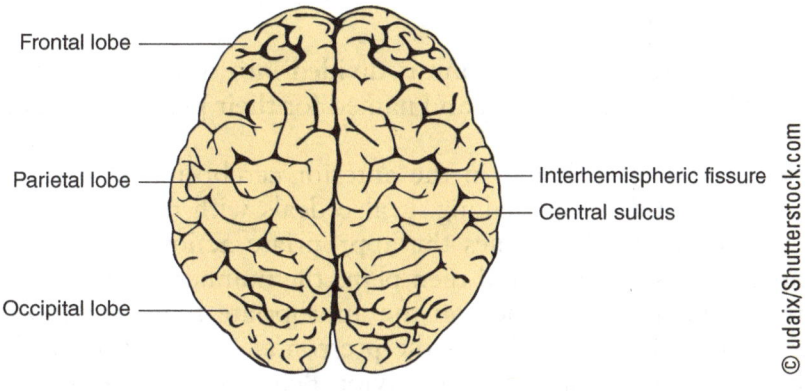

Lateralization of Brain Functions

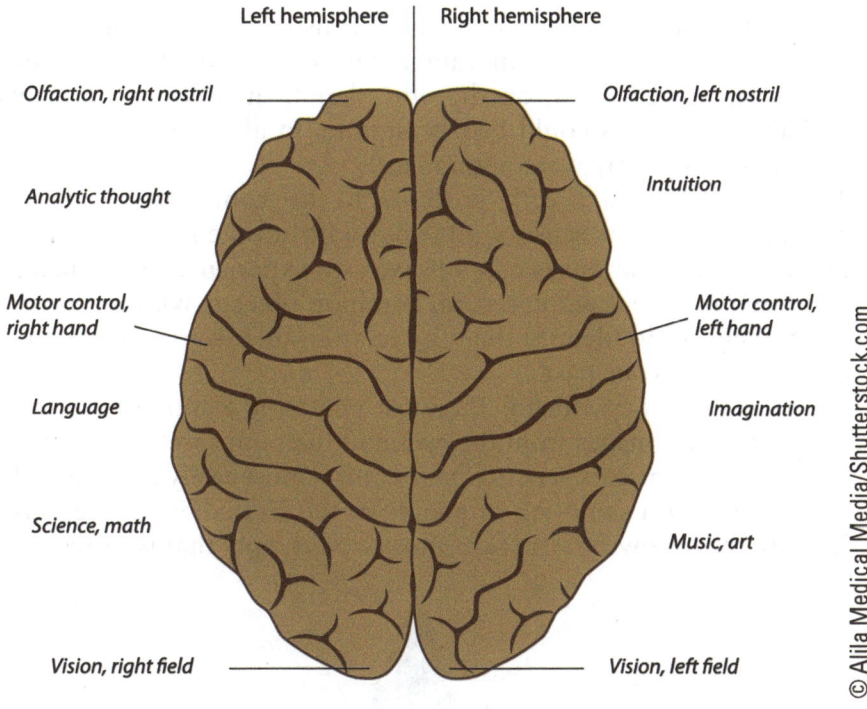

Figure 6.9 (Continued)

Cerebral Hemispheres

The cerebral hemispheres both have similar functions when controlling the stimulation of muscle and the reception and interpretation of sensory information, but they specialize in higher levels of mental activity. The left hemisphere deals with activities requiring analysis. It is well suited for math, philosophy, and language skills such as writing and speaking. The right hemisphere specializes in creative activities, such as music and art. As individuals progress through life, different environments, activities, injuries, etc., can impact the roles of the higher functions of the cerebral hemispheres so that they adapt as necessary.

Lobes of the Cerebrum

Each hemisphere of the cerebrum is divided into four lobes. The lobes are identified by the sulci and gyri and named for their protecting cranial bones.

Frontal lobe the most anterior lobe of the cerebrum that coordinates and stimulates contraction of skeletal muscle, as well as the location that controls personality, behavior, and analysis.

- The frontal lobe, found in the anterior cerebrum, primarily performs analysis and decision making. It also deals with movement. It analyzes a situation and determines the appropriate action. It decides whether movement is required and then plans the number of motor units to activate and stimulates them at the appropriate time. The most anterior portion of the frontal lobe—the prefrontal cortex, deals with analysis that involves responses and behavior. Before mood-altering drugs were invented, some of the axons from the prefrontal lobe were disconnected from the rest of the brain as a means of controlling behavior, especially in dangerous individuals. This procedure, called a prefrontal lobotomy, resulted in a dramatic leveling out of the individual's personality. The person no longer went into rages but also lost the ability to become overly passionate about anything. For some individuals, it became a life-changing benefit. If, on the other hand, the disconnection was too extreme, the results could be devastating. In all cases the results of the prefrontal lobotomy were permanent.

Parietal lobe the central superior lobe of the cerebrum that receives, locates, and interprets sensory information.

- The parietal lobe is directly posterior to the frontal lobe. It receives, locates, and interprets sensory information from the body. The interpretation of these sensations is based heavily on experience. If you placed your hand in a box containing several common objects, without looking you most likely could identify them by relating their sensations with similar experiences in your past.

Occipital lobe the most posterior lobe of the cerebrum that receives and interprets vision.

- The occipital lobe, which is the most posterior section of the cerebrum, receives visual images from the eyes and then interprets what they mean. This process is also experience based and is most easily explained through an example. If a sampling of people in the United States were asked to identify the following structure, most would reply that is a green triangle.

If a brown rectangle is now added to the triangle, the interpretation of many people from temperate climates changes.

When looking at the preceding illustration, a person may now identify it as a pine tree or a Christmas tree. Does it look like a pine tree? Not really, but the general shape is close enough for most to make the association or interpretation based on previous experience. To someone with no previous knowledge of pine trees or Christmas trees, it is merely a green triangle and a brown rectangle. It is this interpretation that is useful to us when seeing part of a total image.

- The temporal lobe is the most lateral of the divisions of the cerebrum. It receives and interprets sound based on experience. It is important to note that sound is not necessarily the same as language, which is processed elsewhere in the brain. The temporal lobe is more concerned with distinguishing between a ticking clock and a gunshot or differentiating between a purr and a growl.

> **Temporal lobe** the most lateral lobe of the cerebrum that receives and interprets sound.

The lobes of the cerebrum are interconnected to provide a constant assessment of the environment and initiate plans to respond to both desires and requirements to maintain safety and homeostasis.

> **Comprehension Check-up:**
> 1. The cerebrum can be divided into left and right sides, known as _____.
> 2. The lobe of the cerebrum that deals with the location and interpretation of sensory information is the _____ lobe.
>
> 1. hemispheres 2. parietal

Cerebellum

The cerebellum is posterior to the brain stem and inferior to the cerebrum. The cerebellum has two major functions. First, it provides significant input to help us maintain our balance. Someone with damage to the cerebellum may appear to be drunk when completely sober. Second, it deals with fine motor coordination. In other words, it makes our movements fluid and precise, rather than jerky.

> **Cerebellum** the structure of the brain that maintains balance and fine motor coordination to cause movements to be fluid.

Brain Stem

Coming out of the center of the brain is the brain stem. As the center of vital functions, the brain stem is critical to our well-being. That is, it controls functions essential to life, such as breathing, blood pressure, heart rate, sneezing, coughing, and gagging. At first it would seem that sneezing and coughing are not as vital to life as breathing or controlling our blood pressure. They are, however, major factors in opening our airway to allow us to continue breathing. The brain stem can be divided into three sections:

> **Brain stem** the structure of the brain that connects the rest of the brain to the spinal cord. Within the brain stem are centers for vital functions such as control of breathing, heart rate, blood pressure, and processes necessary to maintain an open airway.

- The *midbrain* connects the cerebrum to the rest of the central nervous system.
- The *pons* connects the cerebellum to the rest of the central nervous system. It also contains some of the respiratory centers that control the length of inspiration.
- The *medulla oblongata* connects the brain to the spinal cord. It also regulates cardiac activities and some respiratory activities, as well as other vital functions.

Centers for vital functions along with clusters of neurons dealing with other functions are found throughout the brain stem.

Diencephalon

In the center of the brain at the superior end of the brain stem is the *diencephalon*. Because it consists of several structures, the functions of the diencephalon are quite varied.

- The *thalamus* relays sensory information such as touch, pain, temperature, pressure, sight, and sound to multiple locations within the brain to provide quick analysis and interpretation.
- The *hypothalamus* maintains homeostasis of food and water intake, bodily temperature control, hormone levels. It is discussed in greater detail in a later chapter.
- The *pineal gland* is an endocrine gland that acts as our biological clock. It acts as a timer setting daily cycles known as *circadian rhythms*. Examples of rhythmic cycles are patterns that cause us to wake up or go to sleep at the same time every day. If meals are eaten at roughly the same time every day, as those times draw near we become hungry. This is also discussed in more detail in a later chapter.
- The *limbic system* involves numerous structures in the diencephalons and cerebral cortex. It is a functional classification more than an anatomical location. It deals with stress and emotion and also takes an active role in memory and learning.

Associated Structures in the Brain

There are several structures found in the brain that are not involved with its function but that are necessary for maintaining homeostasis of the brain and spinal cord.

Ventricles

Deep within the brain are spaces known as *ventricles*. There are four of them. Within these spaces are ependymal cells that produce cerebrospinal fluid. This fluid exits from the ventricles to surround and protect the brain and spinal cord.

Meninges

The brain and spinal cord are covered by three membranes known as meninges.

> **Meninges** membranes The brain and spinal cord.

- The *dura mater* is a tough, thick membrane attached to the inside of the cranium and vertebral canal. This membrane essentially isolates the central nervous system from its protective structures and helps hold the brain in place.
- The *arachnoid mater* is the middle layer that consists of spiderweb-like fibers named after spiders—arachnids. This layer is filled with cerebrospinal fluid, which is produced in the ventricles. It provides a fluid cushion around the brain and spinal cord.
- The *pia mater* is a thin covering directly on the brain and spinal cord.

Clinical Notes—Stroke

According to the American Stroke Association (ASA):

"A stroke is a disease that affects the arteries leading to and within the brain. It is the No. 5 cause of death and a leading cause of disability in the United States. Stroke can be caused either by a clot obstructing the flow of blood to the brain (called an ischemic stroke) or by a blood vessel rupturing and preventing blood flow to the brain (called a hemorrhagic stroke). A TIA (transient ischemic attack), or "mini stroke", is caused by a temporary clot."

Types of Stroke

Ischemic stroke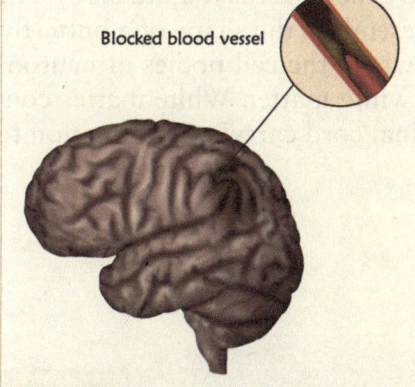

Hemorrhagic stroke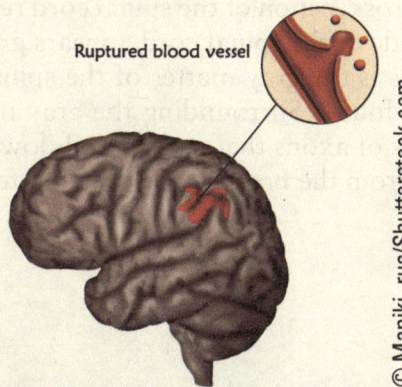

SPOT A STROKE
LEARN THE WARNING SIGNS AND ACT FAST

B E F A S T

BALANCE — LOSS OF BALANCE, HEADACHE OR DIZZINESS

EYES — BLURRED VISION

FACE — ONE SIDE OF THE FACE IS DROOPING

ARMS — ARM OR LEG WEAKNESS

SPEECH — SPEECH DIFFICULTY

TIME — TIME TO CALL FOR AMBULANCE IMMEDIATELY

CALL 911 IMMEDIATELY

Spinal Cord

The medulla oblongata of the brain stem connects the brain to the spinal cord when it exits the skull through the *foramen magnum,* Latin for "large hole," in the occipital bone. Inferior to the skull the spinal cord runs through the *vertebral canals* of the spine. The spinal cord transmits impulses for the stimulation of muscle from the brain and transmits sensory information to multiple areas in the brain. In addition, the spinal cord has the ability to initiate skeletal muscle contraction on its own without waiting for stimulation from the frontal lobe of the cerebrum by causing reflexes to occur. *Reflexes* are rapid responses to excessive sensory input from some area of the body. For example, if a person touches something hot with his or her bare hand, the arm is immediately jerked away as the input of heat and pain to the spinal cord causes stimulation of muscles to withdraw the upper extremity.

Structure of the Spinal Cord

A cross section of the spinal cord reveals two general areas (Figure 6.10). The inside of the spinal cord appears gray and resembles the shape of a butterfly. This is the gray matter of the spinal cord, where the cell bodies of neurons are found. Surrounding the gray matter is white matter. White matter consists of axons that run up and down the spinal cord carrying information to or from the brain to the gray matter.

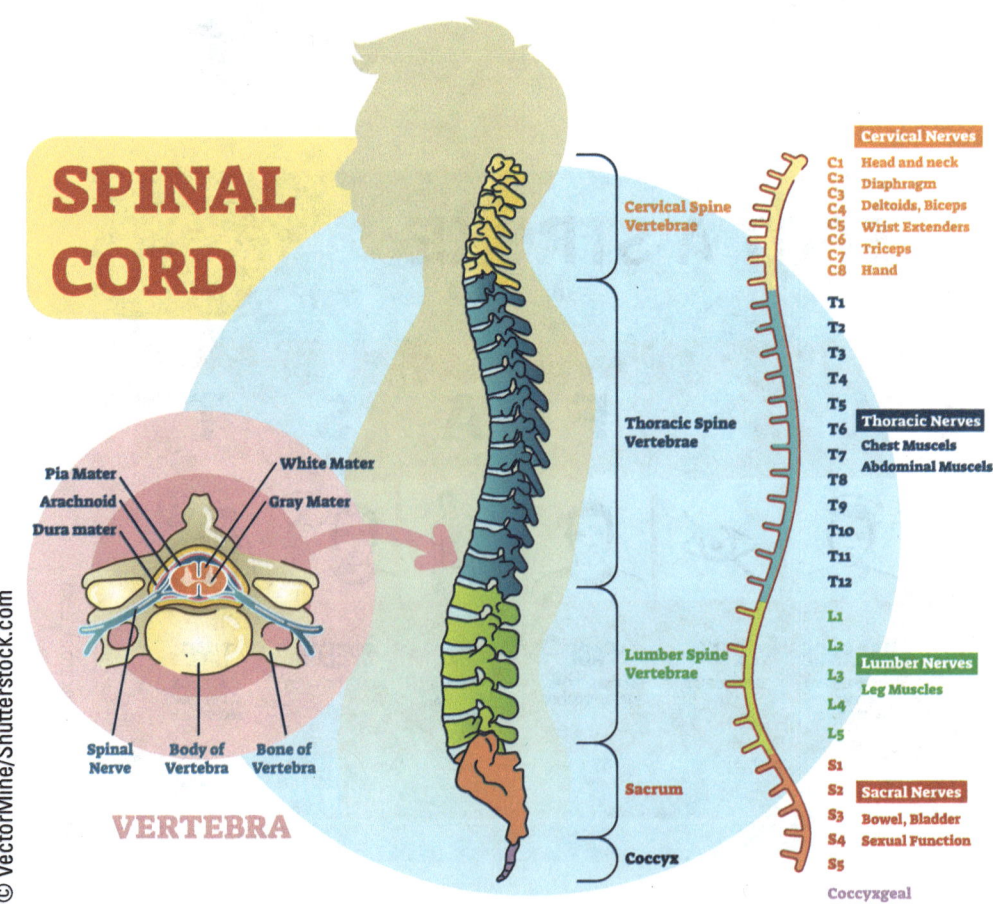

Figure 6.10

Spinal cord

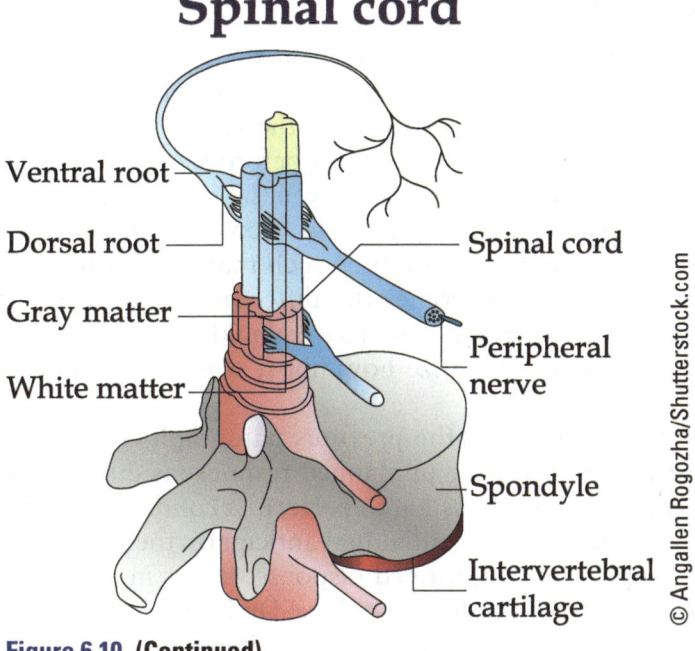

Figure 6.10 (Continued)

Gray Matter of the Spinal Cord

The gray matter of the spinal cord can be divided into three sections known as *horns*:

1. The *anterior (ventral) horn* contains the cell bodies of motor neurons. These neurons form the motor units that stimulate groups of skeletal muscle fibers.
2. The *posterior (dorsal) horn* receives sensory information from the body, primarily from the skin and joints.
3. The *lateral horn* contains cell bodies of autonomic neurons that stimulate or alter the movement of smooth muscles and glands.

Interneurons

In the gray matter of the spinal cord are neurons that connect sensory neurons to motor neurons known as *interneurons*. If sensory information coming into the spinal cord is strong enough to create an action potential in the interneuron, it may also result in the stimulation of skeletal muscle to cause a reflex or response initiated in the spinal cord.

> **Comprehension Check-up:**
> 1. The cell bodies of motor neurons stimulating skeletal muscle are found in the _____ horn of the spinal cord gray matter.
> 2. Sensory neurons are connected to motor neurons in the gray matter of the spinal cord by _____.
>
> 1. anterior 2. interneurons

Reflex Arc and Reflexes

The *reflex arc* consists of a sensory neuron that is connected to an interneuron; the interneuron stimulates a motor neuron to activate a muscle contraction (Figure 6.11). The purpose of the reflex arc is to cause muscle contraction at the level of the spinal cord before there is time for complete analysis and stimulation from the brain. Even though the brain is able to provide a rapid response, originating the muscle contraction in the spinal cord instead could increase the response time such that a slight injury becomes a major catastrophe or even becomes the difference between life and death. For example, a person touching a hot stove would pull his or her hand away before even having time to realize how hot it was. As a result, the burn is minimized.

The response discussed in the previous paragraph is considered an *unconditioned reflex* in which the unplanned response is due to sensory input to the spinal cord. It is also possible to perform a *conditioned reflex* if a pattern is established to which the individual anticipates the need for a response. The classic example of a conditioned response occurs in an experiment performed by Dr. Ivan Pavlov. In his procedure, Dr. Pavlov rang a bell whenever he fed a group of dogs. He discovered that, after time, just ringing the bell caused the dogs to salivate even though food was not yet present.

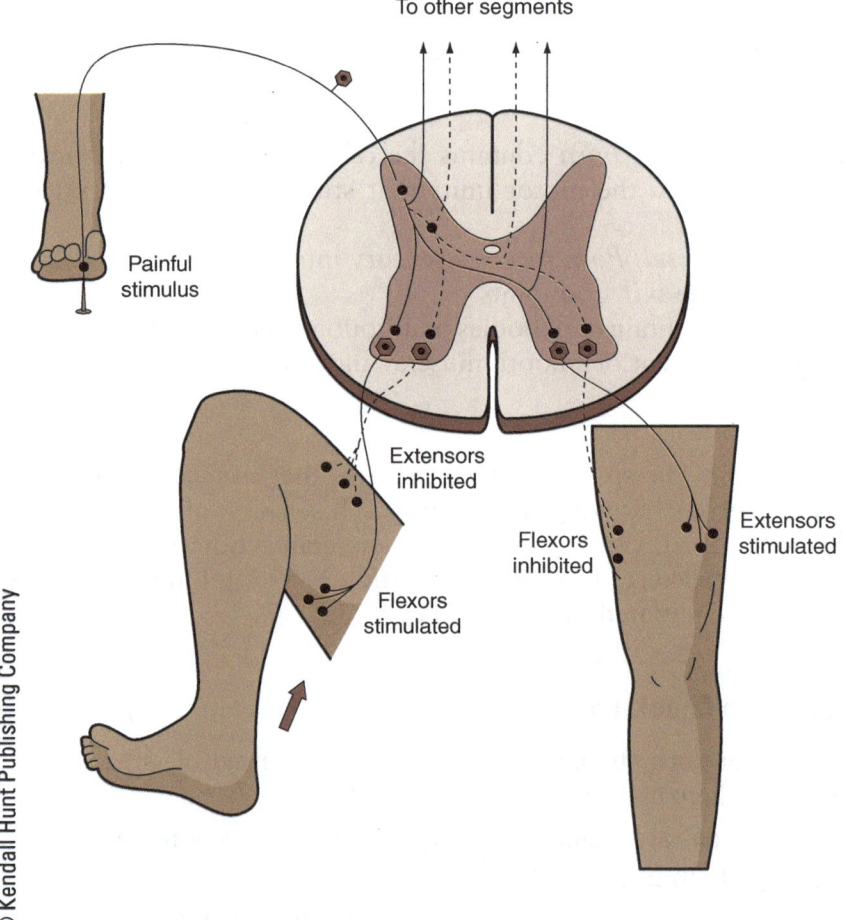

Figure 6.11 The Reflex Arc. Muscle contraction originates in the spinal cord rather than in the brain.

Reflexes may result from the sudden stretching of a tendon. This is commonly tested by a physician when he or she uses a rubber hammer to tap on various tendons on the body and observe the results. Best known is the tapping of the patellar tendon in the knee. Striking just below the patella, the percussion hammer stretches the tendon, causing the impulse from stretch receptors to be transmitted through sensory neurons to the spinal cord. There interneurons create an action potential in motor neurons to the quadriceps femoris, resulting in a *knee jerk reflex*. Another type of reflex occurs when pain sensors are activated. The reflex resulting would be a sudden *withdrawal reflex* from the source of injury. Again the action is initiated in the spinal cord rather than the brain.

Interneurons also connect to motor neurons on the opposite side to counterbalance sudden movements. If an individual suddenly jerks back from danger, losing balance could potentially put that person at risk of falling. By counterbalancing the movement, the individual remains in a stable position.

White Matter of the Spinal Cord

The axons carrying information between the brain and spinal cord in the white matter are often found in bundles called *tracts*, which run to or from a common place. There are two general categories of tracts: those transmitting stimulation from the brain and those carrying information to the brain.

- *Descending tracts* carry action potentials originating in the brain and descending the axons in the spinal cord to motor neurons in the anterior horn to stimulate skeletal muscle. Two examples of descending tracts are:
  - *Corticospinal* or *pyramidal tract*, which is named based on its origin and termination. These neurons begin in the motor cortex of the frontal lobe. Their axons cross to the opposite side of the spinal cord, where they terminate in the anterior horn to stimulate skeletal muscle on the opposite side of the body. In the motor cortex of the frontal lobe of the cerebrum are pyramid-shaped cell bodies of muscle-stimulating neurons forming the pyramidal tract.
  - *Extrapyramidal tract*, which contains axons from neurons that originate in areas of the brain other than the motor cortex. They provide regulation of muscle contraction.
- *Ascending tracts*, which receive action potentials from sensory receptors in the skin or throughout the body. This information is received by neurons in the posterior horn. The axons carry information to the brain for interpretation. An example of an ascending tract is the *spinothalamic tracts*. Again named for their origin and termination, they originate in the posterior horn of the spinal cord and terminate in the thalamus, where sensory information can be distributed to the appropriate locations in the brain for evaluation and action, if necessary.

Nerve Roots

Axons outside the spinal cord that carry information to or from the cord are found in bundles. The bundles of axons inside the vertebral canal are the *nerve roots*. The axons exiting the anterior spinal cord are the *ventral roots*, and those running into the posterior cord are the *dorsal roots*. The cell bodies of these neurons are outside the spinal cord but inside the vertebral canal. When there are groups of neuron cell bodies outside the central nervous system, they are known as *ganglia*. These groups of ganglia are found on the dorsal root and are known as the dorsal root ganglia.

> **Comprehension Check-up:**
> 1. The ability of the spinal cord to initiate contraction as a result of excessive sensory input before the brain has time for complete analysis is known as a _____.
> 2. Axons outside of the spinal cord but inside the vertebral canal form _____.
>
> 1. reflex 2. nerve roots

Clinical Notes—Myotome Testing

Myotome testing is used when an individual is suspected to have suffered neurological damage consistent with a spinal cord injury or spinal nerve root injury. Muscle testing would be performed to specific muscle groups controlled primarily by individual spinal nerve roots. Grading would then be done on a 0 to 2 scale. 0 = no palpable muscle contraction. 1 = slight contraction but not within normal limits. 2 = normal muscle contraction and strength. Strength would be compared bilaterally to assess for normal limits. Using this form of testing would assist in identifying the level and extent of spinal cord or spinal nerve root injury.

© Auttapon Wongtakeaw/Shutterstock.com

Contributed by Dave Smith. © Kendall Hunt Publishing Company

Clinical Skills—Dermatome Testing

Dermatome testing is used to assess the ability for sensory information to be carried through the spinal nerve segments on their way back to the spinal cord and ultimately to the brain. Dermatomes are tested by comparing light touch, pinprick, temperature and/or vibration of identical locations on opposite sides of the body. Testing is done by placing stimulus in the testing location on both sides at the same time. The sensory portion of a spinal nerve root would be considered "intact or undamaged" if both sides felt identical and felt what is known to be "normal" for that area. Sensory testing areas represent spinal nerve roots and are shown in the chart below.

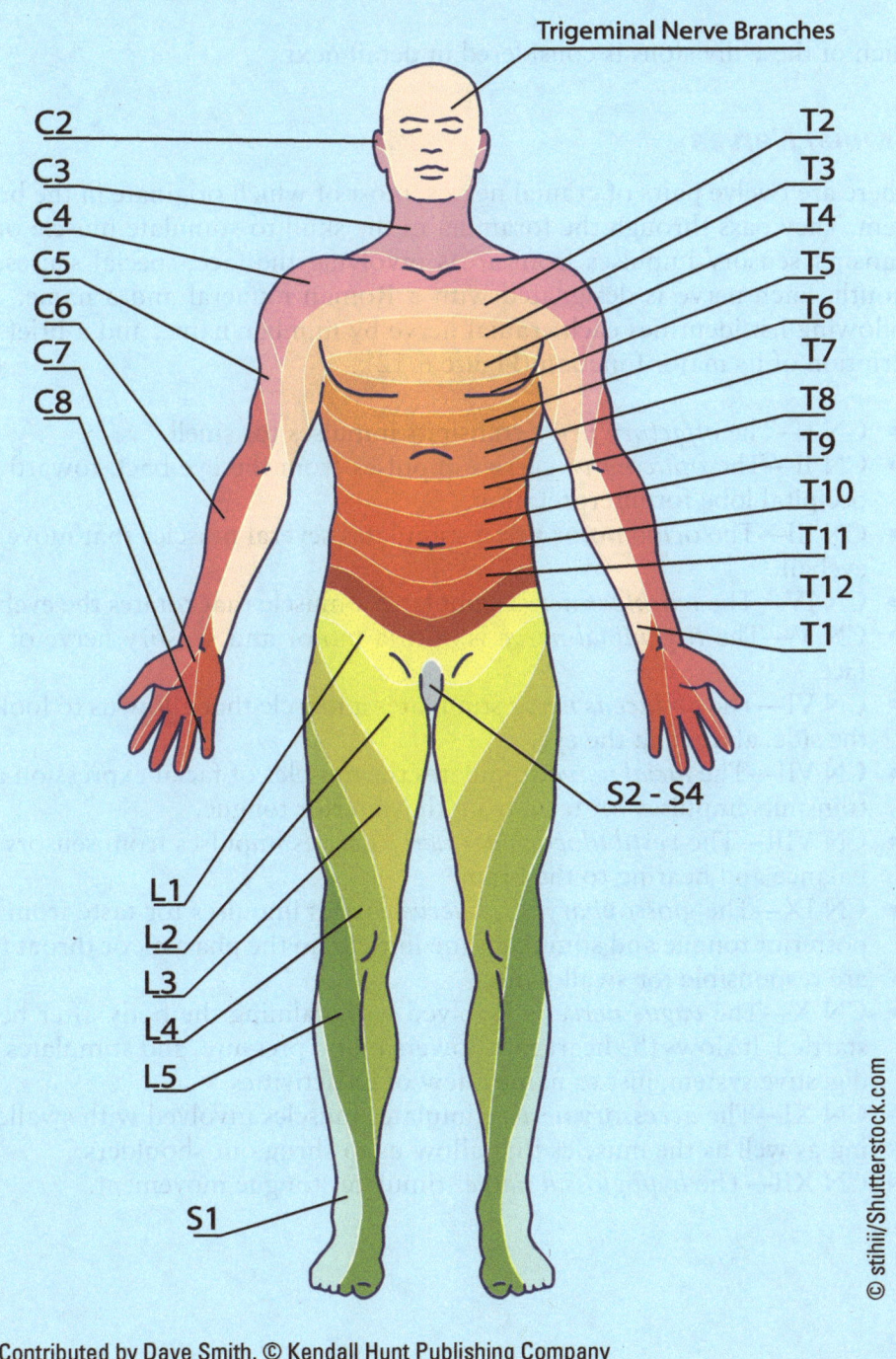

Contributed by Dave Smith. © Kendall Hunt Publishing Company

PERIPHERAL NERVOUS SYSTEM

The peripheral nervous system is located outside the central nervous system. It consists of two major divisions:

1. Cranial nerves exit through foramina (holes) in the skull to stimulate muscle and glands in the head and neck and to transmit sensory input from those areas.
2. Spinal nerves exit between each vertebra to stimulate muscle below the head. They also conduct sensory input from the body to the central nervous system.

Each of these divisions is considered in detail next.

Cranial Nerves

There are twelve pairs of cranial nerves, most of which originate in the brain stem. They pass through the foramina in the skull to stimulate muscle or to transmit sensory impulses from areas involving the face, special senses, or mouth. Each nerve is designated with a Roman numeral and a name. The following list identifies each cranial nerve by number, name, and a brief description of its major function (Figure 6.12).

- CN I—The *olfactory nerve* transmits impulses for smell.
- CN II—The *optic nerve* carries impulses from the eye back toward the occipital lobe for interpretation.
- CN III—The *oculomotor nerve* stimulates several muscles that move the eyeball.
- CN IV—The *trochlear nerve* stimulates a muscle that rotates the eyeball.
- CN V—The *trigeminal nerve* is both a motor and sensory nerve of the face.
- CN VI—The *abducens nerve* stimulates a muscle that allow us to look to the side, abducting the eye.
- CN VII—The *facial nerve* stimulates the muscles of facial expression and transmits impulses for taste from the anterior tongue.
- CN VIII—The *vestibulocochlear nerve* carries impulses from sensors for balance and hearing to the brain.
- CN IX—The *glossopharyngeal nerve* carries impulses for taste from the posterior tongue and stimulates the muscles in the pharynx or throat that are responsible for swallowing.
- CN X—The *vagus nerve* is involved with calming the body after being startled. It slows the heart rate, lowers blood pressure, and stimulates the digestive system, just to name a few of its activities.
- CN XI—The *accessory nerve* stimulates muscles involved with swallowing as well as the muscles that allow us to shrug our shoulders.
- CN XII—The *hypoglossal nerve* stimulates tongue movement.

CRANIAL NERVES

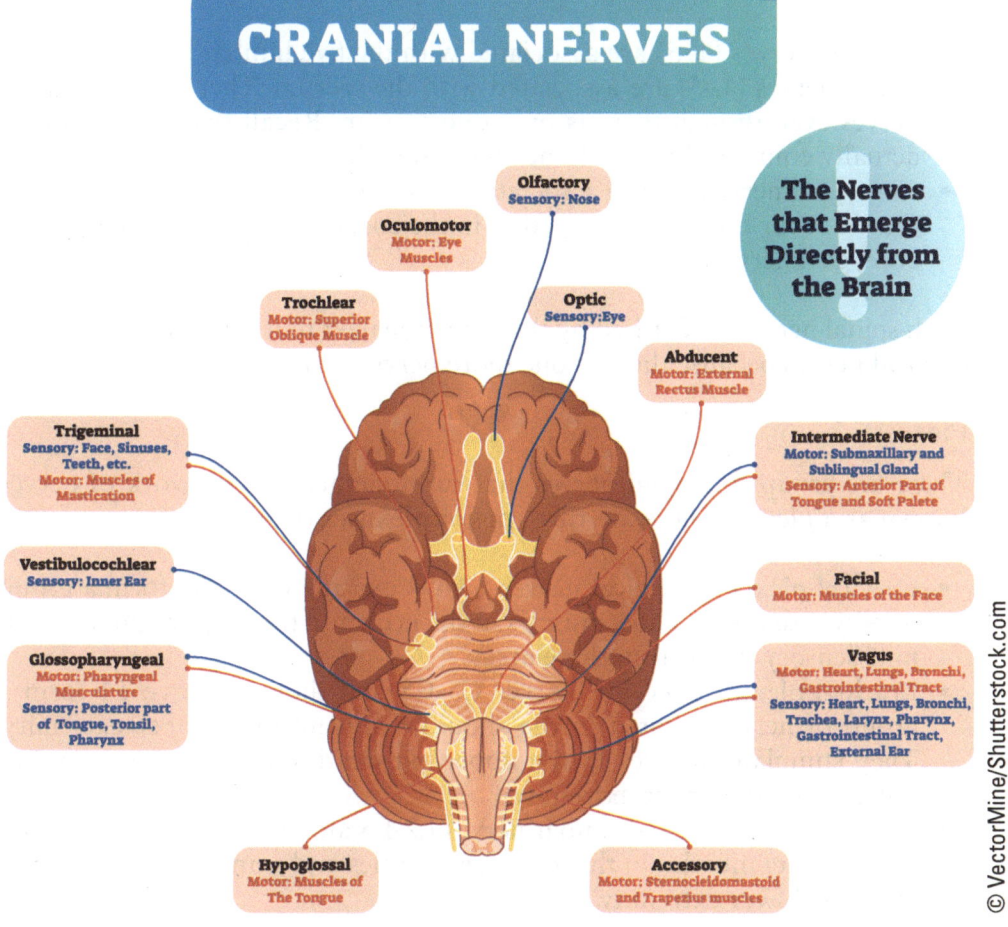

Figure 6.12

Comprehension Check-up:
1. Cranial nerve _____ stimulates the muscles that change facial expression.
2. The sense of smell is detected by cranial nerve _____.

1. CN VII—facial nerve   2. CN I—olfactory nerve

Spinal Nerves

Spinal nerves are formed when the ventral and dorsal roots inside the vertebral canal combine to form a common nerve as they exit from the spinal column. A spinal nerve exits from between each vertebra on both left and right sides. Spinal nerves are associated with each region of the spinal column.

- Cervical nerves (1–8) exit from both sides of the neck. Remember that there are only seven cervical vertebrae; however, there is a pair of spinal nerves exiting between the skull and first cervical vertebra resulting in an extra pair of cervical nerves.

- Thoracic nerves (1–12) are found exiting between the 12 thoracic vertebrae.
- Lumbar nerves (1–5) are associated with the five lumbar vertebrae.
- Sacral nerves (1–5) are associated with the sacrum. There are five sacral nerves even though there is only one sacrum. Recall that the sacrum is actually composed of five bones fused together.
- Coccygeal nerves (1) are associated with the coccyx. Although there are three or four bones fused together in the coccyx, there is only one pair of coccygeal nerves.

Each spinal nerve is identified by the vertebrae next to it. The exception is the cervical nerves because there is one more nerve than vertebrae.

Plexus

There are groups of spinal nerves that run together to a network of nerves known as a *plexus*. There are three major plexuses (Figure 6.14).

- **Cervical plexus**—Cervical nerves C1-C4 exit the neck and form a group of nerves that stimulates muscles in the neck and the diaphragm. It transmits sensory input from these areas.
- **Brachial plexus**—Cervical nerves C5-C8 and thoracic nerve T1 form a group of nerves located in the shoulder and into the upper extremity. They stimulate muscles in the upper extremities and relay sensory impulses from the upper extremity.
- **Lumbosacral plexus**—Lumbar nerves L2-L5 along with sacral nerves S1-S4 stimulate muscles in the lower extremities and transmit sensory input from the lower extremity.

> **Cervical plexus** a group of spinal nerves (C_1–C_5) that innervate structures in the neck.

> **Brachial plexus** a group of spinal nerves (C_5–T_1) that innervate structures in the upper extremity.

> **Lumbosacral plexus** a group of spinal nerves (L_2–S_3) that innervate structures in the lower extremity.

Comprehension Check-up:

1. There are _____ cervical nerves.
2. The muscles of the upper extremity are stimulated by nerves from the _____ plexus.

1. 8 2. brachial

CERVICAL PLEXUS

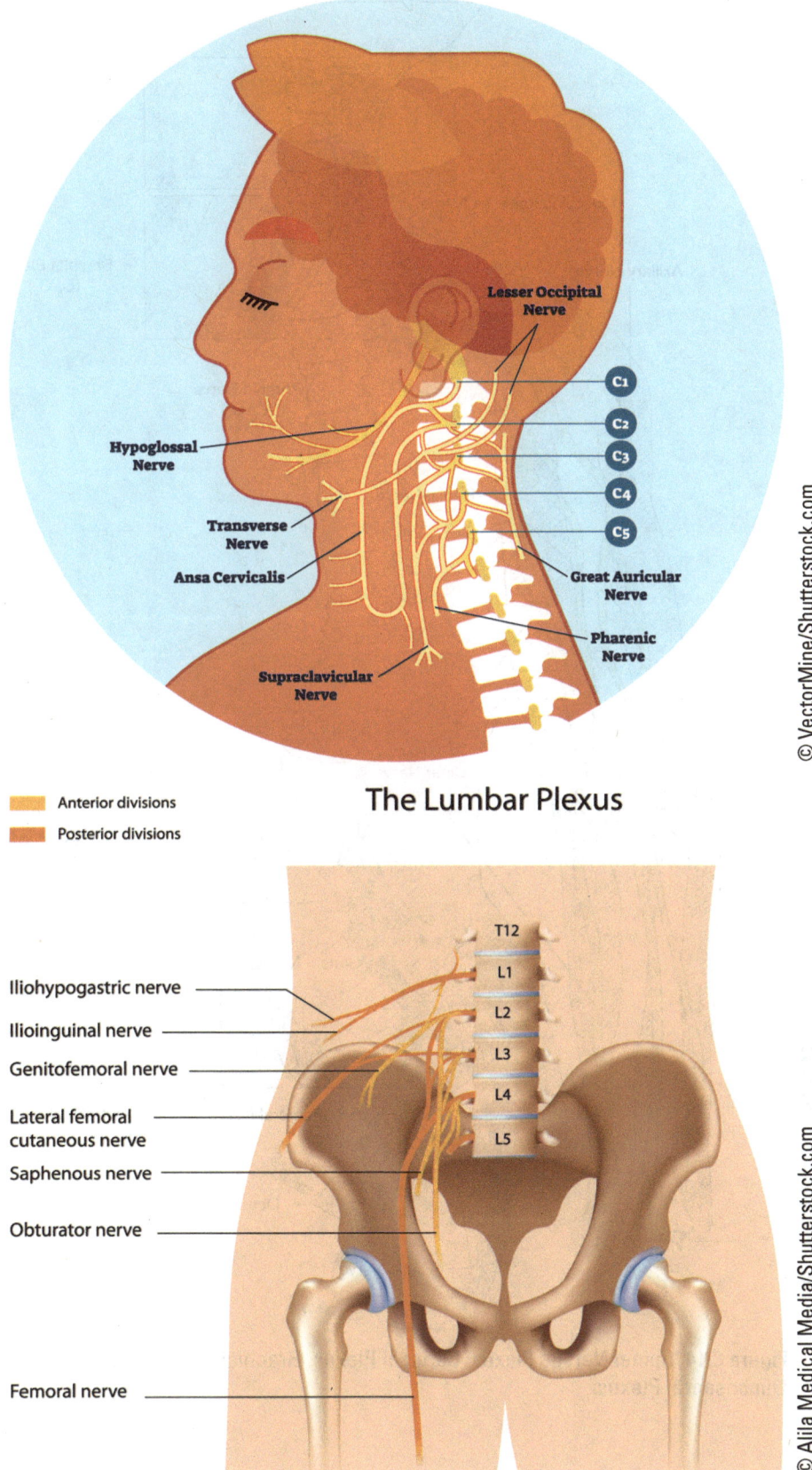

The Lumbar Plexus

Figure 6.13

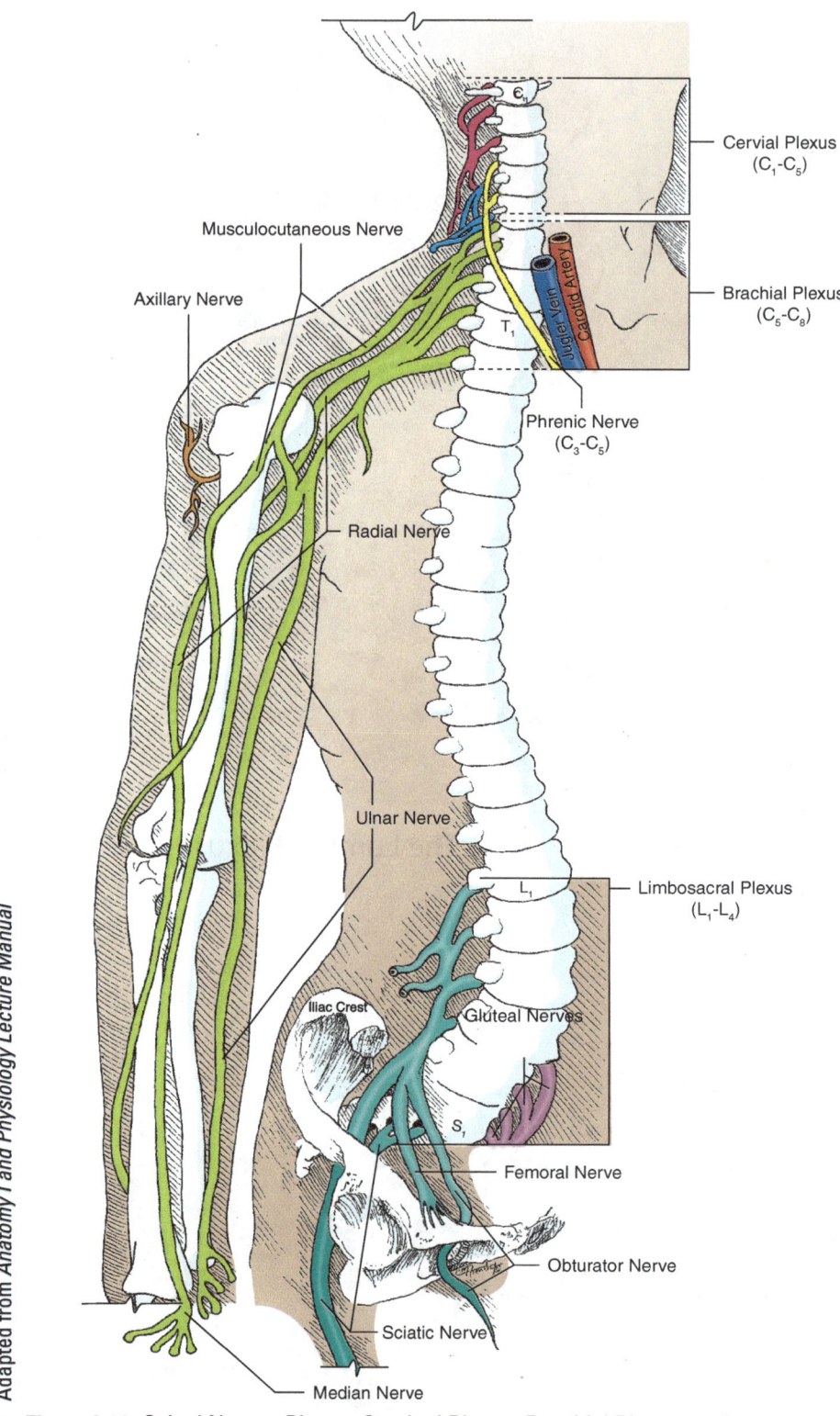

Figure 6.14 Spinal Nerves Plexus: Cervical Plexus, Brachial Plexus, and Lumbosacral Plexus.

Nervous System Responses to Daily Life

The following scenario illustrates some of the involvement of the nervous system in the simple act of taking a sip of hot coffee. The terms discussed in this chapter are highlighted.

My intention is simply to sit down and enjoy my morning cup of coffee. It seems simple enough. Before I pick up my cup, the occipital lobe of my cerebrum is going to collect data from my eyes about the shape, color, and location of the cup. Those parameters are relayed to my frontal lobe to assimilate information from my eyes through CN II about the size of the cup and correlate it with previous data about the weight of a cup of coffee to determine how many motor units it will take to lift the cup. A decision also has to be made concerning the length of time I intend to hold my coffee cup. The frontal lobe of my brain works out a movement plan and the motor cortex begins stimulating motor units. I reach for the cup.

As my fingers grasp the handle, sensors in my skin detect the texture and touch of the cup. The information generated by sensors in my hand passes through sensory neurons to the posterior horn of my spinal cord and is relayed up to the thalamus through my spinothalamic tract. The thalamus distributes the information to various areas of my brain for analysis and decision. The parietal lobe of my cerebrum determines where the sensations originated. As I begin to lift the cup, pressure sensors in my hand monitor the weight of the cup as receptors in the muscles in my upper extremity send information to my brain about the amount of tension being created. The frontal lobe of my cerebrum quickly makes adjustments to my motor unit stimulation to correct for any miscalculation concerning the weight of the cup. My cerebellum is actively making the lifting of my coffee cup smooth and coordinated.

Other areas of my brain are concerned with maintaining my posture so that I don't spill my coffee. As I lift the cup to my mouth, I smell the coffee through CN I and detect the warmth through CN V as it nears my lips. Muscles in my face, stimulated by CN VII, position my lips to drink. I am close to enjoying the taste of coffee when I am accidentally bumped by a family member passing by. The coffee splashes over the edge of the cup onto my leg. The heat from the spilled coffee is sensed by temperature receptors in my thigh. The sensing of heat and pain is rapidly transmitted by sensory neurons to the posterior horn of my spinal cord, where interneurons in the reflex arc pass the action potential to motor neurons in the anterior horn that stimulate muscles in my lower extremity, causing a withdrawal reflex that moves my thigh out of the way just in time to prevent another splash from causing more pain. The impulse from the posterior horn of my spinal cord has also traveled through the ascending spinothalamic tract to the thalamus, where it is distributed to the appropriate areas of my brain. One area is the parietal lobe of the cerebrum, where I interpret this sensation as very unpleasant. As that information is transmitted to the prefrontal area of my frontal lobe, I weigh the consequences of my reactions and decisions and determine I will not allow this incident to negatively affect my day.

Homeostasis—Holding in Balance

The nervous system is active in maintaining the homeostasis of many areas of the body. Of particular importance is the hypothalamus, which controls body temperature, regulates food and water intake, and causes the release of hormones affecting water balance, growth, metabolism, development, and reproduction. Not only is the brain responsible for conscious interpretation, decisions, and interactions, it also, through regulation of the respiration rate, maintains the level of carbon dioxide in our bloodstream as a means of assisting in the control of blood pH within homeostatic range.

There Is No Substitute for Exercise

Exercise relieves stress. Stress causes us to create a sympathetic response, resulting in our "fight-of-flight" system increasing blood pressure, heart rate, and respiration rate. During times of emergency, we decrease our vigilance against foreign invaders because we are more concerned at the moment with immediate survival. By relieving stress, we increase our resistance against disease.

Regular exercise also increases coordination and reflex action. Repeated practice creates conditioned reflexes that allow us to respond without having to think through the same process time after time. For example, a basketball player, after repeatedly practicing shooting a basketball, does not have to think about the position of his feet, the angle of the shot, or the degree of force needed behind the throw of the ball in order to be successful most of the time. The movement becomes programmed as a reflex because of repeated exercise.

The brain also controls the level of muscle tone required to stabilize joints and maintain posture. Exercise improves muscle tone and thereby reduces the risk of injury that could result because of the rapid change of position and movement that occurs throughout the activity.

Clinical Scenario—Cervical Fracture

A 40-year old male suffered injuries when his car was hit by another car coming from the left side in a 4-way traffic. The impact was on the driver side of the car. A 911 was called. It appeared that the victim suffered injuries to the left side of his body. The EMT / paramedic team carefully stabilized the head and neck before moving the patient out of his car. The patient was on placed on the backboard, and secured his neck with a cervical collar, and was transported to the nearest trauma hospital.

Upon examination, the patient has neck trauma as confirmed by an MRI (Magnetic Resonance Imaging), showing cervical fractures to his C4, C5 and C6. The patient has difficulty with his breathing. He is intubated and connected to a ventilator.

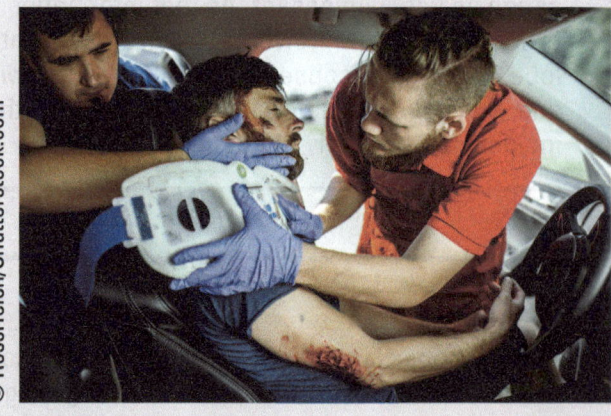
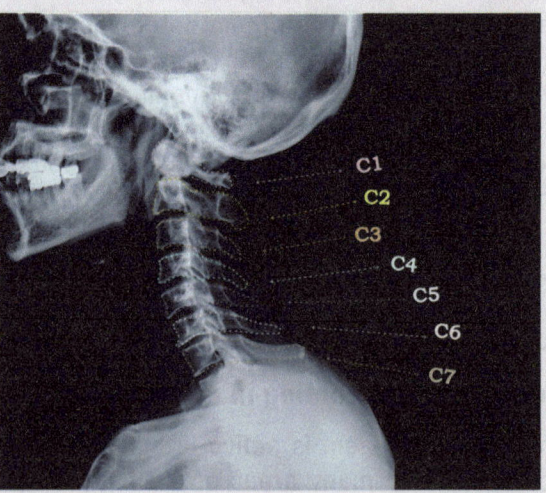

Questions:

1. What are the common signs and symptoms of cervical fracture?
2. What is the effect of a fracture in C2, C3 and C4 on the body?

THE SENSES

ᴸThe nervous system receives feedback or stimuli from the environment or surroundings through its five (5) senses namely: touch or tactile, vision, taste, smell, and hearing. These senses have sensory receptors that convert external stimuli into energy and transmit it into action potential that is delivered to the CNS.

ᴸTactile sensations detect touch, pressure and vibration. The tactile receptors are located in the dermis of the skin, which transmit the messages to the brain. Receptors that react to light touching are called Merkel's discs; receptors that react to pressure and vibration are called Meissner's and Pacinian corpuscles; and receptors that react to deep pressure and stretch are called Ruffini corpuscles.

TACTILE RECEPTORS IN THE SKIN

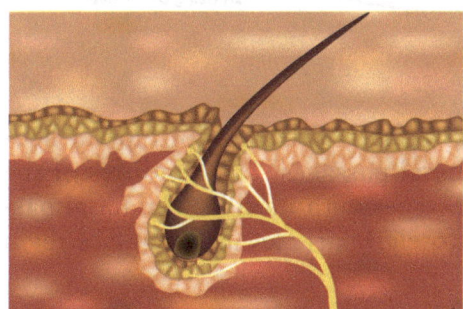

A root hair plexus

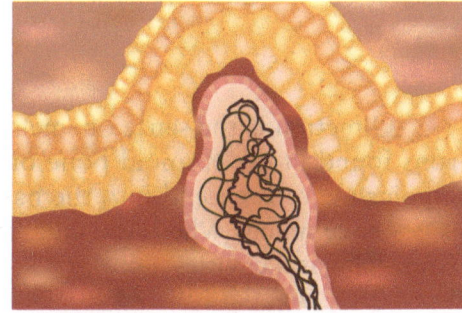

A tactile (Meissners) corpuscle

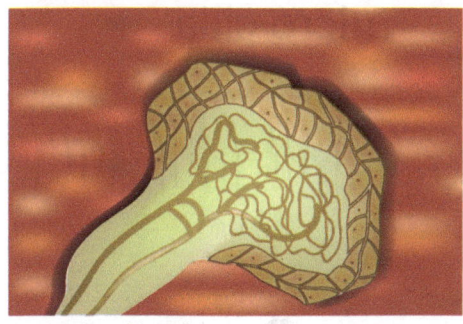

Krause and buld

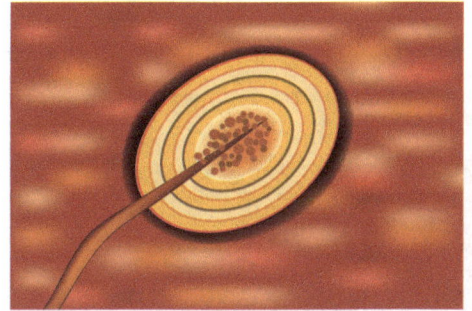

A lamellated (pacinian) corpuscle

© Sakurra/Shutterstock.com

Figure 6.15

ᴸThermal sensations detect changes in heat and cold. The thermal receptors are composed of free nerve endings in the epidermis and within the deep visceral organs. Pain sensations detect pain. Pain receptors, also called nociceptors, are located throughout the body except the brain and eyes. Proprioceptive sensations detect body positioning through the joints and the muscle tension caused muscle contraction.

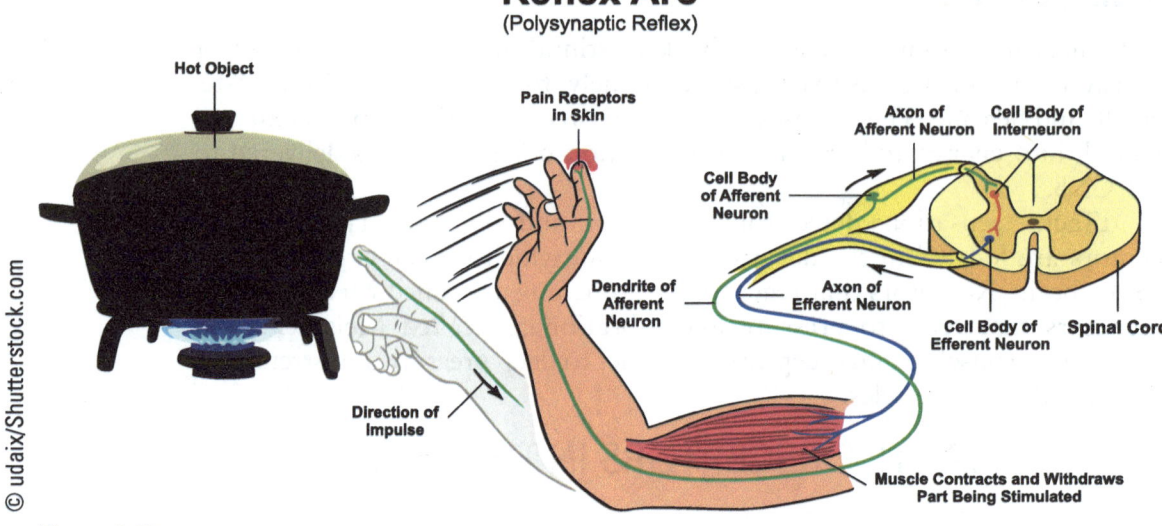

Figure 6.16

Olfaction is the sense of smell. It has olfactory neurons or bipolar neurons found in the ethmoid bone of the nasal passages. The dendrites of the bipolar neurons extend into the nasal passages with its cilia. The nasal passages have mucus where the cilia catch foreign particles and removes them from the nasal cavities.

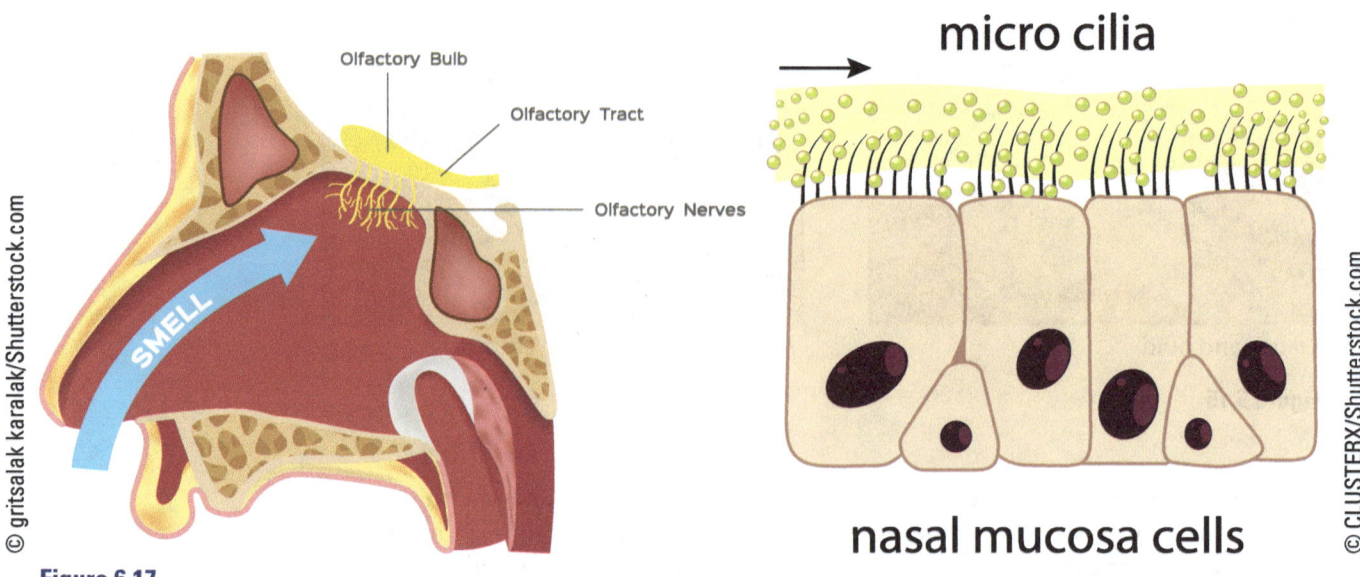

Figure 6.17

Gustation is the sense of taste. The tongue is layered with taste buds and projections called papillae. The taste buds contain gustatory cells that activate when in contact with chemicals from the food. The tastes of the tongue are sweet (glucose), sour (acid), salty (sodium), and bitter (base or alkaline substance).

BASIC TASTES

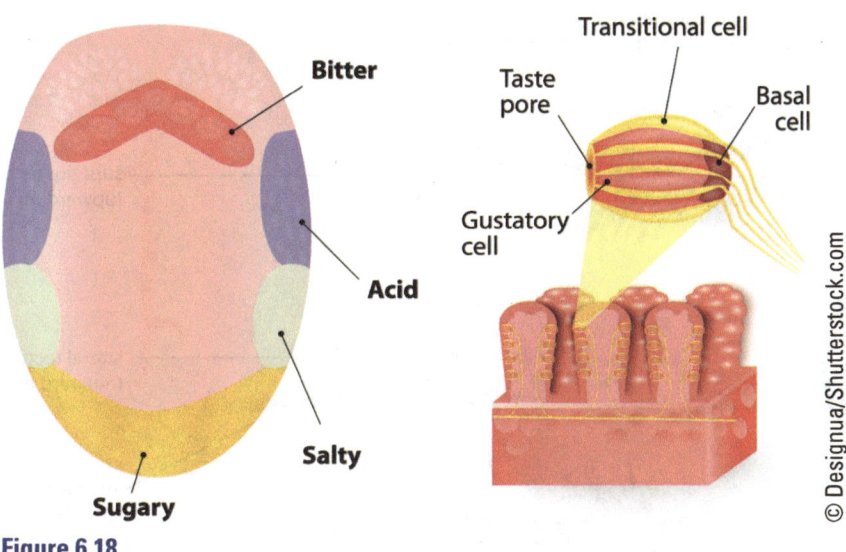

Figure 6.18

Vision

The ability to see is critical to most individuals for normal everyday life. It is interesting to note that vision is the only special sense that can be completely corrected in most cases.

Accessory Structures of the Eye

There are two basic structures outside the eyeball that are essential to vision: the *lacrimal glands*, which produce tears, and skeletal muscles, which cause eye movement. The lacrimal glands are located superior and lateral to the eye. They place tears onto the eye to clean and lubricate it as well as to provide nutrients to the external cornea. Blinking pushes the tears medially, where they drain into the nasal passage. Normal tear production is slow enough that the nasal passage is able to deal with the drainage. Crying, however, causes the system to be overwhelmed. Excessive drainage into the nasal passage causes our nose to run. When tears cannot drain fast enough, they run down the cheeks.

There are also six skeletal muscles that move the eye in all directions (Figure 6.19). The four rectus muscles are located around the eye and enable the eye to turn. For example, when the superior rectus muscle contracts, the eyeball turns upward. The two oblique muscles twist the eye to keep the

---

Lazy Eye (Amblyopia)—Muscles Pulling with Unequal Strength

Occasionally one of the eye-moving muscles is weaker than its antagonist, resulting in what is known as a *lazy eye*. Because the opposing muscles do not pull with equal strength, the eye tends to point in one direction, making it difficult to look the opposite way. Exercises are done to strengthen the weak muscle. If that does not work, the stronger muscle can be medically weakened.

Muscles of the Human Eye

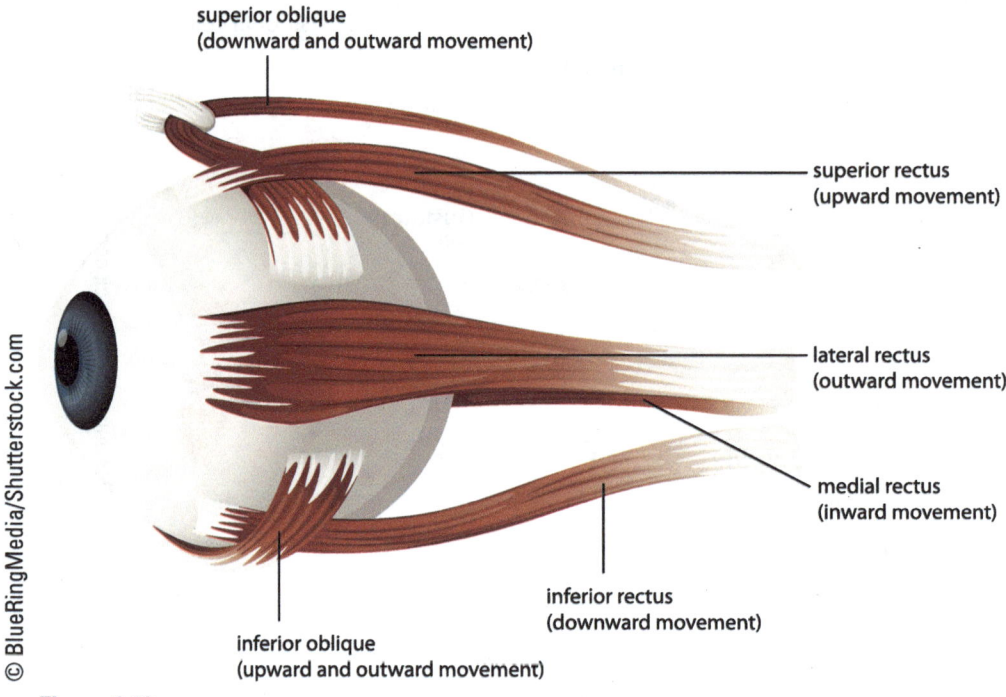

Figure 6.19

image upright within a given range even when our head is not perfectly vertical. These muscles receive their stimulation from CN III, IV, and VI.

Anatomy of the Eyeball

The eyeball is composed of the following structures (Figure 6.20):

- *Sclera*—Thick, tough, white membrane covering most of the external eyeball.
- *Cornea*—Clear anterior covering where light passes into the eye.
- *Choroid*—Located inside the sclera. This layer contains blood vessels and lymphatic vessels. In the anterior portion of the eye the choroid forms the ciliary body, which holds the lens in place and becomes muscles that change the thickness of the lens.
- *Iris*—Colored muscular structure that constricts or relaxes to control the amount of light entering the eye.
- *Pupil*—The opening in the iris for light to enter.
- *Lens*—Focuses light on the retina, in conjunction with the cornea.
- *Retina*—Contains cells that create an action potential in the presence of light. It lines the interior of the eyeball, except the anterior area that controls light entering the eye.
- *Fovea centralis (macula lutea)*—Part of the retina that provides best color vision.
- *Optic disc (blind spot)*—The area where the optic nerve and blood vessels enter the eye. Because this area contains only axons and blood vessels, the eye cannot create an action potential in this location. During a physical examination the physician looks at the optic disc to determine

---

Cornea the clear anterior covering through which light passes into the eye.

Iris the colored muscular structure that constricts or relaxes to control the amount of light entering the eye.

Pupil the opening in the iris for light to enter.

Lens focuses light on the retina, along with the cornea.

Retina contains cells that create an action potential in the presence of light.

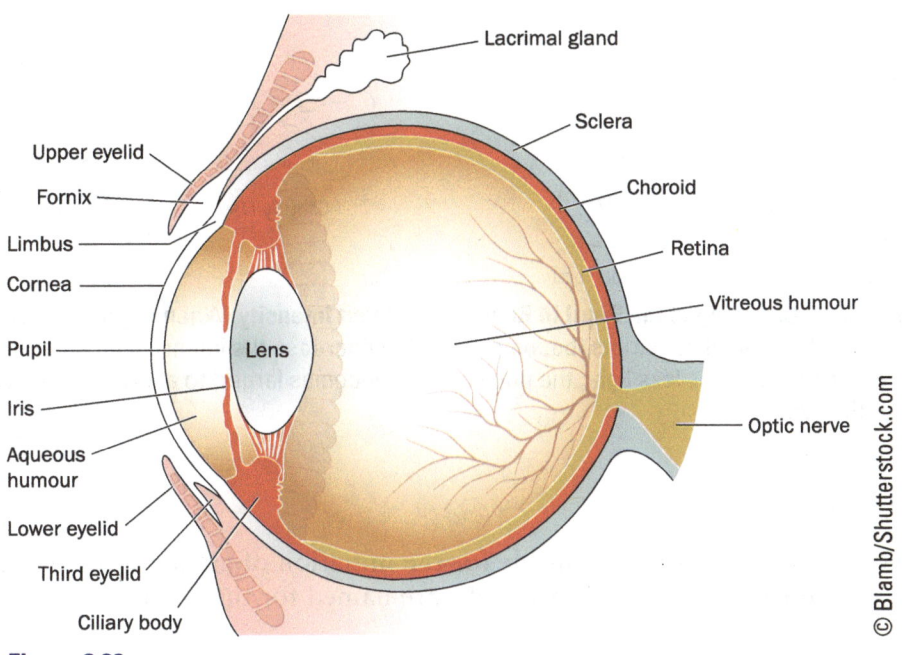

Figure 6.20

whether it is normal. Abnormal optic discs can result from other eye problems or from increased pressure inside the cranium.
- *Vitreous humor*—Clear gel that fills the eyeball in the posterior chamber.
- *Aqueous humor*—Waterlike fluid in the anterior chamber of the eye that provides nutrients to the internal cornea. The pressure of this liquid on the cornea is measured as a possible indication of glaucoma.

Image Formation

For light to focus on the retina, the rays of light must be bent into a smaller image. This is accomplished by two structures: the cornea and the lens. The cornea is curved, causing light entering the eye to be partially focused on the retina. The lens then completes the focusing process. The lens becomes thinner or thicker to accommodate for the distance of the object viewed. When a faraway object is viewed, the lens becomes thinner. The lens is thickened when close objects are viewed. The internal muscles found in the ciliary body, known as *ciliary muscles*, change the thickness of the lens. It is this changing of the lens that allows us to see clearly.

The muscles of the iris relax or constrict to adjust the amount of light entering the eye through the pupil (Figure 6.21). Too much light would overwhelm the cells in the retina, preventing a clear image from being seen. When the light source is low, the iris increases the size of the pupil to allow as much light as possible in to stimulate cells in the retina.

Comprehension Check-up:

1. Except in times of distress when tear production is excessive, tears placed on the eye normally drain into the _____.
2. Light enters the eye through the _____.

1. nasal passage   2. pupil

Figure 6.21 Changing of the Pupil in Response to Light Intensity. When bright objects are viewed, the pupil constricts (becomes small) to decrease the intense light entering the eye. When there is less light, the pupil dilates (becomes larger) to allow more light to enter the eye.

Physiology of Sight

Two types of cells are responsible for the initiation of an action potential in the retina: rods and cones. These cells are named for their shape.

Rods

Rods enable us to see black and white. They require less light to function than do cones. As a result, if we go outside on a moonlit night and there are no streetlights, we will not be able to determine the color of the surrounding objects, although we may be able to distinguish them in black and white.

> **Rods** light-sensitive cells that enable us to see black and white.

Cones

Each cone has the ability to detect one color: red, blue, or green. Cones sensing each color are side by side throughout the retina. When light from a colored object reaches a cone, for example, red light reaches the retina, the cones sensing red cause action potentials to be created. If a purple object is viewed, both red and blue cones are stimulated to cause action potentials. It is the combination of the cones stimulated that allows us to see color variations.

> **Cone** light-sensitive cells that have the ability to detect one color: red, blue, or green.

Retina

The retina converts light to an action potential. The retina consists of a layer of rods and cones covered by a layer of bipolar cells, over which is a layer of ganglion cells. The fovea centralis consists almost entirely of cones with few rods, making it the area of our best color vision. There are fewer cones and an increasing number of rods as the distance from the fovea centralis becomes greater until reaching the lateral margins of the retina, which are

---

Do All Animals Have Color Vision?

Animals that hunt or forage at night, such as dogs, cats, cattle, and deer, do not need the ability to see color. Rather they need to be able to see with as little light as possible. As a result, their eyes possess only rods and they are color-blind. Animals active during the day have the ability to distinguish color very clearly. Most birds fall into this category.

Color Blindness Is from the Mother's Side of the Family

Color blindness is the result of defective cones. It occurs mostly in males and comes from the mother's side of the family. Genes for the development of cones are found on the X chromosome. There are two chromosomes in our DNA that determine our sex. Females have two X chromosomes and males have one X and one Y chromosome. The Y chromosome does not possess the appropriate genes for color detection. Since we receive one sex chromosome from each parent, we always receive an X chromosome from our mother. If the father also provides an X chromosome, then the offspring is female. When the father contributes a Y chromosome, the result is a male. That means if a son is color-blind, it is because the X chromosome he received from his mother had defective genes for color discrimination. In the case of a daughter, since she has two X chromosomes, she has a backup set of instructions. She is not color-blind but is considered a carrier of the defect. She will be color-blind only if both X chromosomes are defective. If she has one defective X, there is a 50/50 chance she will pass it to her offspring.

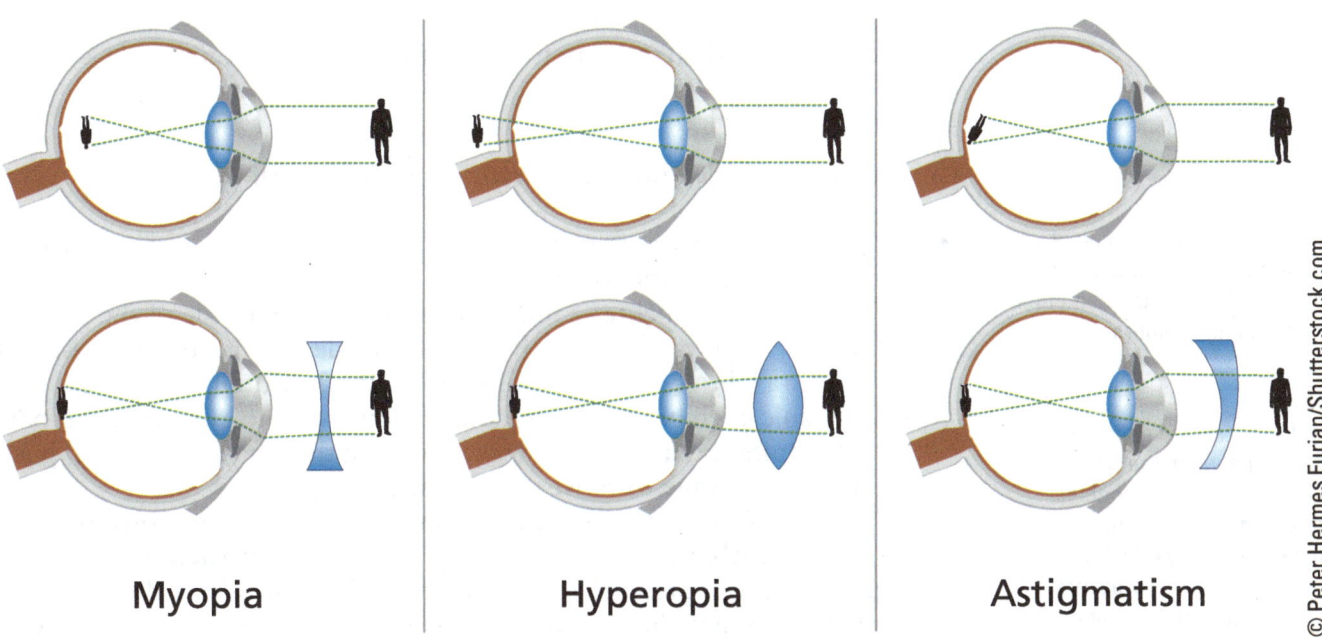

Figure 6.22

virtually all rods. In other words, our peripheral vision can see only black and white.

Physiology of the Retina

In the retina between the layer of rods and cones are two types of bipolar cells, light-on and light-off. When light is not present, light-off bipolar cells are depolarized by rods or cones. The depolarization of light-off bipolar cells inhibits ganglion cells from transmitting impulses to the brain. When light arrives at the rod or cone, a chemical change occurs which results in lighton bipolar cells being stimulated to cause ganglion cells to transmit impulses through CN II to the brain. This action potential carries to the brain both the location and the color of one specific spot of light reaching the retina. The occipital lobe assembles all of the inputs from each eye into an image and then identifies the picture. Because there is a fixed distance between both eyes, the brain is able to correlate the inputs into a three-dimensional image.

> **Comprehension Check-up:**
> 1. Color is detected in the retina by _____.
> 2. Bipolar cells in the retina create an action potential that depolarizes a _____ whose axon transmits the image to the brain.
> 3. Another term for nearsightedness is _____.
>
> 1. cones 2. ganglion cell 3. myopia

Hearing and Balance

Although our hearing may not be as acute as that of some animals, it still is a highly significant factor in normal bodily function and defense. Hearing loss can affect an individual's ability to interact with the environment.

Anatomy of the Ear

The ear can be divided into three sections (Figure 6.23):

- **External ear**—This part of the ear has constant access to the environment. It consists of the following:
  - Auricle—the funnel for sound waves on the lateral side of the head
  - External auditory canal—the tunnel inside the head where sound waves travel to be converted into an action potential
  - Tympanic membrane—Commonly known as the *eardrum*, the part that vibrates because sound waves arriving through the ear canal push on the membrane to cause its movement
- **Middle ear**—This part of the ear transmits the sound vibrations from the tympanic membrane to the cochlea for conversion into action potentials. It also allows air pressure to equilibrate with the inside of the tympanic membrane so that changes in atmospheric pressure do not interfere with the ability to hear.
  - The three bones of the middle ear amplify the vibrations of the tympanic membrane:
    - Malleus (Hammer)
    - Incus (Anvil)
    - Stapes (Stirrup)
  - The auditory tube (eustacian tube) runs from the back of the throat to the inside of the tympanic membrane to allow air pressure to equalize.
- **Inner Ear**—This part of the ear includes the labyrinth, a structure enclosed in bone that contains apparatuses to detect sound, rotation, motion, and gravity. Sound waves are converted by a snail shell–shaped section of the labyrinth known as the cochlea. Rotation is detected by three semicircular canals. Motion and gravity are converted to an action potential by structures found in the vestibule of the labyrinth known as the *utricle* and *saccule*.

External ear the section of the ear from the auricle on the lateral side of the head to the tympanic membrane.

Middle ear the section of the ear between the tympanic membrane and the cochlea. It consists of three bones and the auditory tube for equalization of air pressure on the inside of the tympanic membrane.

Inner ear the innermost part of the ear that consists primarily of apparatuses to detect sound, rotation, motion, and gravity; also known as the *labyrinth*.

Cochlea the snail-shaped section of the labyrinth in the inner ear that converts sound waves into an action potential.

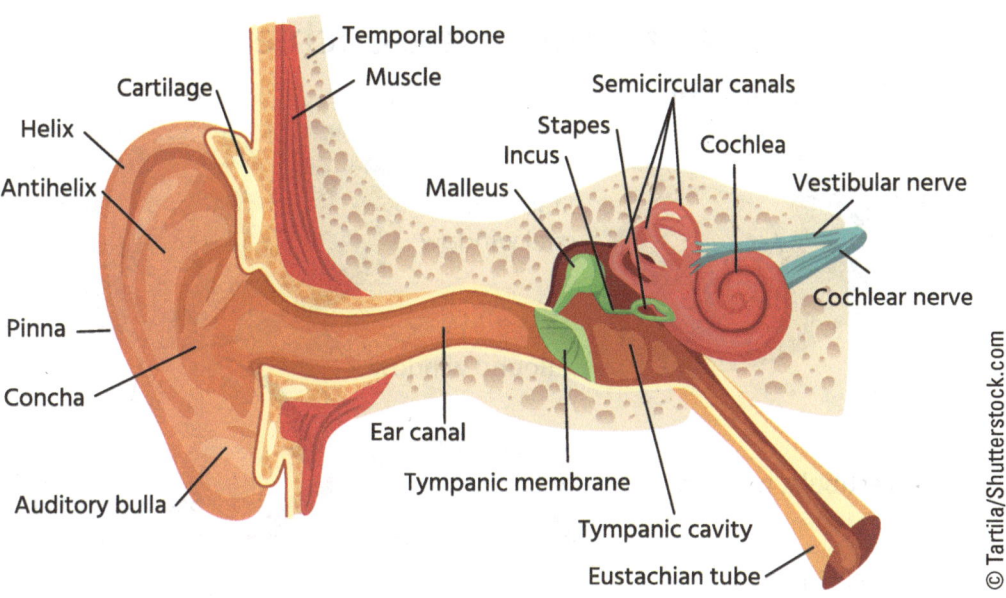

Figure 6.23

Cochlea

To understand the creation of action potentials in response to sound vibrations, it is useful to discuss the cochlea in more detail. If the snail shell shape of the cochlea is unrolled, two general areas can be seen: the spiral organ and a surrounding fluid-filled space (Figure 6.24). The *spiral organ* is in the center of the cochlea and contains the apparatus to create action potentials. The action potential is created by cells that have hairs on their apical surface, known as hair cells. Those hairs are embedded in a rooflike structure known as the *tectorial membrane*. The hair cells and tectorial membrane are surround by fluid. This fluid is called *endolymph,* and it fills the spiral organ. The space outside the spiral organ contains a fluid known as *perilymph*. This space can be divided into the scala vestibule, which is the fluid-filled chamber above the spiral organ, and the scala tympani, which contains perilymph below the spiral organ. The stapes is attached to one end of the scala vestibuli through an area known as the *oval window.*

> **Hair cells** cells in the inner ear have hairs on the apical surface. These create an action potential when the hairs are bent. Those hairs are embedded in several different structures. This allows the structures in the labyrinth to detect sound, rotation, motion, and gravity.

Physiology of Hearing

Following are the steps involved with the creation of an action potential for sound perception:

1. Sound waves are funneled by the auricle into the external auditory canal.
2. The sound waves reach the tympanic membrane, causing it to vibrate.
3. The vibration of the tympanic membrane is transmitted and amplified by the three bones of the middle ear to the oval window of the cochlea.
4. As the stapes pushes in and out on the oval window, it creates waves in the perilymph.
5. The spiral organ rides up and down on the waves in the perilymph.
6. As the spiral organ rides the waves, it causes hairs of the hair cells, which are stuck in the tectorial membrane, to bend.

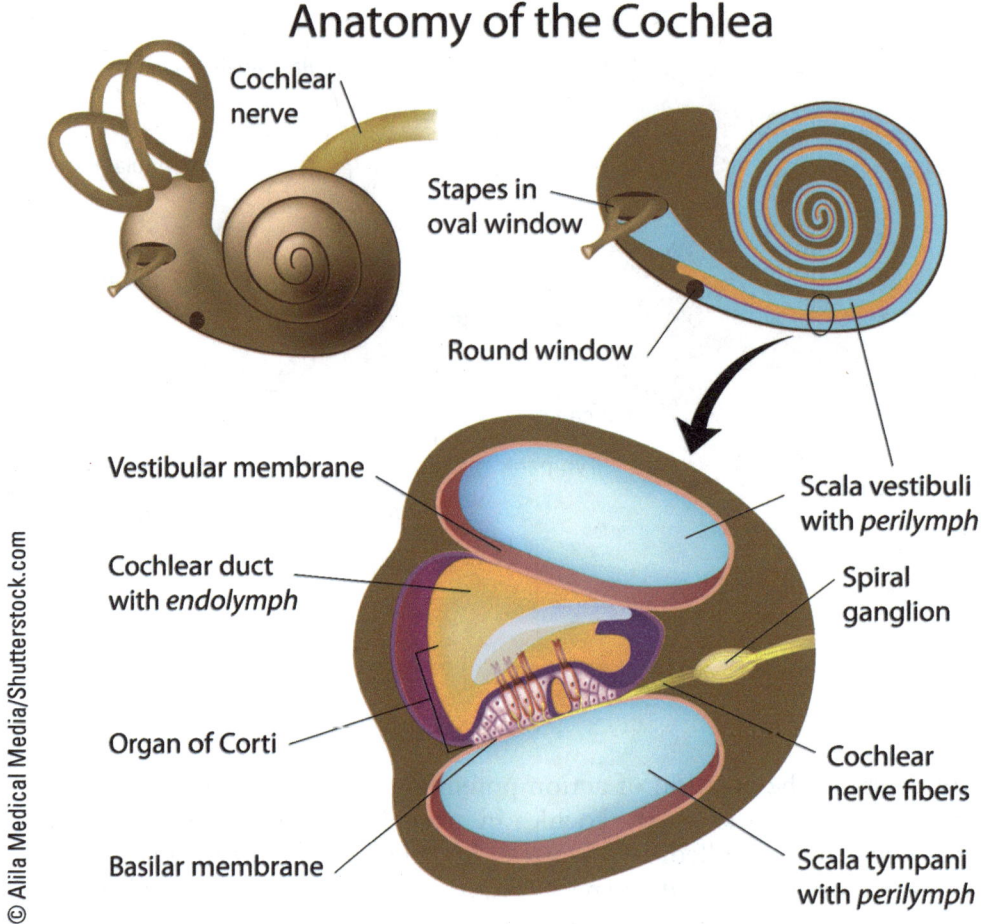

Figure 6.24

7. The bending of the hairs on the hair cells causes the hair cells to create an action potential.
8. The action potential travels down CN VIII toward the brain for interpretation.

Hair Cells

It is the bending of the hairs on hair cells that creates an action potential in the spiral organ of the cochlea. The bending of the hairs causes sodium channels to open, resulting in depolarization of the hair cell. CN VIII is attached to the hair cells to transmit the action potential to the brain for interpretation. By changing the configuration slightly, other structures in the inner ear are able to detect rotation, motion, and gravity.

> **Comprehension Check-up:**
> 1. Sound waves are converted to vibrations at the end of the external auditory canal by the _____.
> 2. Sound vibrations are converted into action potentials in the _____.
>
> 1. tympanic membrane   2. cochlea or spinal organ

Balance

The combined sensations of rotation, motion, and gravity allow our brains to determine our position in the environment. That information can be processed to determine posture, stimulate antigravity muscles, and maintain our balance while moving through our surroundings.

Semicircular Canals

Semicircular canals are composed of three half-circles of fluid-filled tubes that detect rotation. There is a semicircular canal for each plane in each inner ear. On one end of each semicircular canal is a bulge known as an *ampulla*, where receptor cells (cristae) are located. Cristae consist of hairs of the hair cells embedded in a hat-like structure, known as a *cupula*. This apparatus is exposed to the endolymph in a semicircular canal. As the head turns in one direction, the fluid in the semicircular canal in that plane pushes on the cupula. This causes the hairs in the crista to bend, detecting rotation in that particular direction. The action potential for detection of rotation travels through CN VIII to the brain for analysis and to determine the appropriate adjustments necessary to maintain balance.

> **Semicircular canals** three half-circles of fluid-filled tubes that detect rotation through the bending of hairs on hair cells.

Maculae

Inside the vestibule of the labyrinth are two additional structures, the utricle and saccule, collectively known as the maculae that contain hair cells covered by a gelatinous membrane. Imbedded in the gelatinous membrane are crystals of minerals and protein known as *otoliths* (Latin for ear stones). The otoliths add weight to the gelatinous membrane causing it to be pulled downward by gravity. The purpose of the maculae is to detect head position and linear acceleration. As an individual moves forward, the gelatinous membrane drags the hairs of the hair cells backward creating an action potential that informs the brain that forward movement is in progress. Backward movement bends the hairs in the opposite direction. Turning the head to look down causes the gelatinous membrane to be pulled downward by gravity (Figure 6.25). This process also bends hairs on the hair cells, informing the brain of the force of gravity.

> **Maculae** consists of the utricle and saccule, which are hair cells covered by a gelatinous membrane that detects motion and gravity.

Combining the action potentials from the semicircular canals and maculae provides essential information for the brain to determine the appropriate muscle stimulation required to maintain balance.

Adaptation

When a receptor is stimulated, it produces repeated action potentials known as generator pulses, as long as the sensor is activated. Several but not all of the general senses have the ability to achieve adaptation (Figure 6.26). That is, after continually being stimulated some receptors will eventually stop producing generator pulses. This is a very helpful condition. If it were not for adaptation, the touch or pressure of our clothes would be a constant irritation. For example, if you have ever been around someone who just started wearing glasses for the first time they are constantly putting them on and taking them off. They clean them at the slightest speck or smudge. They are intently aware that the glasses are there. But someone who has worn glasses for a long time may not even be aware that they are on their face. Many

> **Generator pulses** repeated action potentials produced by a sensory receptor as long as it is manipulated.

> **Adaptation** after continual manipulation, a sensory receptor may stop producing generator pulses.

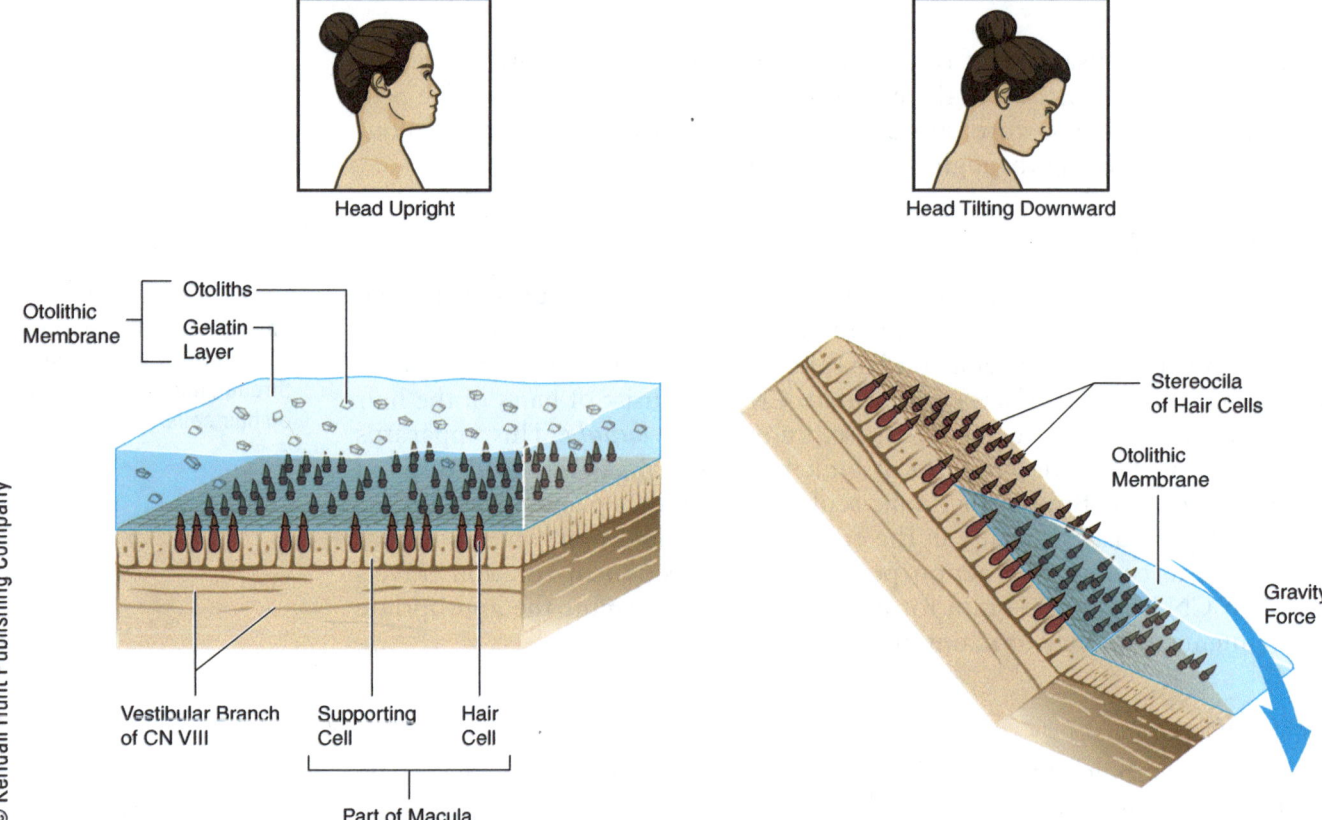

Figure 6.25 Response of the Maculae to Gravity.

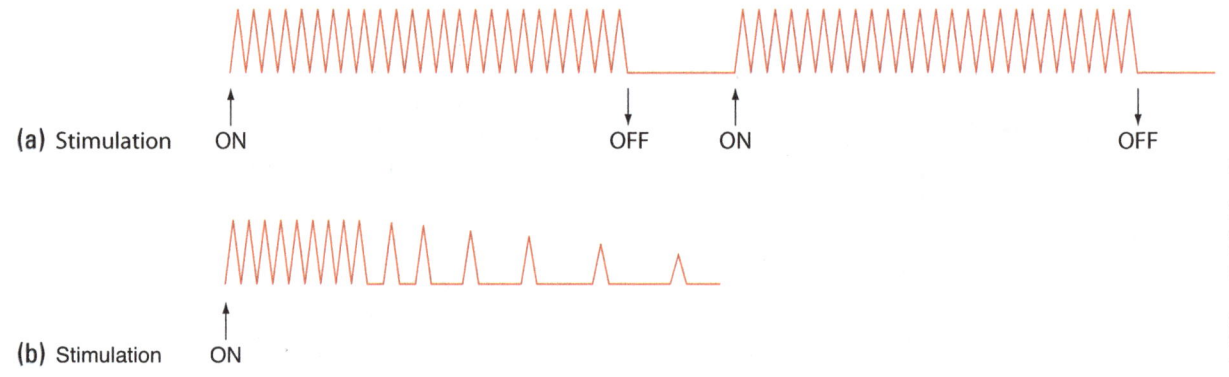

Figure 6.26 (a) Generator pulses are action potentials that are repeated as long as a sensory receptor is stimulated. (b) Adaptation is the termination of generator pulses after continual stimulation of some sensory receptors.

people have spent a considerable amount of time looking for the glasses they were wearing.

Temperature sensors also adapt. When first stepping into a hot shower or hot tub, the water seems to be very hot. It is not long before the water feels comfortable. In the example previously described where one finger was placed in hot water and another in cold, after a minute the finger in the hot water did not sense the water as being so hot, nor did the finger in the cold water feel as cold. When placing both fingers in lukewarm water, the one that had been in the hot water, had heat receptors that stopped producing action potentials. The cold receptors in the same finger were ready to work and were stimulated by the water causing it to feel cool. The opposite condition occurred in the finger placed in cold water. If there was only one sensor for both hot and cold, then placing both fingers in the lukewarm water the temperature would have felt the same.

Not all sensory receptors adapt. Nociceptors for pain and sensors for proprioception need to remain constantly vigilant. Although we might prefer that the pain sensors adapt, the consequences might very well be disastrous if they did. We also need to know the location of all of our bodily structures at all times.

Some of our special senses adapt as well. For example, after eating a piece of candy, a piece of fruit may not taste as sweet and may actually taste sour. Our sense of smell adapts to odors, even when they are quite unpleasant. Adaptation often makes subtle differences in our detections of sensations that often make our lives considerably more pleasant.

Comprehension Check-up:

1. Rotation is detected by the _____.
2. Gravity is sensed in the _____ of the labyrinth.
3. After a receptor has been creating generator pulses while it is stimulated, eventually it may stop the production of impulses by a process known as _____.

1. semicircular canals   2. vestibule or maculae   3. adaptation

CLINICAL TERMS TO REVIEW

Peripheral nervous system, central nervous system
Autonomic nervous system- sympathetic and parasympathetic
Somatic nervous system
Neuron, dendrites, sensory and motor neurons, synapse
Schwann cells, ependymal cells- cerebrospinal fluid (CSF)
Nerve impulse, resting potential, depolarization, repolarization
Cerebrum- frontal lobe, parietal lobe, occipital lobe, temporal lobe

Cerebellum
Brain stem- medulla oblongata. Pons, midbrain
Diencephalon- thalamus, hypothalamus, pineal gland, limbic system
Meninges- dura mater, pia mater, arachnoid mater
Spinal cord- gray matter, white matter
Cranial nerves- CN I through CN XII
Spinal nerves- cervical, thoracic, lumbar, sacral, coccygeal
Plexus- cervical, brachial, lumbosacral

Stroke- ischemic stroke, hemorrhagic stroke, transient ischemic attack (TIA)
Myotome testing, dermatome testing
Cervical fracture- C1, C2, C3, C4, C5
Sensations- tactile, thermal, pain, proprioceptive
Olfaction, gustation, vision, hearing and balance

Test Yourself

Choose the best answer to the following multiple choice questions:

1. There are spaces between the cells of the neurilemma that expose the axon. These spaces are known as
   a. the synaptic cleft.
   b. neurotransmitters.
   c. nodes of Ranvier.
   d. the foramen of Magendie.

2. What type of neuron is a sensory neuron?
   a. multipolar
   b. bipolar
   c. tripolar
   d. unipolar

3. The space between the synaptic knob and the postsynapse is known as
   a. the synaptic cleft.
   b. neurotransmitters.
   c. nodes of Ranvier.
   d. the foramen of Magendie.

4. A ridge on the cerebrum is known as a
   a. fissure.
   b. frenulum.
   c. sulcus.
   d. gyrus.

5. Creative activities such as music and art are a specialty of the _____ hemisphere of the cerebrum.
   a. superior
   b. left
   c. inferior
   d. right

6. The lobe of the cerebrum that performs analysis and decision making along with stimulation of skeletal muscle is the _____ lobe.
   a. occipital
   b. frontal
   c. temporal
   d. parietal

7. The area of the brain stem that connects the cerebellum to the rest of the central nervous system is the
   a. midbrain.
   b. medulla oblongata.
   c. pons.
   d. thalamus.

8. The functional area of the brain that deals with stress and emotion as well as learning is the
   a. limbic system.
   b. thalamus.
   c. pineal gland.
   d. hypothalamus.

9. The thin meningeal covering directly on the brain and spinal cord is known as the
   a. dura mater.
   b. pia mater.
   c. arachnoid mater.
   d. cetateous mater.

10. A sensory neuron connected to an interneuron that can stimulate a motor neuron to cause skeletal muscle to contract forms the
    a. somatic nervous system.
    b. corpus callosum.
    c. central nervous system.
    d. reflex are.

11. Senses that are found throughout the body are considered
    a. special senses.
    b. universal senses.
    c. visceral senses.
    d. general senses.

12. The area sensed by one sensory neuron is known as its
    a. synaptic locus.
    b. receptive field.
    c. insertion.
    d. tactile area.

13. Loss of proprioception in a limb would result in
    a. the loss of sensing body position and identity with the limb.
    b. the loss of the ability to sense pain in the limb.
    c. the inability to block pain in the limb.
    d. the inability to detect cold in the limb, although the ability to sense heat would not be affected.

14. Smell can
    a. block pain as well.
    b. also increase our ability to hear.
    c. cause us to recall significant memory related to a particular smell.
    d. cause an inhibition of reflexes.

15. Chemical receptors for taste on the tongue are found in
    a. gustatory bodies.
    b. taste buds.
    c. glossoreceptors.
    d. nociceptors.

16. The amount of light entering the eye is controlled by the
    a. lens.
    b. cornea.
    c. retina.
    d. iris.

17. Light is focused on the retina by the
    a. iris.
    b. optic disc.
    c. lens.
    d. sclera.

18. The detection of hearing, rotation, forward or backward motion, and gravity are all the result of
    a. the bending of hairs on hair cells.
    b. chemicals attaching to receptors on chemosensitive areas.
    c. pressure created within the ear as air passes over the tympanic membrane.
    d. light refracted through the eardrum onto specific photoreceptors in the inner ear.

19. The three bones that amplify and transfer sound vibrations are found in the
    a. outer ear.
    b. inner ear.
    c. middle ear.
    d. auditory (eustacian) tube.

20. We have semicircular canals
    a. for each plane.
    b. on the left side only.
    c. in the horizontal plane only.
    d. in the vertical plane only.

Chapter 7

The Endocrine System

LEARNING OBJECTIVES

Upon completion of this chapter, you will be able to:

1. Describe the components and functions of the endocrine system and how it maintains homeostasis.
2. Identify the different classifications of hormones and explain their actions and feedback mechanisms.
3. Identify the hormones each endocrine gland secretes and their effects.
4. Describe a clinical example, scenario or disorder related to the endocrine system.
5. Demonstrate how to perform a blood glucose testing.

CHAPTER OUTLINE

Introduction
Endocrine System
Endocrine System Components
- Hormone Action

Endocrine Glands
- Pituitary Gland
- Thyroid Gland
- Parathyroid Glands
- Endocrine Pancreas
- Adrenal Gland
- Gonads: Testes and Ovaries
- Pineal Gland

INTRODUCTION

You have just completed the chapter on the nervous system. Both the nervous system and the endocrine system work together, and their interaction is important in communication—relay of messages. Both have different ways of relaying or transmitting their messages or signals, but they work together to maintain homeostasis.

The endocrine system and the nervous system are connected by the hypothalamus of the brain. The hypothalamus of the brain controls major endocrine glands such as the pituitary gland, while the endocrine system controls and regulates the hormones in the body. The nervous system, through the hypothalamus, stimulates the pituitary gland to secrete or release hormones in the body by way of the bloodstream, which then go to the organs and tissues.

There are about 30 hormones in the body, and each of them has their own specific function. Endocrine glands have no ducts, and are different from exocrine glands, which release hormones through their ducts externally such as sweat glands, mammary glands, and salivary glands.

The importance of learning about the endocrine system is crucial to understanding the diseases associated with an increase or decrease in the secretion levels of specific hormones. Common disorders of the endocrine include: hypothyroidism, hyperthyroidism, hypoglycemia, metabolic disorder such as diabetes, adrenal insufficiency, polycystic ovary syndrome, and gigantism or acromegaly.

According to the American Diabetes Association (ADA), there were 34.2 million people in the United States with diabetes in 2018 and 7.3 million people were undiagnosed. There are about 1.5 million of Americans diagnosed with diabetes every year. Individuals with a low level of insulin can develop diabetes. Insulin, a hormone secreted by the pancreas, helps the cells remove or absorb sugar or glucose, and keeps it in the cell for energy use. However, when the body cannot produce enough insulin, a build-up of sugar in the bloodstream leads to high level of glucose in the blood. There are two types of diabetes: type 1 has a complete absence of insulin, and type 2 has low level of insulin. According to the Centers for Disease Control (CDC), 90%-95% of people with diabetes have type 2.

People with diabetes need to keep their weight, blood pressure, and cholesterol down by exercise and healthy diet. Many of them may also have prescribed insulin shots to keep their blood glucose under control. There is no cure for diabetes. Complications of diabetes are shown below:

CHRONIC COMPLICATIONS OF DIABETES

HEART AND BLOOD VESSEL DISEASES

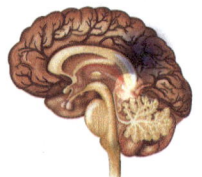
STROKE
BLOCKAGE OF THE BLOOD VESSELS SUPPLYING BLOOD TO THE BRAIN

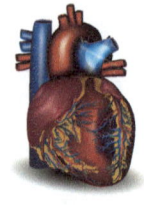
HEART ATTACK
BLOCKAGE OF THE BLOOD VESSELS SUPPLYING BLOOD TO THE HEART

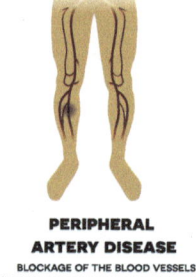
PERIPHERAL ARTERY DISEASE
BLOCKAGE OF THE BLOOD VESSELS SUPPLYING BLOOD TO THE LEGS AND FEET

© Tefi/Shutterstock.com

Figure 7.1

ENDOCRINE SYSTEM

The endocrine system provides slow sustained communication between cells at a distance from each other. The endocrine system relies on the action of endocrine glands, which are ductless glands that transfer their secretions directly into the bloodstream for transport throughout the body (Figure 7.2).

Endocrine cells within the gland produce chemical messages known as hormones that are secreted into the bloodstream. Hormones can have a direct action on cells or may stimulate other glands to release their hormones to accomplish the desired action. These hormones travel throughout the cardiovascular system and then leave the bloodstream to attach to complementary receptors on cells known as target cells (Figure 7.3). When the hormone binds to a cell that has the complementing receptor, then the cell's activity is altered. Unlike the communication of the nervous system, which is very rapid and whose response is very short, the endocrine system moves at the speed of blood flow, but its response may last minutes to hours or even days. The endocrine system is somewhat like writing a letter (hormone) and placing it in the mailbox. It travels throughout the mail system (bloodstream), but it arrives at a specific location with the appropriate mailbox identification (target cell). There is a delay in the letter reaching its destination (that is, it's not instantaneous), but the letter can be repeatedly read and the instructions followed once received. Hormones in the blood are being continuously

> **Endocrine glands** ductless glands that secrete substances into the bloodstream for transport throughout the body.
>
> **Hormones** chemical messages produced by endocrine glands.
>
> **Target cells** cells with complementary receptor sites for a specific hormone that is the intended recipient of the message sent by an endocrine gland.

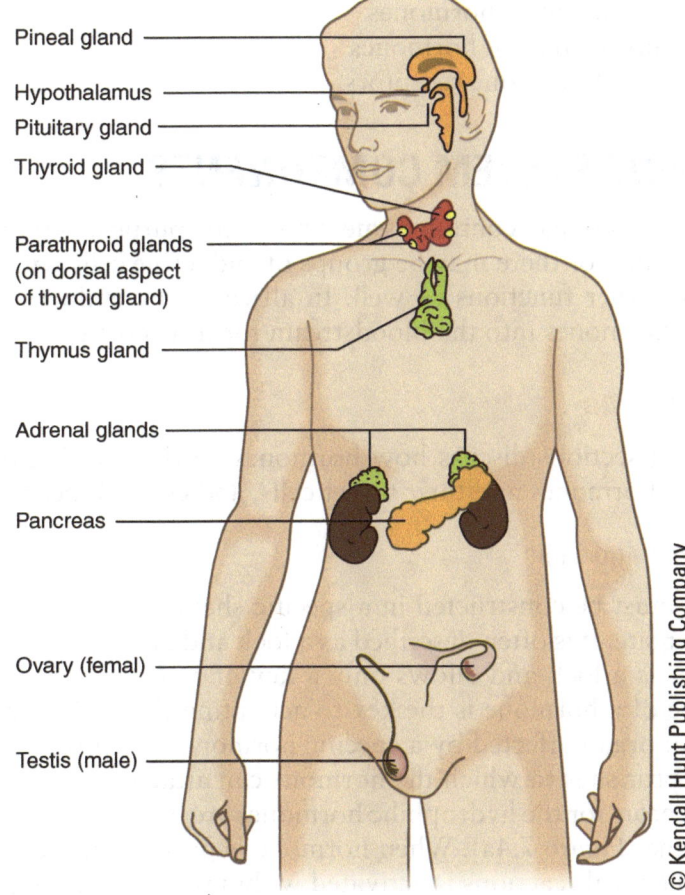

Figure 7.2 Endocrine Glands.

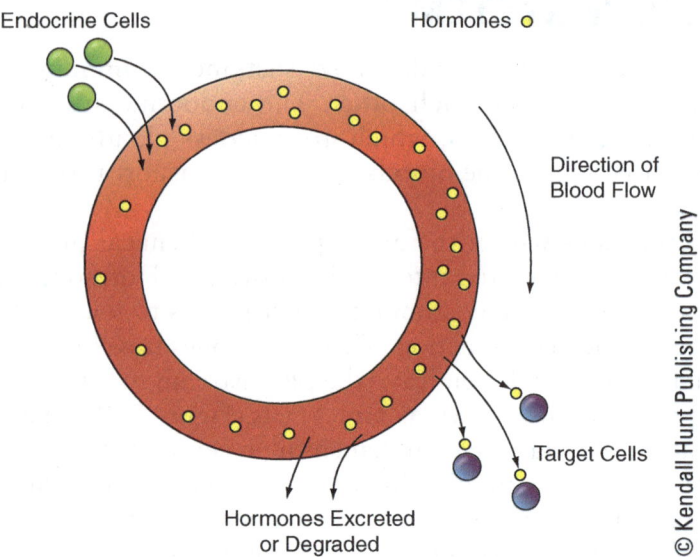

Figure 7.3 Endocrine Cells. Endocrine cells release hormones into the bloodstream. The hormones either attach to target cells or remain in the bloodstream before eventually being excreted or degraded.

degraded and/or secreted and the actual level in the blood reflects a dynamic balance between hormone release and hormone removal. Malfunctions within the endocrine system could be caused by:

- An overproduction of hormones
- An underproduction of hormones
- Insensitivity of hormone receptors

ENDOCRINE SYSTEM COMPONENTS

An endocrine gland may contain tissue whose only purpose is to produce and secrete hormones, or there may be groups of endocrine cells within an organ that perform other functions as well. In all cases the endocrine cells must secrete their hormones into the bloodstream for transport to the target cells.

Hormone Action

The following sections discuss how hormones work, covering the relationship between hormones and their target cells, and control mechanisms.

Target Organs and Cells

A hormone must be constructed in a specific shape to be able to fit exactly on a receptor site. It is often described as a lock and key type of arrangement. The receptor is a lock and allows only a key of a specific shape to operate it. The shape of a hormone is the key to activating the cell's response to the hormone. An organ affected by a specific hormone contains target cells that possess receptor sites to which the hormone can attach.

The receptors for the hydrophilic hormones are on the outer surface of the cell membrane (Figure 7.4a). When hormone is present on the receptor site, a series of chemical reactions is activated within the cell, altering the activity of the cell. Each step in the activating process results in an ever-increasing

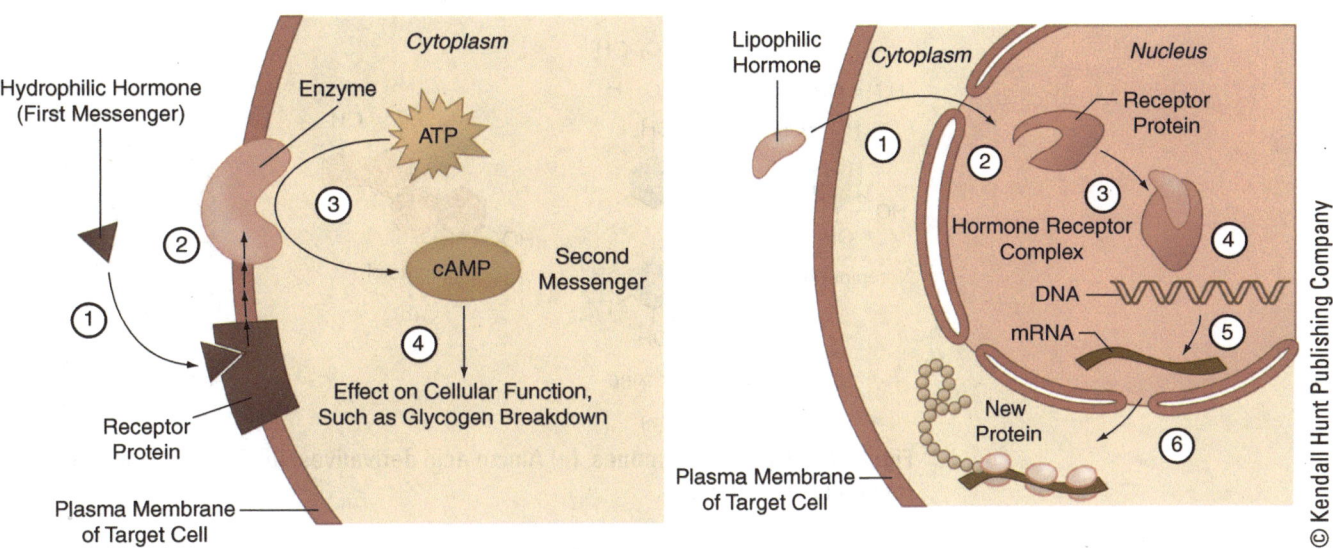

Figure 7.4 Hormone Receptors. (a) External hormone receptor site in the cell membrane. (b). Internal hormone receptor site associated with DNA.

number of responses within the cell. This means that the attachment of a single hormone within the cell results in a rapidly escalating response within the cell, referred to as the *cascade effect*. The external hormones are rapidly degraded, so the response onset is rapid but its response time is relatively short.

Lipophilic hormones pass through the plasma membrane to attach to internal receptor sites (Figure 7.4b). The effect of this type of hormone is to activate a gene in the DNA (deoxyribonucleic acid) that results in a change in the activity of the cell. The degradation of these hormones is relatively slow, so the response from an internal hormone lasts hours or even days.

Chemistry of Hormones

Hormones can be classified as two types: hydrophilic (water-loving or water-soluble) and lipophilic (lipid-loving or lipid-soluble). Phospholipids that make up the plasma membrane form a lipid layer that prevents hydrophilic substances from passing through the cell membrane. Hydrophilic hormones must attach to receptor sites on the outer surface of the target cell. They rely on the chemical action of a second messenger inside the cell that mimics the action of the hydrophilic hormone. Lipophilic hormones can pass through the lipid layer in the cell membrane to directly cause a long-lasting response by activating a gene in the DNA.

There are three categories of hormones based on their chemical composition (Figure 7.5):

- *Amino acid derivatives* are hormones derived from one amino acid, either tyrosine or tryptophan. Amino acid-derived hormones produced by the adrenal medulla (epinephrine and norepinephrine), are hydrophilic. Those made by the thyroid gland (thyroxine and triiodothyrosine) are lipophilic.
- *Steroid hormones* are produced by adding atoms or small molecules to the basic steroid molecule. The basic steroid is derived from cholesterol. Other components of cholesterol are stripped from the basic steroid molecule, then atoms to make the hormone unique are added at specific locations. Steroid hormones are lipophilic.
- *Peptide hormones* are composed of chains of amino acids. They are the most common type of hormones. Peptide hormones are hydrophilic.

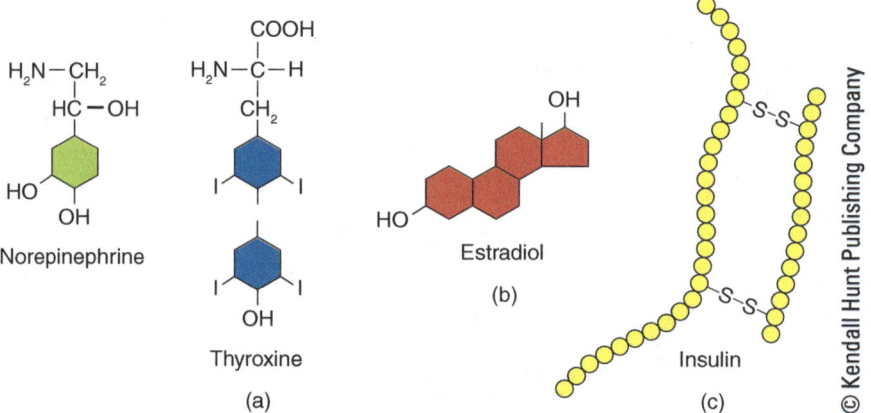

Figure 7.5 Types of Hormones. (a) Amino acid derivatives. (b) Steroid hormone. (c) Peptide hormone.

Hormone Secretion Control

To maintain the reactions caused by hormones within a homeostatic range, the level of each hormone produced must be controlled. Recall from Chapter One that negative feedback is used to control the level of substances in the body. *Negative feedback* is the regulating of a parameter by causing a response that satisfies and turns off the need. Most hormones levels, with one exception, are controlled by negative feedback. When an endocrine gland senses that one of the parameters it maintains is out of homeostatic range, it causes its endocrine cells to release hormones into the bloodstream. When the homeostatic range has been achieved, the need has been satisfied, so the endocrine gland reduces its hormone release. For example, after eating, our blood glucose level increases. If that glucose level exceeds the homeostatic range, the pancreas releases the hormone insulin to cause some of the excess sugar to be removed from the bloodstream and stored in some cells. The result is a decrease in the blood glucose level. When the blood glucose level reaches its desired range, the production of insulin decreases.

There is one exception that occurs during childbirth. In this case *positive feedback* is used: the response increases the release of hormone. Contractions are caused by the hormone oxytocin. To increase the strength of contractions, then it is necessary to release oxytocin at continually increasing levels until birth occurs. Positive feedback occurs for a relatively short period. It is not an ongoing process.

Comprehension Check-up:

1. A cell that has a receptor site for a specific hormone is known as a _____.
2. Hormones derived from cholesterol are _____.
3. Lipophilic hormones attach to receptors _____ the cell.
4. Most hormones are regulated by _____ feedback.

1. target cell 2. steroids 3. inside 4. negative

ENDOCRINE GLANDS

To understand the functions of the endocrine system, it is essential to describe the structure and action of the endocrine glands. Some of the endocrine organs are active participants in other organ systems, such as the stomach, small intestine, and kidneys; these organs are discussed in their respective chapters.

Pituitary Gland

The *pituitary gland* sits in a bowl-shaped region of the sphenoid bone just inferior to the *hypothalamus*. The hypothalamus is the area of the diencephalon of the brain that plays an important role in regulating the functions of many endocrine glands. The pituitary gland is connected inferiorly to the hypothalamus by a stalk known as an *infundibulum*. The pituitary gland is divided into posterior and anterior lobes (Figure 7.6).

Posterior Lobe of the Pituitary Gland

The posterior lobe of the pituitary gland is the most unusual component of the endocrine system because it consists of neurons from the hypothalamus whose terminal ends are in the posterior lobe (Figure 7.7). These neurons produce neurohormones rather than neurotransmitters. When the hypothalamus determines the need for additional hormones from the posterior lobe, neurons whose cell bodies reside in the hypothalamus transmit impulses down their axons in the posterior pituitary gland, where their terminal ends release hormones into the bloodstream. Two hormones are secreted from the posterior lobe:

- Antidiuretic hormone (ADH)—A diuretic is a chemical that causes increased urination. Antidiuretic hormone has the opposite affect in that it causes water retention. If sensors in the hypothalamus determine that the plasma is becoming too concentrated, ADH is released to cause the kidneys to retain additional water from the urine being formed. As a result, the urine becomes concentrated and the plasma is diluted. This is a very important parameter for maintaining the concentration of the plasma.

> **Posterior lobe of the pituitary gland** the pituitary gland is surrounded on three sides by the sphenoid bone and is inferior to the hypothalamus. The posterior section of the pituitary gland consists of neurons from the hypothalamus whose terminal ends are in the posterior lobe. Those terminal ends release hormones rather than neurotransmitters.

> **Antidiuretic hormone** a hormone produced by the posterior pituitary gland that causes the kidneys to retain additional water to dilute plasma that has become too concentrated.

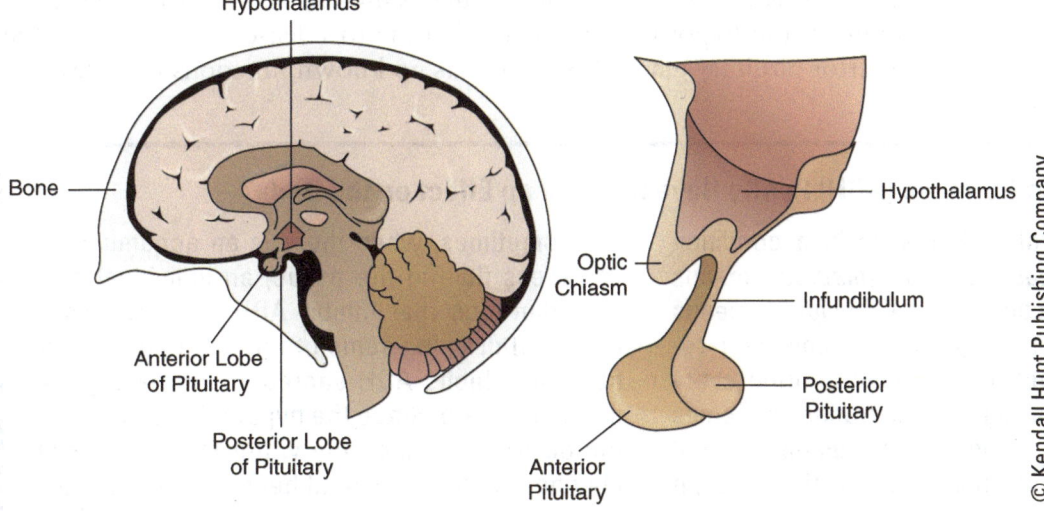

Figure 7.6 The Pituitary Gland. (a) This gland is located inferior to the hypothalamus anterior to the brain stem. (b) It has anterior and posterior lobes, which are connected to the hypothalamus by the infundibulum.

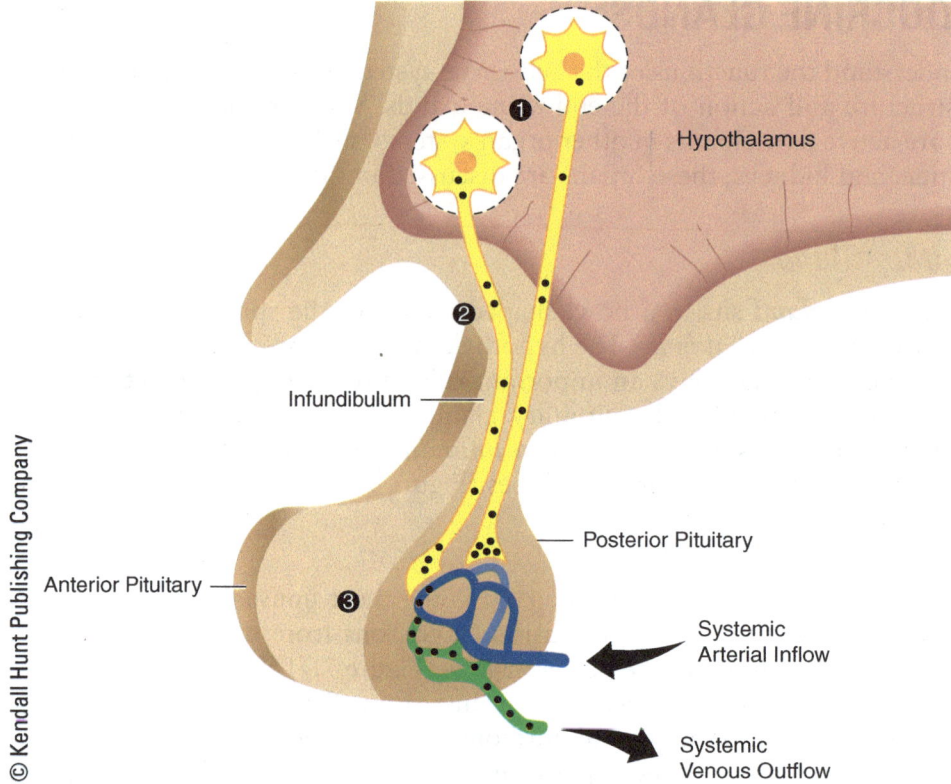

Figure 7.7 The Posterior Lobe of the Pituitary Gland. This is an extension of the hypothalamus, whose nerves terminate in the posterior lobe and produce neurohormones rather than neurotransmitters.

> **Oxytocin** a hormone that affects the female reproductive system; it causes contraction of smooth muscle in the uterus during labor. Synthetic oxytocin may be given intravenously to induce labor. It also causes contraction of smooth muscle in the mammary glands of the breasts, resulting in the release of milk.

- Oxytocin—This hormone affects the female reproductive system. Oxytocin causes contraction of smooth muscle in the uterus during labor. Synthetic oxytocin may even be given intravenously as a means of inducing labor. It also causes contraction of smooth muscle in the mammary glands of the breasts, resulting in the release of milk. Another hormone, prolactin, causes the production of milk, but the milk must not be released as soon as it is produced. Instead, suckling stimulates nursing receptors in the breasts to inform the hypothalamus that it is time to release oxytocin from the posterior pituitary gland. Oxytocin has no known functions in males.

Does Antidiuretic Hormone (ADH) Really Have Much of an Effect on the Body?

Underproduction of ADH results in a condition known as *diabetes insipidus (diabetes* means "increased production of urine"; *insipidus* means "tasteless—without sugar"). This condition results in the production of gallons of urine every day, and the plasma concentration becomes harder to control. Diabetes insipidus has several causes. If hypoproduction of ADH is the cause, it can be treated by replacement therapy with synthetic ADH.

Sometimes, when there is an accumulation of excess fluid in the tissue, an individual may take diuretics that inhibit ADH secretion as a means of draining some of the extra fluid. Alcohol also inhibits ADH secretion, increasing the need to urinate. Since the hypothalamus creates the desire to replace the water loss, the result may be that the individual becomes dehydrated, causing dry mouth the morning after drinking alcohol.

Anterior Lobe of the Pituitary Gland

The anterior lobe of the pituitary gland contains hormone-producing cells but does not release any of the hormones until the hypothalamus instructs it to do so (Figure 7.8). The hypothalamus maintains the homeostasis of several parameters in the body by releasing regulatory hormones into the anterior pituitary. The regulatory hormones may cause those endocrine cells to release their hormones, or they may inhibit the release of hormones from the anterior lobe. Six hormones are produced by the anterior lobe:

- *Human growth hormone* (hGH)—As the name implies, growth hormone causes the body to develop. It also increases metabolism, because the process of growth requires increased energy.
- *Thyroid-stimulating hormone* (TSH)—It stimulates the growth of the thyroid gland and controls the production of thyroid hormones.
- *Follicle-stimulating hormone* (FSH)—FSH is a reproductive hormone that has an action on the gonads in both males and females. Surrounding each egg in the female ovary is a sphere of cells called a *follicle*. FSH stimulates a follicle to release *estrogens* (female sex hormones) and to begin the process of releasing an egg each menstrual cycle. It is also responsible for maintaining the production of sperm in males.
- *Luteinizing hormone* (LH)—In females, LH causes rupture of a mature follicle, resulting in the release of the egg from the ovary by a process known as *ovulation*. Once the follicle is empty, it is converted into the corpus luteum, which secretes estrogens and progesterone. In males, LH (also known as *interstitial cell-stimulating hormone* [ICSH]) causes increased secretion of testosterone from the testes.

> **Anterior lobe of the pituitary gland** the pituitary gland is located on the medial underside of the brain inferior to the hypothalamus. The anterior section of the pituitary gland contains hormone-producing cells, but they do not release any of their hormones until the hypothalamus instructs them to do so through releasing regulatory hormones.

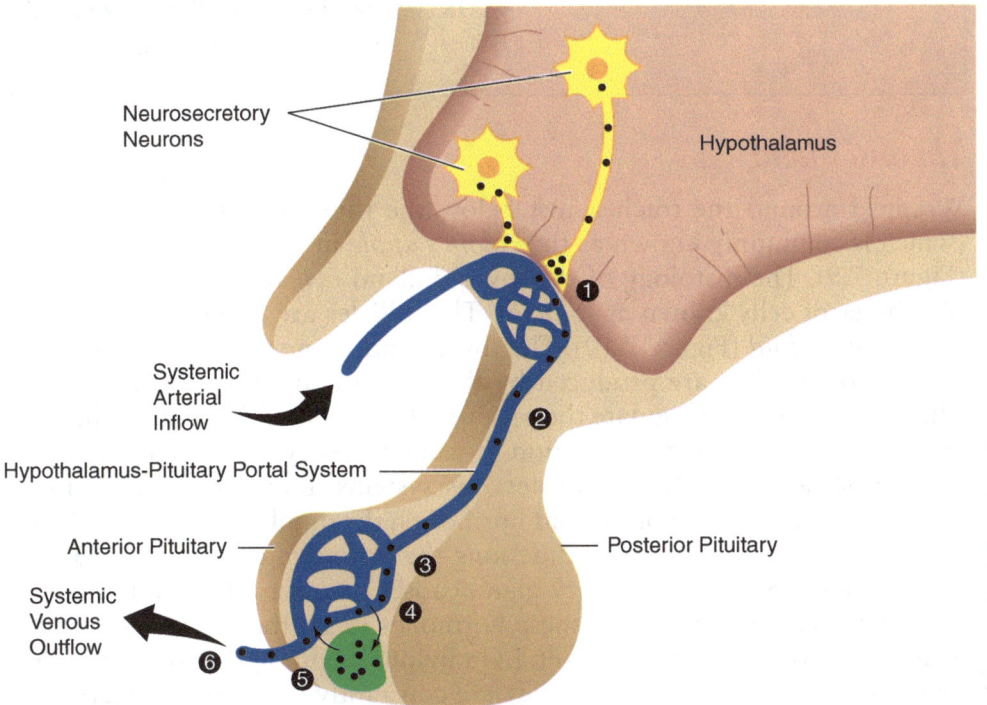

Figure 7.8 The Anterior Lobe of the Pituitary Gland. This has a special group of capillaries, the hypothalamus-pituitary portal system, which transports regulatory hormones from the hypothalamus to the anterior lobe to control the release of hormones from this lobe.

- *Prolactin* (PRL)—PRL regulates mammary gland growth and breast milk production in females. Oxytocin, working in conjunction with prolactin, is necessary for the milk to be released.
- *Adrenocorticotropic hormone* (ACTH)—Although the name seems complicated, breaking it down to its component parts makes it quite descriptive. *Adreno-* refers to the adrenal gland. *Cortico-* is the cortex or outer layer of the adrenal gland. A *tropic* hormone is one that stimulates another endocrine gland. Putting it all together, this hormone stimulates endocrine cells in the cortex of the adrenal gland. This hormone primarily affects the production of cortisol in response to chronic stress by increasing the availability of glucose, lipids, and amino acids to be used as energy sources when coping with a difficult situation.

The hormones produced by the anterior lobe of the pituitary gland affect three general categories: growth, metabolism, and reproduction.

> **Comprehension Check-up:**
>
> 1. Neurons from the hypothalamus whose terminal ends are in the posterior lobe of the pituitary gland secrete _____ rather than neurotransmitters.
> 2. The hormone from the anterior lobe of the pituitary gland that fine-tunes metabolic rate is _____.
> 3. Pituitary hormones that regulate female reproduction include _____, _____, and _____.
>
> 1. neurohormones
> 2. thyroid stimulating hormone—TSH
> 3. oxytocin, FSH, LH, and prolactin

Thyroid Gland

Wrapped around the trachea just below the thyroid cartilage in the neck (which is commonly known as the *Adam's apple* in men is the thyroid gland (Figure 7.9). The histology of the thyroid gland reveals that it is composed of spheres of cells known as *follicles*. The follicles are filled with a material known as *colloid* (Figure 7.10). The thyroid hormone, triiodothyronine (T_3) and thyroxine (T_4), are produced by the follicles and stored in the colloid. These two hormones regulate the production of ATP, and maintain the basal metabolic rate and heat production. In children, they promote the development of skeletal, muscular, and nervous systems. Iodine is required in the formation and function of thyroid hormones; the numbers 3 and 4 in T_3 and T_4 stand for the number of iodine ions the hormone can carry. TSH from the anterior lobe of the pituitary gland causes the thyroid gland to produce T_3 and T4. Hypersecretion of these hormones raises the metabolic rate to a level where weight gain is difficult. In individuals with hyperthyroidism, ATP production may be so high that they are constantly agitated or fidgety. They may have tremors and be very excitable. Hyposecretion of T_3 and T_4 results in decreased ATP production. These individuals often become cold intolerant, feel continually tired, and easily gain weight.

> **Thyroid gland** this gland is located in the anterior neck wrapped around the thyroid cartilage (referred to as Adam's apple in men). The thyroid gland produces hormones to increase basal metabolic rate and also a hormone to decrease blood calcium by causing the excess to be converted into bone.

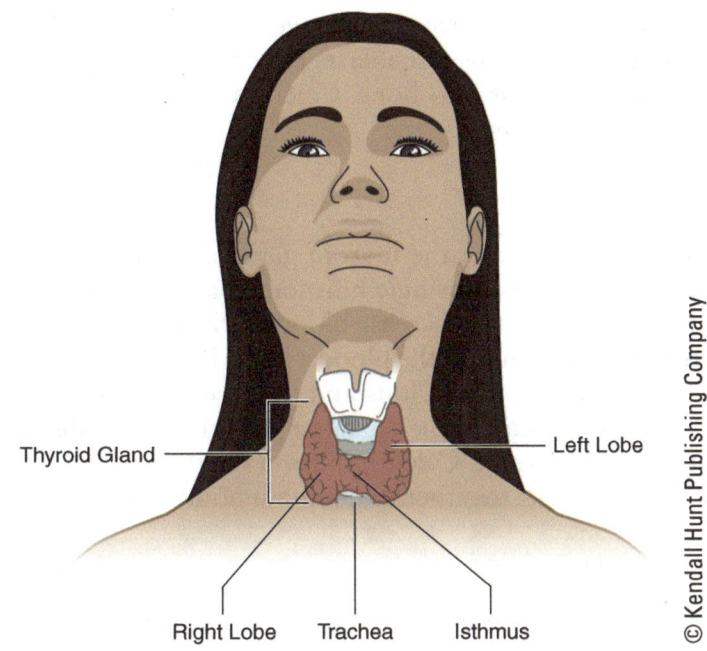

Figure 7.9 Thyroid Gland. The thyroid gland produces hormones T_3 and T_4.

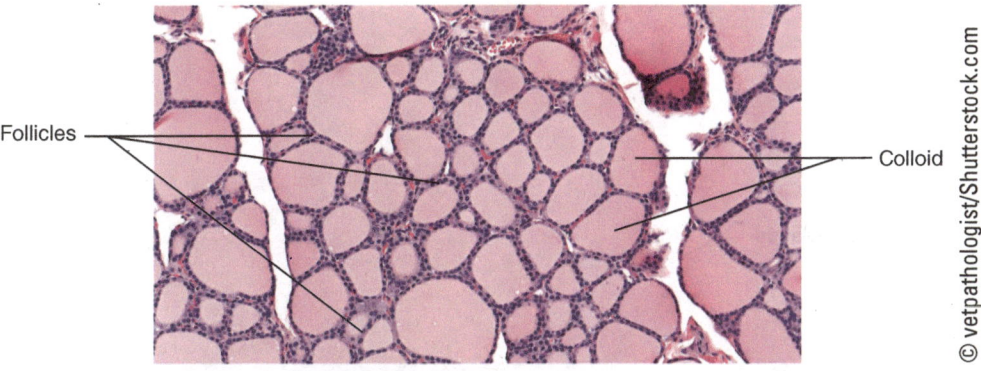

Figure 7.10 Histology of the Thyroid Gland. This gland is composed of spheres of cells known as *follicles* filled with a fluid called *colloid*.

The Importance of Vitamin D

The absorption of calcium into the bloodstream requires activation of vitamin D. Vitamin D has also been shown to help prevent diabetes mellitus (uncontrolled blood glucose), to fight cancer, and to counter autoimmune diseases, although the processes involved are not yet understood. Milk is fortified with vitamin D in the United States as a means of increasing this important vitamin in the American diet. It is the only vitamin we are able to produce by chemical reactions caused by the sun on our skin. Other sources of vitamin D include fish liver oil and eggs.

Goiter is an enlarged thyroid gland that might result from deficiency in iodine intake. It may lead to digestive as well as circulatory problems. To reduce the prevalence of goiter caused by iodine deficiency, many governments require the enrichment of salt with iodine.

Between the follicles of the thyroid gland are endocrine cells that produce another hormone known as *calcitonin* (CT). If the level of calcium in the bloodstream exceeds the homeostatic range, CT has the ability to decrease blood calcium by stimulating osteoblasts to remove excess calcium from the bloodstream and deposit it into bone. Calcitonin also causes the kidneys to excrete excess calcium ions.

Parathyroid Glands

Embedded in the posterior of the thyroid gland are four parathyroid glands (Figure 7.11). They produce a single hormone: parathyroid hormone (PTH). PTH stimulates osteoclasts to increase blood calcium by breaking bone down and depositing the calcium into the bloodstream. PTH also causes the kidneys to decrease excretion of calcium ions and to increase the production of vitamin D_3 (also known as calcitral), the latter stimulates the absorption of calcium from the digestive tract. Calcitonin from the thyroid, and PTH from the parathyroid glands, work together to maintain homeostasis of blood calcium.

Endocrine Pancreas

The pancreas is located inferior and posterior to the stomach in the upper abdomen (Figure 7.12). It is composed of two types of glands: both

> **Parathyroid glands** there are four parathyroid glands embedded in the posterior thyroid gland. They produce a hormone to increase blood calcium by causing bone to be broken down so that its stored calcium can be placed into the blood. It also causes increased absorption of calcium from the digestive system.

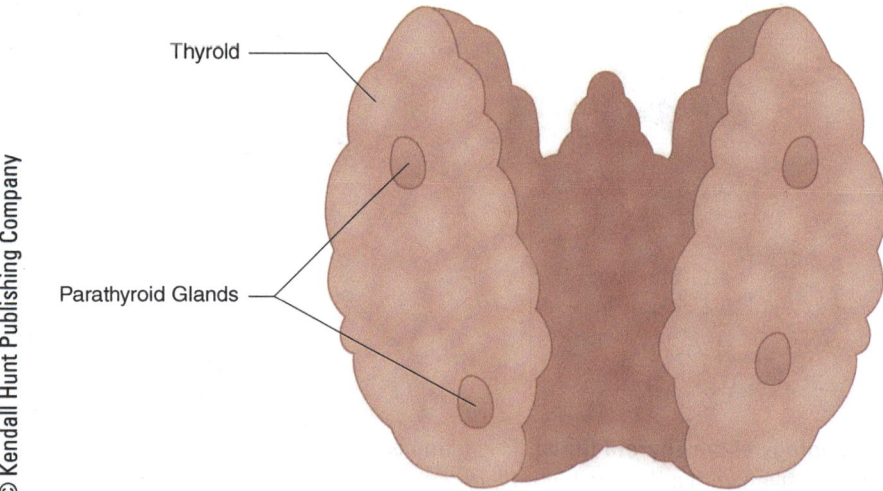

Figure 7.11 Parathyroid Glands. These four glands are embedded in the thyroid gland and produce parathyroid hormone (PTH).

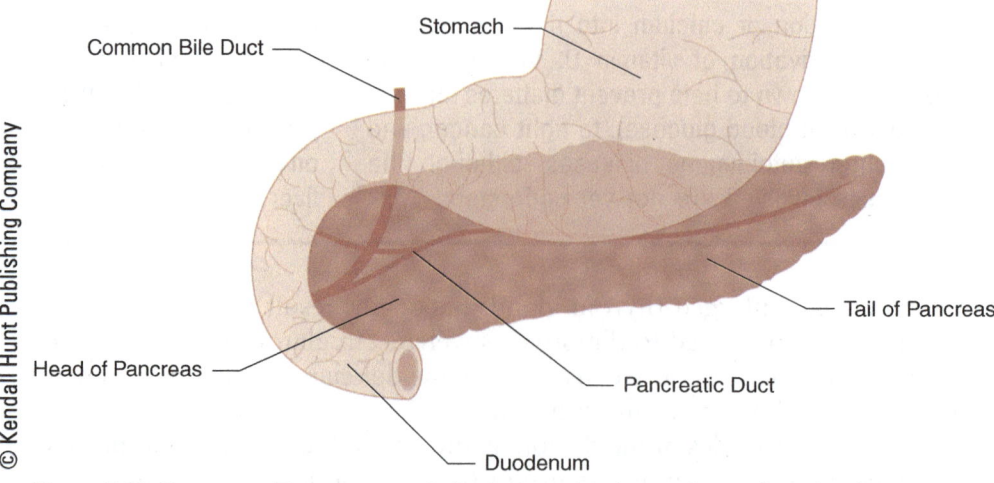

Figure 7.12 Pancreas. The pancreas is both an endocrine and exocrine gland.

exocrine and endocrine. Our concern in this chapter is with the endocrine portion. Scattered throughout the pancreas are groups of endocrine cells known as pancreatic islets (Figure 7.13). Within each pancreatic islet are alpha and beta cells that serve to maintain blood glucose levels within the homeostatic range.

A significant reason for maintaining blood glucose levels within a specific range is that the brain cannot store large amounts of nutrients. Instead, the brain receives its metabolic requirements through the bloodstream. Its primary source of energy to make ATP is blood glucose. If the blood glucose level is out of range, brain function diminishes, and the farther the individual is from homeostasis, the more rapidly it diminishes. This is a constant concern for an individual with diabetes mellitus who must take insulin injections to control blood sugar after eating. If the ratio of food to insulin is not within homeostatic range, the individual may become unconscious in the short term or suffer from other long-term effects, such as decreased vision and decreased blood flow to the extremities and increased risk of cardiovascular disease, stroke, or kidney failure, just to name a few of the potential hazards.

> **Pancreatic islets** within the pancreas, which is located inferior and posterior to the stomach, are groups of endocrine cells known as *pancreatic islets*. These cells produce hormones to maintain blood glucose within its homeostatic range.

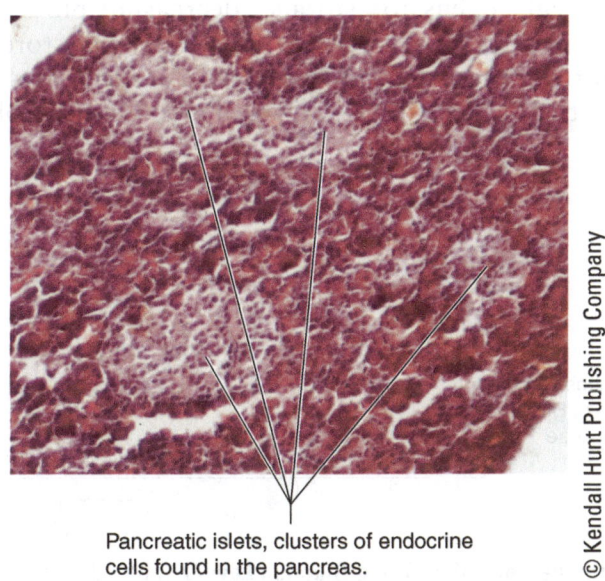

Pancreatic islets, clusters of endocrine cells found in the pancreas.

Figure 7.13 Pancreatic Islets. These are clusters of endocrine cells found in the pancreas.

Diabetes Mellitus—Blood Glucose Out of Control

An individual with diabetes mellitus ("mellitus" means sweet) is unable to produce enough insulin to maintain blood glucose control. There are two types of diabetes mellitus. Type I (insulin-dependent) diabetes mellitus usually occurs when the immune system attacks and destroys beta cells. As a result, affected individuals cannot produce insulin and must take regular injections in order to maintain blood glucose. Type II (non-insulin-dependent) diabetes mellitus typically occurs later in life. In this condition, the beta cells actually produce normal or even above normal levels of insulin. The problem is the target cells, which have a decreased response to insulin. The result is increased blood glucose because the excess can't be moved into storage. The causes of Type II diabetes are not yet well understood, but in about 90% of cases, the affected individuals are obese.

Alpha Cells

Alpha cells respond to low levels of blood sugar. Alpha cells produce the hormone *glucagons,* which causes the liver to increase the release of stored glucose into the bloodstream.

Normally, high nutrient venous blood from the intestine flows to the liver before going to the heart to be pumped to the rest of the body. As this nutrient-rich blood passes through the liver, hepatocytes (liver cells) remove excess glucose and chain it into glycogen molecules for storage as a means of lowering excess blood glucose. If there is a need to increase blood glucose, glucagon alerts the liver of the greater demand. Not only does it allow the release of some of the stored glucose from the liver, it may even cause additional chemical reactions that will increase the ability of the body to raise blood glucose levels.

Beta Cells

Beta cells respond to blood glucose levels above homeostatic values. When blood glucose gets too high, beta cells produce insulin, which facilitates the transport of glucose into liver and muscle cells and turns molecules of glucose into glycogens for storage, decreasing blood levels of sugar. Increased insulin directs the liver and muscle cells to store additional glucose received from the digestive system. It also decreases the breakdown of fat because there is an excessive level of glucose available as a source of energy (Figure 7.14).

> Insulin produced by the endocrine pancreas. It causes increased uptake of glucose by cells as a means of lowering excessive blood glucose levels.

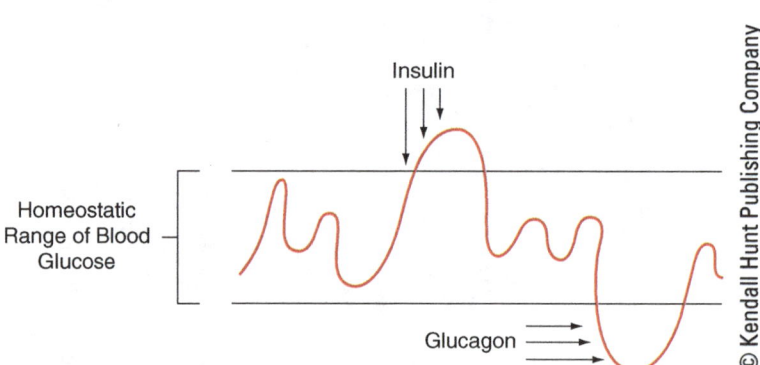

Figure 7.14 Hormone Control of Blood Glucose Levels. If blood glucose levels exceed the homeostatic range, insulin is released to transport glucose into cells to lower blood glucose. If blood glucose drops below homeostatic minimum, glucagon causes the breakdown of glycogen to increase blood glucose.

Comprehension Check-up:

1. The hormones _____ and _____, produced by the thyroid gland, affect metabolic rate.
2. Alpha and beta cells in the _____ islets affect blood glucose.

1. T_3 and T_4. 2. pancreatic

Clinical Skills—Blood Glucose Testing

1. Lay out all equipment in an organized manner.
2. Clean finger with an alcohol prep pad in circular manner.
3. Use a sterile lancet to obtain a drop of blood from the fingertip.
4. If necessary, squeeze finger gently to obtain a large drop of blood.
5. Place glucose strip into the glucometer. Wick up blood with the tip of the test strip.
6. Read the result on the screen, and record numerical reading with mg per dl as units.

© Proxima Studio/Shutterstock.com

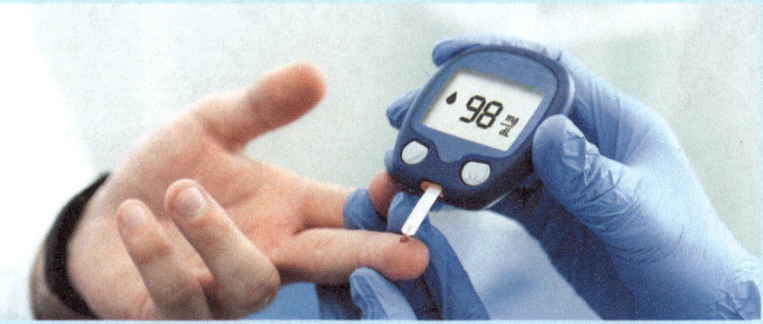

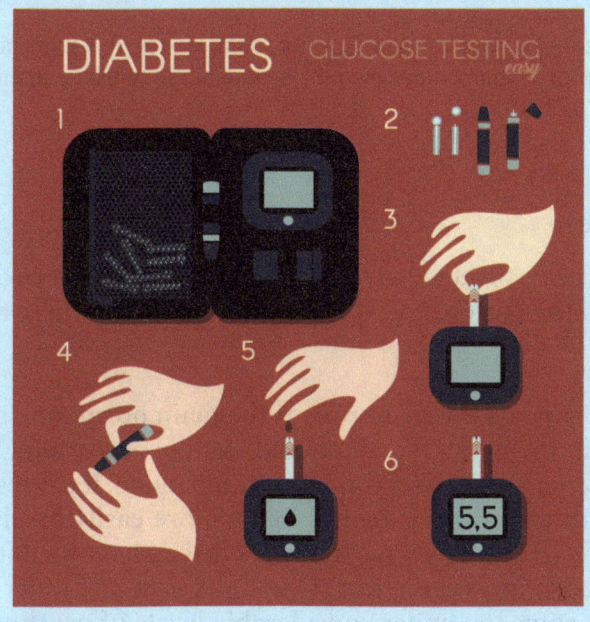

Contributed by David Flint. © Kendall Hunt Publishing Company

Adrenal Gland

The adrenal glands are superior to the kidneys. They are essentially two endocrine glands with one wrapped around the other. The outermost gland is known as the *cortex* and the inner core is referred to as the *medulla* (Figure 7.15).

> **Adrenal glands** located superior to the kidneys, the adrenal glands can be divided into two internal layers. The outer cortex produces steroid hormones. The inner medulla of the adrenal gland is connected to the sympathetic nervous system to cause an extended fight-or-flight response.

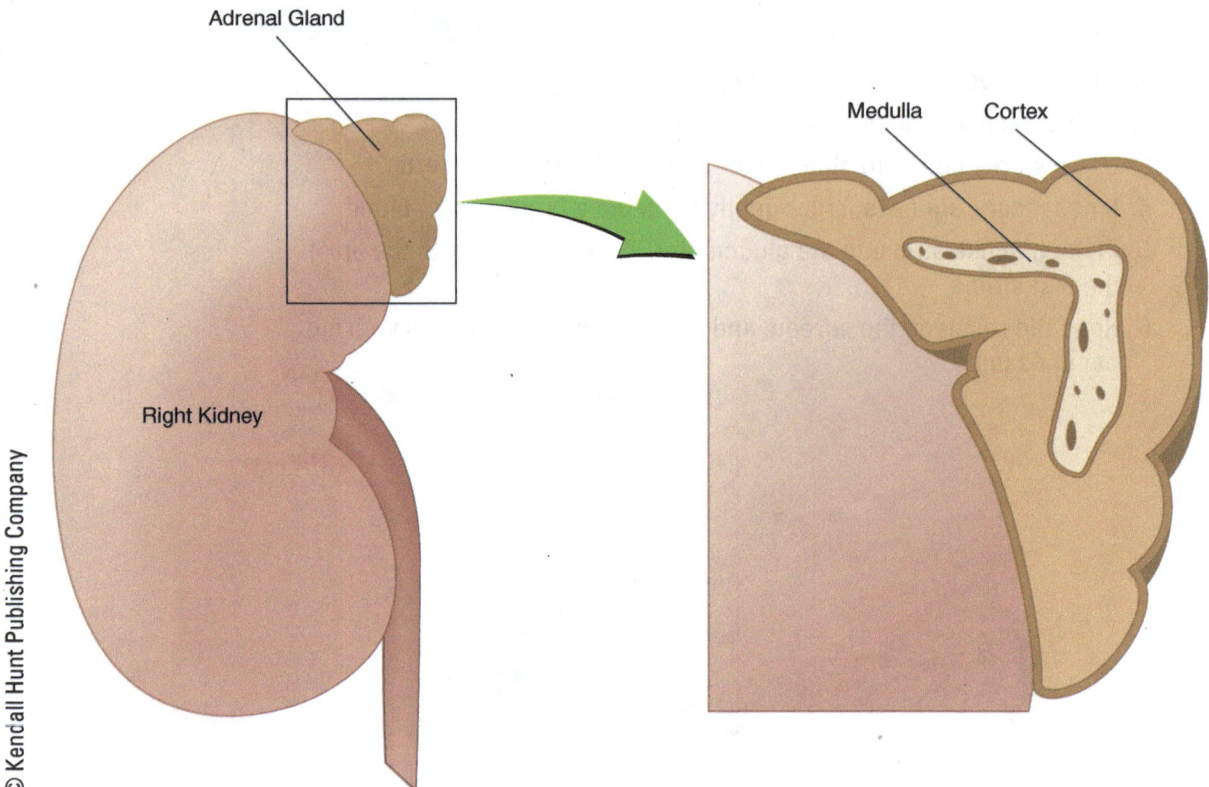

Figure 7.15 Adrenal Gland. The adrenal gland is located superior to the kidney. It is two glands in one: The outer cortex produces steroid hormones, and the inner medulla is connected to the sympathetic nervous system to extend the "fight-or-flight" response.

Adrenal Cortex

The *adrenal cortex* produces steroids referred to as *corticoids*. The suffix *-cortic* refers to the cortex, and *-oids* implies steroids. There are three classes of "corticoids":

- *Mineralocorticoids*—*Aldosterone* is the main hormone. It causes the kidney to retain the mineral sodium and excrete potassium.
- *Glucocorticoids*—*Cortisol* is the major hormone. It causes the conversion of protein to glucose and makes lipids more easily available for coping with chronic stress. Excessive levels of cortisol have an anti-inflammatory and immunosuppressive effect. Cortisol secretion is increased by stress. As a result, someone under considerable levels of psychological pressure may, because of increased cortisol, suppress the immune system and increase his or her risk of disease.
- *Gonadocorticoids*—These include the sex hormones estrogens and androgens. Both estrogens (female hormones) and testosterone (one of the androgens, male hormones) are produced in the adrenal glands of both males and females. The level of these hormones is fairly insignificant with the exception that one of the androgens, dehydroepiandrosterone (DHEA), affects females. Acting like testosterone, it is responsible for sex drive and pubic and axillary hair in females. Because the testes in males produce significantly higher levels of testosterone, the affect of the gonadocorticoids on men is minimal.

Adrenal Medulla

The *adrenal medulla* is innervated by the sympathetic nerve as a means of extending the "fight-or-flight" response when under threat or extreme stress. Recall that the sympathetic nervous system prepares the body for action. It increases blood flow to muscle and increases heart rate and blood pressure. It opens air passages. It slows digestion so that more blood can be diverted to muscle for fighting or running than to the intestine for absorption of nutrients.

In order for the sympathetic nervous system to perform its function, it must continually stimulate the organs it affects. In addition to the short-term effects on the target organs (muscles, heart, blood vessels), the sympathetic nervous system stimulates the adrenal medulla to release the epinephrine (adrenalin) and norepinephrine (noradrenalin) into the bloodstream. These hormones can cause relatively long-lasting responses without the need for constant nerve stimulation. In this manner the endocrine system can complement the sympathetic nervous system to maintain the fight-or-flight response for an extended period.

> **Testes** organ in the male reproductive system responsible for production of sperm to provide half of the chromosomes required by the offspring and also for the secretion of hormones responsible for male characteristics. The testes are located in the scrotum, the tissue sac that remains cooler than the rest of the internal body.

Gonads: Testes and Ovaries

Not only do reproductive organs produce sperm or eggs, but they also produce hormones that communicate with the rest of the body to best prepare it for reproduction.

Testes

In males, between tubules in the testes, where sperm production occurs, are endocrine cells known as interstitial cells, or *Leydig cells*. These cells produce testosterone (Figure 7.16), which is responsible for sex drive. There are a

> **Testosterone** produced by the testes, is responsible for sex drive. Other effects caused by testosterone that become evident in the male during puberty, such as facial, pubic, and armpit hair growth, lowering of the voice, and increased bone growth and muscle mass.

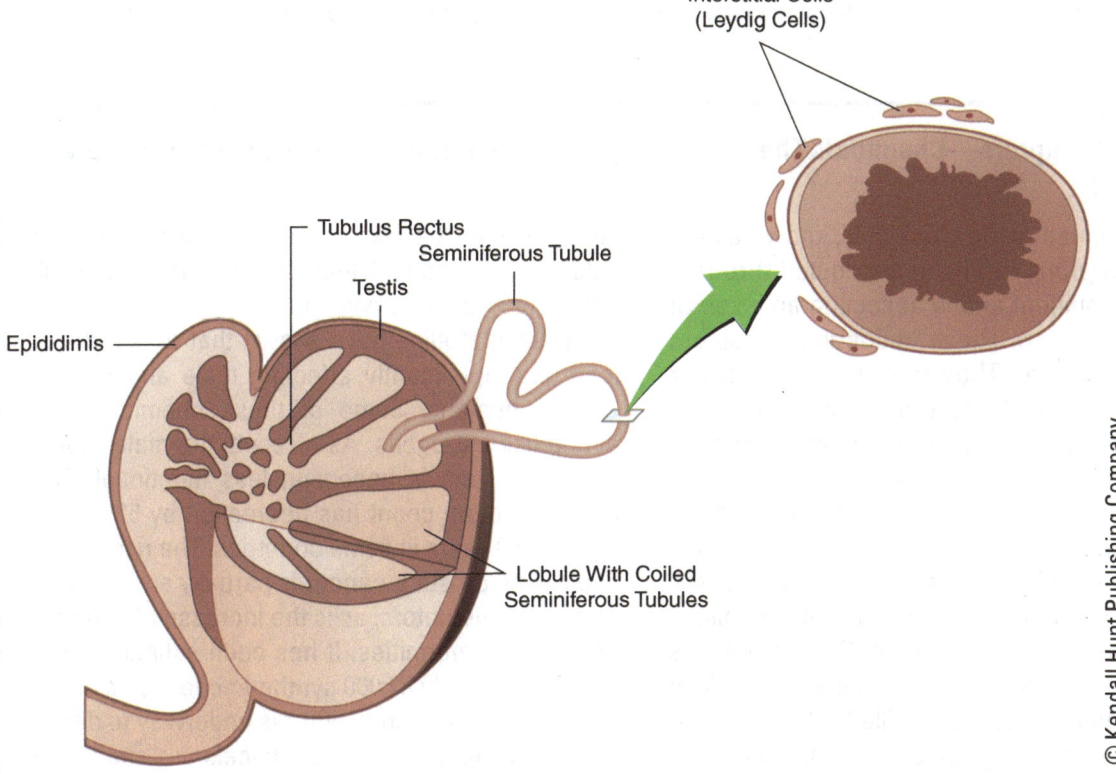

Figure 7.16 In the testes, seminiferous tubules produce sperm and Leydig cells produce testosterone.

> **Ovaries** located in the pelvic cavity, the ovaries release ova at regular intervals, providing the female's contribution of chromosomes required by the offspring. The ovaries also produce hormones responsible for assisting in the regulation of the reproductive cycle and providing female characteristics.

number of other effects caused by testosterone that become evident in males during puberty, such as facial, pubic, and axillary hair growth; lowering of the voice; and increased bone growth and muscle mass.

Ovaries

The ovaries, found in the pelvic cavity of females, produce two types of reproductive hormones (Figure 7.17): estrogens and progesterone. *Estrogens* prepare the woman's body to receive a fertilized ovum (egg). In other words, these hormones cause changes in various organs in the woman's body to allow her to provide maximum support of the fetus. During puberty, estrogens cause growth

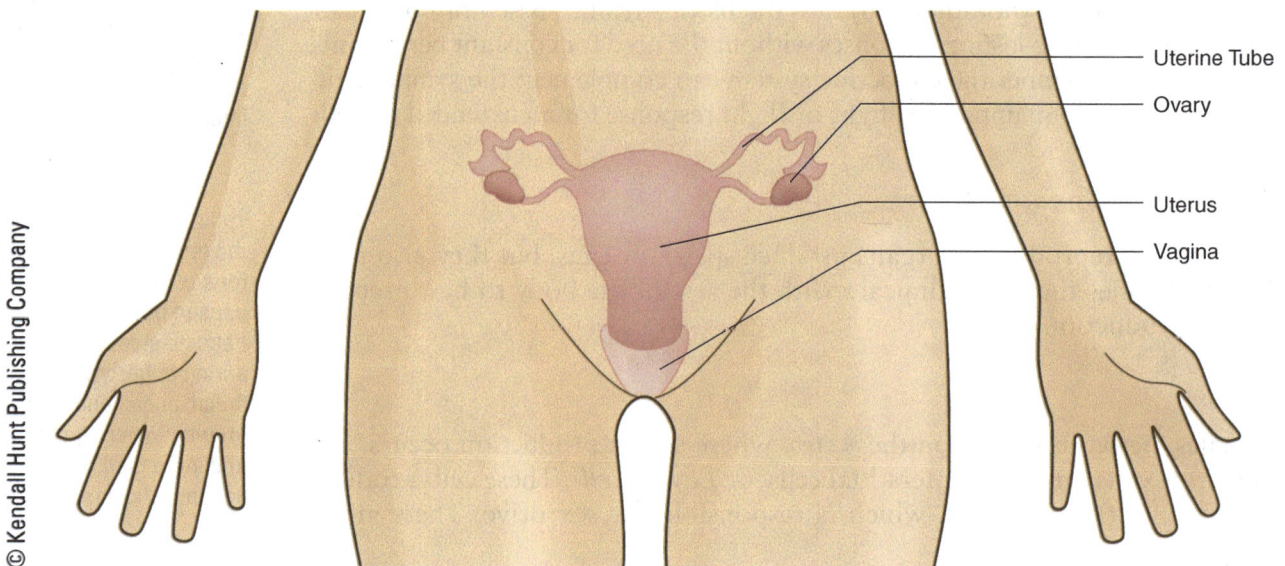

Figure 7.17 Ovaries. The ovaries are located in the female's pelvic cavity and produce ova and two hormones: estrogens and progesterone.

Endocrine Disruptors—Chemicals That Can Change Our Endocrine System Even in Extremely Small Doses

A group of human-made chemicals, known as *endocrine disruptors*, has been distributed throughout the environment and is becoming an issue of increasing concern. These endocrine disruptors come in many forms. They may be insecticides or herbicides, chemicals used in plastics, food additives, detergents, or even automobile exhaust. As these chemicals are absorbed into our body, they are able to cause changes in the production of hormones.

The amount of an endocrine disruptor required to alter our endocrine system is sometimes miniscule, such as one part per billion. That quantity is often hard to comprehend. Think of it this way. Take 100,000 railroad tank cars each filled with 10,000 gallons of water and empty all of them into one location to form a lake. Add one gallon of the endocrine disruptor into the lake and you have produced a solution strong enough that drinking of this water will alter your endocrine system.

Recent studies have shown that these chemicals are significantly affecting male animals, including humans. Some of these chemicals are acting like estrogens. As a result, the male reproductive system is becoming less functional. The average sperm count has decreased by 50% over the past 50 years in some countries. The rise in testicular and prostate cancer is partially attributed to endocrine disruptors, as is the increase of male reproductive deformities. It has been estimated that there are roughly 87,000 synthetic chemicals in our environment. A serious effort is underway to determine how many of those chemicals are endocrine disruptors.

of pubic and axillary hairs and increased development of breasts and reproductive genitalia. They also stimulate deposition of fat in the breasts and thighs.

After the egg has been released from its surrounding sphere of cells, the follicle is converted into an endocrine structure that produces progesterone. Essentially progesterone maintains the uterus in its prepared state by keeping the lining of the uterus thick and ready to support a fetus. It also prevents contraction of the smooth muscle in the wall of the uterus until the birthing process begins. The functions of the reproductive hormones are discussed in more detail in Chapter 13.

Pineal Gland

The pineal gland is found deep in the center of the brain in the diencephalon (Figure 7.18). It releases melatonin, a hormone that regulates the daily cycles known as *circadian rhythms*. There are chemicals in the brain referred to as *clock proteins* that must be set every day. It is the function of melatonin to "'set the clock" by the periods of daylight and dark. Some individuals are especially sensitive to the decreased daylight of the winter months because of decreased levels of melatonin. It may cause mood changes and decreased clarity of thinking. Individuals working on "'swing shifts" who work every other week on day shift and work evenings the other weeks often experience continual tiredness, difficulty concentrating, and increased health problems because their biological clock is unable to establish regular settings. Those who travel also find increased tiredness after crossing several time zones in a single day. Their normal day/night cycle, and therefore their melatonin secretion, is disrupted, and they experience a condition known as *jet lag*. For this reason, some individuals have found taking melatonin supplements while changing work schedules or traveling minimizes the effects of changes in circadian rhythms.

> **Progesterone** produced by the corpus luteum in the ovary. It maintains the uterus in its prepared state by keeping the lining of the uterus thick and ready to support a fetus. It also prevents the contraction of the smooth muscle in the wall of the uterus until the birthing process begins.

> **Pineal gland** located deep in the center of the brain, this gland acts as a biological clock, establishing daily cycles known as circadian rhythms.

> **Melatonin** produced by the pineal gland. It synchronizes the body's cycles with periods of daylight and dark.

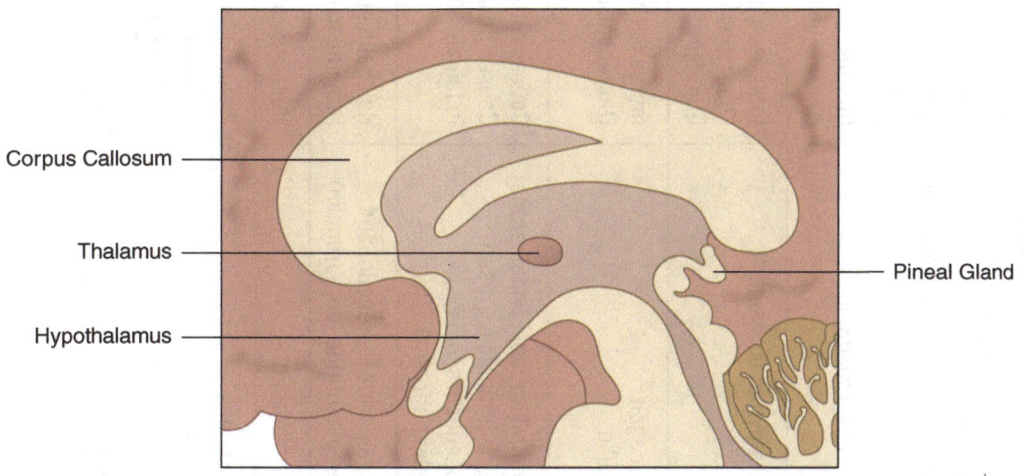

Figure 7.18 Pineal Gland. The pineal gland is located between the cerebrum and cerebellum posterior to the brain stem.

Comprehension Check-up:

1. Steroids produced by the adrenal gland are referred to as _____.
2. The hormones that prepare the woman's body to support a developing fetus are _____.

1. corticoids 2. estrogens

Chapter 7 The Endocrine System

Endocrine Gland	Hormone	Effect
Posterior pituitary gland	Antidiuretic hormone Oxytocin	Causes additional water retention by the kidneys to dilute plasma when it becomes too concentrated. Causes uterine contraction and milk release.
Anterior pituitary gland	Growth hormone Thyroid-stimulating hormone Follicle-stimulating hormone Lutenizing hormone (interstitial cell-stimulating hormone) Prolactin Adrenocorticotropic hormone	Causes body development and increased metabolism, increases the release of T_3 and T_4 from the thyroid gland. Female—Causes the ovary to prepare to release another ovum. Male—Stimulates sperm production in the testes. Female—Causes ovulation and formation of the corpus luteum. Male—increases testosterone secretion. Causes milk production. increases the production of cortisol in response to stress.
Thyroid gland	T_3 and T_4 Calcitonin	Increases basal metabolic rate; promotes development of skeletal, muscular, and nervous systems. Decreases blood calcium by causing the excess to be converted into bone.
Parathyroid gland	Parathyroid hormone	Increases blood calcium by causing (1) the breakdown of bone to add bone calcium to the blood, (2) increased absorption of calcium from the digestive system, (3) decreased calcium loss through urination.
Endocrine pancreas Alpha cells Beta cells	Glucagon insulin	Increases blood glucose by causing the release of stored glucose into the bloodstream. Decreases blood glucose by facilitating the transport of excess blood glucose into storage.
Adrenal cortex	Mineralcorticoid (aldosterone) Glucocorticoid (cortisol) Gonadocorticoids (dehydroepiandrosterone)	Causes retention of additional sodium and excretion of potassium. Causes an increase in the availability of glucose, lipids, and amino acids during periods of stress. In females this hormone acts like testosterone to increase sex drive.
Adrenal medulla	Epinephrine (adrenalin) and norepinephrine (noradrenalin)	Released in conjunction with the sympathetic nervous system to increase blood pressure and heart rate and to increase blood flow to muscle for an extended period.
Testes in the male	Testosterone	Responsible for sex drive and male physical characteristics.
Ovaries in the female	Estrogens Progesterone	Responsible for preparing the female's body to receive and support a fertilized ovum. Inhibits uterine contraction and maintains the viability of the functional layer of the uterus throughout pregnancy.
Pineal gland	Melatonin	Establishes circadian rhythms (daily cycles) in the body.

Homeostasis—Holding in Balance

The endocrine system plays an enormous role in maintaining homeostasis. The following is a list of parameters controlled by endocrine glands. (Not all of these control systems have been discussed yet.)

- Water balance-controlled by the hypothalamus/pituitary gland
- Growth-controlled by the hypothalamus/pituitary gland
- Basal metabolic rate-controlled by the hypothalamus/pituitary gland and thyroid gland
- Blood glucose-controlled by the hypothalamus/pituitary gland, adrenal glands, and endocrine pancreas
- Development-controlled by the hypothalamus/pituitary gland and testes or ovaries
- Reproduction-controlled by the hypothalamus/pituitary gland, testes or ovaries, and adrenal gland
- Blood calcium-controlled by the thyroid and parathyroid glands
- Plasma volume-controlled by the adrenal gland, kidneys, and heart
- Red blood cell production-controlled by the kidneys
- Digestive secretions-controlled by the stomach and duodenum

Each endocrine gland provides a vital factor in maintaining homeostasis. If a single endocrine gland malfunctions, medical intervention must be provided or severe consequences will result.

There Is No Substitute for Exercise

Regular, sustained exercise improves the ability of the endocrine system to maintain homeostasis of many parameters in the body. For example, exercise increases basal metabolic rate, allowing us to become or remain trim. The resulting decreased level of visceral fat allows our endocrine system to more efficiently control blood glucose. Increased visceral fat resulting from a lack of exercise and a failure to burn excess calories often results in insulin resistance, leading to heart disease and stroke.

Exercise causes additional deposits of calcium in bone to strengthen them. It is especially important that children maintain a high level of activity as bones grow and develop.

Clinical Scenario—Hypoglycemia

A 42-year-old semiconscious male is experiencing a seizure. The following were observed:

Chief Complaint (CC) = Dizzy, Semiconscious

Level of Consciousness (LOC—AVPU) = Patient responds to pain on the AVPU scale.

Glasgow Coma Scale (GCS) = 8

A—Airway = Open

B—Breathing = Respiratory Rate (RR) = 38

C—Circulation = Cool and clammy, Heart Rate (HR) = 140, Blood Pressure (BP) = 80/50

A—Allergies = none

M—Medications = none

(continued)

P—Past History = none
L—Last oral intake = none
E—Events = Working in his garden then passed out.
Physical Exam (PE) = Skin cool and clammy, no eye response, incomprehensible words, localized response to pain, no sign of trauma
Pulse oximetry (SP02) = 88
Blood Glucose Level (BGL) = 37
Temperature = 96°F

In this scenario, first take note of the objective vital signs such as the heart rate, respiratory rate, blood pressure, and temperature. Are these values higher or lower than normal? It will be important in this course to know normal vital signs. Note objective signs such as pulse oximetry and blood glucose.

Next, determine which of the 11 body systems are most affected. What things stand out so that there appears to be a constellation of signs and symptoms that point to a specific issue? Then test your hypothesis.

Note that the heart rate and the respiratory rate are fast indicating a sympathetic (fight or flight) response. Blood pressure, blood glucose, pulse oximetry, and temperature are low. The cardiovascular, respiratory, nervous, and endocrine systems are all affected. Many of these signs appear to be some kind of shock. Shock is a lack of blood perfusion to the vital organs. Is there some kind of treatment that could reverse some of these trends?

Two interventions that could be applied immediately are to provide oxygen and some kind of sugar. Sugar in the form of a paste applied to the inside of the mouth below the tongue if the patient s conscious. If the patient is unconscious, then an IV route will provide dextrose, a kind of sugar. As it turns out, our patient improved greatly after treatment with sugar. After oral glucose gel was administered, the patient became fully conscious and was able to communicate to the Health Care workers. The other vital signs soon normalized as well. Our patient, with a history of diabetes mellitus, was found to be suffering from hypoglycemia, low blood sugar. Normal values for blood glucose are between 70 and 110 mg per dl.

Contributed by David Flint. © Kendall Hunt Publishing Company

CLINICAL TERMS TO REVIEW

Pituitary gland, antidiuretic hormone (ADH), oxytocin
Human growth hormone (HGH), thyroid-stimulating hormone (TSH), follicle-stimulating hormone (FSH), luteinizing hormone (LH), prolactin (PRL), adrenocorticotropic hormone (ACTH)

Thyroid gland, parathyroid gland
Pancreatic islets, alpha cells, beta cells, insulin, glucagon
Adrenal cortex—mineralocorticoids, glucocorticoids, gonado-corticoids
Adrenal medulla

Testes, ovary, progesterone, testosterone
Pineal gland, melatonin, circadian rhythms
Blood glucose testing, glucometer, hypoglycemia, hyperglycemia
Level of consciousness (LOC)

Test Yourself

Choose the best answer to the following multiple choice questions:

1. Hormones that cannot pass through the plasma membrane but must attach to receptor sites on the outside of the cell are
   a. hydrophilic.
   b. multipolar.
   c. lipophilic.
   d. hydrophobic.

2. Control of most hormone levels in the bloodstream is by
   a. positive feedback.
   b. response to blood pressure.
   c. complexes formed after intake of vitamins.
   d. negative feedback.

3. The relationship between a hormone and its receptor site is often compared to a
   a. lock and key.
   b. bat and ball.
   c. hand and glove.
   d. chisel and hammer.

4. The ability of a hormone to affect a target cell depends on its
   a. size.
   b. shape.
   c. color.
   d. ability to bend.

5. In order for the anterior lobe of the pituitary gland to release its hormones, it must
   a. receive conscious decisions to do so from the frontal lobe of the cerebrum.
   b. wait for the body temperature to spike to signal release.
   c. receive releasing hormones from the hypothalamus.
   d. strip extra atoms off the basic steroid molecule.

6. In order to be able to absorb calcium into the bloodstream, we need vitamin
   a. A.
   b. B.
   c. C.
   d. D.

7. Why do the mammary glands require two hormones for milk secretion?
   a. One causes milk to be produced, and the other causes its release in response to nursing.
   b. One is hydrophilic and adds liquid to milk, whereas the other is lipophilic and adds milk fat.
   c. One affects the mother, causing her to produce and release milk, whereas the other affects the baby, increasing his or her desire to nurse.
   d. One hormone is produced to allow nursing of one child to continue during pregnancy of another, whereas the other hormone is produced only when the woman is not pregnant.

8. An endocrine gland is defined as a
   a. spherical gland that contains fluid for the production of steroids.
   b. gland that secretes its chemical messages through a duct into the target organ.
   c. ductless gland that secretes hormones directly into the bloodstream.
   d. group of cells that relies on the rhythmic pulsations of the bloodstream to cause release of its secretions out of the organ in which it resides.

9. Which endocrine gland is concerned with the control of blood glucose?
   a. posterior pituitary gland
   b. anterior pituitary gland
   c. endocrine pancreas
   d. thyroid gland

10. Which endocrine gland is concerned with growth, metabolic rate, and reproduction?
    a. posterior pituitary gland
    b. anterior pituitary gland
    c. endocrine pancreas
    d. thyroid gland

The Cardiovascular System and Blood

Chapter 8

LEARNING OBJECTIVES

Upon completion of this chapter, you will be able to:

1. Describe the components and functions of the cardiovascular system and how it maintains homeostasis.
2. Describe the anatomy of the heart and discuss the cardiac cycle and conduction system.
3. Compare and contrast the anatomical and functional differences among blood vessels in the body and describe how they supply blood to and from the heart.
4. Distinguish between pulmonary, systemic, portal, and fetal circulation.
5. Describe the characteristics and functions of blood and how it helps maintain homeostasis.
6. List and describe each component of blood.
7. Explain the ABO blood typing system and the Rh factor.

CHAPTER OUTLINE

Introduction

The Cardiovascular System
- Structure of the Cardiovascular System
- Overview of the Cardiovascular Function

The Heart
- Location and Surface Anatomy
- Layers of the Heart Wall
- Pericardium
- Heart Chambers
- Major Blood Vessels Associated with the Heart
- Heart Valves
- Blood Supply to the Heart
- Conduction System of the Heart
- Electrocardiogram
- Function of the Heart and Cardiac Cycle

Regulation of Heart Function

Blood Vessels
- Types of Blood Vessels
- Anatomy of Blood Vessels
- Differences among Blood Vessels
- Arteries
- Major Arteries
- Veins
- Major Veins
- Capillaries

The Circulation
- The Pulmonary Circulation
- The Systematic Circulation
- Venous Return
- Portal Circulation
- The Fetal Circulation

CHAPTER OUTLINE (CONTINUED)

Functions of Blood

Composition of Blood

- Erythrocytes (Red Blood Cells)
- Leukocytes (White Blood Cells)
- Platelets (Thrombocytes)
- Blood Cell Formation
- Complete Blood Cell Count (CBC)
- Plasma

Hemostasis

- Vascular Spasm
- Formation of a Platelet Plug
- Coagulation
- Inappropriate Clotting

Blood Types

- ABO System

INTRODUCTION

If you have reached this far in Anatomy and Physiology, give yourself a pat on your back. You're halfway through this course. You will be learning in this chapter the cardiovascular system. This is the center of clinical anatomy and physiology because it deals with circulation, heart, and blood.

The cardiovascular system performs a vital function because it carries oxygen and nutrients that nourish the different parts of the human body. It also removes the metabolic waste products from the body and carries them to specific body organs for elimination, such as the lungs, liver, and kidneys. Circulation or the movement of the blood throughout the body is made possible by the blood vessels such as the veins and arteries. The cardiovascular system also helps regulate the body temperature and maintains fluid balance throughout the body.

You will also learn about the composition of blood, especially the cellular elements. White blood cells attack foreign invaders such as bacteria and viruses. The central organ of the cardiovascular system is the heart. The fetal heart beat starts around 5 weeks of pregnancy, and can be heard or detected by ultrasound around 8 weeks of pregnancy. The human heart beats roughly over 3 billion times if a person lives until 80 years old.

Cardiovascular exercise is important to keep the heart healthy and fit. A healthy lifestyle includes a healthy diet, exercise, and staying away from smoking and heavy alcoholic drinks.

© Tefi/Shutterstock.com

Figure 8.1

Figure 8.2

THE CARDIOVASCULAR SYSTEM

The cardiovascular system provides a means to transport substances to virtually every cell in the body. The process is similar to our use of rivers to move items from one location to another. A large variety of items can be loaded onto a ship, barge, or container and sent down the river to another location. In the same way, substances are transported to the appropriate locations anywhere in the body—only the cardiovascular system is a continuous loop.

Structure of the Cardiovascular System

The cardiovascular system consists of a pump, the heart, and blood vessels that carry blood throughout the body and vessels through which blood passes. Arteries transport blood away from the heart. Arteries branch into capillaries, which are so thin that nutrients, gases, and wastes can diffuse through their walls between the blood and the surrounding cells. Blood from the capillaries drains into veins, which return blood to the heart. There are two divisions of the cardiovascular system: the systemic circulation, which transports blood to all of the body except the lungs, and the pulmonary circulation, which moves blood to the lungs (Figure 8.3). Associated with the cardiovascular system is the lymphatic system, which drains excess tissue fluid passing out of the capillaries back into the bloodstream. The lymphatic system is discussed in Chapter 10.

Overview of the Cardiovascular Function

The cardiovascular system is responsible for the transport of nutrients, gases, and wastes among cells. It transports hormones from endocrine cells to target cells. Defensive cells can move from one location to another throughout the bloodstream. Heat is conducted from active cells through blood to the skin, where it can be released into the environment.

> **Arteries** vessels that transport blood away from the heart.

> **Capillaries** blood vessels composed of a single layer of squamous epithelium through which nutrients, gases, and wastes exchange between blood and cells throughout the body.

> **Veins** vessels that bring blood back to the heart.

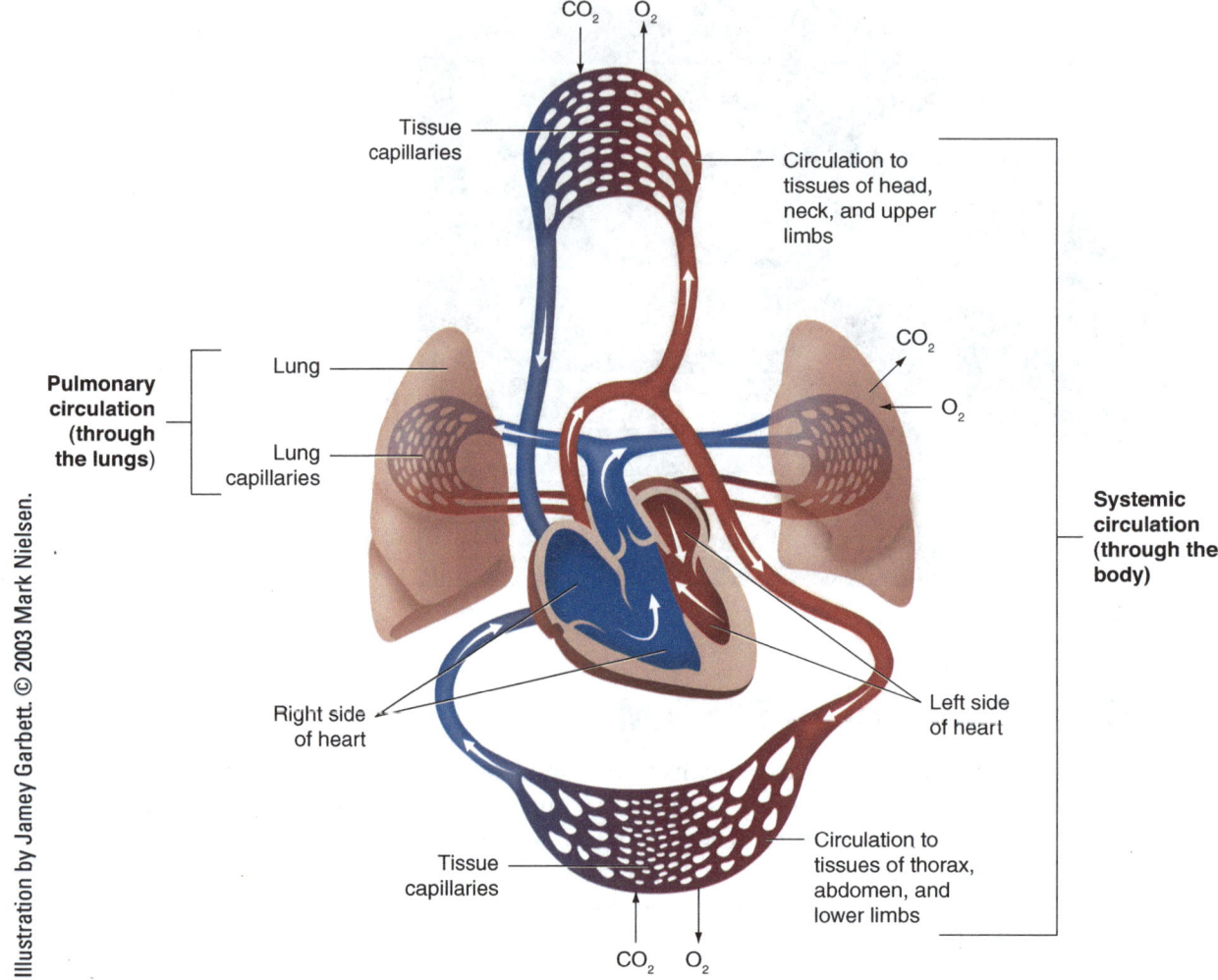

Figure 8.3 General Layout of the Cardiovascular System.

THE HEART

Circulation of blood throughout the body requires the pumping action of the heart. Beating 70 to 80 times per minute, it will pump more than 7,500 liters (about 2,000 gallons) of blood per day and will beat more than 100,000 times every 24 hours, pushing blood through roughly 60,000 miles of blood vessels.

Location and Surface Anatomy

The heart is located deep in the thoracic cavity in an area known as the mediastinum. This section in the center of the thoracic cavity contains all other structures found in the thoracic cavity other than the lungs. The heart is primarily located between the sternum and the thoracic vertebrae, with its inferior edge resting on the diaphragm. The ventricles come to a point, the apex, which lies in the left side of the chest (Figure 8.4). The atria are superior to the ventricles and expand as blood flows into them (Figures 8.5 and 8.6). Because of the comparatively thin walls of the atria, they appear to be much smaller than the ventricles even though they eject the same volume of blood. The ventricles, which pump blood to maintain circulation throughout the body, are the bulk of the heart.

> **Mediastinum** the section of the thoracic cavity between the two lung cavities. It contains all organs found in the thoracic cavity other than the lungs.

> **Apex** the ventricles of the heart form a point known as the *apex*.

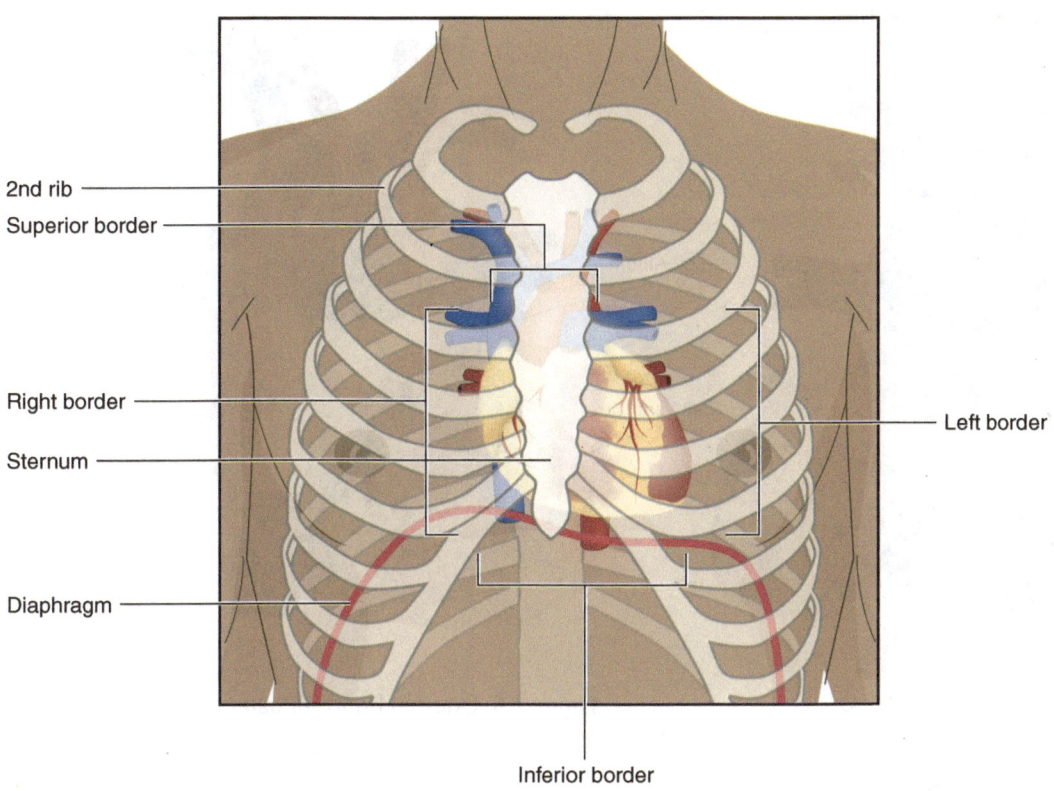
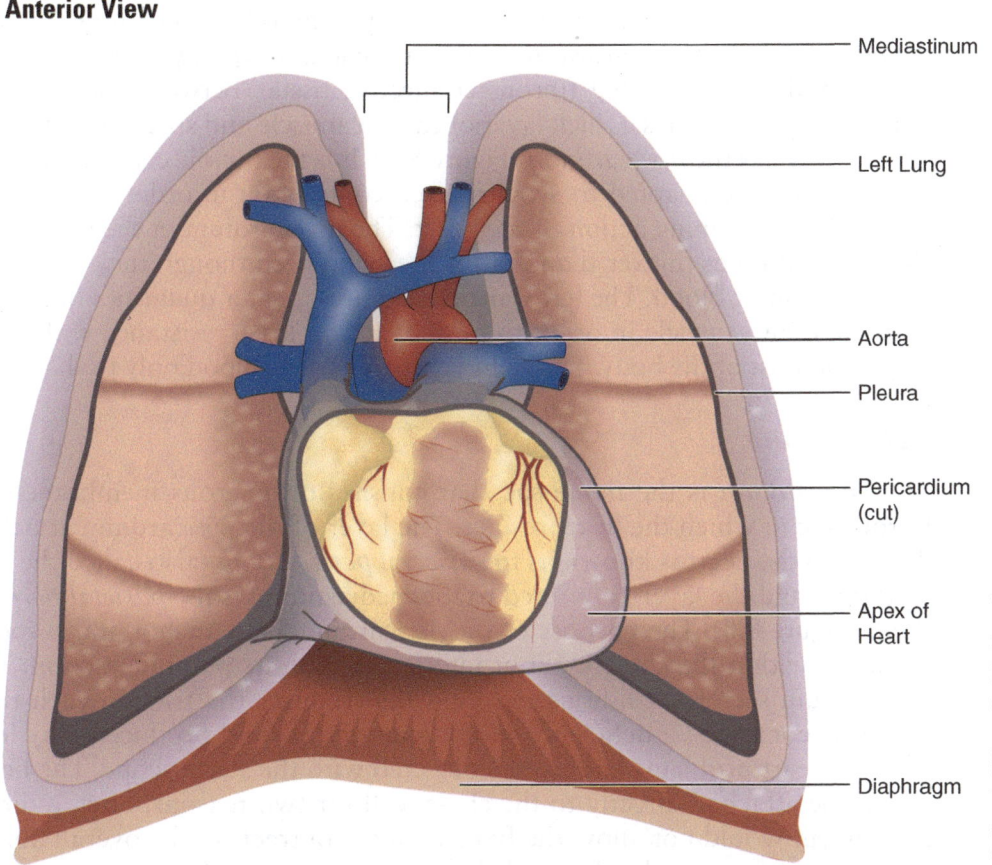

Figure 8.4 Placement of the Heart in the Thoracic Cavity. The heart is located slightly to the left of the center of the chest between the two lungs.

Illustration by Jamey Garbett. © 2003 Mark Nielsen.

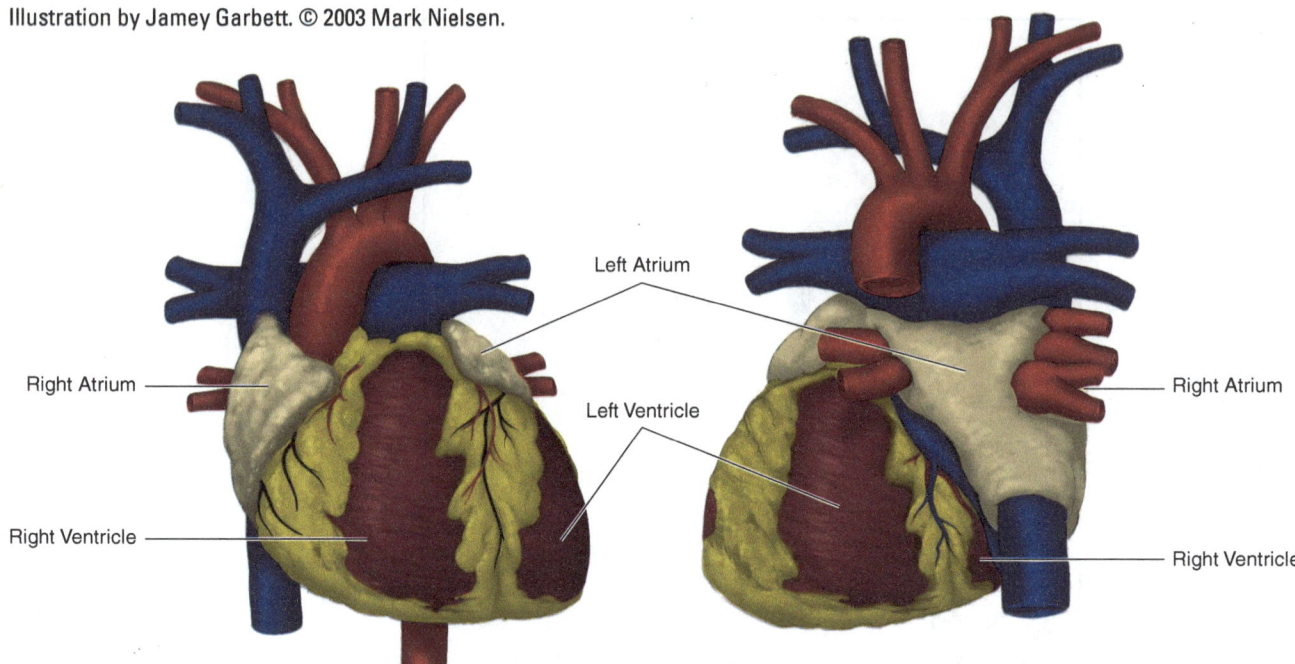

Figure 8.5 External Heart—Anterior View.

Figure 8.6 External Heart—Posterior View.

Layers of the Heart Wall

The interior surface of the heart is covered by a smooth epithelium known as the *endocardium*. The *epicardium* is the epithelium covering the heart's outer surface and the underlying fibrous connective tissue. Between the endocardium and epicardium is a thick layer of cardiac muscle and special conduction fibers known as the *myocardium* (Figure 8.7). It is the muscle cells of the myocardium that are adapted to synchronized contraction in order to maintain the rhythmic pumping action of the heart. The myocardium of the left ventricle is several times thicker than that of the right, even though they pump the same amount of blood. The left ventricle requires greater quantities of muscle than the right ventricle in order to push against a large resistance within the organs of most of the body. The right ventricle pushes blood only to the lungs.

Pericardium

The *pericardium* is the tissue sac that consists of a serous membrane and fibrous sac in which the heart resides. The layer of the pericardium attached directly to the heart is known as the *visceral pericardium* (also called the *epicardium*), and the layer connected to the mediastinum is the *parietal pericardium* (Figure 8.8). The purpose of the pericardium is to anchor and suspend the heart within the chest without interfering with its pumping action. The pericardium secretes a lubricating fluid that reduces the friction each time the heart contracts; a deficiency in the quantity of fluid produced may lead to some heart problems and can be detected with the stethoscope. The heart cannot be attached directly to the chest wall for two reasons. First, a direct attachment would not allow the fixed area to contract, so the overall ability of the heart to pump blood would be decreased. Second, if the heart was suddenly jarred, for example, by someone landing from a jump or from falling, the rapid displacement of the heart could cause tearing, with potentially fatal results. By placing the heart within this tissue sac, the pericardium can act as a shock absorber as well as an attachment that allows free movement.

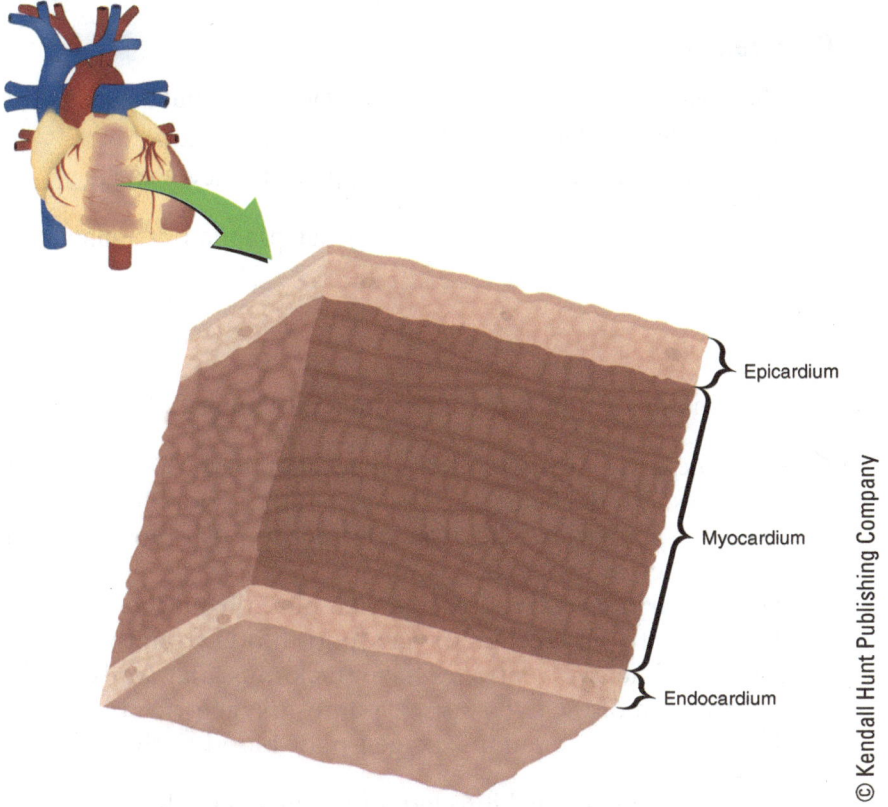

Figure 8.7 Layers of the Heart. The heart consists of three layers: endocardium (inner lining), myocardium (cardiac muscle and conducting fibers), and epicardium (outer covering).

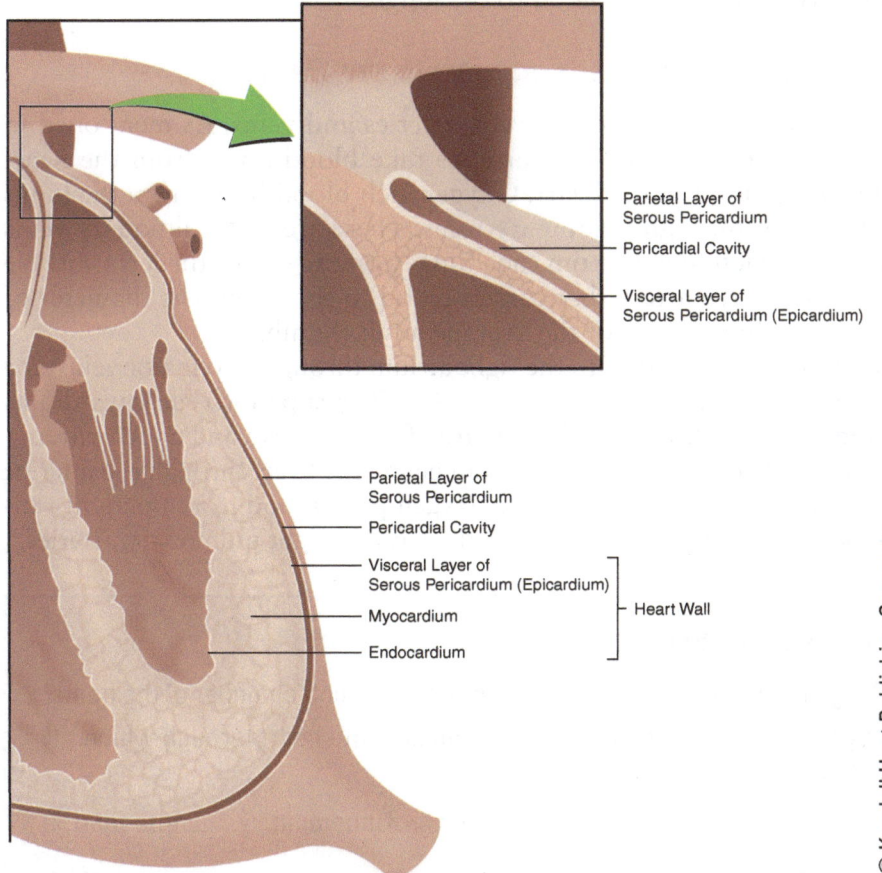

Figure 8.8 Pericardium.

Left atrium the chamber of the heart that receives oxygen-rich blood from the lungs.

Right atrium the chamber of the heart that receives oxygen-poor blood from the body.

Left ventricle the chamber of the heart which pumps oxygen-rich blood to the body.

Right ventricle the chamber of the heart which pumps oxygen-poor blood to the lungs.

Septum the wall that separates the left and right sides of the heart.

Pulmonary veins vessels that bring oxygen-rich blood from the lungs to the left atrium.

Aorta the artery that carries oxygen-rich blood from the left ventricle to the body.

Superior vena cava large vein that returns oxygen-poor blood from the upper half of the body to the right atrium.

Inferior vena cava large vein that returns oxygen-poor blood from the lower half of the body to the right atrium.

Pulmonary trunk (artery) the artery that carries oxygen-poor blood from the right ventricle to the lungs.

> **Comprehension Check-up:**
> 1. The area of the thoracic cavity where everything but the lungs, including the heart, can be found is known as the _____.
> 2. The tissue sac that holds the heart in place is known as the _____.
>
> 1. mediastinum   2. pericardium

Heart Chambers

The heart consists of four separate chambers (Figure 8.9).

- Atria—The atria are two thin-walled chambers that receive blood. As blood flows into the heart, the atria expand to pool blood until it is ready to enter the ventricles. The atria contract to push blood into the ventricles while they are relaxed.
  - The left atrium receives oxygen-rich blood from the lungs.
  - The right atrium receives oxygen-poor blood from the body.
- Ventricles—The two ventricles have thick muscular walls for pumping
  - Contraction of the left ventricle pumps oxygen-rich blood to the body.
  - The right ventricle pumps oxygen-poor blood to the lungs.

The heart can be divided into left and right sides. The separating wall between the two sides of the heart is known as the septum. The left side pumps oxygen-rich blood throughout the body, except the lungs, whereas the right side transports oxygen-poor blood to the lungs.

Major Blood Vessels Associated with the Heart

It is important to differentiate between arteries and veins. As mentioned in the previous section (Figure 8.9), arteries take blood away from the heart and veins bring blood to the heart. Oxygen-rich blood flows from the lungs to the left atrium through pulmonary veins. There are two pulmonary veins draining into the left atrium from each lung. Coming out of the left ventricle is the aorta, the large artery that carries blood to the body. The diameter of the aorta is roughly the size of the average adult thumb.

Oxygen-poor blood flows to the right atrium through two major veins: the superior vena cava and the inferior vena cava. The superior vena cava drains blood from above the heart, that is, primarily from the head and upper extremities. Blood returns to the right atrium from below the heart through the inferior vena cava. The pulmonary trunk takes oxygen-poor blood from the right ventricle to the left and right lungs through the left and right pulmonary arteries.

> **Comprehension Check-up:**
> 1. Oxygen-poor blood from the body returns to which chamber of the heart?
> 2. Oxygen-rich blood returns from the lungs to the heart through which vessels?
>
> 1. right atrium   2. pulmonary veins

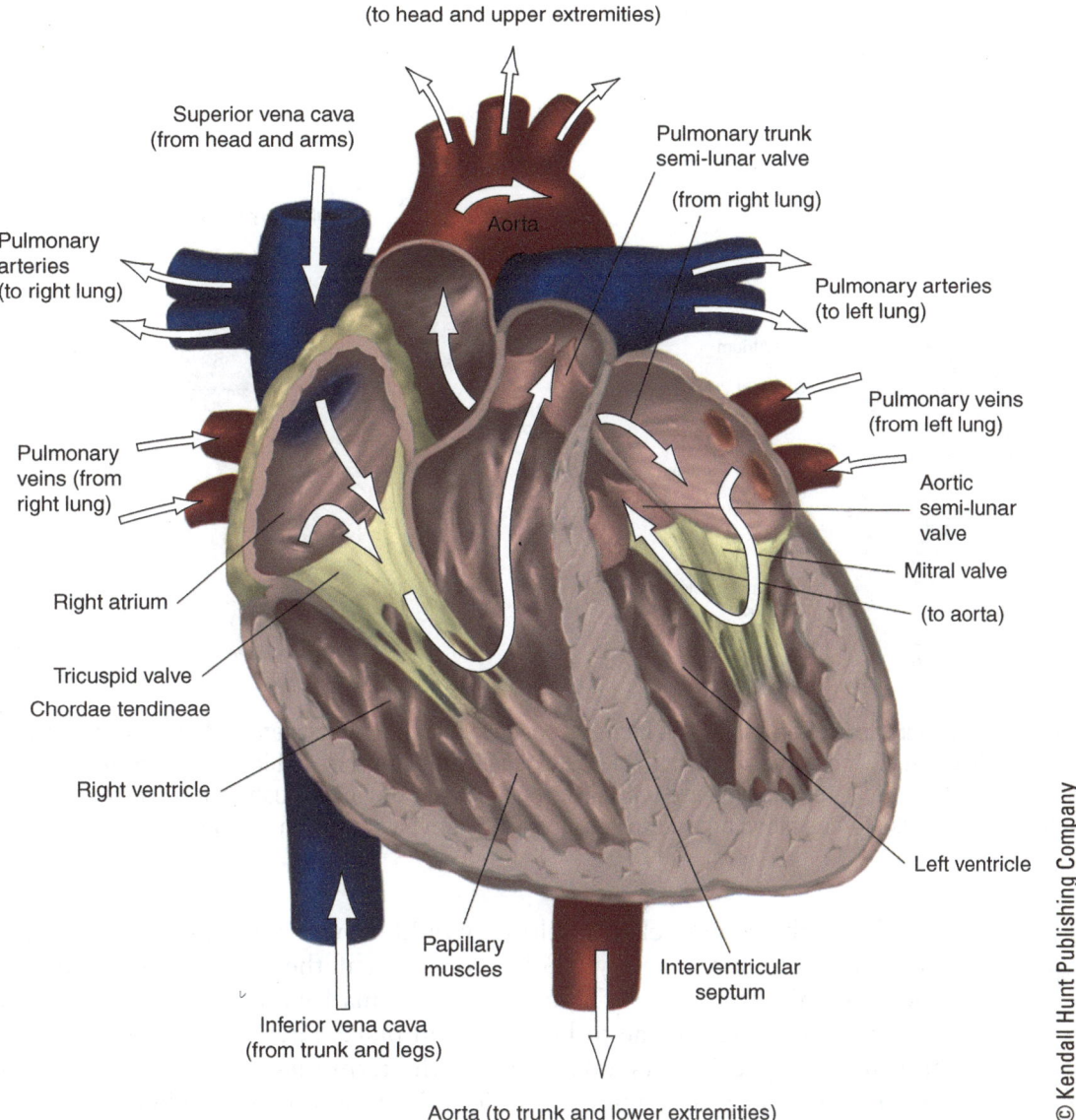

Figure 8.9 Internal Structures of the Heart.

Heart Valves

The atrioventricular valves are one-way valves that allow blood to flow from each atrium into only its corresponding ventricle. As blood from the contracting ventricle is pushed back toward the atrium, it pushes on the underside of the valve, causing it to billow out like a parachute. The valve is actually composed of two or three parachute-like leaflets that billow out to meet across the opening between the atrium and ventricle to block the flow of blood (Figure 8.10). As a result, the only exit for blood from the ventricles is through the aorta on the left and through the pulmonary trunk on the right. The valve between the left atrium and left ventricle is known as the bicuspid (mitral) valve. It has two leaflets. Between the right atrium and right ventricle is the tricuspid valve, which has three leaflets. These valves are so large that they have connective tissue cords, known as *chordae tendineae*, holding their leaflets in place, much like the cords of a parachute. If these chordae tendineae rupture, the valve will not close tightly and may lead to blood reflux into the atrium.

Bicuspid (mitral) valve the two-cusp one-way heart valve between the left atrium and left ventricle.

Tricuspid valve the three-cusp one-way heart valve between the right atrium and right ventricle.

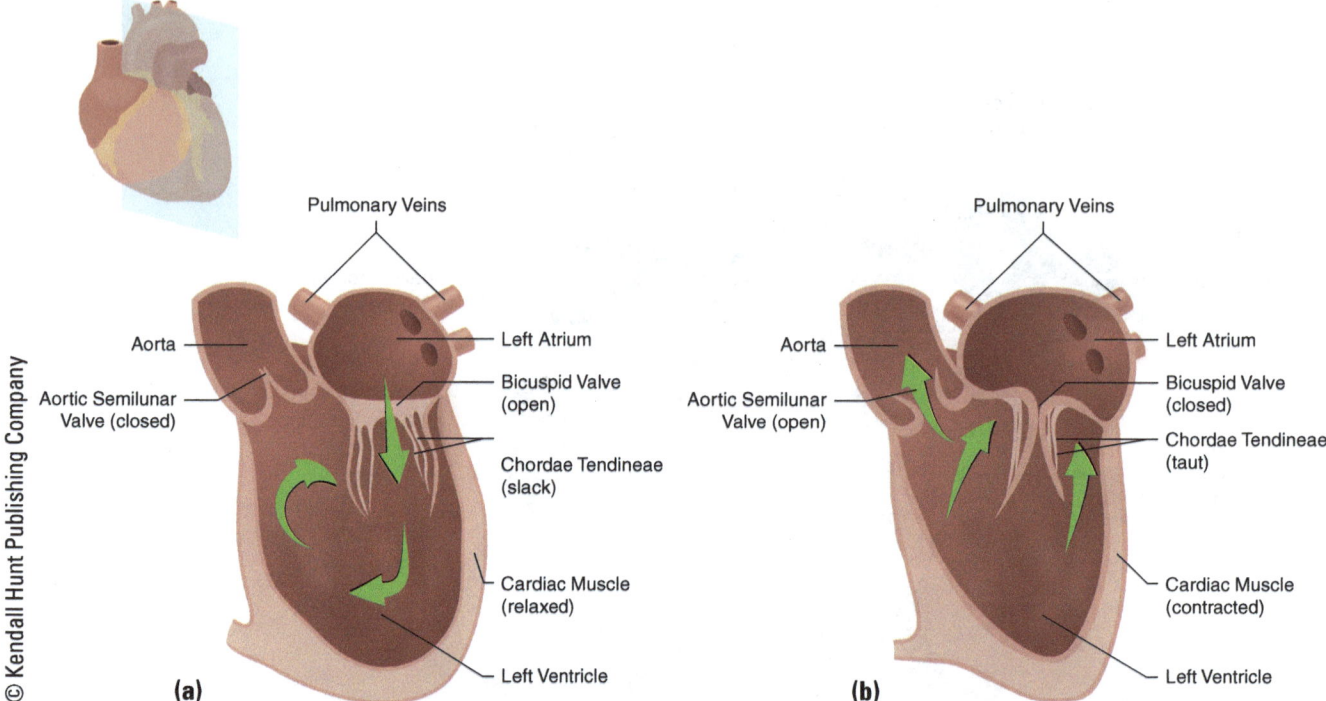

Figure 8.10 Movement of Atrioventricular Valves. (a) The bicuspid valve is open during relaxation of the left ventricle as blood flows in from the left atrium. The aortic semilunar valve is closed by the pressure of blood in the aorta so that it cannot backflow into the left ventricle. (b) The bicuspid valve is forced closed as blood is pushed against it by the contracting left ventricle, which also forces blood against the aortic semilunar valve, causing it to open.

> **Aortic semilunar valve** the one-way heart valve located in the outflow of the left ventricle leading to the aorta to prevent backflow into the ventricle. It is composed of three half-moonshaped cusps.

> **Pulmonary semilunar valve** this one-way heart valve is composed of half-moon cusps and controls direction of blood flow from the right ventricle into the pulmonary trunk.

> **Coronary arteries** vessels that carry oxygen-rich blood to the myocardium when it is relaxed.

When the heart relaxes, blood would flow back into the ventricles from the pulmonary artery/trunk and aorta, if there were not additional one-way valves in place. These valves are small enough that they do not require chordae tendineae to hold them in place. Each valve possesses three half-moonshaped leaflets, giving rise to the term *semilunar valves*, that seal the arteries to prevent backflow of blood into the ventricles when they relax. There is an aortic semilunar valve and a pulmonary semilunar valve.

When the heart valves close, sound waves created can be heard through a stethoscope placed on the patient's chest on either side of the superior and inferior ends of the sternum (Figure 8.11). There are two sounds heard close together. The first sound is created by the closing of the mitral and tricuspid valves as the ventricles begin their contractions. The second sound is made by the closing of the semilunar valves on the aorta and pulmonary trunk as the ventricles relax. Contraction of the ventricles occurs between those two sounds. After the second sound has occurred, the heart is relaxed.

Blood Supply to the Heart

Blood needs to flow into ventricular muscle during relaxation because the myocardium is too thick for exchange of substances in the heart chambers by simple diffusion. In order to do so, our heart has arteries on its external surface to carry oxygen-rich high-nutrient blood to the myocardium during relaxation. These arteries are known as coronary arteries (Figures 8.12 and 8.13). They branch off of the aorta as it exits the left ventricle. A left coronary

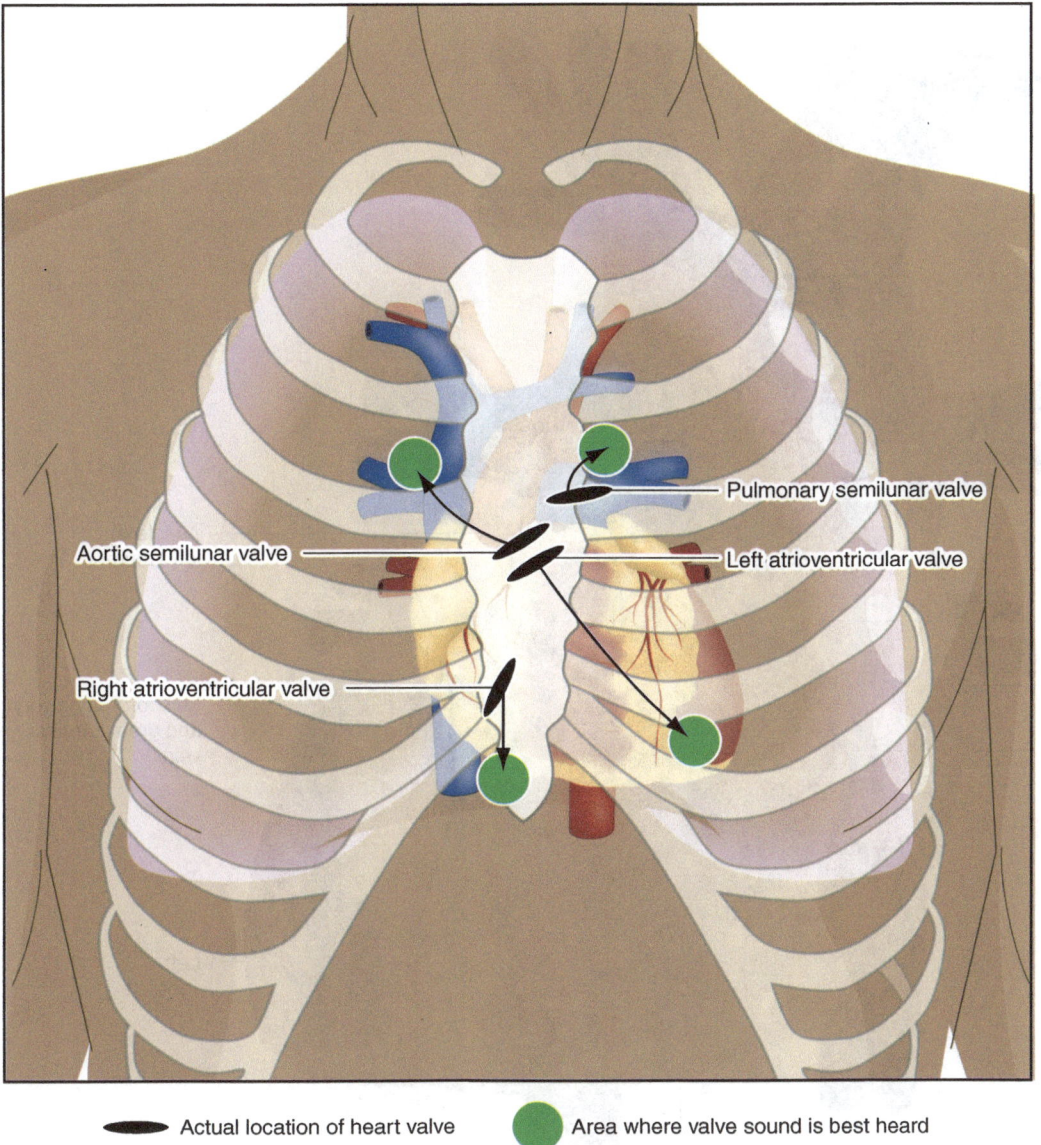

- ● Actual location of heart valve
- ● Area where valve sound is best heard

Figure 8.11 Heart Valves. Location of heart valves and where they are best heard on the anterior chest.

artery supplys blood to the left ventricle and anterior heart, and a right coronary artery supplys the right ventricle and posterior heart with blood. Once the myocardium has exchanged nutrients, gases, and wastes with blood passing through the capillaries, blood flows through coronary veins to be drained through the coronary sinus into the right atrium.

Conduction System of the Heart

Cardiac muscle has the ability to cause itself to contract. It does not wait for stimulation from the brain like skeletal muscle does. Within the heart are specialized cardiac muscle cells that create and transmit the action potential throughout the heart. Those areas that have the ability to create an action potential are known as *nodes* (Figure 8.14). The node that depolarizes at the fastest rate lies in the myocardium of the right atrium near the superior vena cava and is called the *sinoatrial (SA) node*. It is commonly referred to as

> **Coronary veins** vessels that receive oxygen-poor blood from the myocardium and return it to the right atrium.

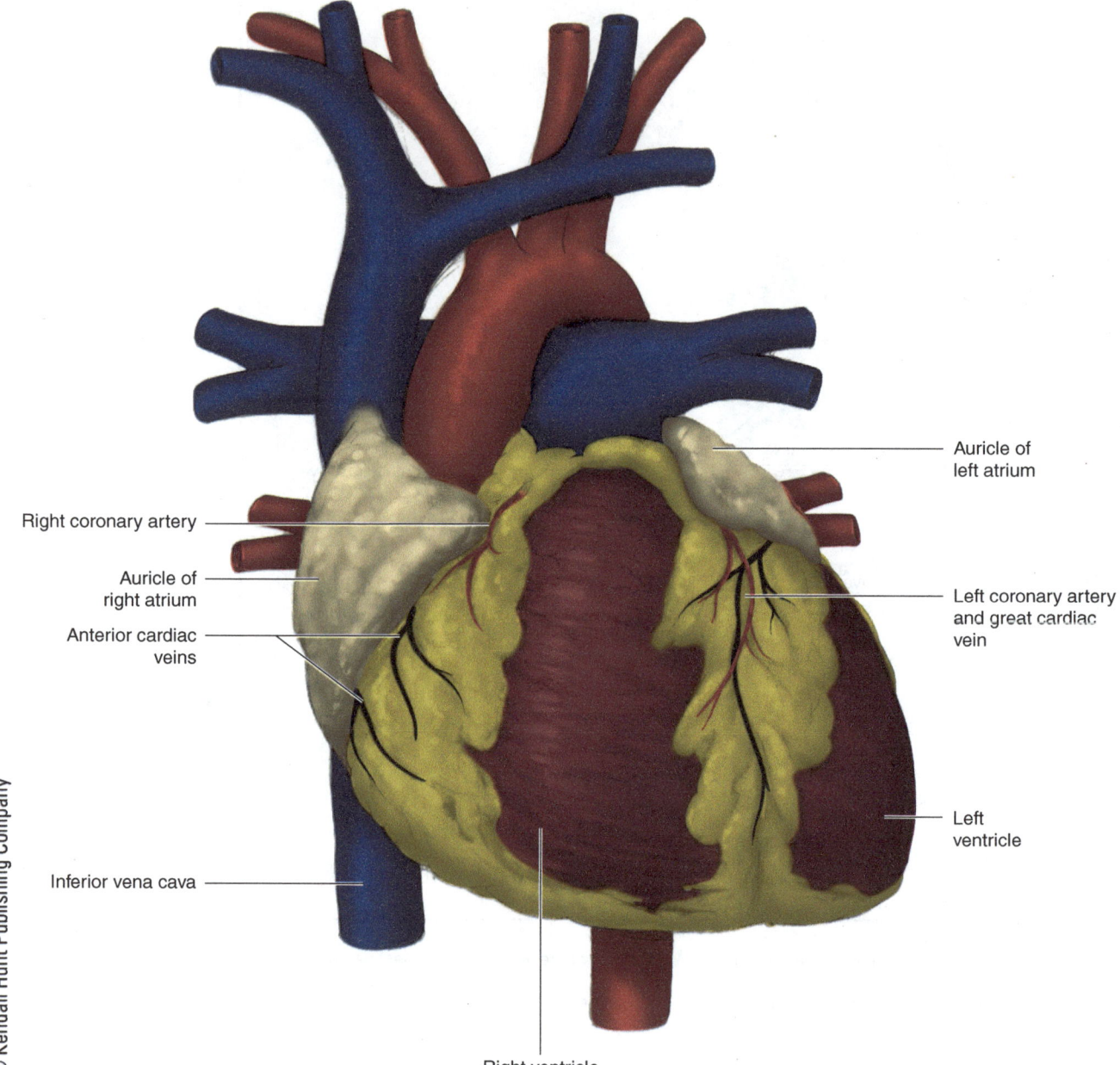

Figure 8.12 Coronary Arteries and Veins on the Anterior Heart.

Why Does External Compression on the Sternum Move Blood through a Non-beating Heart?

In instances when the heart has stopped contracting, pressure on the sternum can squeeze the heart enough to push blood from the ventricles into the arteries. The heart valves ensure the direction of blood flow. As a result, external compression of the heart, as occurs during cardiopulmonary resuscitation (CPR), is able to circulate enough blood through an individual's body to maintain life until medical intervention can potentially cause the heart to again beat spontaneously.

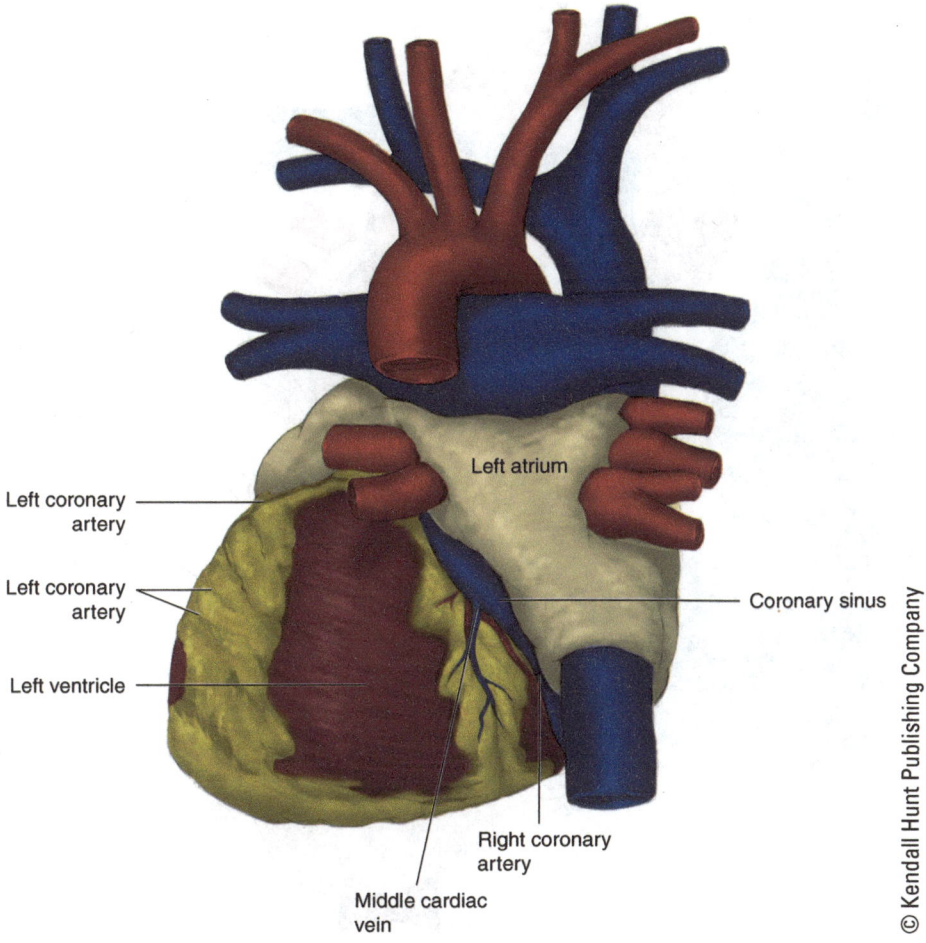

Figure 8.13 Coronary Arteries and Veins on the Posterior Heart.

the *pacemaker of the heart*. Its action potential travels across both atria and through specialized conduction fibers to another node located in the septum of the lower right atrium; this node is called the *atrioventricular (AV) node*. From the AV node the action potential enters the ventricles. The AV node also causes a slight delay in the transmission of the action potential to allow the atria time to contract first and push a majority of the blood into the relaxed ventricles before stimulating their contraction. After the delay has occurred, the action potential travels via the AV bundle to the ventricular septum, forming the Bundle of His. It then splits, and heads toward each ventricle along specialized conduction cells known as *Purkinje fibers*. These specialized conduction cells rapidly carry the stimulus to contract to the ventricular muscle.

Comprehension Check-up:
1. The _____ side of the heart pumps oxygen-poor blood to the lungs.
2. The vessel that carries blood out of the left ventricle is the _____.
3. The heart valve between the left atrium and left ventricle is known as the _____ valve.
4. Arteries that supply blood to the heart muscle during relaxation are known as _____ arteries.

1. right 2. aorta 3. mitral 4. coronary

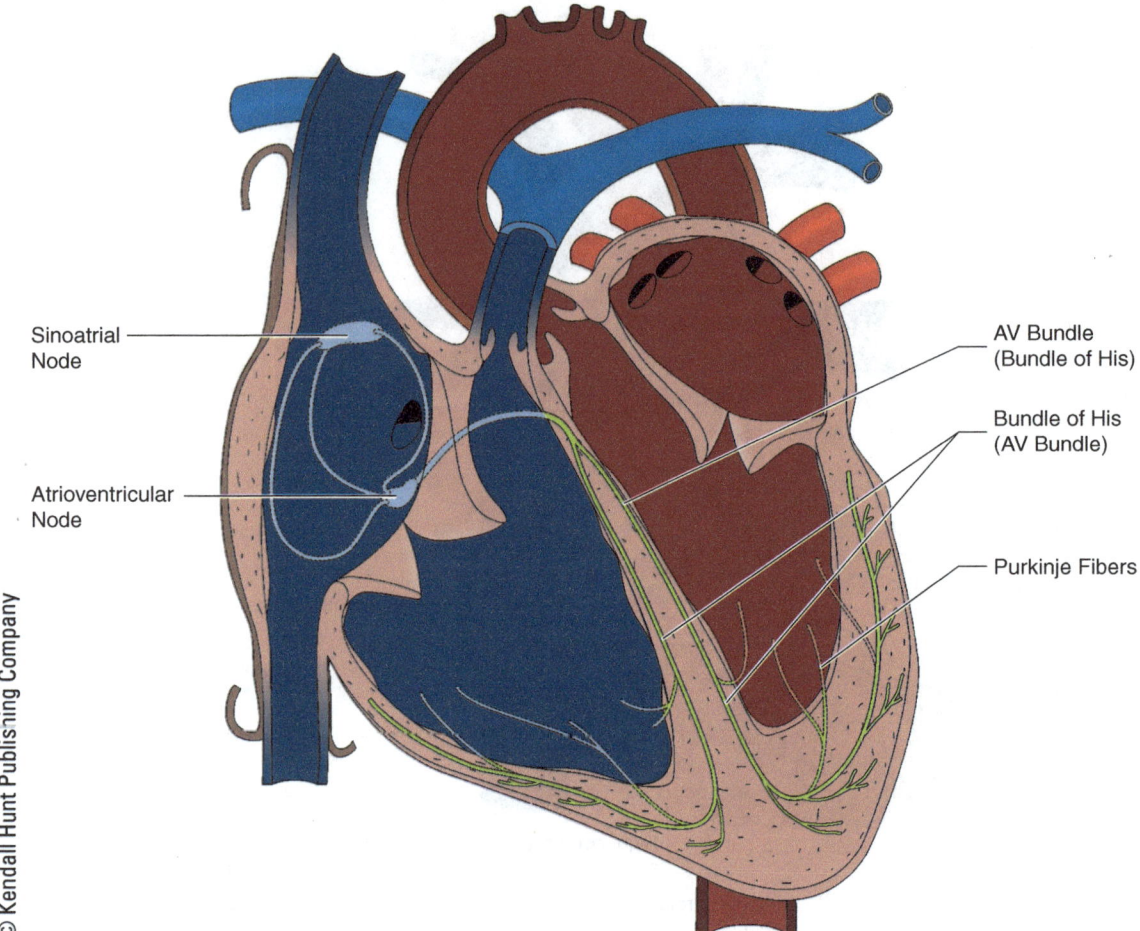

Figure 8.14 Conduction System of the Heart. The cells of the sinoatrial (SA) node depolarize at a faster rate than anywhere else in the heart. The action potential is transmitted from the SA node to the atrioventricular (AV) node, where it enters the ventricles through the AV node. The action potential is delayed until the atria have time to contract. After that time, the action potential is transmitted through the Purkinje fibers to the ventricles to stimulate contraction.

Electrocardiogram

The exchange of ions that occurs during depolarization and repolarization of the heart creates an electrical charge monitored on the surface of the body as an *electrocardiogram (ECG)* (Figure 8.15). Skin electrodes can detect the ECG either on the limbs or chest wall. In some countries *cardio* is spelled

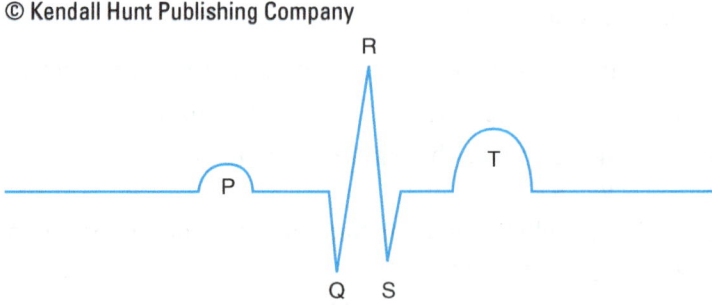

Figure 8.15 Electrocardiogram of One Cardiac Cycle.

with a "k," resulting in the equivalent term *EKG*. The electrocardiogram is composed of three major waves. They are as follows:

- The P wave is a small increase in the electrical potential. It indicates the depolarization of the atria followed by atrial contraction.
- The QRS complex is a large, rapid increase and decrease in the electrical potential signifying the depolarization of the ventricles, resulting in their contraction.
- The T wave is a large, slow increase in electrical potential. It indicates the repolarization of the ventricles, which occurs as the ventricles relax.

By analyzing the electrocardiogram, a physician is able to determine whether there are abnormalities in the conduction of the heart. It is a very useful tool in the diagnosis of heart disease.

> **Comprehension Check-up:**
> 1. The "pacemaker of the heart" is the ____.
> 2. The wave of the electrocardiogram that signifies depolarization of the ventricles is the _____.
>
> 1. sinoatrial node 2. QRS complex

Function of the Heart and Cardiac Cycle

The capability of any pump is determined by the volume of a substance it can move per minute. The heart is also evaluated as a pump when determining cardiac function. The cardiac cycle is the series of events that occur from the start of one heart contraction to the start of the next.

Cardiac Function

To determine the ability of the heart to maintain homeostasis as a pump, it is useful to define some terms involved with the process.

- Ventricular systole is the term used to designate that the ventricles are contracting.
- Ventricular diastole signifies that the ventricles are relaxing.
- *Stroke volume* is the amount of blood ejected by each chamber during one contraction.
- *Cardiac output* is the total amount of blood ejected by each chamber per minute.
- Heart rate is the total number of contractions made by the heart within 1 minute, which is normally approximately 70 to 80 beats per minute.

Stroke Volume

The stroke volume of both sides of the heart is normally the same. If this is not the case, an imbalance of the cardiovascular system occurs. Stroke volume is roughly 70 ml per beat in the average adult while at rest. Determining stroke volume requires some sophisticated equipment and is, if necessary, determined by a cardiologist.

> **Ventricular systole** term used to designate that the ventricles are contracting.
>
> **Ventricular diastole** term used to signify that the ventricles are relaxing.
>
> **Heart rate** total number of contractions made by the heart within 1 minute.

Cardiac Output

Cardiac output is a parameter for determining the ability of the heart to work as a pump. There are two factors whose product determines the cardiac output. The first is stroke volume. Once the amount of blood ejected per heartbeat has been determined, one can calculate the total volume ejected per minute, which depends on the number of contractions that occur (Cardiac Output = Stroke Volume × Heart Rate). This measurement is easily obtained by counting the number of pulses felt in the wrist in 1 minute. The normal, relaxed adult heart pumps the equivalent of all of our blood volume every minute, approximately 5,000 ml. For example, if an individual has a stroke volume of 70 ml per beat with a heart rate of 72 beats per minute, that person's cardiac output is (70 ml/beat) × (72 beats/min) = 5,040 ml/min.

Cardiac Cycle

The *cardiac cycle* (Figures 8.16 and 8.17) begins when the heart is at rest. All four chambers are relaxed. The mitral and tricuspid valves are open, and both semilunar valves are closed. The cardiac cycle is as follows:

1. The SA node creates an action potential.
2. The action potential spreads across the atria. This is seen as the creation of a P wave on the EKG.
3. The atria contract, pushing blood into the relaxed ventricles.
4. The action potential travels through conduction fibers to the AV node at the same time it is spreading across the atria.
5. After a slight delay—enough for the atria to contract—the action potential travels through the AV bundle, down the septum, and along the Purkinje fibers to the ventricular muscle. The QRS complex is seen on the EKG.
6. The ventricles contract (systole).
7. The contraction of the ventricles forces the closing of the mitral and tricuspid valves. The first heart sound is heard.
8. Increased pressure within the ventricles forces the semilunar valves open as blood is pushed into the aorta and the pulmonary trunk.
9. The ventricles repolarize, creating the T wave on the ECG.
10. The ventricles relax (diastole). With no pressure against them, the mitral and tricuspid valves open. Backpressure from the arteries forces the aortic and pulmonary semilunar valves to close again. The second heart sound is heard.

What Causes Congestive Heart Failure?

Normally both sides of the heart have the same stroke volume. If, however, the left ventricle is unable to maintain its normal stroke volume because of heart disease, the imbalance can have serious side effects. If the right ventricle is functioning normally, then more blood enters the left ventricle than it is able to pump out. As a result, blood backs up into the left atrium. The backup continues through the pulmonary veins, causing blood to pool in the lungs. The excess blood in the lungs results in an accumulation of fluid within pulmonary tissue (pulmonary edema), causing congestion. This problem is known as *congestive heart failure*. Along with giving drugs to strengthen the heart, diuretics can be administered. Increased urine formation helps draw some of the excess fluid out of the lungs and provides some relief from the congestion.

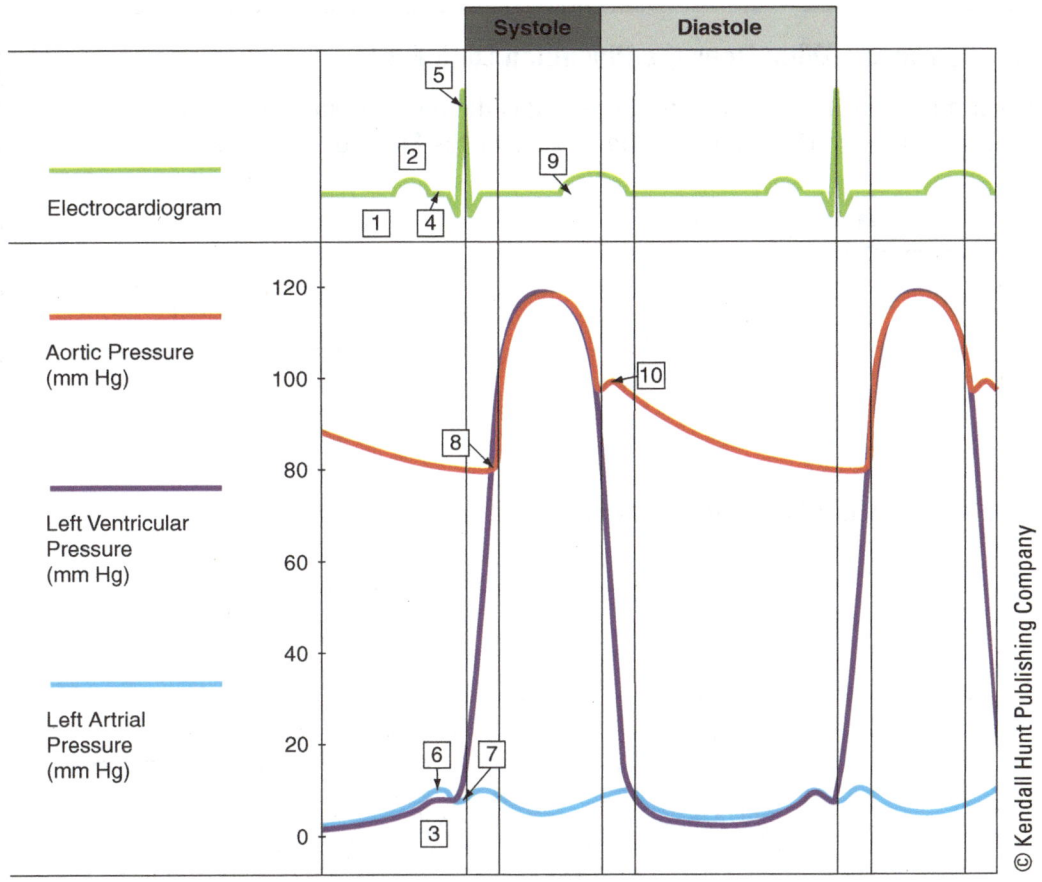

Figure 8.16 Cardiac Cycle—Electrocardiogram and Systemic Pressure.

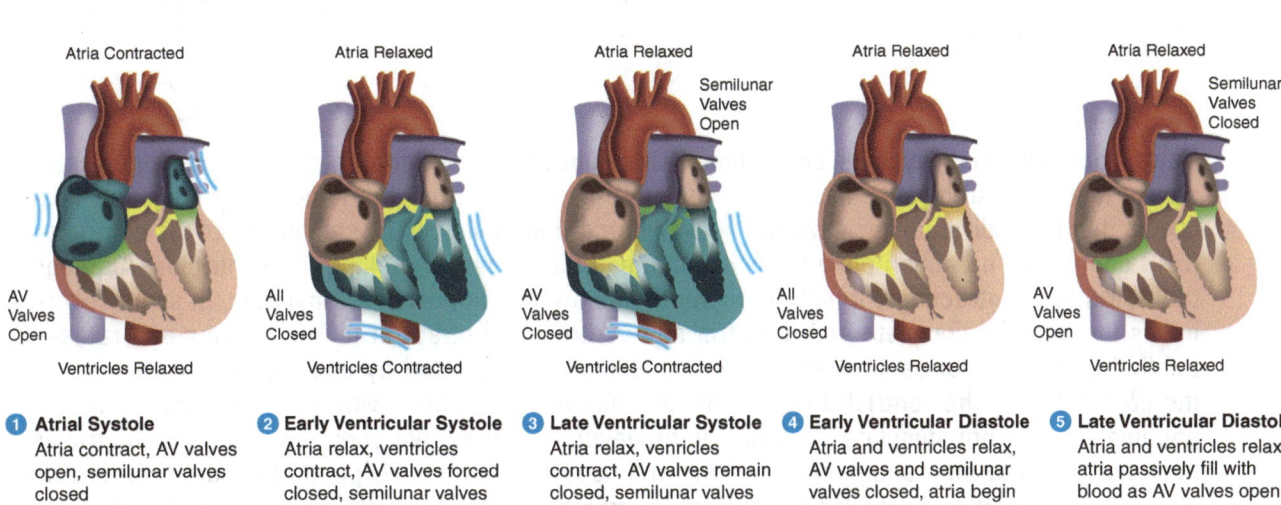

Figure 8.17 Cardiac Cycle

Abnormal Electrocardiograms—What Does the Physician Look For?

Heart disease is a major health issue in the United States and is the number one cause of death. We will discuss a few of the more common irregular heartbeats seen on the EKG. For comparison, a normal EKG is given:

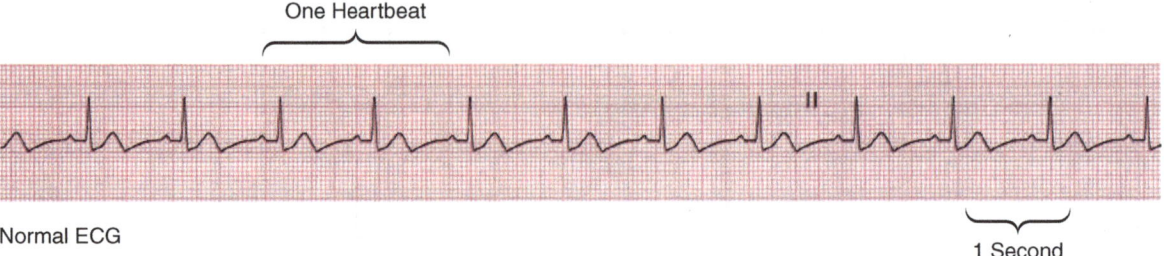

Normal ECG

1. Tachycardia is an abnormally fast heart rate of greater than 100 beats per minute.

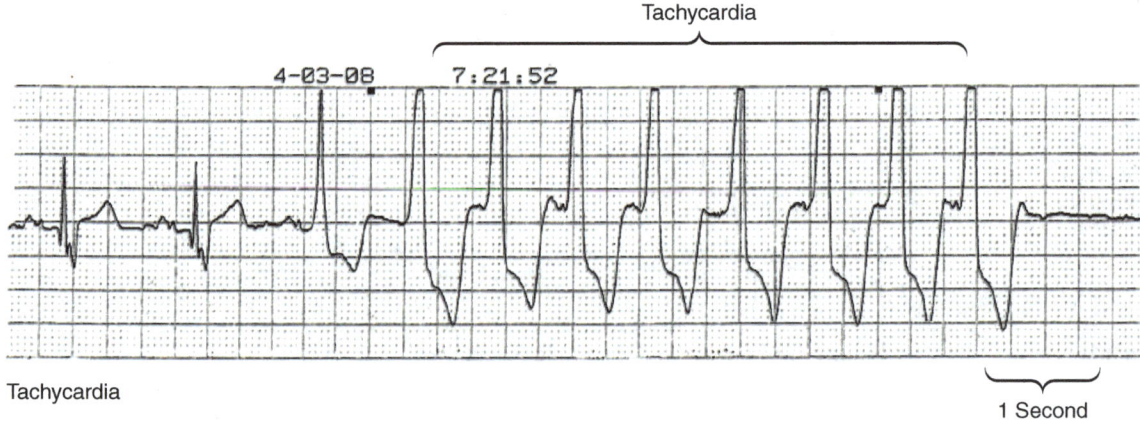

Tachycardia

2. Bradycardia is a heart rate less than 60 beats per minute.

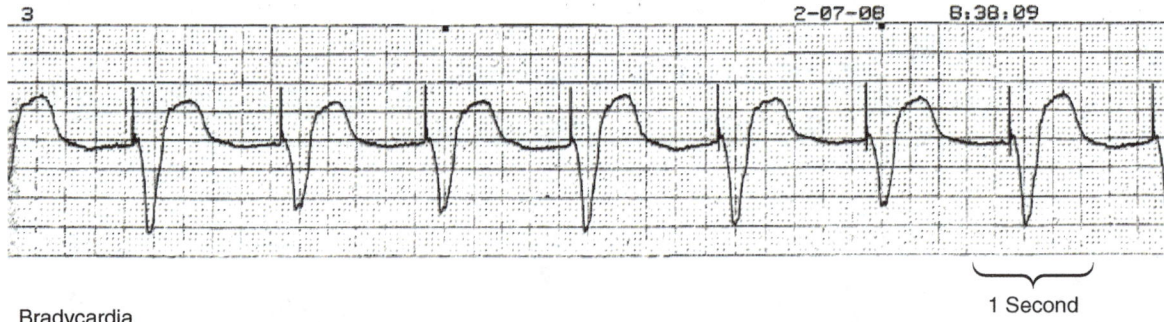

Bradycardia

3. Heart block is the inability of the action potential originating in the sinoatrial node to reach the ventricles or to provide sufficient stimulus to initiate a QRS complex from the AV node. Notice in the illustration that this individual has several P waves that are not followed by stimulation of the ventricles (QRS complex). The action potential is sometimes blocked from reaching the ventricles. Although their atria are contracting normally, their ventricles are too slow and not rhythmic. In some severe cases the block is complete and the atria and ventricles contract at different times and rates—totally independent of each other. These symptoms are alleviated by implanting a pacemaker. If the contraction of the ventricles occurs on time and is in sequence with the atrial stimulation, the pacemaker does not stimulate the heart. If, however, the atria have contracted and the ventricles have not received their action potential within a specified period, the pacemaker delivers its own stimulus, causing the ventricular contraction to remain coordinated with the atrial contraction.

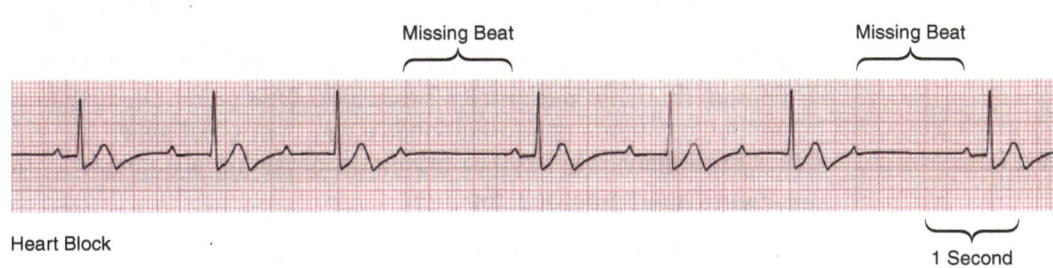

Heart Block

4. Premature ventricular contraction (PVC) occurs when the ventricles do not wait to receive the action potential from the SA node but instead contract early. This premature contraction does not allow time for blood to fill the ventricles. As a result, the ventricular contraction ejects only a small amount of blood. Once the PVC occurs, the ventricle waits until the next stimulus from the SA node. If a pulse is being taken when a PVC occurs, it feels like a skipped beat because the early contraction did not eject enough blood to feel the pulse. PVCs feel like a flutter in the chest. Occasional PVCs occur in most people. They are not a cause for alarm unless there are many of them, particularly in sequence. They are often caused by stress and too much caffeine. If you notice PVCs increasing, it would be suggested that you find a constructive form of stress relief and decrease your caffeine intake.

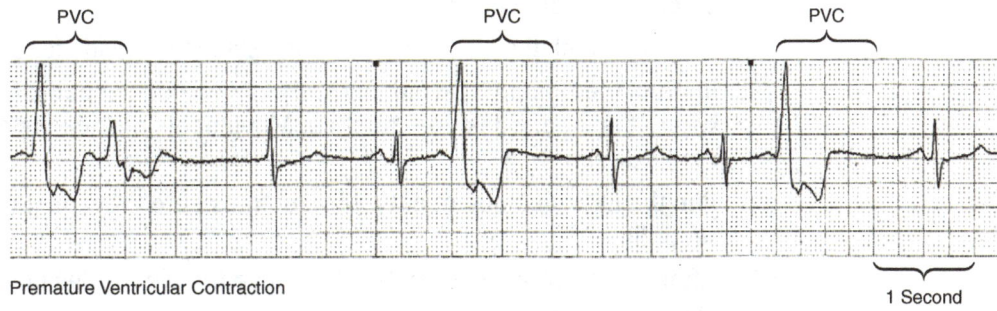

Premature Ventricular Contraction

5. Ventricular fibrillation (V-fib) is one of the most common causes of death due to heart disease. During fibrillation the cardiac muscle cells contract independently of each other. Think of it this way. Imagine a Viking ship with 30 rowers side by side is moving swiftly across the Baltic Sea. As long as all 60 Norsemen row together, the movement is swift and directed. Now suppose they all get into an argument. Each decides to row at his own pace regardless of what the others are doing. Coordination is lost, and the boat has no direction. That is similar to fibrillation. The action potential to the ventricles is no longer causing coordinated contraction. Instead, each cardiac muscle cell contracts without waiting for stimulation from its neighbor. As a result, the ventricles quiver but do not contract as a unit. An insignificant amount of blood is moved. If nothing is done to aid this individual, the brain will begin to die within 4 to 6 minutes. For this reason, cardiopulmonary resuscitation (CPR) can be lifesaving. Once the heart starts fibrillating, it rarely exits this condition spontaneously. To end the fibrillation, a strong electrical charge is used to defibrillate the heart. Once the fibrillation has ended, the opportunity is now available for the SA node to again pace the heart. The sooner a fibrillating heart can be defibrillated, the greater the chances are this condition can be corrected. For this reason, defibrillators are placed in many public areas. Medical expertise is not required to operate them; rather, they are designed to be used by the general public. These units will defibrillate only if the ECG detected by the apparatus registers ventricular fibrillation.

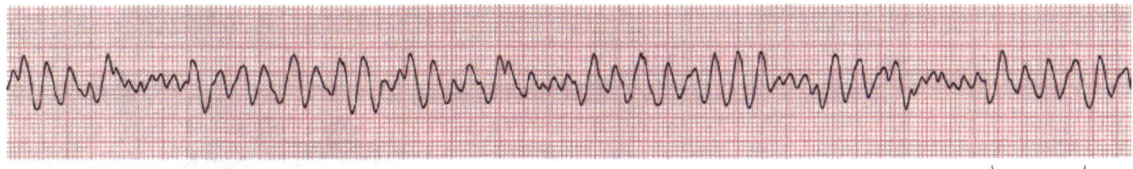

Ventricular Fibrillation

Clinical Skills—EKG

A simple way to evaluate heart function would be to perform an electrocardiogram (ECG). To perform a 12-lead ECG on a patient is easy and gives tremendous amounts of information is a short period of time.

The following steps are critical to obtaining a quality recording of the patient's heart electrical function.

1. Obtain order from doctor.
2. Gather EKG equipment with electrodes and supplies such as alcohol wipes, gloves, and so on.
3. Introduce yourself to the patient, and briefly explain the procedure in simple terms.
4. At all times, protect the patient's privacy by closing the door, closing the curtains, and providing appropriate covering or clothing to the patient.
5. Place the patient in a semi-recumbent position, or flat on the bed.
6. Using alcohol wipes, clean the areas of the patient's skin where the electrodes would be attached.
7. Place or attach the electrodes using the placement guide for a 12-lead EKG.
8. Instruct the patient to avoid any body movements, especially the areas where the electrodes are attached. Movements may cause artifacts in the EKG recording.
9. Run the EKG machine to get a recording. Ensure that you get quality recording.
10. After the recording is completed, remove the electrodes gently to avoid skin abrasions.
11. Thank the patient. Maintain patient's privacy.
12. Document the procedure, and submit EKG report to the physician.

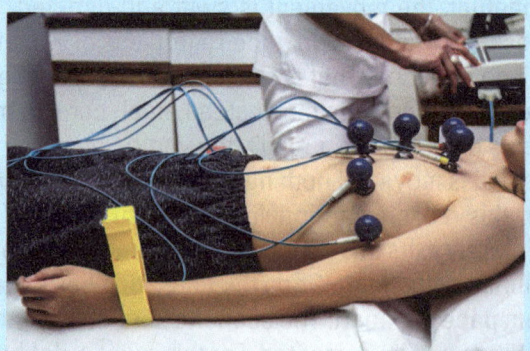

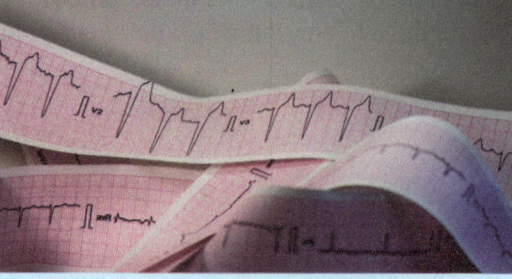

Contributed by Eugene Demekhin. © Kendall Hunt Publishing Company

REGULATION OF HEART FUNCTION

The heart achieves a rhythmic contraction as a result of intrinsic stimulation originating from the SA node. The rate and contractility (intensity of contraction) can be altered by several extrinsic factors, however. The autonomic nervous system can cause significant changes. The sympathetic nervous system increases blood flow to muscle for fighting or running by raising heart rate and by increasing the intensity of the contraction resulting in higher blood pressure. The parasympathetic nervous system calms the body after an emergency and decreases heart rate and the stroke volume. There are sensors that determine blood pressure and chemical composition of the blood. If these receptors inform the brain that the blood pressure, oxygen content, carbon dioxide level, or pH of the blood are out of homeostatic range, then a change in heart rate and stroke occurs to correct the problem. There are also hormones that affect rate and contractility. Epinephrine (adrenalin), which extends the sympathetic nervous system response, increases heart rate and contractility. Exercise increases the need for nutrients to produce ATP, it causes a greater demand for oxygen and it results in a stronger need to remove accumulating carbon dioxide from metabolism, causing the heart rate and stroke volume to increase. Over time, regular exercise can increase the stroke volume, even at rest.

> **Comprehension Check-up:**
> 1. The term for relaxation of the heart is _____.
> 2. A measure of the amount of blood ejected by the heart in one minute is known as _____.
>
> 1. diastole 2. cardiac output

BLOOD VESSELS

In order for cells to survive in a complex organism, they must be within a short distance of a source for nutrients and oxygen and have a means to remove carbon dioxide and wastes. Blood vessels are the conduit through which transport occurs.

Types of Blood Vessels

Arteries become smaller and more branched as they become more distal from the heart. Eventually, branched arteries become capillaries through which exchange of nutrients, gases, and wastes can occur between the bloodstream and cells. The capillaries then combine into larger vessels that form veins to return blood to the heart.

Anatomy of Blood Vessels

The layers of the blood vessels are referred to as *tunics* (Figure 8.18).

- *Tunica intima*—This is the innermost lining of all blood vessels. It consists of an exceedingly thin layer of simple squamous epithelium and the underlying fibrous connective tissue.

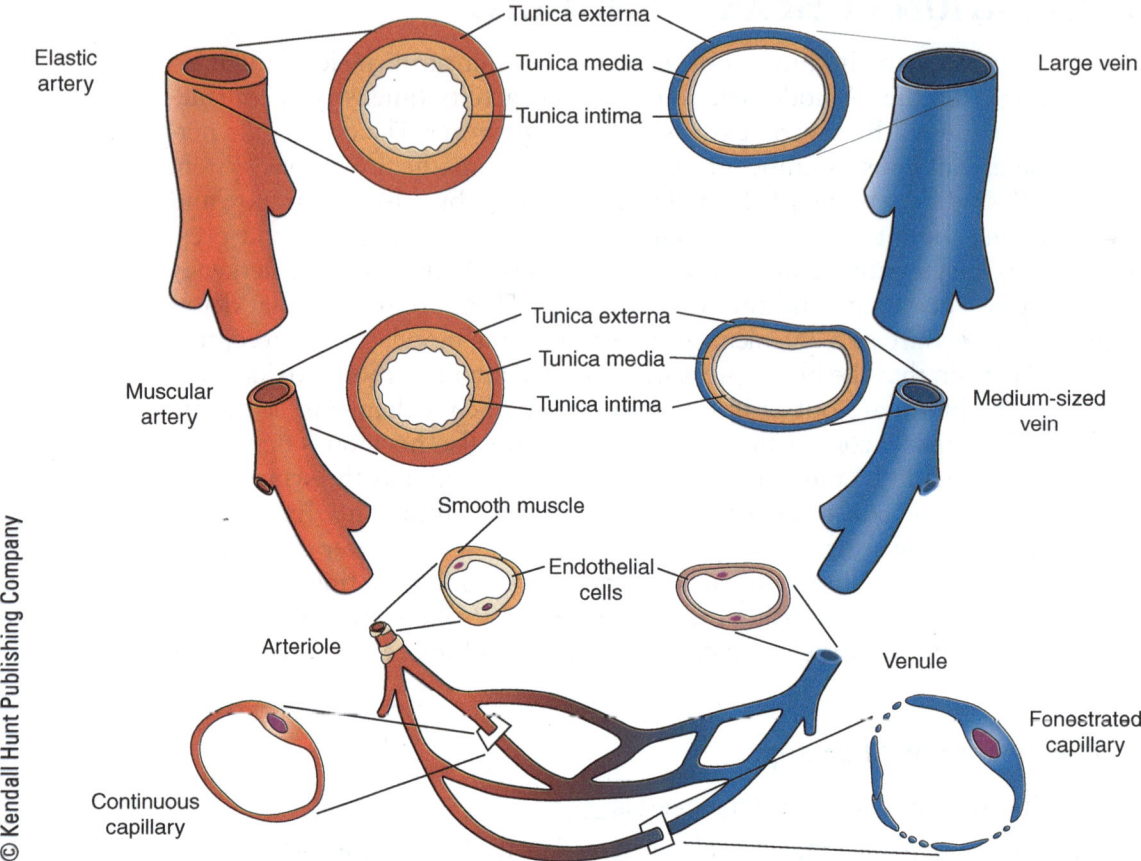

Figure 8.18 Anatomy of Blood Vessels.

- *Tunica media*—This middle layer contains reinforcing collagen fibers, elastic fibers to allow stretching, and smooth muscle spiraled around the blood vessel. Contraction of the smooth muscle causes the blood vessel to become smaller in diameter.
- *Tunica externa*—This tough outer covering contains collagen fibers that not only reinforce the blood vessel but may also attach it to surrounding tissue.

Differences among Blood Vessels

Blood vessels are not rigid pipes like those used in the plumbing in a house. Some have the ability to change their diameter, meaning they can alter blood flow. That control may occur within the blood vessel itself, or it may be in response to brain stimulation, especially in times of crisis.

Large arteries contain higher quantities of elastic fibers in the tunica media than do other vessels. They are able to stretch every time the heart pumps blood into the aorta. Arteries function at higher pressures than capillaries or veins, so their tunica externa contains more collagen than that of the other vessels.

Because capillaries consist of simple squamous epithelium, diffusion of gases, nutrients, and wastes between blood and surrounding cells can occur. Capillaries form a branching bed of vessels that run throughout most tissue. Rarely are cells, except for those in the epithelia, more than one or two layers away from access to their source of nutrients, gases, and waste.

Veins have thinner walls than arteries do. They do not typically handle high pressures and therefore have less reinforcement than arteries. Because they are not involved in controlling the distribution of blood throughout the body, they do not have as much smooth muscle in their tunica media as do arteries.

> **Comprehension Check-up:**
> 1. The lining of blood vessels, composed of simple squamous epithelium and the underlying fibrous connective tissue, is known as the tunica _____.
> 2. Exchange of nutrients, gases, and wastes between the bloodstream and cells occurs through which type of blood vessel?
>
> 1. intima 2. Capillaries

Arteries

There are three types of arteries through which blood passes as it flows away from the heart.

- Elastic arteries stretch with every contraction of the heart. They are the largest of the blood vessels. As the heart relaxes, the elastic arteries provide tension on the blood, holding the blood pressure constant. As blood continues to flow onward, the elastic arteries continue to decrease in diameter to maintain pressure until the next contraction of the heart occurs.
- Muscular arteries contain additional smooth muscle. These vessels can change their diameter to alter blood flow to regions of the body.
- Arterioles are wrapped with smooth muscle to act as valves to control local blood flow to a single capillary bed.

The smooth muscle in the arteries remains partially contracted at virtually all times. This allows the blood vessel diameter to become larger as the muscle relaxes or smaller by increasing the level of contraction. For example, if you were being chased by a wild animal, it would be of critical importance that blood flow to your skeletal muscles would increase so they would have greater access to nutrients and oxygen. Your sympathetic nervous system would cause the smooth muscle in vessels carrying blood to skeletal muscle to relax so blood could reach active muscle more easily. At the same time, areas of your body not essential for fighting or running, such as your digestive system, would receive less blood as the autonomic nervous system causes smooth muscle in those blood vessels to constrict. Changes in the diameter of the arteries may be the result of stimulation of smooth muscle in the walls of the vessels by the brain or by release of hormones, or it may be the result of chemical changes, such as levels of oxygen or carbon dioxide or alterations of pH in the tissue.

Major Arteries

The following is a list of the major arteries of the body (Figure 8.19). The aorta, which can be divided into three sections: (the ascending aorta, aortic

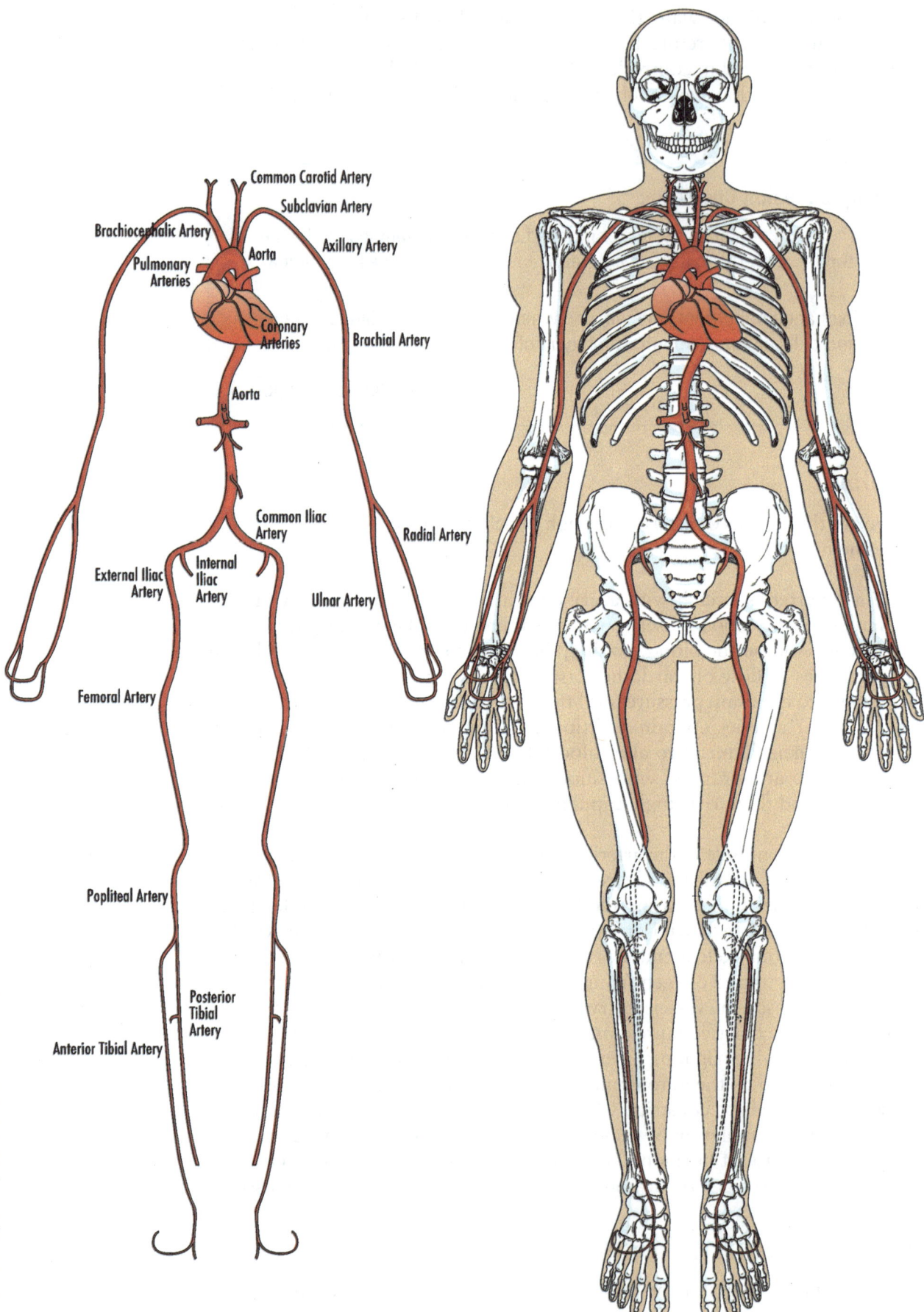

Figure 8.19 Major Arteries—Anterior View.

arch, and descending aorta), carries all oxygen-rich blood from the left ventricle to the systemic circulation.

- Ascending aorta—Exiting from the left ventricle, the aorta carries all of the oxygen-rich blood for distribution throughout for the body. Immediately upon exiting, two coronary arteries branch off the ascending aorta to supply blood to the heart itself.
- Aortic arch—The aorta curves into an arc as the main vessel starts to descend. Three branches off the aortic arch supply blood to the upper body.
  - Brachiocephalic artery—This single vessel supplies blood to both the right side of the head and the right upper extremity; it is the first branch off the aortic arch. Shortly after branching off the aorta, the brachiocephalic artery divides into the right common carotid artery and the right subclavian artery. The right common carotid artery supplies blood to the head. It splits into the internal carotid artery, which supplies blood to the brain and some other internal structures, and the external carotid artery, which flows to the skin and muscles of the head and neck. The right subclavian artery provides blood to the right upper extremity.
  - Left common carotid artery—This is the middle of the three arteries branching off the aortic arch. It supplies blood to the left side of the head. Like the right common carotid artery, the left common carotid artery splits into internal and external carotid arteries.
  - Left subclavian artery—This is the third branch off the aortic arch. It supplies blood to the left upper extremity.
- Descending aorta—The descending aorta carries blood to the lower body. It can be divided into two sections.
  - Thoracic aorta—The thoracic aorta resides in and supplies blood to the chest wall and thoracic organs other than the heart.
  - Abdominal aorta—Located below the diaphragm, the abdominal aorta supplies blood to the abdominal wall, viscera, and lower extremities.
- Arteries to the brain—There is a special arrangement of blood vessels to the brain called the *circle of Willis* (Figure 8.20). Arterial blood from both internal carotid arteries provides input to this vascular circle. Three branches on each side exit from the circle of Willis, carrying blood to the brain. They are the anterior, middle, and posterior cerebral arteries. The advantage of the circle of Willis is that blockage of any one location on the circle does not prevent blood flow to the remaining areas. Many individuals have varying anatomy of these vessels, but as long as the cerebral arteries are intact as they emerge from the carotids and divide within the brain, the blood vessels may anastomose and provide the nutrients and oxygenation necessary for survival.
- Arteries to the upper extremities are as follows:
  - Axillary artery—The subclavian artery passes inferior to the clavicle. As the artery continues, a section passes through the axilla (armpit) to become known as the axillary artery.
  - Brachial artery—As the artery to each upper extremity enters the upper limb, it becomes known as the *brachial artery.*
  - Radial artery—At roughly the elbow, the brachial artery splits to follow the two bones of the forearm. The lateral artery is the radial artery. At the wrist, the radial artery passes superficial to the distal end of the radius. Its pulsations are easily detected on the anterior aspect of the wrist and are often used to determine heart rate.
  - Ulnar artery—This is the medial artery in the forearm.

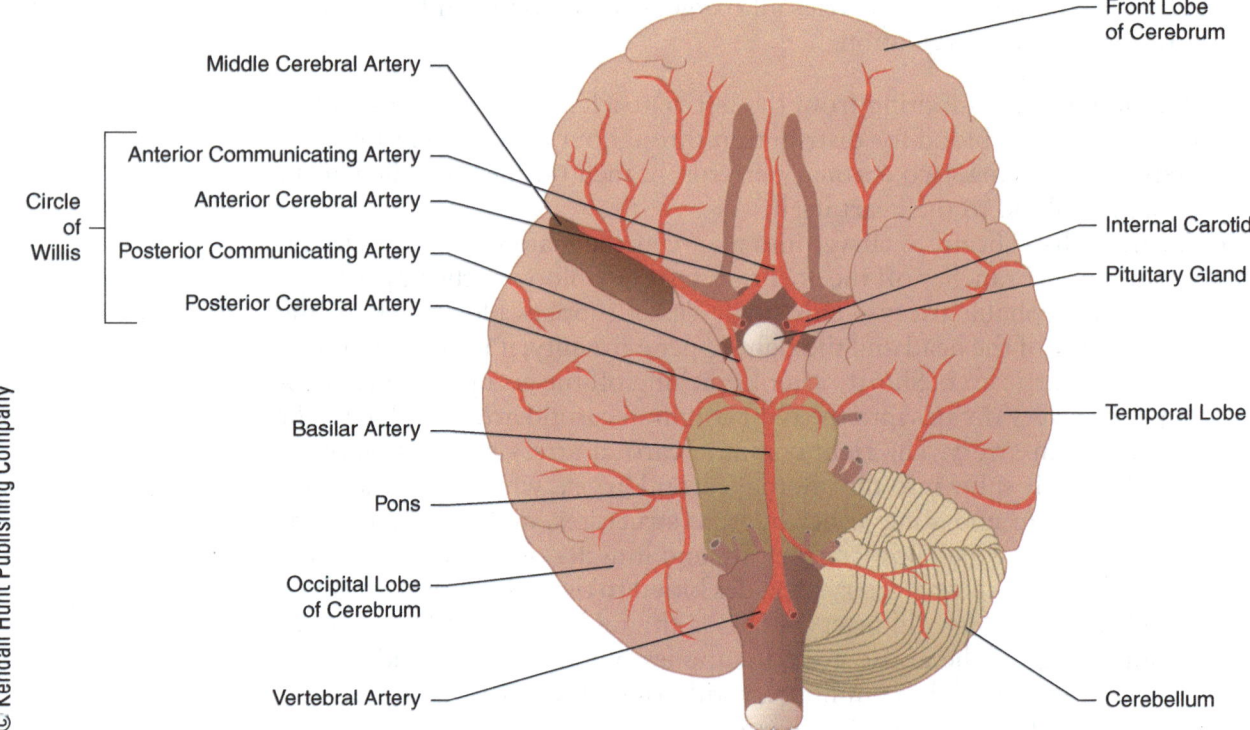

Figure 8.20 Arterial Supply to the Brain.

- Arteries to the abdomen are as follows:
  - Celiac trunk—This single vessel is the first ventral branch off the abdominal aorta inferior to the diaphragm. It splits into three branches:
    - Gastric artery, which provides blood to the stomach
    - Splenic artery, which supplies blood to the spleen
    - Hepatic artery, which provides blood to the liver
  - Superior mesenteric artery—This single vessel supplies blood to the all of the small intestine and some of the upper colon.
  - Renal arteries—Each kidney receives oxygen-rich blood from a renal artery emerging from the aorta.
  - Gonadal arteries—An artery on each side runs to the ovaries or testes.
  - Inferior mesenteric artery—This single branch supplies blood to the lower colon.
  - Common iliac arteries—The abdominal aorta divides into two branches as it enters the pelvic region. Each common iliac artery branches into the following two arteries:
    - Internal iliac artery, which provides blood to the organs in the pelvis
    - External iliac artery, which exits the abdomen to supply blood to the anterior abdominal wall and lower extremities
- Arteries of the lower extremities are as follows:
  - Femoral artery—This artery provides blood to the thigh and is the continuation of the external iliac artery.
  - Popliteal artery—As the femoral artery enters the posterior knee it becomes known as the popliteal artery.
  - Anterior and posterior tibial arteries—Below the popliteal artery the vessel divides into these arteries, which supply blood to the leg.
  - Dorsalis pedis artery—As the anterior tibial artery enters the dorsal side of the foot (top of the foot in the anatomical position), it becomes

the dorsalis pedis artery. This artery is important as a location for testing for the presence of a pulse. If an individual's lower extremity has been placed in a splint or cast, it is critical that a pulse be detectable in the dorsalis pedis artery after the device has been applied. Lack of a pulse indicates that the blood flow to the lower extremity has been compromised.

> **Comprehension Check-up:**
> 1. The type of blood vessels that acts like a valve to control blood flow through a bed of capillaries is known as a(n) _____.
> 2. Blood flows to the head through the _____ artery.
>
> 1. arteriole 2. carotid

Veins

There are two types of vessels through which blood passes as it returns to the heart. As capillaries converge to form larger vessels, they empty into *venules*. Diffusion of nutrients, gases, and wastes between the bloodstream and cells does not occur through venules. They combine into veins to carry blood back through ever larger veins toward the heart.

Veins in the extremities have one-way valves (Figure 8.21). They not only prevent backflow of blood into the capillaries, they assist with moving

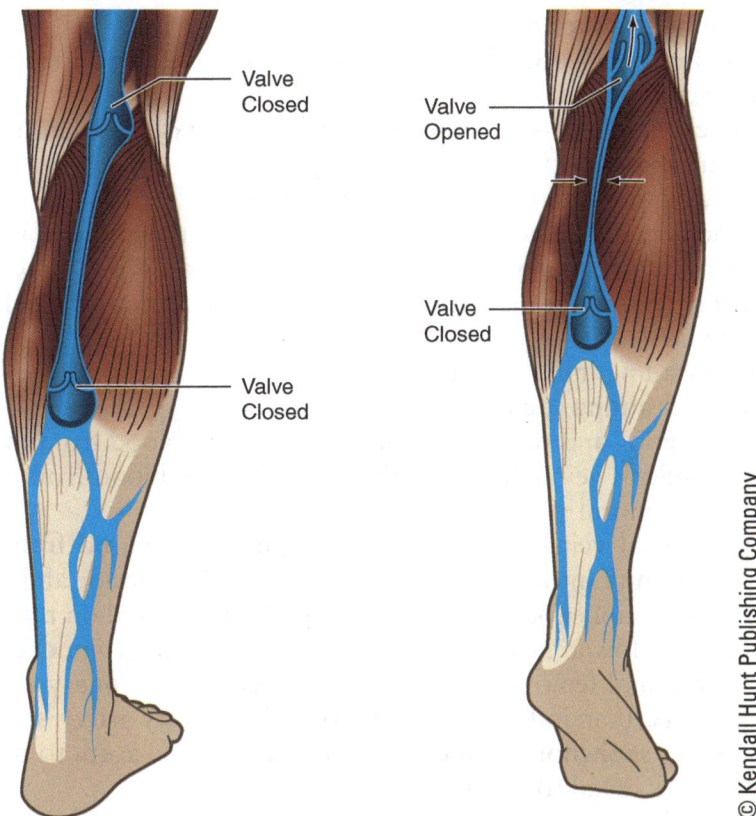

Figure 8.21 Venous Valves. Venous valves in the extremities assist blood flow back to the heart. As muscles contract and compress deep veins, blood is forced through the next one-way valve, moving blood toward the heart.

blood back to the heart. Deep veins are found between muscles. As a person exercises and muscles shorten, the muscles compress these deep veins, forcing blood through the next valve. As a result, muscle contractions continue moving blood upward, making it less strenuous on the heart. This is another advantage of exercise.

Major Veins

Most of the major veins correspond to arteries in the same area of the body. Blood drains into the right atrium from the body through the superior and inferior venae cavae (Figure 8.22).

- Veins draining superior to the heart:
- Veins draining from the head:
  - Internal jugular veins—drain blood from the brain
  - External jugular veins—drain blood from the skin and muscles of the head
- Veins draining each upper extremity:
  - Radial vein—removes blood from the lateral side of the forearm
  - Ulnar vein—drains blood from the medial side of the forearm
  - Brachial vein—a deep vein that drains the arm into the axillary vein
  - Cephalic vein—a superficial vein that removes blood from the lateral upper limb into the axillary vein
  - Basilic vein—drains the medial upper extremity into the proximal end of the brachial vein
  - Median cubital vein—located between the distal ends of the cephalic and basilic veins and found in the anterior crease of the elbow; is a common site for blood withdrawal
  - Axillary vein—found in the axilla and receives blood from the upper extremity
  - Subclavian vein—receives blood from the axillary vein and external jugular vein
- Veins draining into the superior vena cava:
  - Brachiocephalic vein—receives blood from the subclavian veins and the internal jugular vein; whereas there is only one brachiocephalic artery (found on the right side of the body), there is a brachiocephalic vein on each side
- Veins draining inferior to the heart:
  - Veins draining each lower extremity:
    - Anterior and posterior tibial veins—drain blood from the leg
    - Popliteal vein—found in the knee; empties blood from the leg into the femoral vein
    - Femoral vein—located in the thigh; drains blood from the lower extremity into the external iliac vein in the inferior abdomen
    - Great saphenous vein—a superficial vein located on the medial side of the lower extremity from the ankle to the groin; drains into the proximal femoral vein; commonly used for coronary artery bypass grafts (If an individual has a blocked coronary artery, sections of the great saphenous vein can be grafted from the aorta to the coronary artery distal to the blockage.)
  - Veins draining the abdomen:
    - External iliac vein—receives blood from the lower extremity
    - Internal iliac vein—drains blood from the pelvic cavity

Chapter 8   The Cardiovascular System and Blood   231

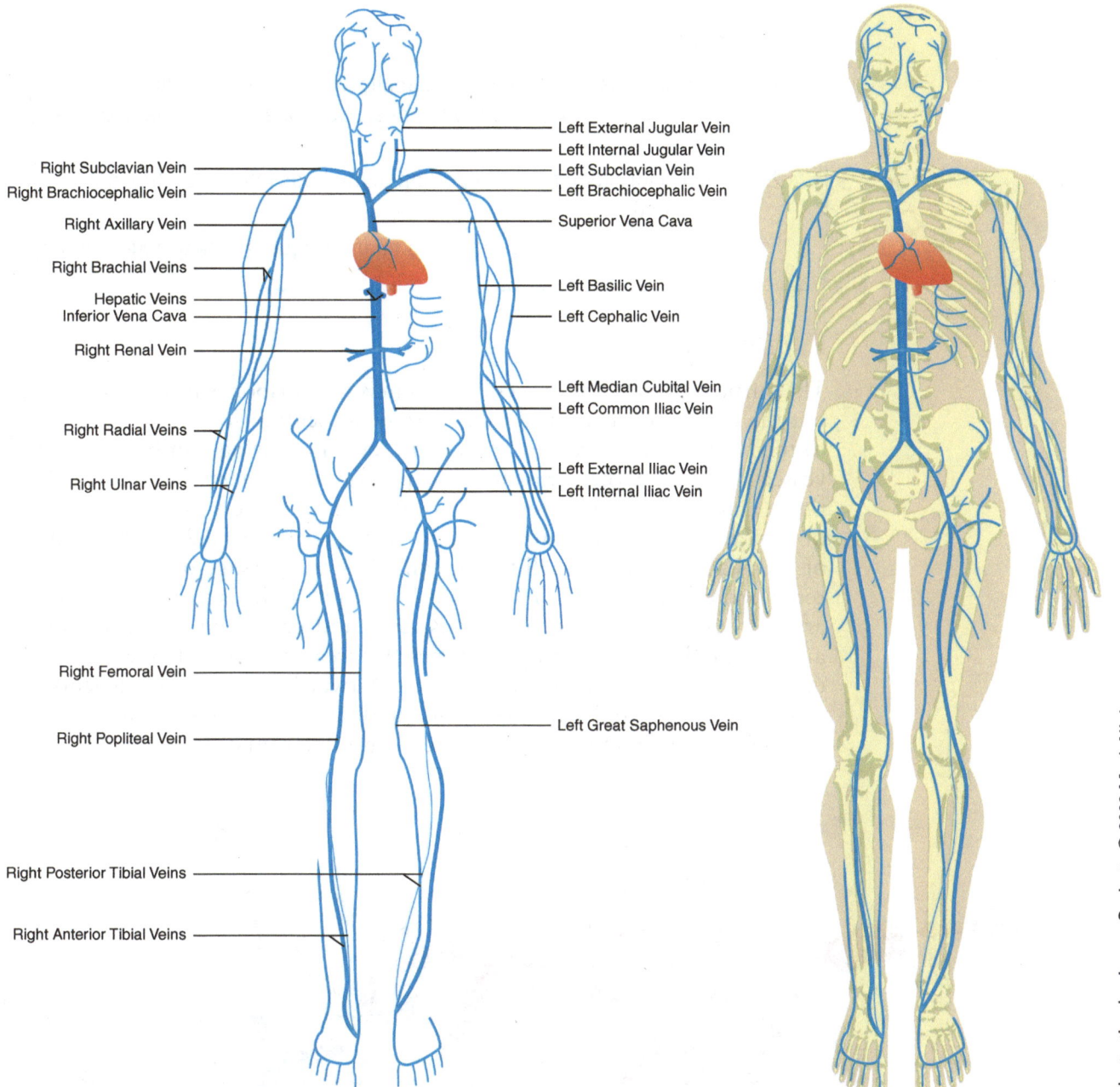

Figure 8.22 Major Veins—Anterior View.

- Common iliac vein—results from the convergence of the external and internal iliac veins emptying blood into the abdominal inferior vena cava
- Renal veins—drain blood from the kidneys, which are posterior to the peritoneum (Their venous drainage is not connected to the hepatic portal system.)
- Hepatic portal system—carries blood from the abdominal gastrointestinal organs to the liver
- Left and right hepatic veins—drain blood from the liver into the inferior vena cava
- Inferior vena cava—drains blood into the right atrium

Comprehension Check-up:

1. Venous blood is prevented from backflowing into the capillary bed and also assists returning blood to the heart because veins in the extremities have _____.
2. Oxygen-poor blood from the portion of the body inferior to the heart drains into the _____.

1. one-way valves   2. inferior vena cava

Capillaries

The purpose of the cardiovascular system is to transport substances throughout the body, so it is critical that there be a location for access of each cell to the bloodstream. Arteries branch into ever-increasing numbers of smaller vessels. Eventually they branch into capillaries so numerous that virtually all tissue is not more than a few cells away from these vessels, with the exception of avascular epithelial tissue and cartilage (Figure 8.23). The diameter of each capillary is so small that it can allow the passage of only one single red blood cell at the time. The network of capillary beds forms an enormous area of contact between the bloodstream and cells. The collective surface area of capillaries is roughly 1.5 times the area of a basketball court, about 650 square meters (7,000 square feet). Outward diffusion of nutrients and oxygen from the capillaries to the surrounding cells and inward movement of carbon dioxide and wastes occurs freely through thin capillary walls.

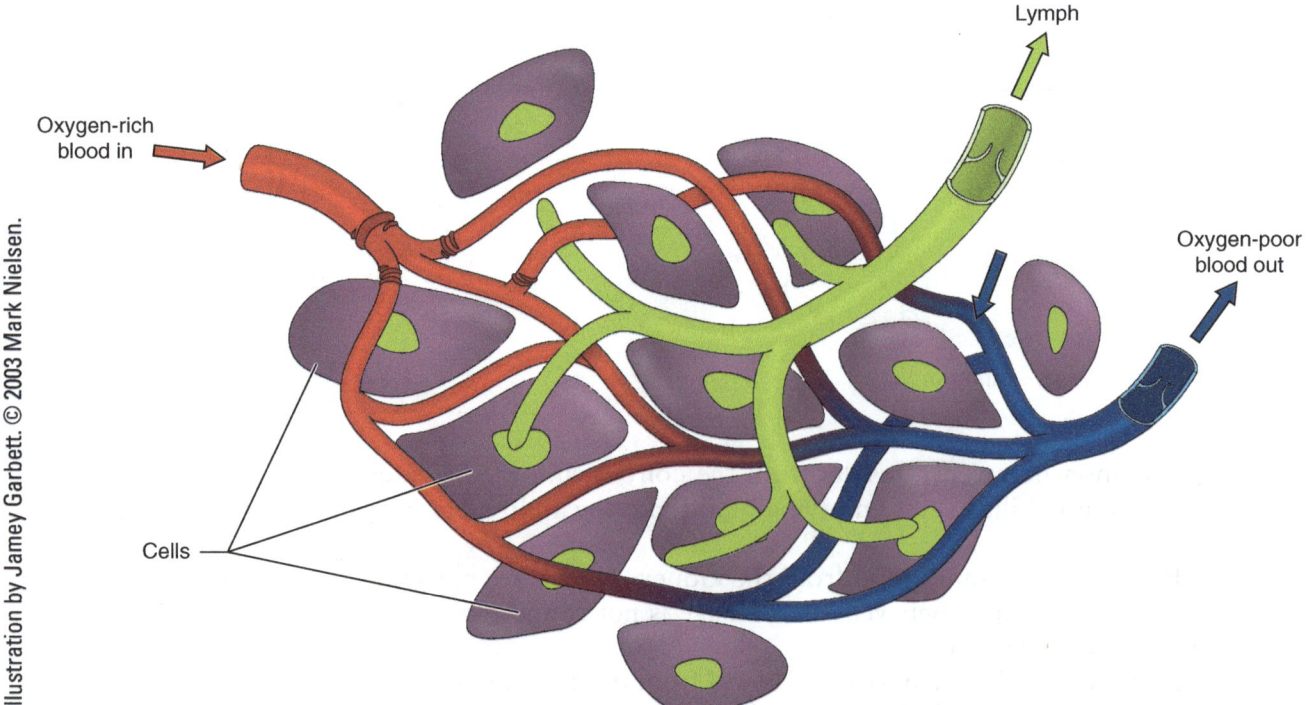

Figure 8.23 Microcirculatory Structure. An arteriole branches into capillaries, which pass through tissue to exchange gases, nutrients, and wastes between the cells and blood. The capillaries recombine to form a venule as blood flows back toward the heart. A lymphatic vessel collects excess tissue fluid (discussed in the next chapter). As oxygen-rich blood ➡ flows through the proximal capillaries, oxygen diffuses into the cells. As more diffusion occurs, blood oxygen levels decrease ➡ so that by the time the capillary blood reaches the distal end it is oxygen-poor. ➡

THE CIRCULATION

The right side of the heart receives oxygen-poor blood from the body and transports it to the lungs and back, forming what is known as pulmonary circulation. The left side moves oxygen-rich blood to everywhere else and back to the heart via the systemic circulation (Figure 8.24).

Pulmonary circulation the transport of blood from the heart to the lungs and back.

Systemic circulation the transport of blood from the heart to everywhere in the body and back, except for the lungs.

The Pulmonary Circulation

After the pulmonary trunk exits the right ventricle, it splits into two pulmonary arteries carrying oxygen-poor blood to the lungs. Arterial branches

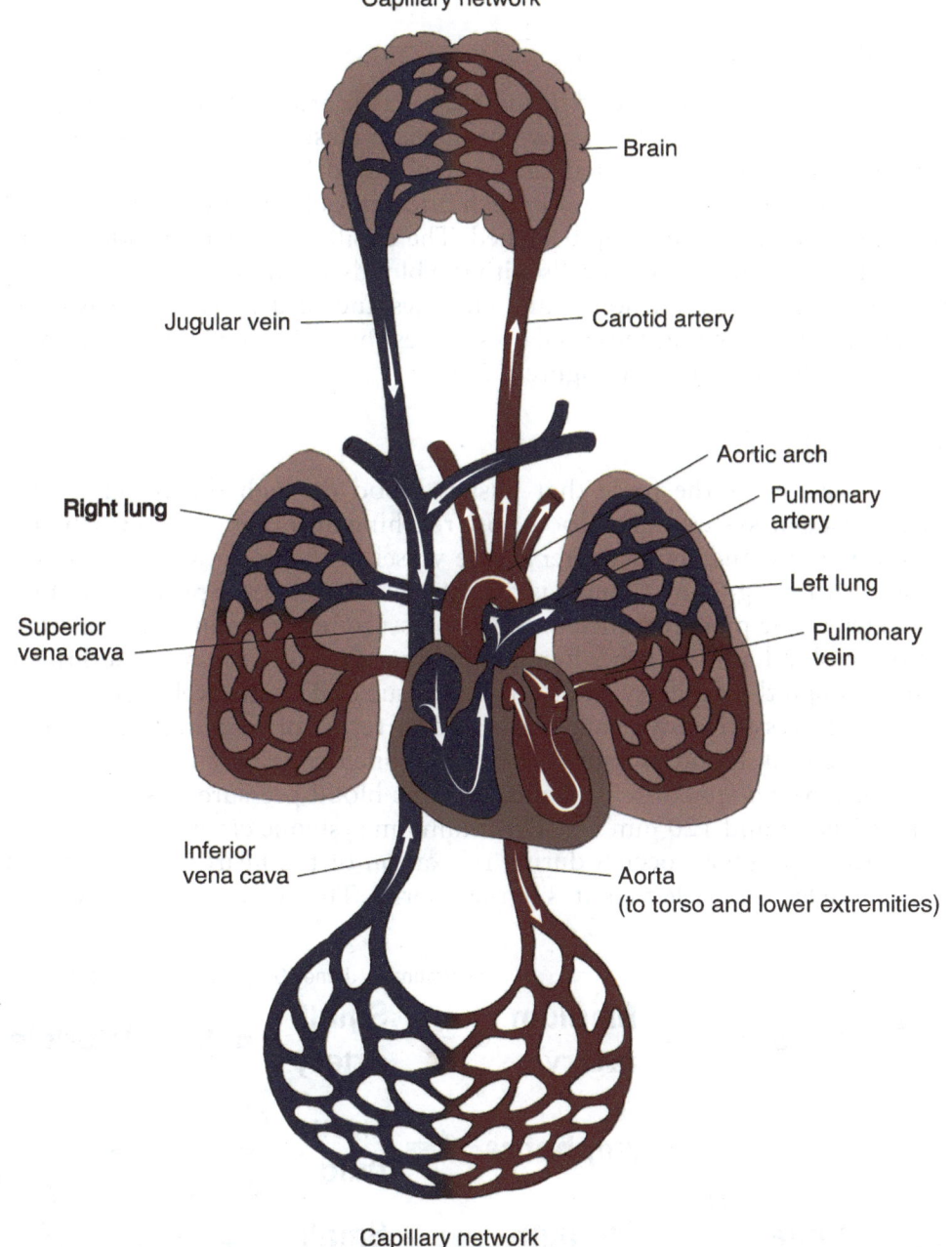

Figure 8.24 Pulmonary and Systemic Circulation. Pulmonary circulation transports blood from the heart to the lungs and back to exchange gases. Systemic circulation transports blood from the heart to the body and back, carrying nutrients, gases, and wastes to the appropriate organs or tissues.

form ever-increasing numbers of branches until terminating in capillary beds wrapped around alveoli (air sacs) in the lungs. The alveoli are composed of simple squamous epithelium, as are the capillaries. The diffusion of gases, due to concentration differences, occurs easily between the capillary and alveolar membranes as these are so thin. Normally the concentration of oxygen in the alveoli is greater than that in the capillaries. Oxygen diffuses into the bloodstream. Carbon dioxide from cellular metabolism is greater in the capillaries than in the alveoli, so it diffuses into the air sacs to be exhaled into the atmosphere. The end result is that the blood exiting the lungs and flowing to the left atrium through the pulmonary veins contains increased levels of oxygen and less carbon dioxide than does the blood entering the lungs.

The Systemic Circulation

The left ventricle pumps blood through the elastic arteries that maintain pressure even when the heart relaxes. From these elastic arteries, blood passes through muscular arteries that can control blood flow to regions throughout the body. Muscular arteries branch into arterioles that can control local blood flowing through a capillary bed. The capillary beds tremendously increase the surface contact of cells with the bloodstream as a means of accessing nutrients, gases, hormones, and enzymes and of disposing of wastes by diffusion. Blood exiting the capillaries passes through venules and veins as it returns to the right atrium (Figure 8.25).

Blood Pressure

> **Blood pressure** the force created by the ventricles that pushes blood through vessels throughout the body.

Blood pressure is the force that pushes blood through the vessels of the cardiovascular system. The blood flow reaching an organ is affected by the blood pressure and the diameter of the vessel in the same way the amount of water running from a faucet is directly related to water pressure and the diameter of the pipe. Insufficient blood pressure results in inadequate circulation of blood. On the other hand, high blood pressure can force additional fluid through the capillary walls into the tissue and cause swelling.

Blood pressure is typically measured in terms of maximum and minimum. The peak blood pressure is reached while the heart is contracting and is termed *systolic pressure.* A healthy systolic blood pressure of systemic circulation is around 120 mm Hg. The minimum systemic circulation pressure, or *diastolic pressure,* occurs during relaxation of the heart and is different within the left ventricle than it is in the arteries. The aortic valve closes when

Illustration by Jamey Garbett. © 2003 Mark Nielsen.

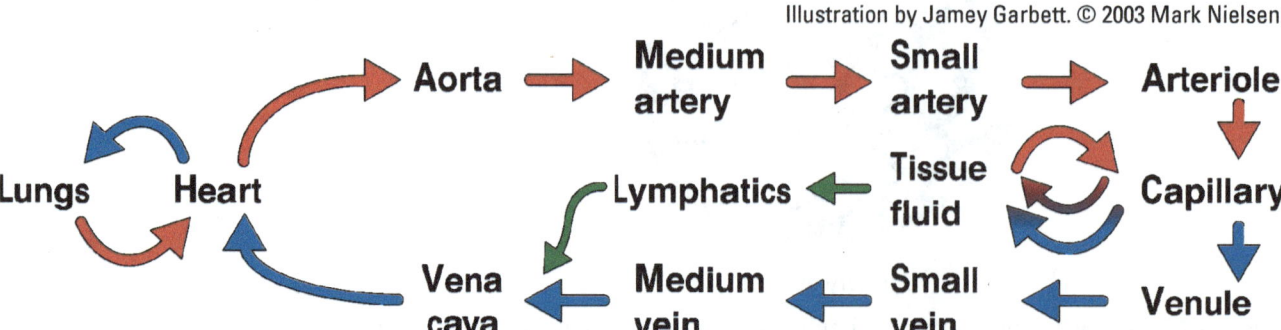

Figure 8.25 Schematic Representation of the Cardiovascular System. Oxygen-Rich Blood, Oxygen-Poor Blood, Lymphatic Fluid.

the left ventricle relaxes to maintain the diastolic pressure at a higher level so that blood can continue to be pushed through the vessels. During relaxation, the pressure within the left ventricle drops to 0 mm Hg while the arterial pressure typically remains at 70 to 80 mm Hg.

> **Clinical Notes—Hypertension**
>
> Hypertension, high blood pressure, is a common problem for many people. The concern with hypertension is that additional pressure could cause an artery to rupture. Rupture would result to excessive blood loss and disrupt the blood flow to organs in the body. There are weak areas in our blood vessels that can burst due to the additional pressure. A weak area in the blood vessel is called aneurysm. An aneurysm on the aorta could lead to rapid fatal blood loss. If an aneurysm breaks in the brain, blood loss occurs within the brain and causes increased intracranial pressure (ICP), in addition to low oxygen content of the blood. This is one cause of a stroke.
>
> Causes of hypertension include: stress, obesity genetic predisposition, smoking, and hormone imbalance. Hypertension has serious consequences, so it is important to check one's blood pressure regularly. If one's blood pressure is consistently and abnormally high, lifestyle changes are needed such as changing one's diet, increasing exercise, and stopping smoking and alcohol intake. Uncontrolled and unstable hypertension may need prescribed medications from the physician.

Venous Return

Venous return refers to the amount of blood returning to the right atrium. Several factors can alter blood flow through veins. The most significant factor contributing to venous return is the amount of blood volume. Blood vessels, primarily arterioles, can adjust their diameter in an attempt to maintain blood pressure. If the blood volume is low, it becomes difficult for the cardiovascular system to constrict enough to maintain both blood pressure and adequate blood flow. If insufficient blood returns to the heart, blood pressure drops. This situation may occur with blood loss, but it could also be an effect of dehydration.

An additional factor is that large veins can distend to pool blood when large volumes are not essential. The process of holding blood not required at the time is known as *venous capacitance*. If there is a sudden need for an increase in venous return and higher blood pressure, the smooth muscle in the veins can be to stimulated to constrict and some of the reserve blood added to the cardiovascular circulation.

> **Comprehension Check-up:**
>
> 1. The exchange of oxygen and carbon dioxide with the atmosphere occurs through the _____ circulation.
> 2. The force that pushes blood through vessels is known as _____.
> 3. The quantity of blood that flows into the right atrium is referred to as the _____.
>
> 1. pulmonary   2. blood pressure   3. venous return

Portal Circulation

There are some special instances when the flow of blood from a specific area passes through only one location—a portal—in order to ensure it is able to provide its function for the entire body. There are two portal systems we will discuss: the hepatic portal circulation and hypothalamus-pituitary portal circulation.

The Hepatic Portal Circulation

All venous blood from the abdominal viscera takes a diversion through the *hepatic portal circulation* before returning to the heart (Figure 8.26). Nutrient-rich venous blood from the digestive tract, spleen, and pancreas flows through the hepatic portal vein to the liver. From there it flows into the inferior vena cava to allow the liver to alter the contents of the blood before making those substances available to the entire body. The liver may remove some excess nutrients for storage or detoxify some potentially hazardous substances absorbed through the digestive system. It can remove foreign

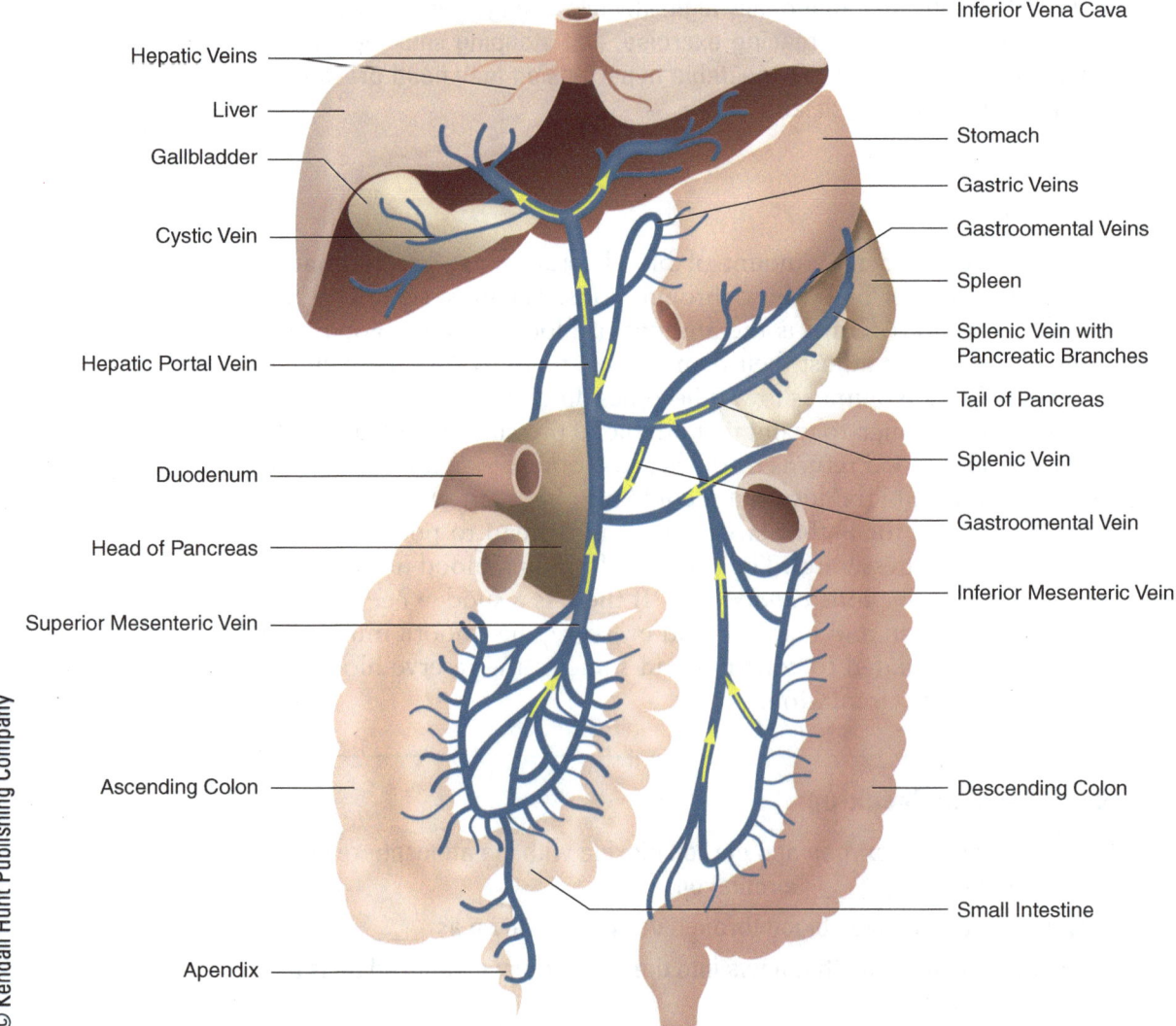

Figure 8.26 Hepatic Portal System. Venous blood from the abdominal organs flows to the liver, where the blood is altered before being transported to the heart to be pumped throughout the body.

organisms and dead red blood cells. The liver may add some of its own substances as well, such as clotting factors needed to stop blood loss if a blood vessel is broken and albumin acting as a thickener for the plasma. The system of veins draining blood from the abdominal organs into a single vessel—the hepatic portal vein—is referred to as the *hepatic portal circulation*.

Hypothalamus-Pituitary Portal Circulation

The hypothalamus controls the release of hormones from the pituitary gland into the bloodstream. The anterior lobe of the pituitary gland produces hormones but does not release them until receiving regulatory hormones from the hypothalamus. It is essential that the circulation of blood between the hypothalamus and pituitary gland provide ready access of the regulatory hormones to endocrine cells in the pituitary gland. There is a system of blood vessels that links the hypothalamus and the anterior pituitary in the brain. This system of capillaries emptying into a large vein is called the hypothalamus-pituitary portal circulation. The *hypothalamus-pituitary portal circulation* guarantees access of its regulatory hormones to the anterior lobe (Figure 8.27).

The Fetal Circulation

Some variations in the circulatory system occur prior to birth because the fetus is exchanging gases, nutrients, and wastes through the placenta rather

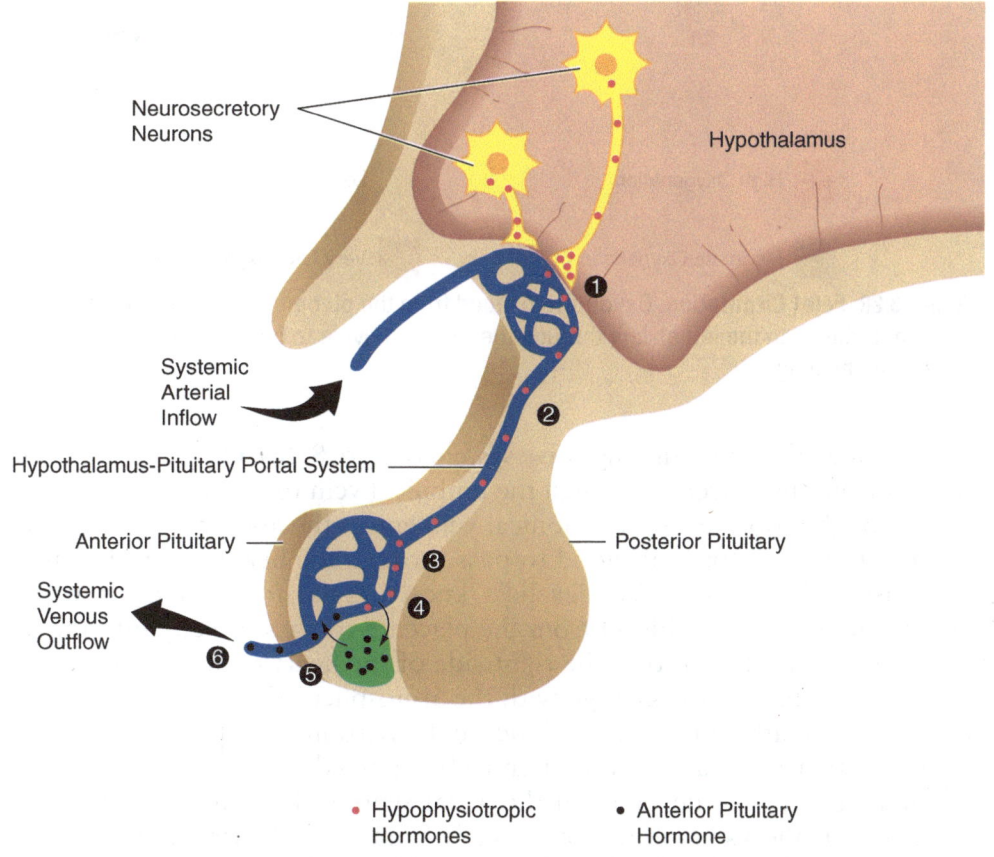

Figure 8.27 The Hypothalamus-Pituitary Portal System. Capillaries from the hypothalamus carry blood to the anterior lobe of the hypothalamus.

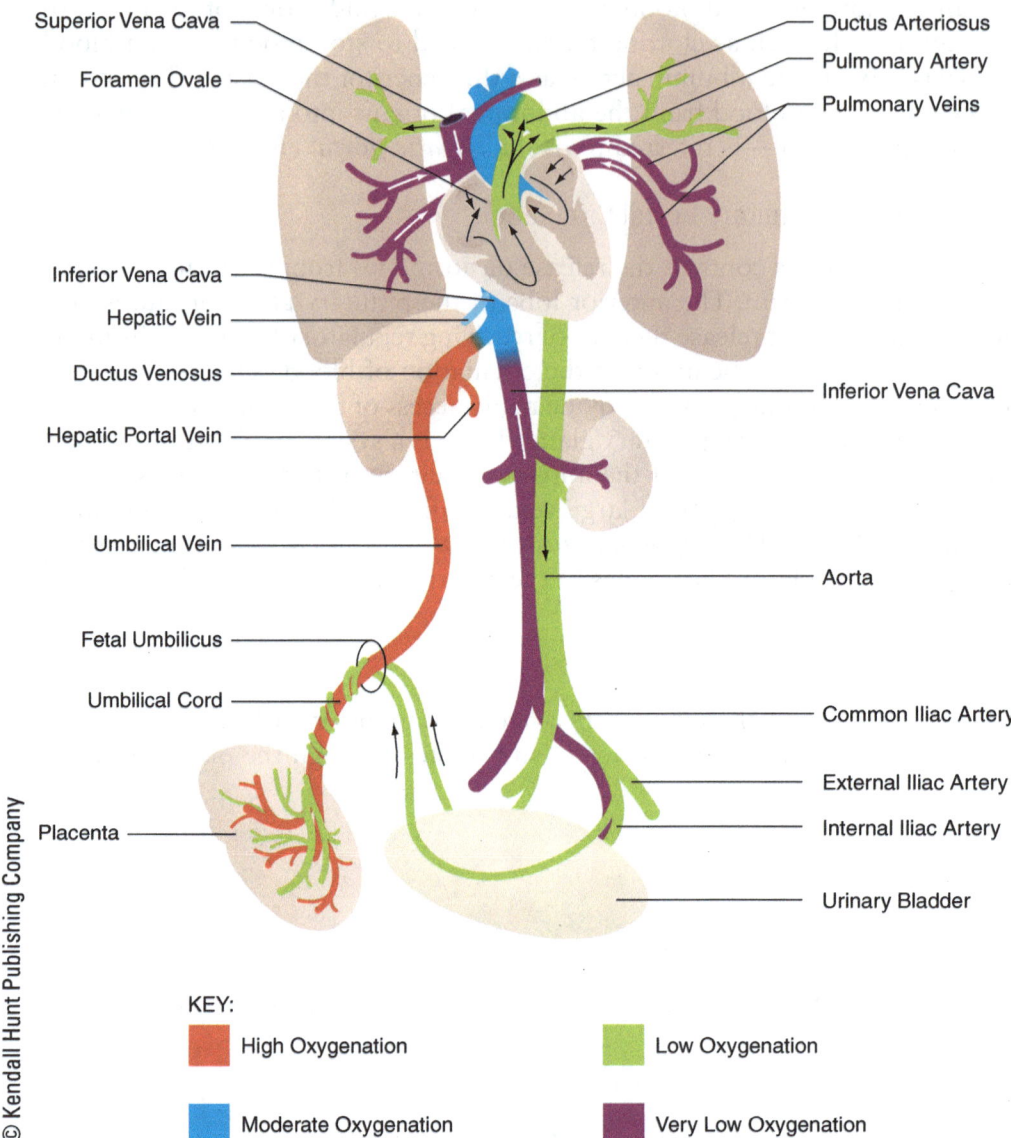

Figure 8.28 Fetal Circulation. Oxygen-rich blood from the placenta flows directly from the right atrium into the left atrium through the foramen ovale to be circulated to the body, bypassing the lungs.

than through the lungs and digestive system (Figure 8.28). Oxygen-rich blood returns from the placenta through the umbilical vein to the developing fetus. Even though blood from the umbilical vein enters the abdomen of the fetus, it is not necessary for that blood to pass through the hepatic portal system. A bypass of the portal system develops, known as the *ductus venosus,* which allows the oxygen-rich blood from the placenta to pass directly into the inferior vena cava. Recall that the right side of the heart pumps oxygen-poor blood to the lungs for exchange with the atmosphere. The fetus, however, is not able to breathe air in its amniotic fluid environment. There is no reason for blood that is already oxygen-rich to be pumped from the right ventricle to the lungs. A hole develops in the atrial septum, known as the *foramen ovale,* which allows blood to flow directly from the right atrium into the left atrium. From there the blood can be pumped by the left ventricle to the body. Some of the blood entering the right atrium does flow into the right ventricle, however, and is pumped into the pulmonary trunk. A connection, or shunt—the ductus arteriosus—develops between the pulmonary trunk and the aorta, allowing the blood pumped by the right ventricle to enter the aorta.

Prior to birth a tissue flap grows down from the top of the atrium. After the child is born and takes his or her first breath, the resistance to blood flow in the lungs changes dramatically. Left atrial pressure increases suddenly, pushing the tissue flap over the foramen ovale to close the hole between the two atria. If the flap does not completely seal the foramen ovale, oxygen-poor blood will continue to enter the left atrium from the right, causing both oxygen-rich and oxygen-poor blood to mix in the left atrium and resulting in the baby appearing blue. After birth the ductus venosus and ductus arteriosus normally close on their own.

Comprehension Check-up:

1. All venous blood from the abdominal viscera flows to the _____.
2. Blood entering the right atrium of the fetus is able to pass directly into the left atrium through a hole in the atrial septum known as the _____.

1. liver 2. foramen ovale

Homeostasis—Holding in Balance

Because the cardiovascular system provides access of virtually every cell in the body through transport of nutrients, gases, hormones, and the removal of wastes, it is essential for the maintenance of homeostasis throughout the body. If a blood vessel becomes blocked or if the heart malfunctions, there is a major alteration of body function because homeostasis of many parameters cannot be held within a specific range.

The homeostatic control of blood volume and blood pressure is partially controlled by the heart. Stretch receptors in the atria determine whether the volume of blood entering is too great. If volume is excessive, the atria produce the hormone atrial naturietic peptide, which results in the decrease in blood volume. Less blood volume results in the lowering of blood pressure as well.

There Is No Substitute for Exercise

A person who is physically active can, over time, increase stroke volume. As the demand for a greater blood supply increases with activity, the heart compensates by increasing its output per beat. An advantage of a larger stroke volume is the need for a slower heart rate to maintain a normal cardiac output when at rest. In other words, when the active person rests, his or her heart has more time than the heart of an inactive person to rest between contractions.

When an individual is active, the contracting muscles compress veins in the extremities that lie between muscle groups. Each contraction of muscle results in venous blood being pushed through the next one-way venous valve. As a result, blood is circulated back to the heart more efficiently, preventing accumulation of fluid in the hands and feet.

Stress relief, along with improved circulation resulting from exercise, decreases blood pressure. Lower blood pressure decreases the risk of stroke.

> **Erythrocytes** blood cells containing primarily hemoglobin which are responsible for carrying oxygen and a small amount of carbon dioxide throughout the body. Also called *red blood cells*.

> **Leukocytes** blood cells responsible for defense of the body. Also called *white blood cells*.

> **Platelets (thrombocytes)** blood components that are active in the stoppage of blood loss resulting from blood vessel damage.

> **Plasma** the liquid component of blood that transports substances throughout the body.

FUNCTIONS OF BLOOD

Blood is composed of cells, fragments of cells, and liquid. There are two types of blood cells. Red blood cells, known as erythrocytes, carry oxygen and carbon dioxide efficiently in the bloodstream. White blood cells, known as leukocytes, provide a constantly mobilized defense throughout the body. Platelets, also known as thrombocytes, are fragments of cells active in the prevention of blood loss when a vessel is damaged. Plasma, the liquid component, carries substances such as oxygen, carbon dioxide, clotting factors, and other chemicals from one location to another along with blood cells and cellular fragments.

COMPOSITION OF BLOOD

To view the major components of blood, a small sample is obtained in a thin tube. If the tube is placed in a centrifuge and rotated for about five minutes, the red and white blood cells will separate from the liquid plasma (Figure 8.29). This technique is commonly used to determine the percent of red blood cells, known as the *hematocrit*. It is a measure of the oxygen-carrying capacity by the blood. Generally blood contains about 40% to 45% red blood cells and about 1% white blood cells; the rest is plasma. The plasma is composed primarily of water and contains substances such as electrolytes, nutrients, proteins, and hormones dissolved in the fluid.

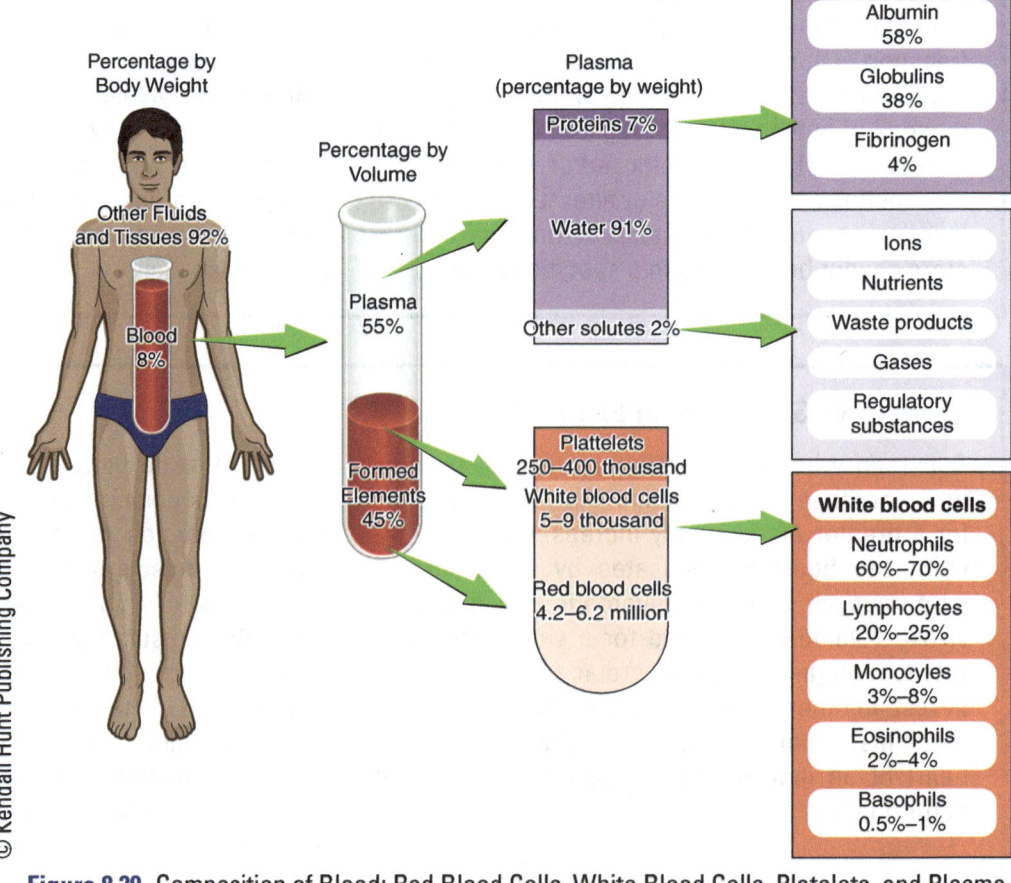

Figure 8.29 Composition of Blood: Red Blood Cells, White Blood Cells, Platelets, and Plasma.

> **Comprehension Check-up:**
>
> 1. The percentage of red blood cells in whole blood is referred to as the _____.
> 2. The liquid found in blood is known as _____.
>
> 1. hematocrit 2. plasma

Erythrocytes (Red Blood Cells)

The primary purpose of erythrocytes (*erythro* means "red") is to capture and transport oxygen and a small portion of carbon dioxide throughout the body. Oxygen and carbon dioxide are carried in the red blood cell by molecules known as *hemoglobin*. To maximize the red blood cell's oxygen-carrying capacity, as it matures, its organelles and nucleus are extruded from the cell, making additional room for hemoglobin. Because the mature erythrocyte is a plasma membrane filled with hemoglobin, it is referred to as a *corpuscle* rather than a cell, although the term *red blood cell* is commonly used.

The production of erythrocytes occurs in red bone marrow. The red blood corpuscle is a biconcave disc; that is, it is caved in on two sides (Figures 8.30 and 8.31). These indentations increase surface contact for increased diffusion of oxygen and also allow the red blood cell to be more flexible and to fold when passing through exceedingly small capillaries.

Leukocytes (White Blood Cells)

Leukocytes (*leuko* means "white") provide defense. Several different types are available to counteract various forms of potential invaders or abnormal body cells (Figure 8.32). White blood cells can be divided into two distinct groups: *granulocytes,* which are cells containing visible granules when the cells are stained, and *agranulocytes,* in which granules are not visible with the standard stain.

Illustration by Jamey Garbett. © 2003 Mark Nielsen.

© Chadsikan Tawanthaisong/Shutterstock.com

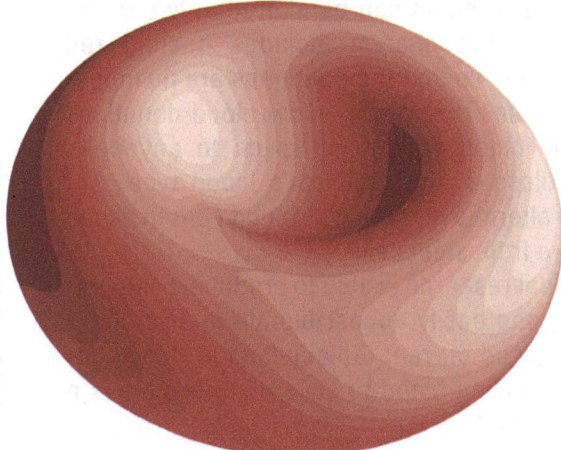

Figure 8.30 Erythrocyte. The erythrocyte is a biconcave disc.

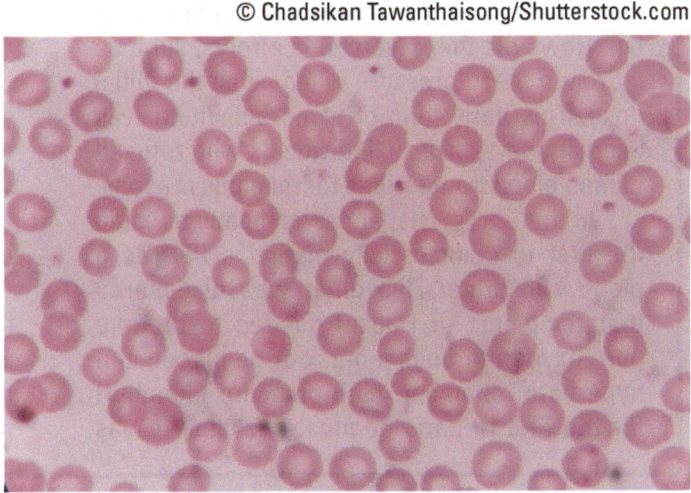

Figure 8.31 Red Blood Corpuscles Microscopic View.

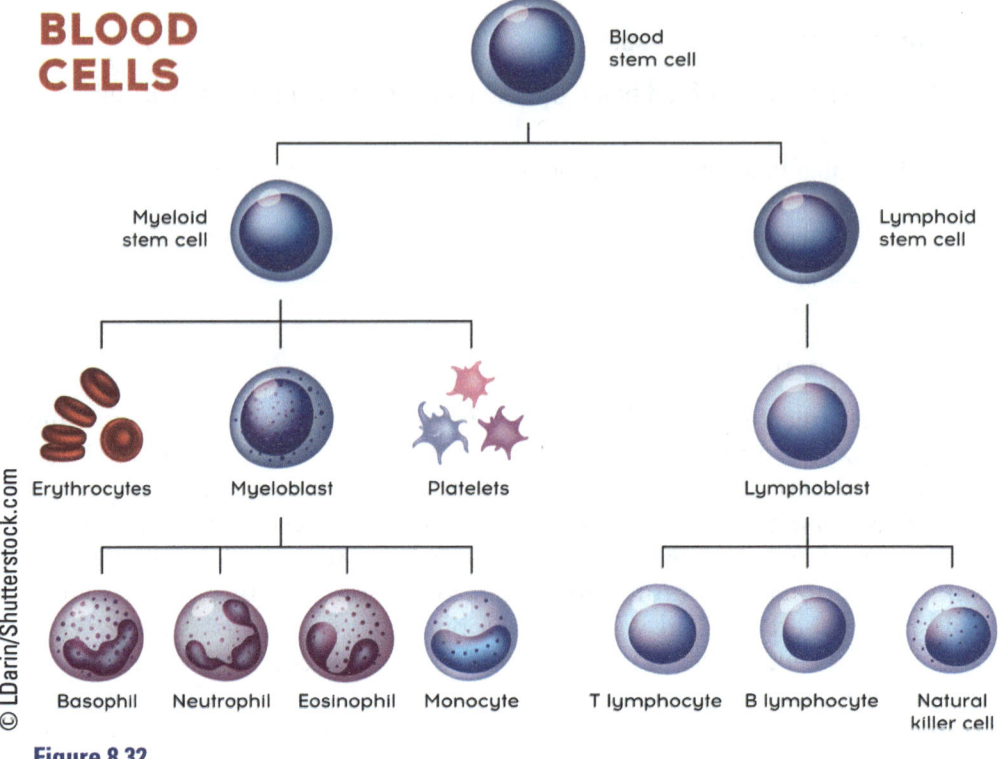

Figure 8.32

Blood Substitutes—Hope When Blood Donations Are in Short Supply

Blood substitutes have been an intriguing area of research for the past 30 years. Early pioneers such as Dr. Robert Geyer at Harvard University Medical School and Dr. Leland Clark at Children's Hospital in Cincinnati, Ohio, determined that an inert chemical similar to the coating used for nonstick cookware (Teflon) had the ability to carry oxygen. This chemical, called *perfluorocarbon* (PFC), was able to deliver oxygen to cells, acting as a substitute for hemoglobin. However, there were side effects, such as a drop in blood pressure when given the initial dose. This early form of blood substitute was used in a few cases in individuals who refused human blood transfusions for religious reasons.

In the 1980's a Japanese company became the first to commercially market a PFC as a blood substitute: Fluosol. It was discovered that recipients of Fluosol needed to breathe pure oxygen through a mask or be placed in a hyperbaric chamber (high-pressure chamber) for the PFC to carry adequate oxygen. It was also found that Fluosol depressed the immune system.

Currently other artificial blood substitutes are available. A new generation of PFC, known as *Oxyfluor*, is being investigated. Other alternatives for blood substitute include altered human hemoglobin, the production of artificial red blood cells, and the use of purified hemoglobin from cattle.

Human blood can be altered in two ways. Ongoing investigation includes coating red blood cells so that their antigens are no longer detectable. That would allow blood donations of any blood type to be given to anyone—regardless of the recipient's blood type. Another alternative is to grow blood cells in tissue culture from stem cells.

Until research is able to produce a safe, reliable, and easily accessible alternative, there is simply no current replacement for blood donation. Giving blood to save another's life is a priceless gift.

> **Causes of Abnormal Quantities of Red Blood Cells**
>
> There are two general categories of abnormal quantities of red blood cells: anemia and polycythemia.
>
> Anemia is the lack of sufficient red blood cells. There are numerous causes:
>
> - Nutritional deficiencies, such as a lack of sufficient iron for production of hemoglobin
> - Kidney malfunction, resulting in a failure to stimulate red blood cell production
> - Blood loss from injury or disease processes
> - Failure of the red bone marrow to produce sufficient red blood cells as a result of chemical toxins, radiation, or excessive exposure to x-rays.
>
> The overproduction of red blood cells is termed *polycythemia*. One cause is a tumor-like condition in the red bone marrow that results in the production of too many red blood cells. Polycythemia interferes with blood viscosity, blood flow, and oxygen supply to the organs.
>
> It is also possible to have an abnormally high quantity of red blood cells that is not a cancerous condition but results from the reduced ability to obtain oxygen as a result of chronic respiratory disease. In this case it is not the bone marrow that is malfunctioning, but rather a means of compensating for the increased need to be able to hold as much oxygen as possible when the availability is decreased.

Granulocytes

There are three types of granulocytes identified by the color of the stain taken up by their granules. In each of these cells is a nucleus that appears to be broken into pieces connected by thin strands. No other cells in the body have this unusually shaped nucleus. The types of granulocytes are as follows:

- *Eosinophils* (*phil* means "love"; these cells "love" eosin) absorb the acidic red stain eosin, causing these cells to be filled with red granules (Figure 8.33). They are active when allergens are present or when the body is invaded by parasites.
- *Basophils* contain granules colored by a blue alkaline stain (Figure 8.34). These cells produce histamine and heparin. Histamine causes pores in the capillary walls to enlarge, leaking fluid and clotting factors into the surrounding tissue. It may flood the area with fluid to wash away irritating substances or dilute toxins. It may act as a protective measure by allowing the formation of a clot to imprison invaders as other defensive cells rush in to attack. It is released during allergic reactions. For example, if a person is sensitive to pollen in the air, histamine may be released in the nasal passage and eyelids, causing the nose to run and the eyes to become itchy. The individual may counteract these symptoms by taking an antihistamine. The function of heparin is not well understood. It apparently increases the removal of fat particles from the bloodstream after eating a fatty meal. Heparin is used pharmacologically to block the coagulation of blood.
- *Neutrophils* contain unstained granules that have a neutral pH (Figure 8.35). Neutrophils are the most common of the white blood cells. They are phagocytes; that is, they engulf foreign cells—solid objects like dirt, dust, or splinters that may enter the body—and they remove debris, such as dead or damaged cells remaining after defense of the body has ended.

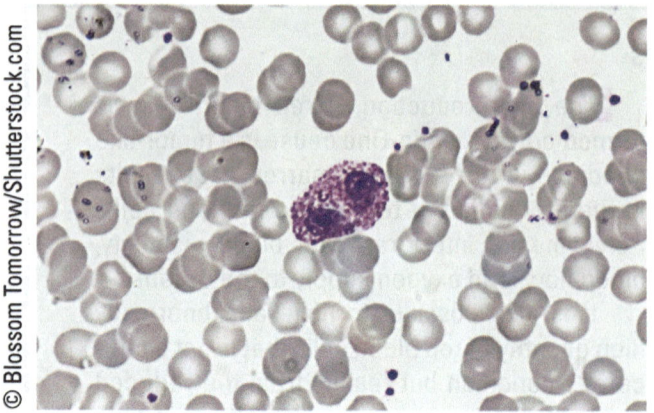

Figure 8.33 Eosinophil.

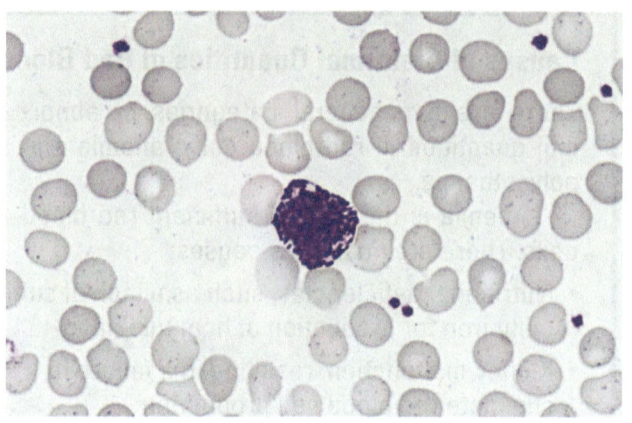

Figure 8.34 Basophil.

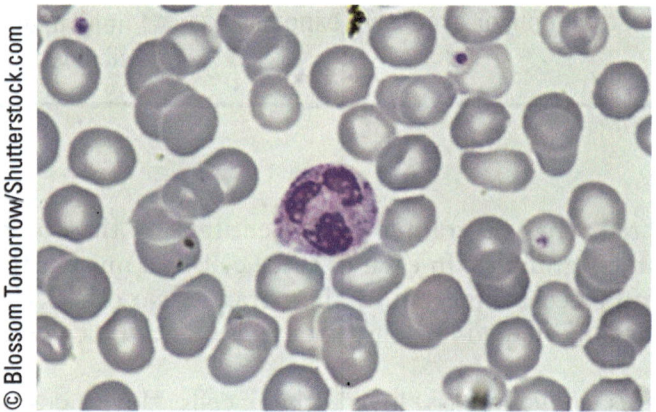

Figure 8.35 Neutrophil.

Agranulocytes

Agranulocytes not only are devoid of stained granules but also have a non-segmented nucleus. There are two types:

- *Monocytes* are the largest of the white blood cells (Figure 8.36). When a monocyte leaves the bloodstream, it enlarges to become a macrophage, which is a very large defensive cell that lives primarily in the skin, the lungs, and the small intestines.
- *Lymphocytes* are the smallest of the leukocytes. Their cytoplasm is filled almost entirely by its nucleus (Figure 8.37). Lymphocytes have the ability to recognize specific foreign organisms and substances. Lymphocytes also control immunity. They are discussed in more detail in Chapter 10.

Platelets (Thrombocytes)

Platelets are active in stopping blood loss from a broken blood vessel. Platelets are fragments (Figure 8.38) that break off from very large cells called *megakaryocytes*. Platelets function in the process of terminating blood loss, known as *hemostasis* (discussed later).

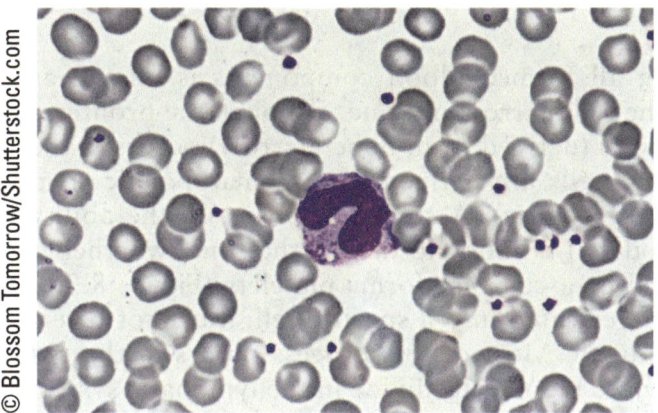
Figure 8.36 Monocyte.

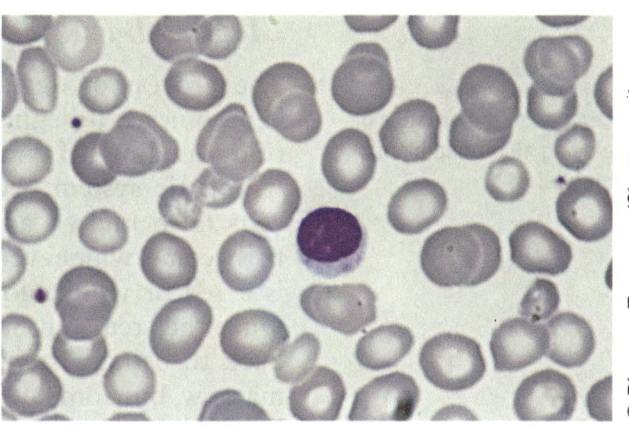
Figure 8.37 Lymphocyte.

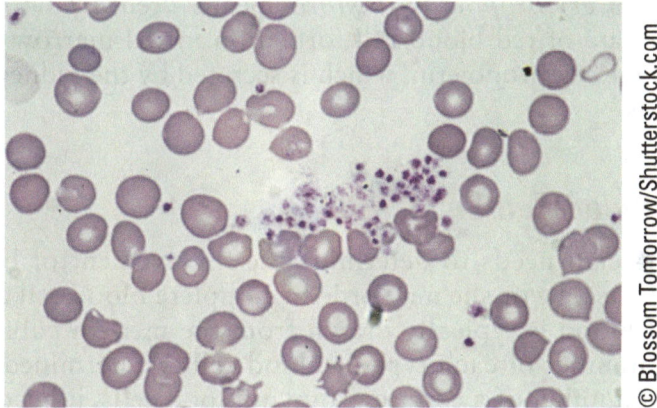
Figure 8.38 Platelet.

Comprehension Check-up:

1. White blood cells that do not contain granules are known as _____.
2. Most oxygen is carried in the blood by _____.

1. agranulocytes   2. red blood cells or hemoglobin

Abnormal White Blood Cells

Many diseases affect the quantity of white blood cells. Two result in abnormal leukocytes: infectious mononucleosis and leukemia.

Infectious mononucleosis is a contagious viral disease that results in pronounced fatigue, sore throat, and a slight fever. This disease causes the production of abnormally shaped lymphocytes. Other than the increase in lymphocytes there is no other increased formation of additional leukocytes.

Leukemia is the cancerous overproduction of immature white blood cells. Not only do these excessive quantities of leukocytes crowd out red blood cells and platelets, but they are also produced so rapidly that they are immature and incapable of defending the body. Contrary to what would seem to be the typical outcome, as this disease progresses, infection becomes a significant concern.

Blood Cell Formation

The process of developing all formed blood components is known as hemopoiesis. It begins in the red marrow of bone, although the proliferation (dividing) of lymphocytes (derived from bone marrow) occurs in the lymphatic system. All blood cells are derived from hematopoietic stem cells (also known as hemocytoblasts). A stem cell possesses the ability to become whichever blood cell is most needed at the time, whether it is a red or white blood cell or a cell that forms platelets (Figure 8.39). Once the stem cell has differentiated into a specific cell, it cannot revert back to being a stem cell again. Most of this process occurs in the red marrow, but some of the white blood cells divide in the lymphatic system as well.

Red blood cells survive an average of 120 days. They die off and are replaced at the rate of approximately 2 to 3 million per second by a process known as *erythropoiesis* (*erythro* means "red"; *poiesis* means "to make"). The rate of red blood cell formation in red marrow is controlled by the hormone erythropoietin, which is secreted by the kidneys to increase erythropoiesis.

> **Hemopoiesis** the process of developing all formed blood components.

Complete Blood Cell Count (CBC)

When a physician needs to determine the involvement of blood cells in fighting a disease, he or she may order a complete blood cell count (CBC). A sample of blood is typically drawn from the median cubital vein, and the average number of each type of blood cell is determined. If there are an abnormal number of a certain group of blood cells, it becomes a major clue in the determination of the cause of the ailment. Elevated eosinophils may indicate severe allergies or a parasite infection. Abnormally high levels of neutrophils, monocytes, and lymphocytes often indicate a bacterial infection. Hematocrit is also determined along with the content of hemoglobin in those red blood cells. A decreased percentage of red blood cells or abnormally low hemoglobin levels often result in fatigue, cold intolerance, and sometimes shortness of breath. The CBC is relatively easy to obtain and is a comparatively inexpensive method to diagnose the patient's condition.

> **Comprehension Check-up:**
> 1. The term for red blood cell production is _____.
> 2. Severe allergies or a parasite infection is indicated the CBC shows an elevation in _____.
>
> 1. erythropoiesis 2. eosinophils

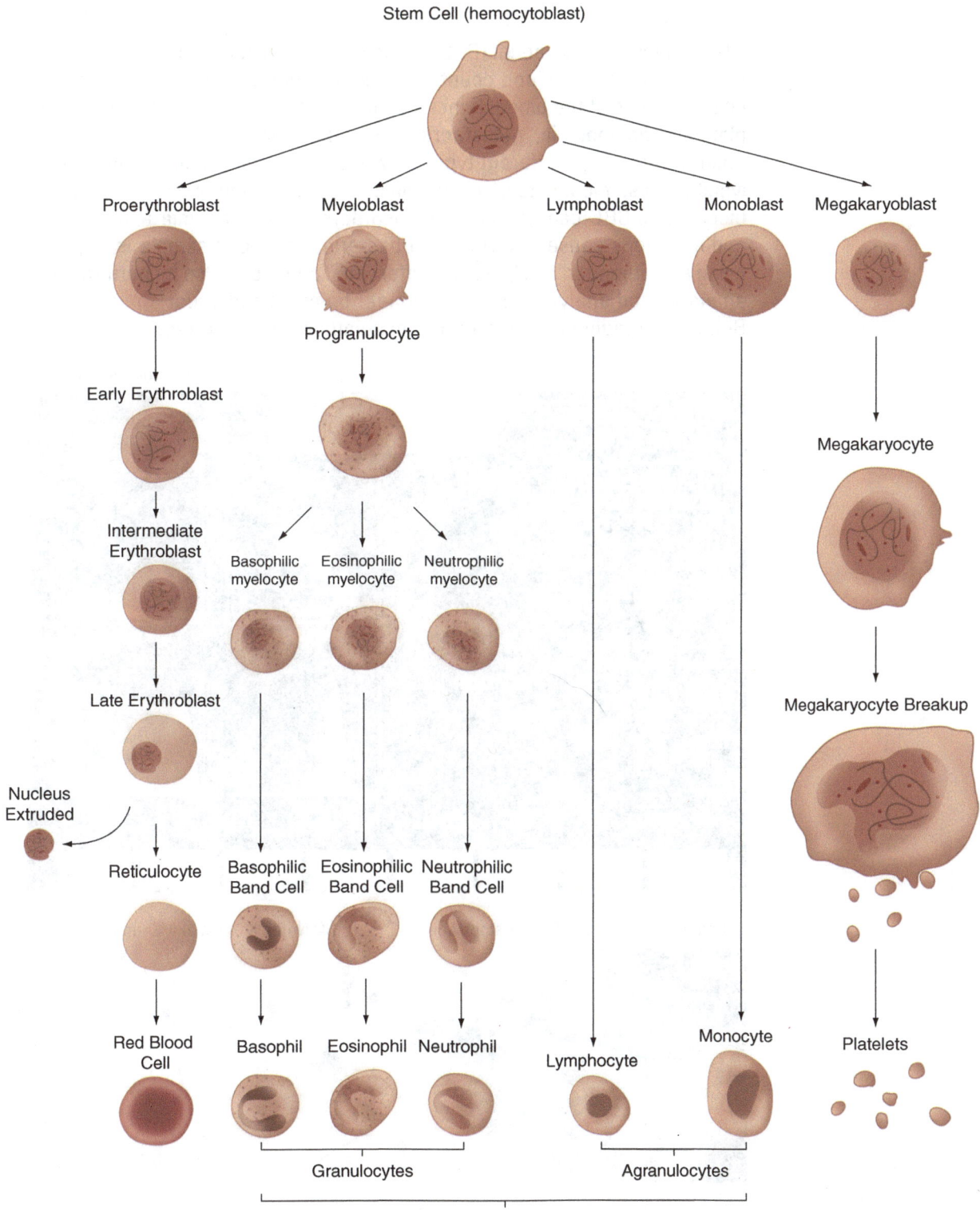

Figure 8.39 **Formation of Blood Cells and Platelets.** All blood cells differentiate from hematopoietic stem cells.

Clinical Notes—Complete Blood Count (CBC)

The complete blood count (CBC) is one of the most commonly ordered blood tests. The complete blood count is an examination of the blood cells or cellular elements of the blood. It includes the red blood cells, white blood cells, platelets, hemoglobin, hematocrit, and differential count. The differential count refers to the different types of white blood cells such as neutrophils, lymphocytes, monocytes, eosinophils, basophils, and possible abnormal blood cells. CBC can detect blood disorders such as anemia and leukemia and infections caused by virus and bacteria. The blood sample is analyzed using an automated analyzer. The differential count can also be performed by an automated analyzer, or a blood smear be examined under a microscope. Below is a reading of CBC report from an automated analyzer.

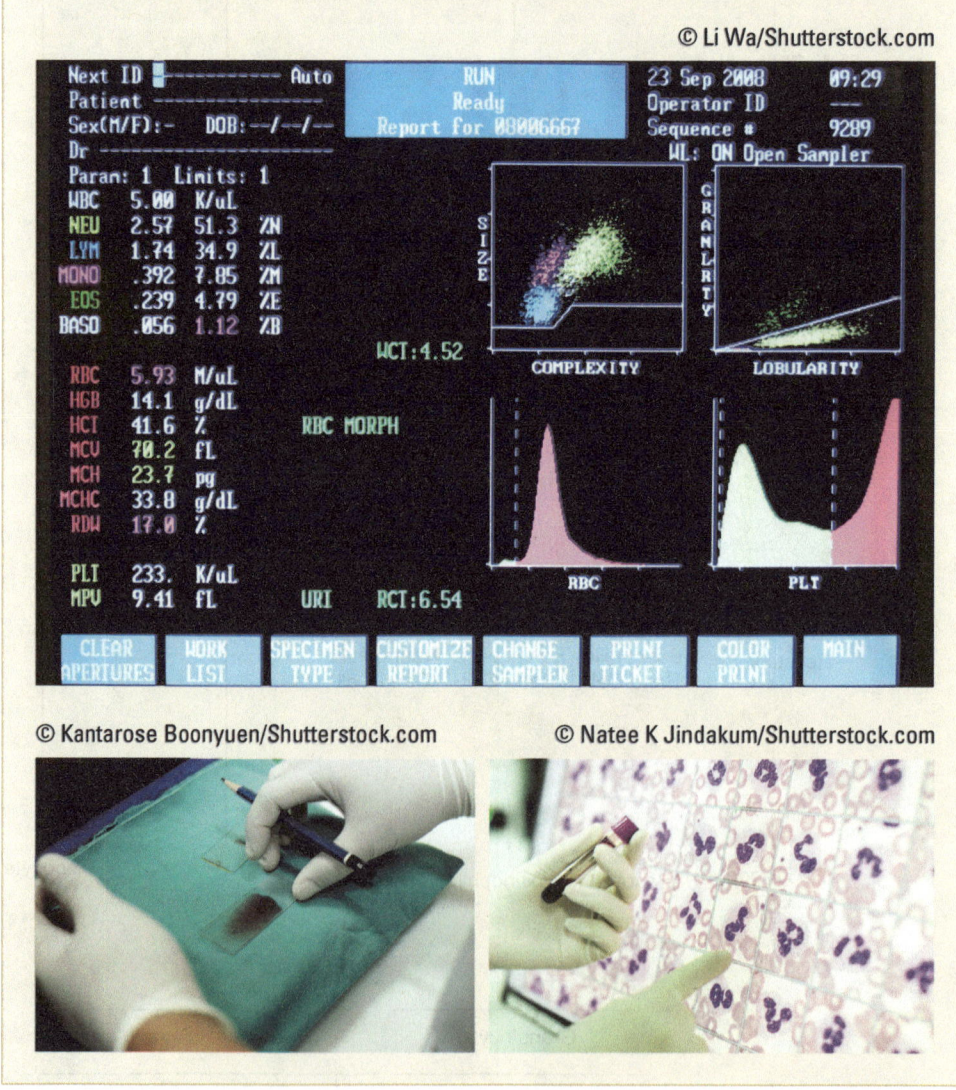

Plasma

Plasma, the liquid portion of blood, is composed mostly of water, with enough sodium chloride to form a 0.9% sodium chloride solution (0.9 grams of sodium chloride per 100 ml of water). If there is a deviation from this level, the homeostasis of the entire body may be affected. Plasma more concentrated than 0.9% sodium chloride draws water out of the body's cells by osmosis, causing them to shrink and, especially in the brain, to malfunction. The individual may become mentally confused and irritable. If the condition worsens, coma and death could ensue. Plasma diluted to a concentration of less than 0.9% sodium chloride causes cells to swell, resulting in brain cell malfunction similar to that which occurs when cells shrink.

Incorporated into the plasma are the following:

- Plasma proteins:
  - A protein known as albumin increases the ability of the plasma to draw water into the bloodstream by osmosis. Water is drawn in to maintain the balance of fluid between the bloodstream and tissue.
  - Globulins may be used for defense or for the transport of lipids or other substances. Antibodies, a type of globulin, act as chemical tags that identify foreign or abnormal cells to be destroyed by body defenses. They may also neutralize toxins produced by foreign organisms.
  - Clotting factors, also a type of globulin, provide a set of checks and balances that activate if a clot needs to be formed.
  - Fibrinogen is activated to form a clot after all clotting factors have indicated a clot is appropriate.
- Water, the major component of plasma, dissolves or transports all other substances and carries heat.
- Electrolytes or ions essential for communication and metabolism, such as sodium, potassium, calcium, and phosphate, are dissolved in the plasma.
- Nutrients include carbohydrates, lipids, amino acids, and vitamins being transported to cells for metabolism and other cellular functions.
- Metabolic wastes, such as urea from the removal of nitrogen from protein, ketoacids from the metabolism of fat, and bilirubin from the catabolism of heme are transported to the kidneys for removal from the bloodstream.
- Respiratory gases include oxygen and carbon dioxide. Oxygen is needed by cells for aerobic metabolism, and carbon dioxide (CO_2) is given off as a by-product of aerobic adenosine triphosphate (ATP) production. CO_2 is transported from the cells to the lungs to be exhaled from the body.
- Hormones are produced as chemical messages by endocrine glands.

Plasma with the clotting factors removed is known as *serum*. This may be obtained when a sample of blood is allowed to clot. The liquid remaining is serum.

Several systems within the body maintain the components of the plasma within their homeostatic range. Failure to do so could potentially have very serious affects on the body's ability to function and may even result in death.

> **Albumin** a blood protein that increases the ability of the plasma to draw water into the bloodstream by osmosis so that the balance of fluid between the bloodstream and tissue is maintained.

> **Globulins** blood proteins that may be used for defense or for the transport of lipids or other substances.

> **Antibodies** Y-shaped particles that attach to specific antigens. Once these antibodies are fixed to a foreign cell's antigens, other defensive cells go into action to destroy the marked cell. Antibodies may also neutralize toxins and agglutinate red blood cells of the wrong type.

> **Clotting factors** a type of globulin. They provide a set of checks and balances that activate if a clot needs to be formed.

> **Fibrinogen** after all clotting factors have indicated a clot is appropriate, fibrinogen is activated to form a clot.

Comprehension Check-up:

1. The concentration of sodium chloride in plasma is normally _____.
2. The main protein in the plasma that increases osmosis into the bloodstream is _____.

1. 0.9% sodium chloride   2. albumin

HEMOSTASIS

> Hemostasis process by which blood loss is stopped.

Blood loss must be kept to a minimum. Hemostasis is the process by which blood loss is stopped. When a blood vessel is broken, three basic steps occur to terminate blood flow out of the cardiovascular system:

- Vascular spasm
- Formation of a platelet plug
- Coagulation (blood clotting)

Each of these processes is discussed individually (Figure 8.40).

Homeostasis

Broken Vessel

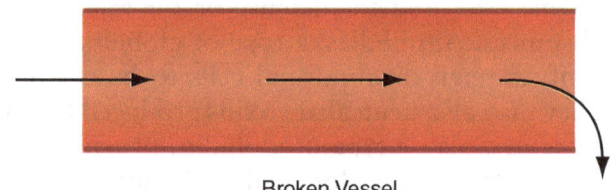

(a) Smooth muscle spasms to reduce opening

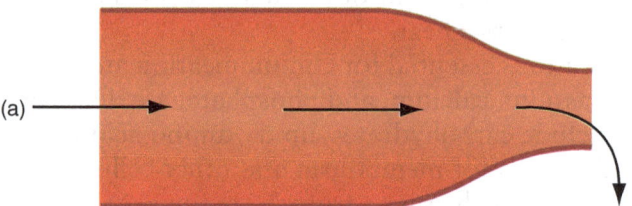

(b) Platelets form a plug

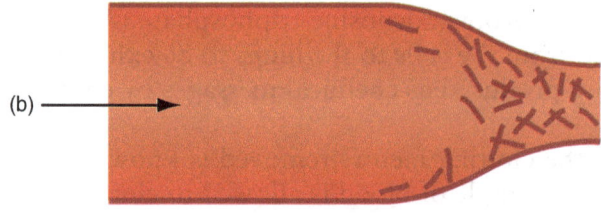

(c) Clotting factors activate fibrinogen to fibrin to form a clot

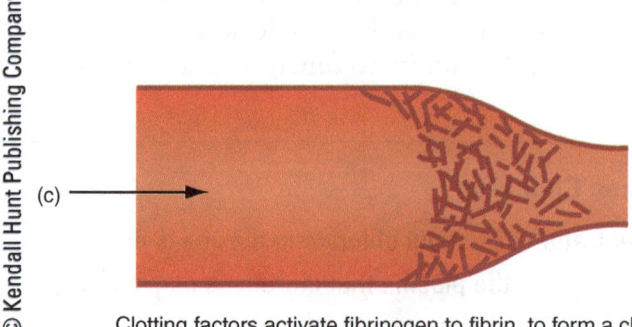

Figure 8.40 Hemostasis. (a) Smooth muscle in the broken vessel spasms decreases the diameter of the vessel to reduce the size of the opening. (b) Platelets attach to the sides of the broken vessel to form a plug. (c) Clotting factors in the plasma are activated and convert fibrinogen to fibrin, which forms the clot to seal the opening.

Vascular Spasm

Wrapped around every blood vessel except for capillaries is smooth muscle. When a blood vessel is damaged, the trauma causes the smooth muscle at the site of the wound to contract. This automatic tension reduces the size of the vessel and decreases blood flow out of the damaged area.

Formation of a Platelet Plug

As blood passes out of the opening in a damaged vessel, platelets are activated by the damaged tissue to become sticky. Platelets attach to the edge of the broken vessel and to each other to form a netlike structure over the opening of the vessel. The net of platelets also possesses the ability to contract, causing it to become more compact and thereby further reducing the size of the opening.

Coagulation

Recall that the plasma contains clotting factors. They are a set of inactive chemicals that must all be activated to cause coagulation (clot formation). It is critical that the plasma remain liquid except when a break occurs in a blood vessel. Therefore, there is a set of checks and balances to ensure the formation of a clot is appropriate. The broken vessel causes activation of one of the clotting factors. One factor is able to activate hundreds of molecules of the next factor that is in line. The process continues activating clotting factors in sequence until all clotting factors in the series activate. If all criteria are met and the final clotting factor in the sequence is activated, fibrinogen is converted into fibrin and the clot is formed. When fibrin is formed, it sticks to the platelet net to form a clot and seal the opening to terminate blood loss. If any factor in the sequence does not activate, the process stops at that point and the conversion to fibrin does not occur.

> **Coagulation** clot formation.

Inappropriate Clotting

Blood clots may also occur in intact vessels. These clots may occur because the lining of the blood vessels has become roughened or narrowed, usually brought on by the presence of a cholesterol-containing ridge known as *plaque*. Chemicals on the plaque may activate platelets to become sticky and initiate the clotting process.

Another type of inappropriate clotting may result from slow-moving blood. Venous valves, particularly in the lower extremities, may become nonfunctional, allowing blood to pool. These distended vessels are referred to as *varicose veins*. Because blood cannot be pumped back to the heart effectively by these deep veins as the surrounding muscles contract, the pooled blood begins to clot. If a clot forms and is free floating, it is considered an *embolus*. Of great concern is when the embolism lodges and becomes a fixed clot, known as a *thrombus*. If there is blockage to the brain or heart, a stroke or heart attack could occur, which could lead to death.

The opposite situation can occur if an individual is born without one of the clotting factors. Because one of the factors determining whether a clot forms is missing, the final stage of clot formation is never reached. This genetically caused condition is known as *hemophilia*. A slight injury can result in significant blood loss. It is essential that individuals with hemophilia carry the missing clotting factor as an injection so that it can be administered at the appropriate time.

> **Comprehension Check-up:**
> 1. The term for a fixed clot is _____.
> 2. The blood components that become sticky and form a net over the opening in a broken vessel are _____.
>
> 1. thrombus   2. platelets or thrombocytes

BLOOD TYPES

If an individual's hematocrit is significantly reduced, a transfusion of additional blood from a donor may be necessary. When blood transfusions were first attempted in the 1800's, the donor and recipient were placed side by side and a catheter was used to carry blood directly from the donor to the recipient. Sometimes the outcome was successful, and sometimes the recipient died. It was later discovered that the cause for success or failure was due to the type of *antigens* (substances that are capable of binding with antibodies and triggering an immune response) in the blood, that is, the proteins attached to the outside of the red blood cell membrane. To explain the results of these direct transfusions, it is necessary to discuss the ABO system of blood typing.

ABO System

There are two common types of antigens attached to the red blood cell membrane, labeled A or B. If neither type of antigen is present, the blood is designated as type O (Figure 8.41). Genetics determines that an individual always produces the same antigens on the red blood cell membranes throughout life. For example, those producing type A red blood cells make only type A antigens and no other type. In the plasma are *antibodies* (protein molecules that react with and destroy the specific antigens) that are designed to destroy the wrong type. Each antibody is capable of holding several red blood cells together, creating clumps of the wrong type—a process called *agglutination*. (Figure 8.42). For example, individuals who produce type A red blood cells have antibodies in their plasma that cause type B red blood cells to agglutinate. It is the antibodies in the plasma of the recipient that determines the type of blood that is acceptable in a transfusion.

Hemophilia Affects Primarily Males and Comes from the Mother's Side of the Family.

The cause for this ailment has to do with sex chromosomes. Of the 46 chromosomes humans normally have, 2 of them determine an individual's sex. A female has two X chromosomes and the male has one X and one Y chromosome. Genes for clotting factors are found on the X chromosome. The male has only one X chromosome, which he receives from his mother. The Y chromosome can come only from his father. Because he has no alternative source of the necessary genetic information, the man with the defective X chromosome is a hemophiliac. If a woman has a defective X chromosome from one parent, she has another X as a backup, so she is typically not a hemophiliac. The only way she would lack clotting factors is if both X chromosomes had the same defect. Although a woman with a defective X chromosome may not exhibit the traits of hemophilia, there is a 50/50 chance her sons will have those traits. Hemophiliac males are advised to receive genetic counseling because all of their daughters would be carriers of the trait.

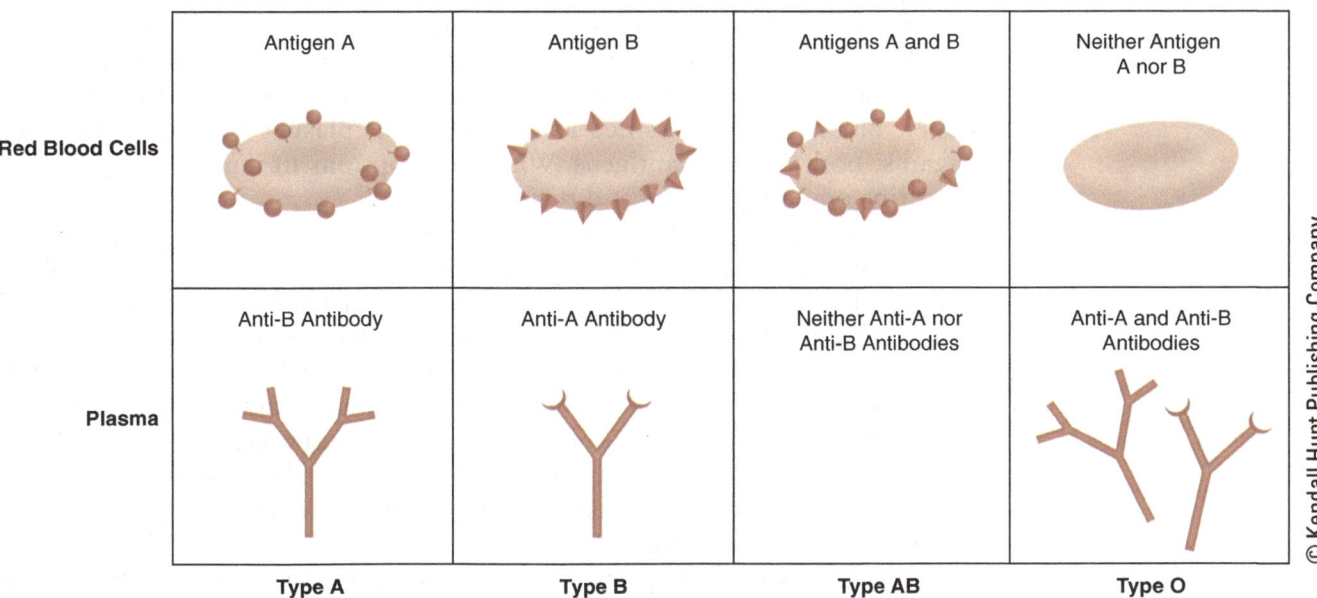

Figure 8.41 Red Blood Cell Antigens. Antigens on the red blood cell identify its type. Antibodies in the plasma attach to antigens of other blood cell types causing those blood cells to clump together (agglutinate).

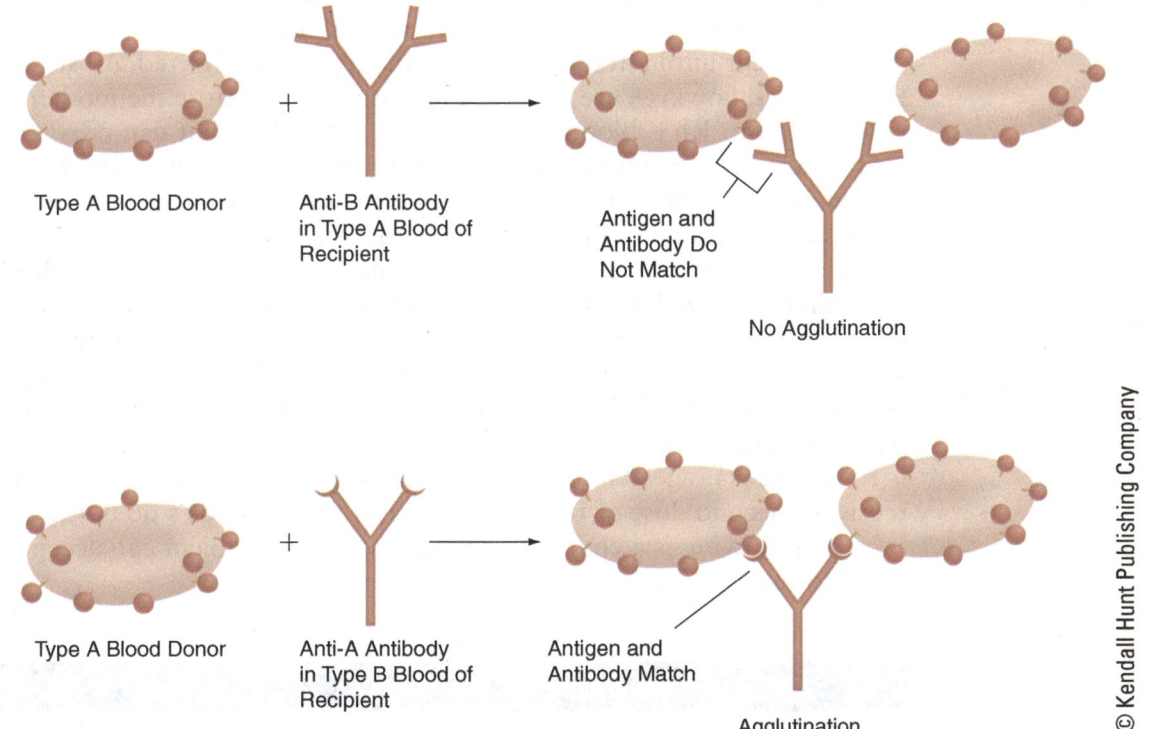

Figure 8.42 Action of Antibodies. When a person with type A red blood cells receives a donation of type A red blood cells, the type B antibodies do not cause agglutination. If someone with type B red blood cells receives a donation of type A red blood cells, the type A antiserum causes agglutination of those type A red blood cells.

Some individuals produce both A and B antigens. Individuals with type AB blood do not have antibodies against either type because they would destroy their own cells if they did. There are others who have neither antigens. Their blood type is classified as type O. Their plasma possesses antibodies against both A and B antigens. With those options in mind it is possible to describe potential donors and recipients.

There is an additional antigen, called *Rh factor* (also known as *factor D*), that is not part of the ABO designation but that may be found on red blood cells. Individuals with this additional tag on their red blood cells have a "+" added to their blood type. If this additional antigen is not present, then a "–" follows the ABO designation. Unlike those with type A, B, or O blood, which possesses antibodies against differing types in their plasma by about age two, Rh– negative individuals do not make D (Rh) antibodies unless exposed to Rh– positive blood. Their D antibody-producing cells must be activated by the presence of Rh+ blood before agglutination would occur. The first time of an Rh– individual receives Rh+ blood, it would not result in agglutination. Any subsequent exposure to Rh+ blood would cause agglutination in this individual. This exposure occurs in two possible ways. If an Rh– patient receives a transfusion of Rh+ blood, D antibodies will be formed. More commonly is a case where an Rh– woman is carrying a fetus whose father is Rh+. There is a 50/50 chance the fetus would be Rh+. During pregnancy the mother's blood does not mix with the fetus's blood. After birth, as the placenta tears away from the uterine wall, some of the capillaries in the placenta break and expose the mother to the baby's Rh+ blood. The first Rh+ baby is not affected by D antibodies because the exposure of the mother to Rh+ blood does not occur until after the baby has already been delivered. All subsequent Rh+ fetuses, however, are at risk of agglutination because the mother's body is sensitized to Rh+ blood and produces antibodies that can cross the placenta and cause agglutination of the fetus's blood, resulting in severe anemia. It is important that D antibodies not be formed in the mother's plasma if at all possible. Antibodies known as *RhoGAM* can be injected into the mother. These antibodies attach to the Rh antigens to prevent the Rh- mother from becoming sensitized against Rh+ blood. It would be too risky to obtain a sample of fetal blood prior to birth to determine its type, so all women who are Rh- receive an injection of RhoGAM fairly late in pregnancy to prevent the formation of D antibodies should some exposure to Rh+ blood occur. After birth a sample of the baby's blood is taken to determine its type. If the baby is Rh-, there is no need to do anything further. If the baby is Rh+, the mother receives a second dose of RhoGAM to ensure D antibodies do not form.

The following is a chart describing whether a recipient can accept a specific blood type from a donor:

		Donor Type			
		A	B	AB	O
Recipient Type	A	Yes	No	No	Yes
	B	No	Yes	No	Yes
	AB	Yes	Yes	Yes	Yes
	O	No	No	No	Yes

Type O– is the universal donor type. These individuals are able to give to all other blood types because there are no antigens on the red blood cells. A person with type AB+ blood is the universal recipient because no antibodies exist in the blood against other types.

> **Comprehension Check-up:**
> 1. The proteins on the red blood cells that distinguish one type from another are known as _____.
> 2. _____ in the plasma of a blood recipient cause the wrong type of blood cells to agglutinate (stick together in clumps).
>
> 1. antigens 2. Antibodies

Homeostasis—Holding in Balance

The control of blood loss resulting from injury or trauma is an important factor in maintaining a constant blood volume. It is critical that blood form a clot when needed, but it is just as important that plasma remain liquid when a blockage would be inappropriate. Let either situation fail and homeostasis of blood volume and pressure is lost, meaning other systems dependent on the flow of blood for transporting essential substances throughout the body are unable to perform their functions.

Constantly vigilant is a circulating force of white blood cells ready to attack invaders to prevent them from overwhelming our organ systems, which could result in a loss of homeostasis. Blood does not contain the entire supply of white blood cells. The majority of them are stored in the spleen and lymphatic nodes. What circulates is a relatively small amount of leukocytes specialized at recognizing and removing anything that may be a potential threat. Although white blood cells are not themselves maintaining homeostasis, they are defending against foreign invaders, which could prevent other systems from keeping parameters within a constant range.

There Is No Substitute for Exercise

When an individual exercises, his or her blood volume decreases for several reasons. Blood pressure increases during exercise, causing additional plasma to filter into the tissue. Sweating moves some of the tissue fluid to the surface of the skin to cool the body. By performing regular, sustained exercise, the body is able to more efficiently compensate for this change in blood volume, allowing the active individual to increase stamina.

As discussed in previous chapters, exercise helps relieves stress. With less stress, the immune system is more active than when in "fight-or-flight" mode. As a result, our white blood cells and other defense mechanisms found in the blood are more active in reducing the risk of foreign invasion.

Clinical Notes—ABO and Rh Blood Type

Blood typing is a simple blood test to determine a person's blood type—whether it is A, B, AB, or O. Antigens in the blood determine the blood type. Type A has the A antigen. Type B has the B antigen, Type AB has the antigens A and B, and type O has neither A nor B antigens. The importance of knowing the blood type is primarily for blood transfusion whether the person is the donor, or the recipient. It requires 100% accuracy to transfuse blood with the correct blood type.

A person who has a blood type A must receive blood from a donor with the same blood type, Type A. If a patient with blood type B receives blood from a type A donor, the patient's immune system will identify the type A antigens from the donor as foreign, and will form anti-A antibodies that will destroy the A antigens of the donor. This will cause transfusion reaction and can cause a fatal illness on the patient.

Blood Transfusion Guide for ABO

A type O person can donate blood to any patient because the blood has no antigens. A type O person can only receive type O blood from other individuals.

A type A person can donate blood to a type A or type AB patient. A type A person can only receive blood from type A or type O individuals.

A type B person can donate blood to a type B or type AB patient. A type B person can only receive blood from type B or type O individuals.

A type AB person can donate blood to a type AB patient only. A type AB person can receive blood from individuals with any blood type.

Blood Transfusion Guide for Rh (Rhesus) Factor

Rh is an antigen on the red blood cells found in at least 85% of the population. It can cause incompatibility in blood transfusions. A person whose blood is Rh-positive (Rh+) has Rh antigens. The person can receive blood from either Rh positive or Rh-negative blood. A person whose blood is Rh-negative has no Rh antigens. An Rh-negative person can only receive blood from another Rh-negative person.

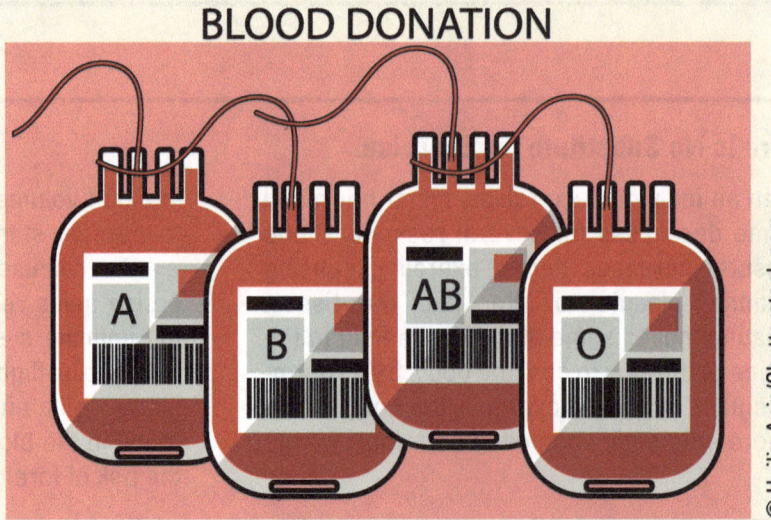

Clinical Skills—ABO and Rh Blood Typing

1. Wash hands. Layout all materials for blood typing.
2. Mix gently the anti-A antiserum (blue), anti-B antiserum (yellow) reagents, and anti-D (Rh) antiserum.
3. Place a drop of each antiserum separately in a glass slide.
4. Disinfect skin of fingertip with an antiseptic to help prevent infection.
5. Prick the fingertip with a sterile lancet.
6. Squeeze the fingertip to obtain drops of blood.
7. Place a drop of blood in the anti-A antiserum, another drop of blood in the anti-B antiserum, and one in the anti-D antiserum.
8. Use three applicator sticks. Use different applicator stick to mix each one of the three antiserum–blood mixtures. Let this mixture stand for 2 minutes.
9. Read the clumping or agglutination. If there is clumping in anti-A antiserum, it is blood type A; if there is clumping in the anti-B antiserum, it is blood type B. If there is clumping on both, it is Type AB; and if there is no clumping on both, it is type O.
10. Record results. Dispose used materials in biohazard container. Wash hands.

© Tino Bandito/Shutterstock.com

© Matej Kastelic/Shutterstock.com

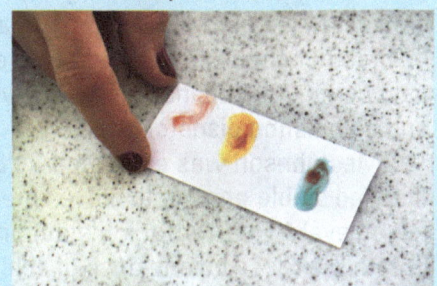

Blood types

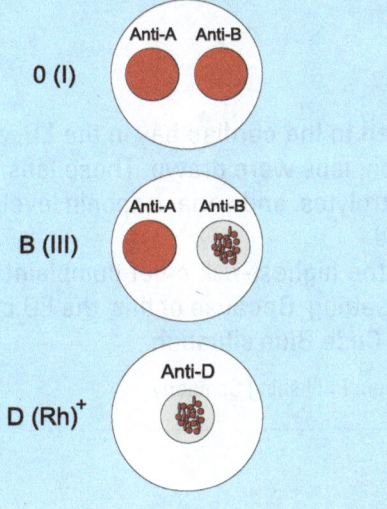
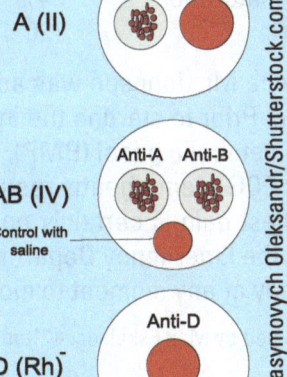

© Gerasymovych Oleksandr/Shutterstock.com

Clinical Scenario—Chest Pain

Mr. Johnson is a 64-year-old male with a history of hypertension, esophageal reflux disease, and hyperlipidemia. Mr. Johnson arrives at the emergency department via EMS complaining of his first ever chest pain. Mr. Johnson was working in his garden. He states that the pain started off as minimal, which he first thought as just a case of his esophageal reflux "acting up." When the pain continued and increased, his wife called 911 and asked the paramedics and ambulance dispatched.

Mr. Johnson was evaluated by EMS at his home. When the patient was deemed stable enough to move safely, he was transported to Henrys' Hospital, a Level 1 Heart Attack Institution. Mr. Johnson was only slightly short of breath. He received one dose of sublingual nitroglycerine by EMS and had no real change in his chest pain.

His vital signs on presentation to the ED are:

BP	150/100
Temp	98.4
O$_2$ saturation	94%
HR	120

Once Mr. Johnson was admitted to the cardiac bay in the ED, an IV was started. Prior to starting the infusion, labs were drawn. These labs include a basic metabolic panel (BMP), electrolytes, and serial troponin levels. A diagnostic ECG is emergently requested.

Chest pain is certainly one of the highest-risk chief complaints you will see in the Emergency Department setting. Because of this, the ED crew must be ready at any moment to move a Code Blue situation.

Contributed by Max Eskelson. © Kendall Hunt Publishing Company

CLINICAL TERMS TO REVIEW

Arteries, capillaries, veins
Myocardium, endocardium, pericardium
Right atrium, right ventricle, left atrium, left ventricle, sinoatrial node, pacemaker
Superior vena cava, inferior vena cava pulmonary artery, pulmonary vein, coronary arteries
Tricuspid valve, pulmonary semilunar or pulmonic valve, bicuspid or mitral valve, aortic valve

Electrocardiogram (ECG), bradycardia, tachycardia, premature ventricular contraction (PVC), ventricular fibrillation (VF), heart block
Stroke volume, cardiac output, heart failure
Pulmonary circulation, systemic circulation, blood pressure, venous return, hepatic portal circulation
Red blood cells, white blood cells, platelets

Neutrophils, lymphocytes, monocytes, eosinophils, basophils
Hemoglobin, hematocrit
Plasma proteins—albumin, globulin, fibrinogen, antibodies, clotting factors
Platelet plug, coagulation
Blood typing, ABO system
Chest pain, complete blood count (CBC)

Test Yourself

Choose the best answer to the following multiple choice questions:

1. The layer of the heart that consists of cardiac muscle and specialized conduction fibers is known as the
   a. endocardium.
   b. pericardium.
   c. myocardium.
   d. epicardium.

2. Oxygen-rich blood is returned from the lungs to the left atrium through the
   a. pulmonary veins.
   b. pulmonary trunk.
   c. pulmonary arteries.
   d. pulmonary septum.

3. The purpose of the heart valves is to
   a. cause oxygen to flow into blood and carbon dioxide to flow out.
   b. ensure the flow of blood is in only one direction through each ventricle.
   c. cause the action potential to travel in only one direction through the heart.
   d. keep the lungs from collapsing as a result of insufficient blood.

4. The action potential that causes the ventricles to contract reaches the ventricles through the
   a. sinoatrial node.
   b. hepatic portal circulation.
   c. coronary arteries.
   d. atrioventricular node.

5. The wave of the electrocardiogram that indicates the depolarization of the atria is known as the
   a. P wave
   b. QRS complex
   c. T wave
   d. U wave

6. In order to calculate the cardiac output of the heart, if I know the amount of blood ejected by the heart per beat, I can determine the output if I know the
   a. height and weight of the individual.
   b. age in days divided by 37.9.
   c. size of the arc produced when rotating the upper extremity plus abdominal circumference at the navel.
   d. heart rate.

7. One complete heartbeat from the start of atrial contraction followed by ventricular contraction and relaxation to the start of the next atrial contraction is referred to as the
   a. life cycle.
   b. cardiac cycle.
   c. systolic phase of the heart.
   d. Krebs cycle.

8. The layer of blood vessels that contains smooth muscle and elastic fibers is known as the
   a. tunica externa.
   b. tunica media.
   c. tunica submucosa.
   d. tunica intima.

9. Arteries that control blood flow to regions of the body are known as _____ arteries.
   a. elastic
   b. diversionary
   c. muscular
   d. regulatory

10. Oxygen-poor blood is pumped out of the right ventricle to the lungs through the
    a. pulmonary veins.
    b. pulmonary portal system.
    c. pulmonary trunk.
    d. alveolar duct.

11. The most common type of leukocyte that functions as a phagocyte is a
    a. lymphocyte.
    b. eosinophil.
    c. basophil.
    d. neutrophil.

12. The leukocyte that has the ability to recognize a specific disease organism and also to control immunity is the
    a. lymphocyte.
    b. eosinophil.
    c. basophil.
    d. neutrophil.

13. The inability to form clots because of a genetic lack of a clotting factor is known as
    a. hemotocrit.
    b. hemophilia.
    c. hemostat.
    d. homunculus.

14. The production of red blood cells is known as
    a. erythropoiesis.
    b. leukocytopenia.
    c. erythroblastoma.
    d. hemolysis.

15. Platelets are pieces of the cytoplasm from
    a. basophils.
    b. megakaryocytes.
    c. erythrocytes.
    d. monocytes.

16. Determining the quantity of blood cells in a blood sample is known as a
    a. Hematocrit.
    b. blood smear.
    c. complete blood cell count.
    d. comprehensive blood spectrum.

17. Universal donors are able to give blood to any other common type because
    a. they have such a large quantity of red blood cells it does not matter if some are destroyed.
    b. they have no antibodies in their plasma.
    c. they have no antigens on their red blood cells.
    d. they produce a chemical that destroys antibodies in someone else's blood.

18. New blood cells are formed from
    a. transformed heart cells.
    b. recycled blood cells reassembled by the spleen.
    c. stem cells in red bone marrow.
    d. blood cells to dividing in the bloodstream.

19. A ridge that forms inside a blood vessel that narrows the size and sometimes causes clot formation is known as a(n)
    a. coarctation.
    b. embolus.
    c. plica.
    d. plaque.

20. What determines the type of blood someone can take for transfusion?
    a. the age and sex of the recipient
    b. the antibodies in the donor's blood
    c. the hematocrit
    d. the antibodies in the recipient's blood

Chapter 9

The Respiratory System

LEARNING OBJECTIVES

Upon completion of this chapter, you will be able to:

1. Describe the components and functions of the respiratory system and how it maintains homeostasis.
2. Summarize pulmonary ventilation and gas exchange.
3. Identify lung volumes and capacities and explain how they are measured.
4. Describe how respiration is regulated and how it responds to exercise.
5. Demonstrate how to perform a pulmonary function testing.

CHAPTER OUTLINE

Introduction
Functions of the Respiratory System
Organization of the Respiratory System
Components of the Respiratory System
- The Nose
- The Pharynx
- The Larynx
- The Trachea
- The Bronchi
- The Bronchioles
- The Alveoli
- The Lungs
- The Pleura

Respiratory Physiology
- Pulmonary Ventilation
- Gas Exchange
- Control of Respiration
- Respiratory Response during Exercise

INTRODUCTION

If you have reached this far in the anatomy and physiology class, then you are ready for this especially important chapter on the respiratory system. The respiratory system is extremely vital to keeping all body systems functional. Learning about the respiratory system, the respiratory structures and their functions, the process of gas exchange, and the control of respiration is essential to the practice of any profession in the health sciences. Any damage, blockage, or injury to this system would affect a person's breathing, cause hypoxemia, and may completely block the delivery of oxygen to the body and brain cells. This can lead to respiratory arrest and death in about 5 minutes. With no oxygen supply to the brain cells, brain damage or brain death can occur. A respiratory arrest can lead to cardiac arrest, too. Cardiac arrhythmias can result from this critical condition leading to cardiopulmonary arrest.

One of the major symptoms of COVID-19 infection is difficulty of breathing. Shortness of breath, or dyspnea, is also observed in pneumonia, asthma, heart disease, and chronic obstructive pulmonary disease (COPD). There is tightness of chest and a feeling of air hunger—a feeling of not being able to breath in sufficient air almost like drowning.

A respiratory arrest can lead to cardiac arrest requiring cardiopulmonary resuscitation (CPR) and defibrillation.

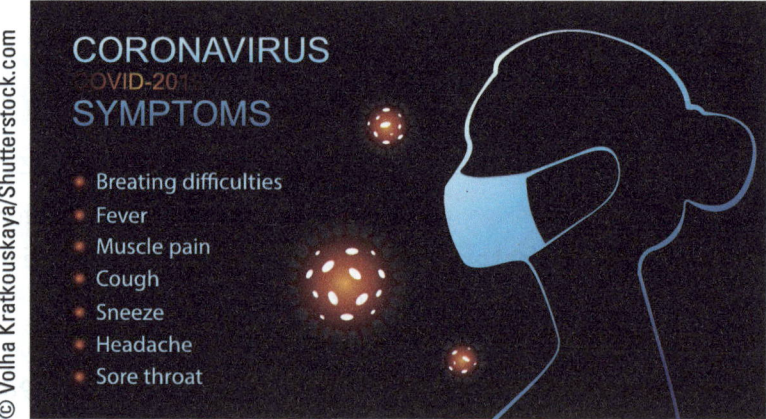

Figure 9.1

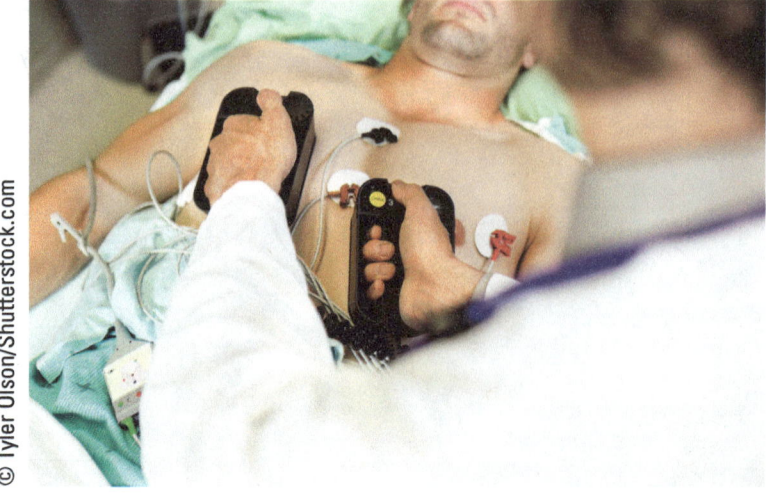

Figure 9.2

Organisms that consist of single cells or only a few cells can exchange gases with the environment by simple diffusion. Complex organisms composed of trillions of cells, like humans, require a system by which gases can be transported throughout the body and made available to virtually every cell. The respiratory system provides access of oxygen to the bloodstream, and carbon dioxide is able to exit the bloodstream to be exported out of the body.

FUNCTIONS OF THE RESPIRATORY SYSTEM

Although a primary function of the respiratory system is the movement of air in and out of the body, it also provides several other functions as well, including:

- Inhalation (inspiration), which is the drawing air into the lungs to provide an exchange of air from which the blood is able to obtain oxygen
- Exhalation (expiration), which is the expulsion of most of the air contained in the lungs to remove carbon dioxide produced during metabolism from the body
- Participation in pH balance by the retention or exhalation of additional carbon dioxide, resulting in the change in the acid content of the blood
- Movement of blood as the vacuum created in the chest during inspiration causes the vena cavae to expand, drawing more blood into the thoracic cavity (Expiration compresses these large vessels to add force, which pushes blood into the right atrium.)
- Vocalization, which is the production of sound that results from the air exhaled from our lungs traveling over the vocal cords, causing them to vibrate and thereby allowing us to speak or sing
- Assistance in the control of blood pressure by the lungs producing angiotensin-converting enzyme (ACE), which results in an increase in blood volume that consequently raises blood pressure
- Movement of air past the olfactory receptors, giving us the sense of smell
- Breath-holding increases abdominal pressure, which assists during urination, defecation, and childbirth

> **Inhalation (inspiration)** the inward movement of air into the lungs.
>
> **Exhalation (expiration)** the outward movement of air from the lungs.

ORGANIZATION OF THE RESPIRATORY SYSTEM

The respiratory system can be anatomically divided into the upper and lower respiratory tracts (Figure 9.3). The upper respiratory tract begins at the nose and ends above the larynx. Other than the nasal passage, the upper respiratory tract is the common passageway for both food and air. The lower respiratory tract starts at the larynx and terminates in alveoli, where gases can exchange with capillary blood. The lower respiratory tract is concerned exclusively with the transport of gases.

The respiratory system can also be divided into the conduction portion, which consists of the air passages through which gases pass on their way to the respiratory portion, and the alveoli, through which gases exchange with blood. If all of the alveoli were spread out side by side, they would equal an area roughly the size of a tennis court.

Pulmonary circulation refers to the division of the cardiovascular system that transports blood from the right ventricle to the lungs so that gases can exchange with the atmosphere and then travel back to the left atrium. *Systemic circulation* is the division of the cardiovascular system that moves blood from the left ventricle to the tissues, where the diffusion of these gases can occur between the bloodstream and cells and then return the blood to the right atrium.

> **Upper respiratory tract** begins at the nose and ends above the larynx. Other than the nasal passage, the upper respiratory tract is the common location for both food and air.
>
> **Lower respiratory tract** starts at the larynx and terminates in alveoli, air sacs, where gases can exchange with capillary blood. The lower respiratory tract is concerned exclusively with the transport of gases.

Chapter 9 The Respiratory System

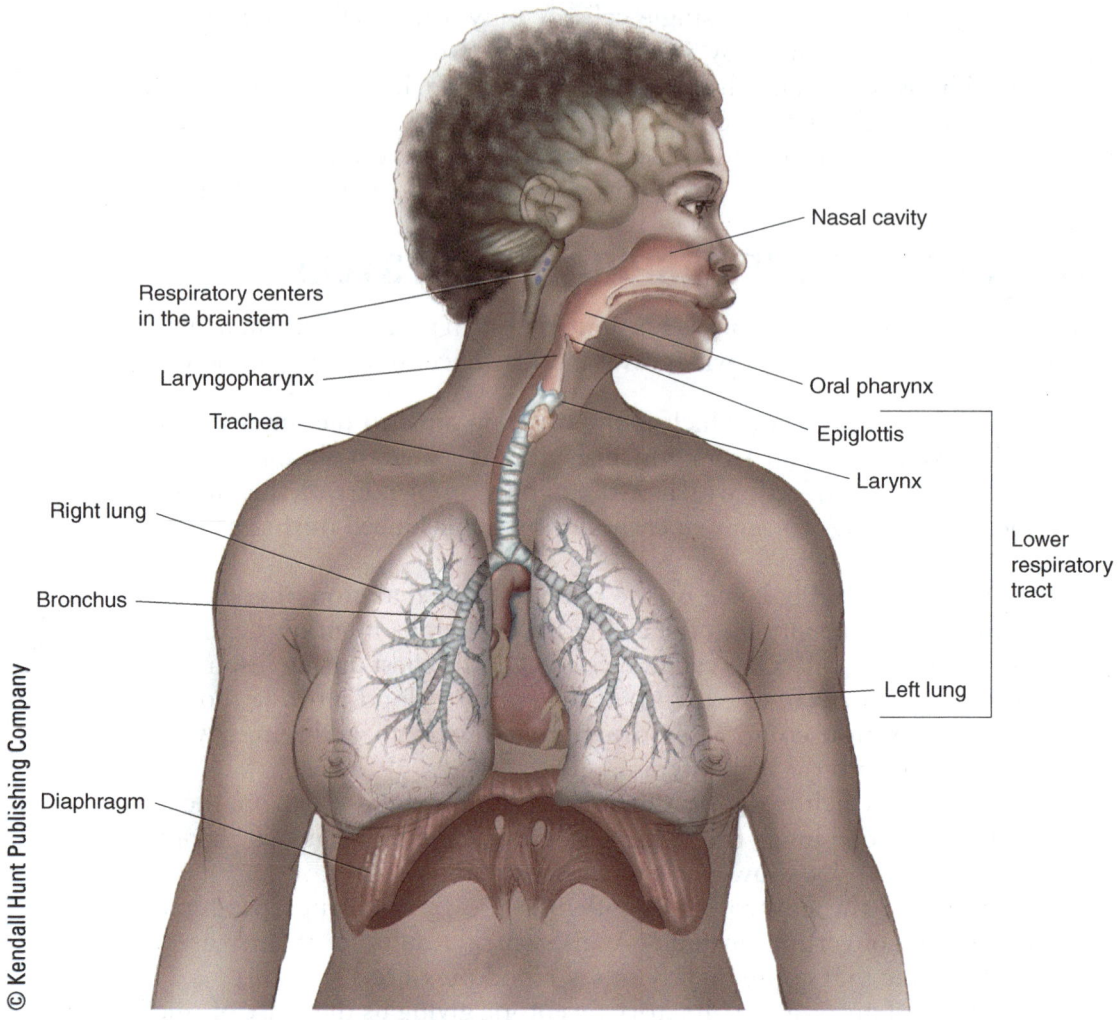

Figure 9.3 Major Organs of the Respiratory System.

Comprehension Check-up:

1. The term for drawing air into the lungs for exchange of gases is _____.
2. Exchange of gases between the atmosphere and blood occurs in the lungs in capillary covered air sacs known as _____.

1. inhalation or inspiration   2. alveoli

COMPONENTS OF THE RESPIRATORY SYSTEM

Our discussion of the respiratory system begins with the upper respiratory tract. It is lined with pseudostratified columnar epithelium and mucus-producing cells. Particles in the air stick to the mucous lining the air passages. Cilia on the outer surface of the pseudostratified columnar epithelium move those particles out of the airway to prevent contamination of the lower respiratory tract. The components of the upper respiratory tract are (Figure 9.4):

- The nose and nasal cavity
- The pharynx

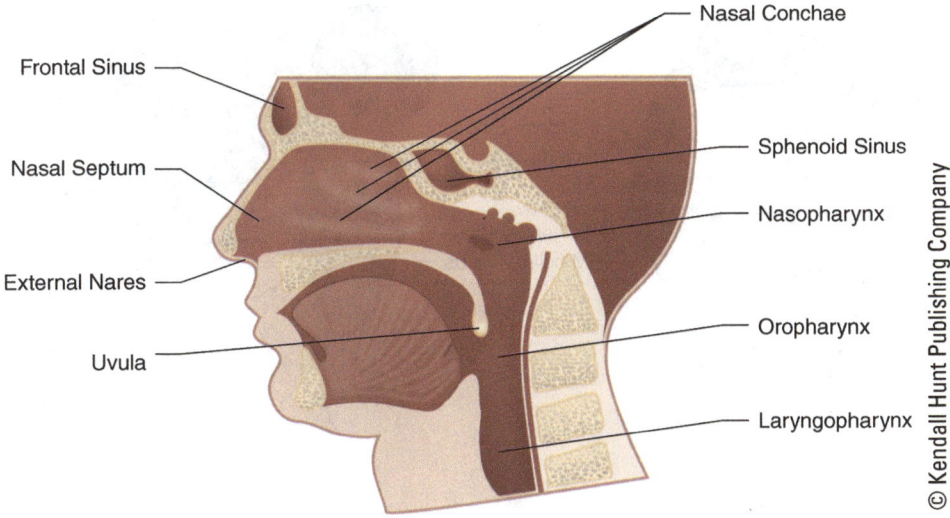

Figure 9.4 The Upper Respiratory Tract.

The lower respiratory tract begins at the split of the larynx (which passes air) from the esophagus (which transports food). The larynx, trachea, and bronchi are lined with ciliated pseudostratified columnar epithelium. The lower respiratory tract contains:

- The larynx
- The trachea
- Bronchi
- Bronchioles
- Alveoli

Sneezing occurs when the upper respiratory tract is irritated. Substances that need to be removed from the lower respiratory tract are ejected by coughing.

The Nose

Air enters the nose through two holes known as *external nares*. Hairs in the entrance to the nasal cavity prevent large objects from entering. The nose is divided into two sides by the *nasal septum*. Posterior to the nose is a space known as the nasal cavity. Located in the mucosa covering the roof of the nasal cavity are olfactory receptors detecting chemicals in the air as it passes through the passageway. *Nasal conchae* are bony projections that protrude into the nasal cavity; they are so named because they resemble conch shells. The blood-filled tissue covering these nasal conchae warm, moisten, and filter air passing through the nasal cavity on its way to the lower respiratory tract.

Associated with the nasal cavity are the four paranasal sinuses. They are spaces within facial bones that resonate sound and lighten the weight of the skull (Figures 9.4 and 9.5). These sinuses are:

- Frontal sinus
- Maxillary sinus
- Sphenoid sinus
- Ethmoid sinus

> **Nasal cavity** the space posterior to the nose.

> **Paranasal sinuses** associated with the nasal cavity are spaces within facial bones that resonate sound and lighten the weight of the skull.

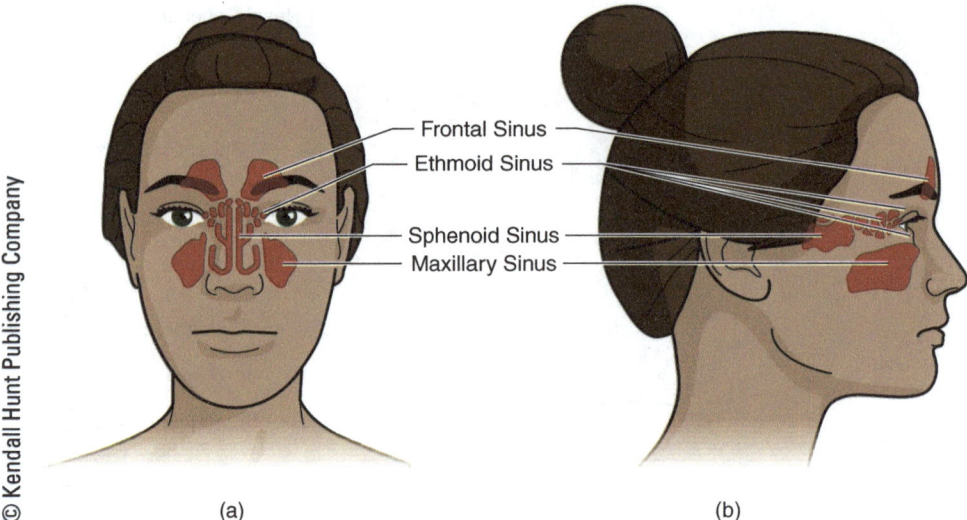

Figure 9.5 Paranasal Sinuses. (a) Frontal view. (b) Sagittal view.

The Pharynx

Posterior to the nasal cavity and mouth is a common passageway for food and air known as the pharynx. It can be divided into three sections:

- The *nasopharynx* is posterior to the nasal cavity.
- The *oropharynx* is posterior to the mouth.
- The *laryngopharynx* in commonly referred to as the throat.

The pharynx is ringed by three sets of tonsils to defend against invasion or foreign organisms and substances such as dust, dirt, and pollen. The oropharynx and laryngopharynx are passages for food as well as for air. There is a tissue flap in the posterior oral cavity called the *uvula*; it projects from the soft palate and elevates when we swallow to prevent food and beverage from entering the nasopharynx.

> **Pharynx** posterior nasal and oral cavities that are a common passageway for food and air.

> **Comprehension Check-up:**
> 1. In the nasal cavity are bony projections covered with epithelium that warm, moisten, and filter air. These projections are called _____.
> 2. The common passageway for food and air found in the posterior oral cavity is known as the _____.
>
> 1. nasal conchae 2. oropharynx

> **Larynx** conducts air into the trachea and also contains the vocal cords.

> **Vocal cords** as air passes between the partially closed vocal cords, sound is created, which allows humans to speak. They may close tightly to provide protection against foreign objects moving toward the trachea.

The Larynx

The pharynx splits into two separate passageways: one for air—the larynx, and the other for food—the esophagus. The larynx, commonly referred to as the voice box (Figure 9.6), is surrounded by the *thyroid cartilage,* whose anterior surface can be seen particularly well in males—called the *Adam's apple*. The larynx has two functions: (1) it conducts air into the trachea, and (2) it contains the vocal cords responsible for the creation of the sound of our voice.

The vocal folds, or vocal cords, are attached to the thyroid cartilage side by side on their anterior end (Figure 9.7). On the posterior end, each cord

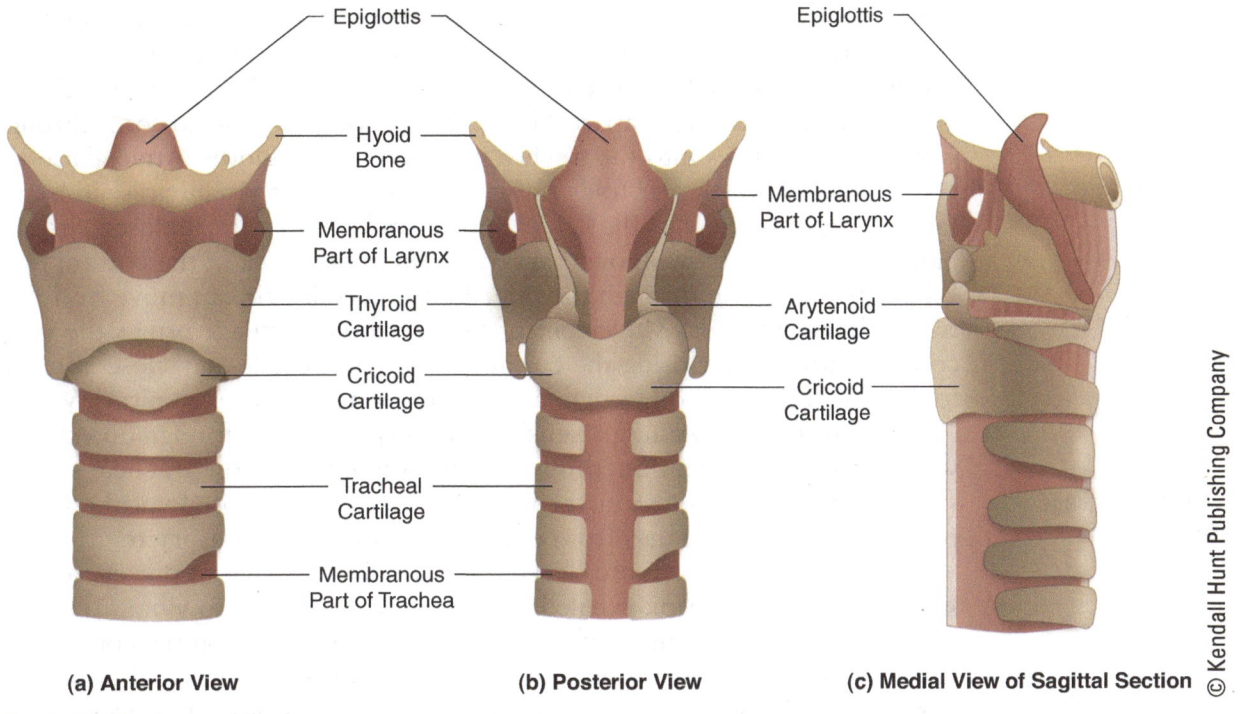

Figure 9.6 Anatomy of the Larynx.

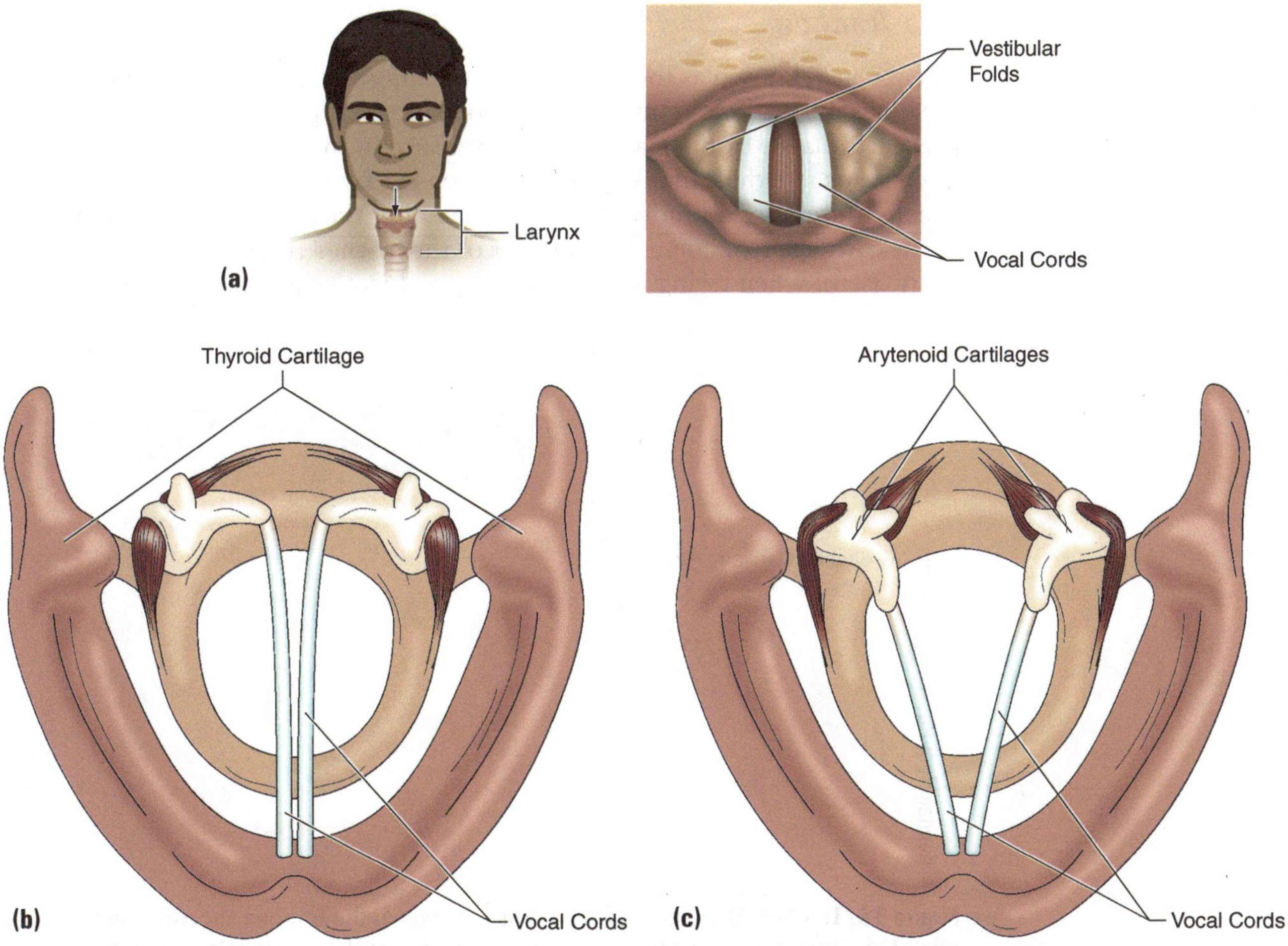

Figure 9.7 Vocal Cords. (a) Superior view also showing vestibular folds. (b) Open; vestibular folds not shown. (c) Closed; when muscle rotates the arytenoids, cartilages move the vocal cords together.

attaches to rotating cartilages known as *arytenoid cartilages*. When stimulated by a motor neuron from the brain, muscles rotate the arytenoid cartilages, causing the vocal folds to be taut and brought together. When the muscles relax, the vocal cords move apart. The opening between the vocal cords through which the air passes into the trachea is known as the glottis. Attached to the vocal cords are vestibular folds (false vocal cords) that visually resemble curtains. A superior view of the larynx looks similar to a pair of closed curtains hanging in front of a window that have been pulled back at the bottom only. Like light streaming through the pulled back curtain, air can move freely between the vocal folds. Movement of the arytenoid cartilage brings the vocal cords together, like closing the curtains. The vocal cords may close abruptly to prevent substances other than air from entering the lower respiratory tract.

The larynx is exceedingly sensitive to the presence of foreign objects and will respond to prevent anything but air from entering the trachea. During swallowing, the epiglottis (a cartilage flap) covers the closed vocal cords to prevent food or beverages from entering the lower respiratory tract.

When the vocal cords are closed but not tightly enough to prevent all passage of air between them, exhaled air can cause these folds to vibrate, creating sound. The intricate control of the moving air and tension on the cords along with mouth and tongue position allows us to vocalize. The sound created by our vocal cords is weak. As the sound passes through the nasal passages and paranasal sinuses, it resonates to produce the full voice we typically expect to hear.

> Epiglottis a flap of cartilage that covers the windpipe while swallowing.

The Trachea

The trachea, known as the windpipe, is the major conduit for air on its way to the lungs. Its walls contain C-shaped cartilages that provide enough rigidity to prevent collapse of the trachea when pressure in the thoracic cavity increases to cause us to exhale air. Attached between the two ends of the "C" is the trachealis muscle (smooth muscle), which can alter the distance between the ends of the cartilage to change the diameter of the trachea (Figure 9.8). This is especially useful when trying to expel a foreign object or mucus from the lung. By

> Trachea the major conduit for air to the lungs.

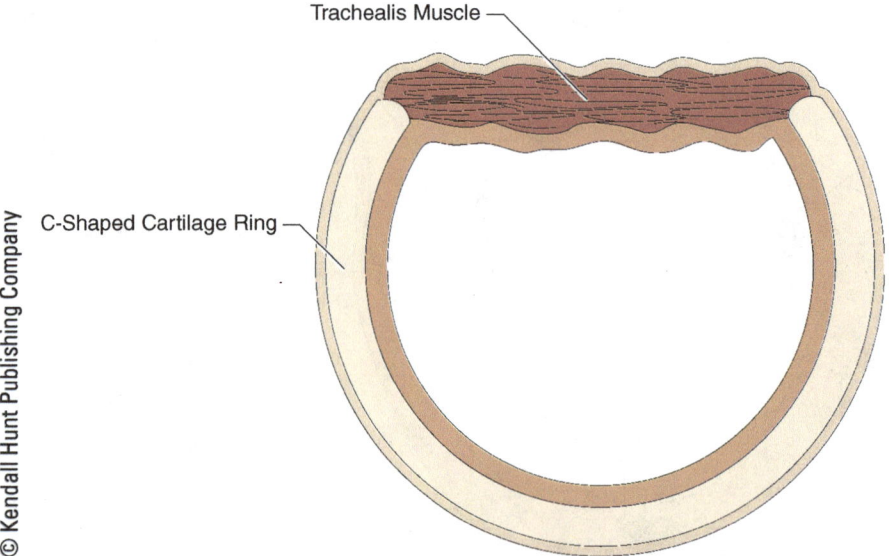

Figure 9.8 Trachea. Superior view of a cross section of the trachea showing the C-shaped ring of cartilage supporting the air passage, with the trachealis muscle connecting the ends of the cartilage. Contraction of this muscle reduces the diameter of the airway.

decreasing the diameter of the trachea when coughing, we create additional pressure as the air rushes out the tube to force unwanted material from the passageway. The trachea, like the upper respiratory tract, is lined with ciliated pseudostratified columnar epithelium and mucus-producing cells.

> **Comprehension Check-up:**
> 1. Cartilage in the larynx that rotates to tighten the vocal cords to allow us to speak is the _____ cartilage.
> 2. The trachea remains rigid during expiration because its walls contain _____.
>
> 1. arytenoid   2. C-shaped rings of cartilage

The Bronchi

The trachea splits toward the two lungs as primary bronchi. Inside the lungs, each primary bronchus divides into secondary then tertiary bronchi, which, like the trachea, are also supported by cartilage rings. Branching continues to even smaller passageways, all of which contain at least some supporting cartilage to prevent collapse of the airway during exhalation. This branching resembles an inverted tree, hence the term bronchial tree (Figure 9.9). The bronchi are also lined by ciliated pseudostratified columnar epithelium and coated with mucus. Particles in the air stick to the mucous coating and then the cilia move back and forth to transport the foreign material out of the air passage.

Bronchi branches of the airway that are visually similar to the branches of a tree.

Bronchial tree the air passages within the thoracic cavity resemble an inverted tree; with the trachea resembling the trunk and the bronchi being the branches.

The Bronchioles

As the bronchi divide further down the bronchial tree, the cartilage rings are gradually replaced by smooth muscle. Their names are changed to bronchioles at the point where cartilage rings are completely absent. Instead, additional smooth muscle spirals around these air passages that constrict or dilate to alter air flow. The *terminal bronchioles* are the smallest branches for conduction of air to areas where gases exchange with the blood. There are bronchioles which possess a few alveoli, known as respiratory bronchioles, because they are able to allow some gas exchange between air and blood, which open to alveolar ducts. The *alveolar duct* is a tube that provides open access to alveoli clustered around it, similar to grapes around a stem. Think of the alveolar duct as a hallway in a motel and the rooms on either side of the hallway with all of the doors open as alveoli. Air enters the alveolar duct and can be distributed to all of the alveoli attached to it.

Bronchioles air passageways at the end of the bronchi. They are encircled by smooth muscle, which can constrict or dilate to alter air flow to distal alveoli.

The Alveoli

The alveoli are air sacs covered by capillaries through which gases can diffuse to allow the exchange of gases between the atmosphere and blood (Figure 9.11). Each alveolus is a sphere composed of simple squamous epithelium, referred to as *type I alveolar cells*. The capillaries attached to each alveolus are also constructed of simple squamous epithelium, allowing rapid exchange of gases between the two membranes. Diffusing into the blood from the alveoli is oxygen. Diffusing out of the blood into the alveoli are carbon dioxide and water vapor.

Alveoli air sacs covered by capillaries through which gases can diffuse, allowing the exchange of gases between the atmosphere and blood to take place.

Chapter 9 The Respiratory System

Bronchial Tree UP-Close

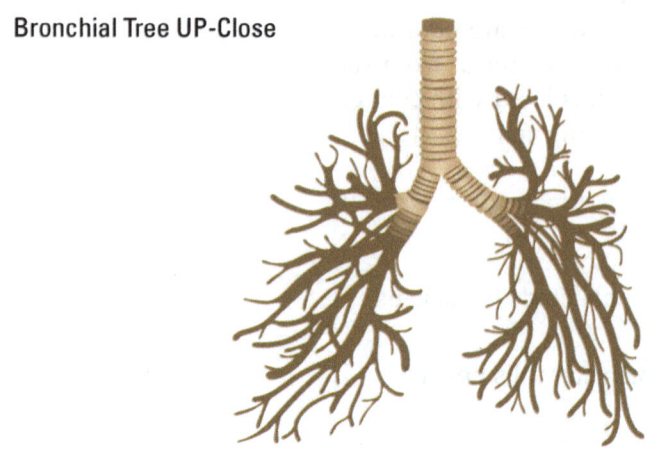

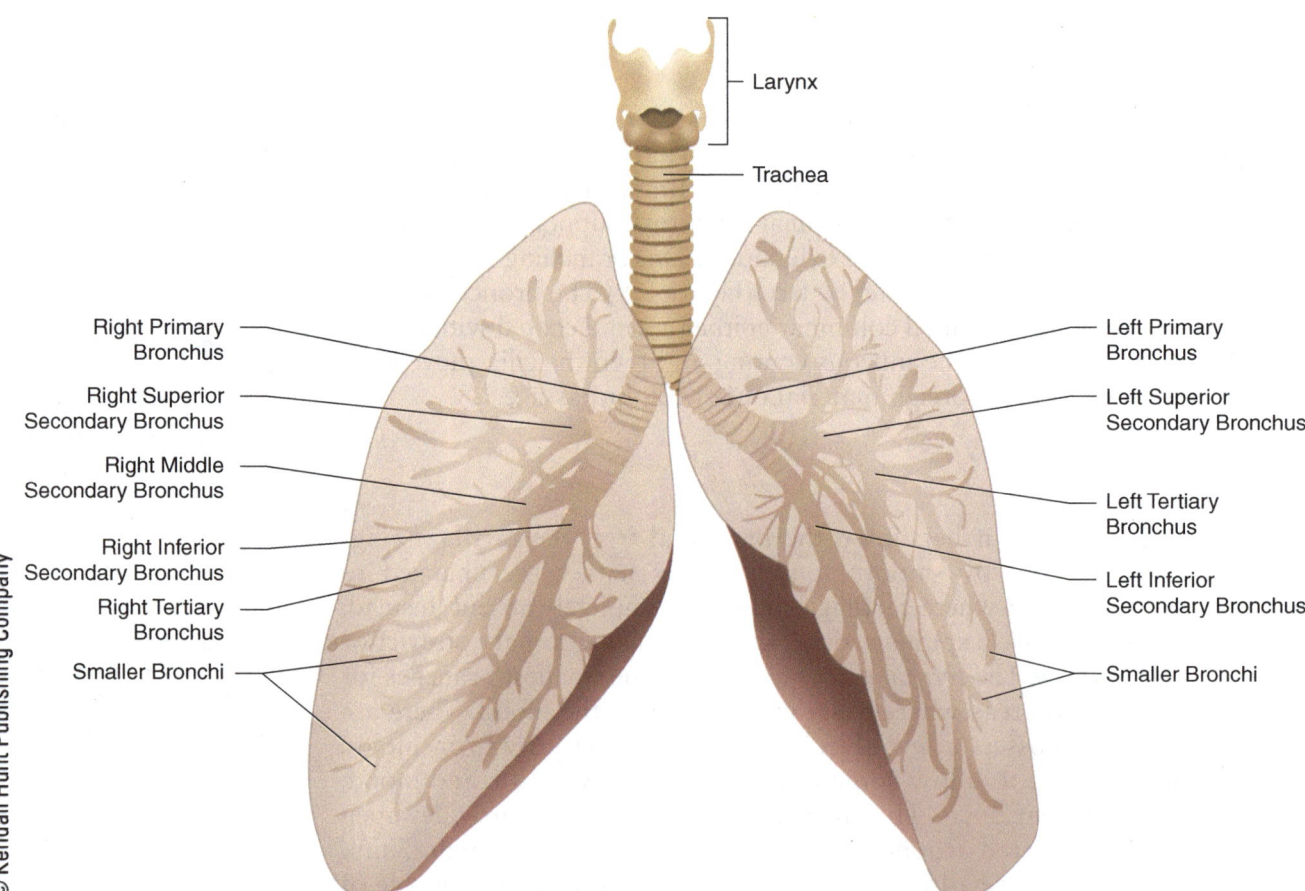

Figure 9.9 Bronchial Tree. The primary bronchi divide into secondary then tertiary bronchi and then to smaller branches.

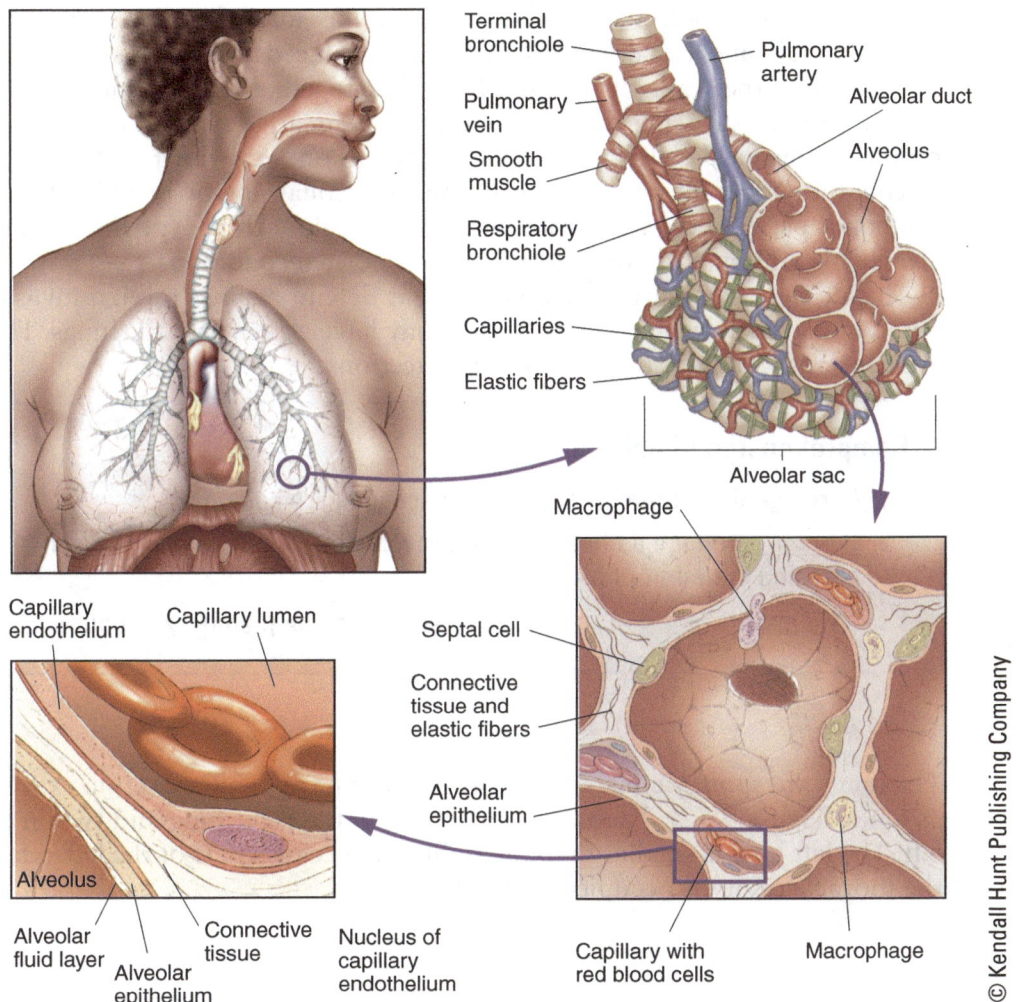

Figure 9.10 Terminal Bronchioles. The terminal bronchioles branch into respiratory bronchioles, which open to alveolar ducts, around which are clustered alveoli. The alveoli are covered with capillaries for exchange of gases.

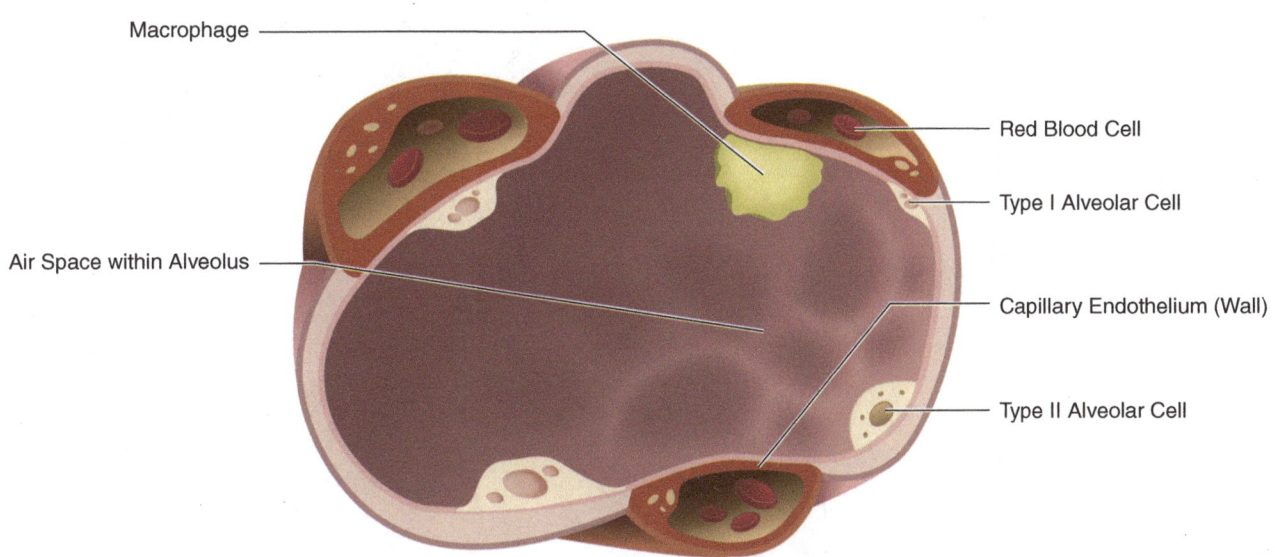

Figure 9.11 The Alveolus. Type I alveolar cells form the walls of the alveolus, and type II alveolar cells produce a chemical to break hydrogen bonding, which would collapse the alveolus during exhalation.

Recall that water molecules can stick together because of hydrogen bonding. When an individual exhales and the alveoli are reduced in size, hydrogen bonding can cause the walls of the alveoli to stick together. To prevent hydrogen bonding from increasing the effort required to breathe, *type II alveolar cells* produce a chemical called pulmonary surfactant onto the inner surface of the alveoli; the surfactant prevents the alveoli from collapsing by breaking hydrogen bonds between the water molecules. As a result of these type II cells in the alveoli, inspiration is almost effortless under normal conditions.

Inside the alveoli are macrophages that defend against foreign invasion and remove particles, such as dust, that may have gotten into the lungs. They are constantly vigilant to prevent lung infections.

> **Comprehension Check-up:**
> 1. Cartilage-lined branches of the air passages are known as _____.
> 2. _____ produce a chemical that prevents hydrogen bonding from increasing the effort of breathing.
>
> 1. bronchi   2. Type II alveolar cells

Lungs spongy, cone-shaped organs that provide a location for the exchange of gases between the atmosphere and the blood.

Lobes distinct sections of each lung.

The Lungs

The lungs are spongy cone-shaped organs that have distinct sections known as lobes (Figure 9.12). There are three lobes in the right lung and two in the left. The left lung is smaller than the right lung because part of the left side of the thoracic cavity is occupied by the heart.

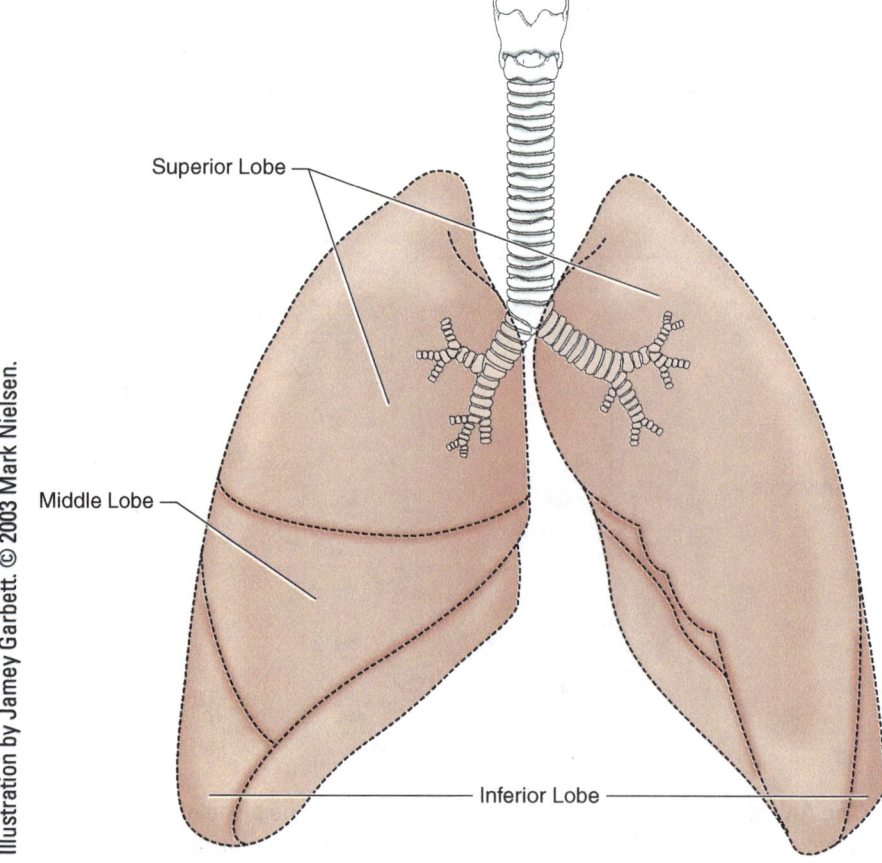

Figure 9.12 Lobes of the Lungs.

The Pleura

The pleura is a serous membrane that lines the thoracic cavity and covers the lungs (Figure 9.13). Imagine, for example, that you place an inflated balloon into a large beaker and then push your fist into the balloon. It is one continuous balloon that contacts both the sides of the beaker and your fist. The pleura are designed in similar fashion. As the lungs develop in utero, they grow into the existing pleura, resulting in one continuous membrane covering the internal thoracic cavity and the lungs. Attached to the wall of the chest are the *parietal pleura*. The *visceral pleura* envelop the lungs. Between the two layers of pleura is a water-like pleural fluid that allows the lungs to slide against the chest wall. This fluid also causes each lung to remain in contact with the parietal pleura so that the lung moves with the wall of the thoracic cavity and diaphragm. It is this contact that results in the movement of air in and out of the lungs. The physiological process is discussed later in this chapter.

> **Pleura** serous membranes that line the thoracic cavity and cover the lungs.

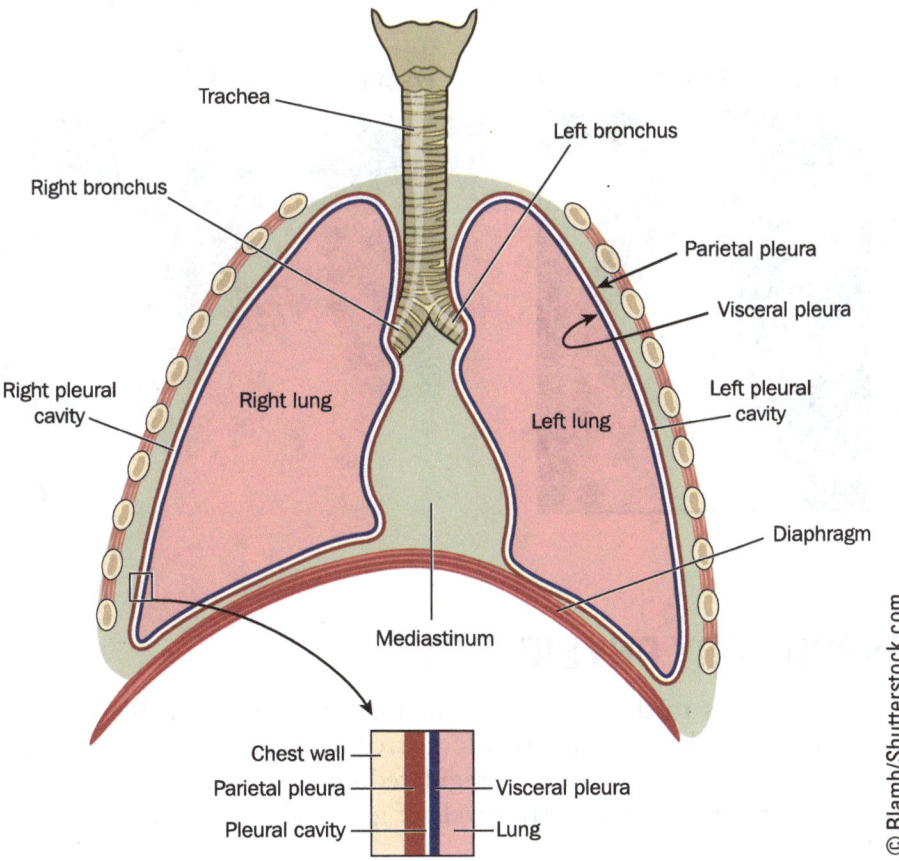

Figure 9.13

Comprehension Check-up:

1. The right lung is divided into _____ lobes and the left lung into _____ lobes.
2. The lining attached to the wall of the thoracic cavity is known as the _____.

1. 3, 2 2. parietal pleura

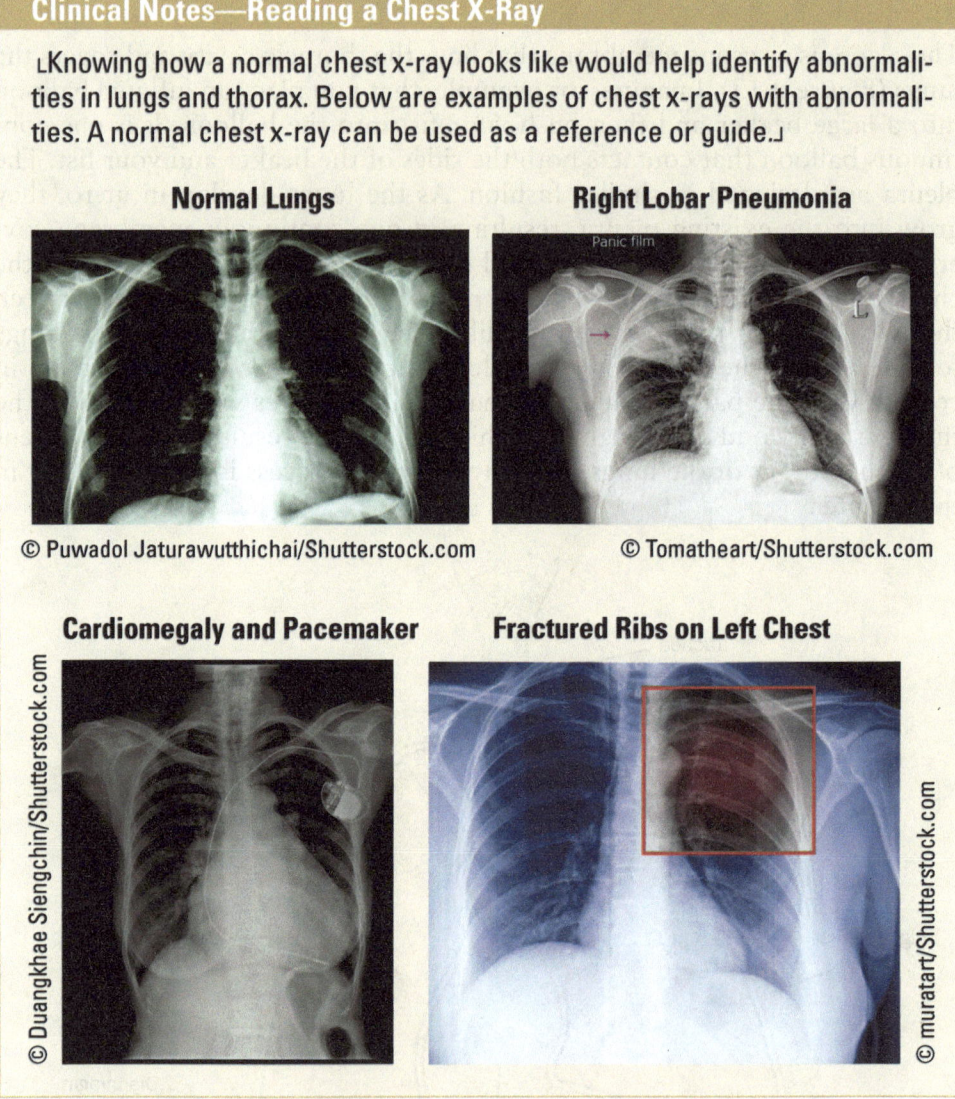

RESPIRATORY PHYSIOLOGY

Respiration the collective functions involved with the movement of gases between the atmosphere, bloodstream, and cells.

Respiration refers to the collective functions involved with the movement of gases between the atmosphere, bloodstream, and cells. The processes involved in respiration are carefully regulated alterations of air movement to allow us to maintain homeostatic levels of gases within the bloodstream.

Pulmonary Ventilation

The exchange of gases between the environment and cells begins with movement of air in and out of the lungs, where those gases have access to the bloodstream through the alveoli. This process of moving air between the lungs and atmosphere is referred to as pulmonary ventilation.

Pulmonary ventilation the process of moving air between the lungs and atmosphere.

Inhalation and Exhalation

The inward movement of air into the lungs is referred to as *inhalation* or *inspiration*. *Exhalation* or *expiration* is outward movement of air from the lungs.

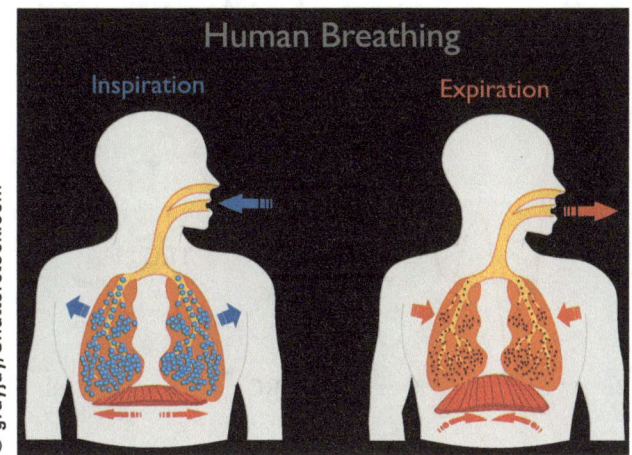 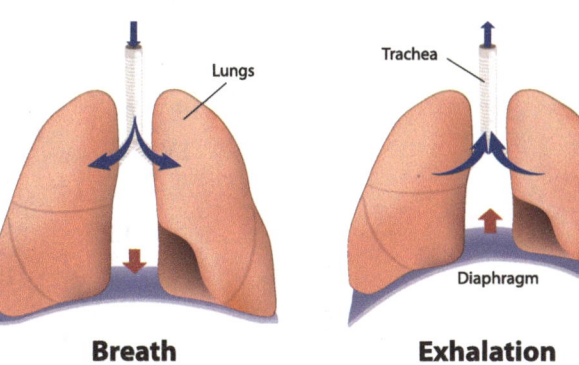

The diaphragm functions in breathing

Figure 9.14

When we inhale, we create a vacuum inside the chest that allows air to move in to fill the void. Recall that the lungs adhere to, yet slide against, the chest wall and diaphragm. When we are relaxed, the contraction of the diaphragm partially expands the lungs, creating a vacuum in the alveoli that draws air in through our air passages (Figure 9.14). When the diaphragm relaxes, an increase in alveolar pressure is created, pushing some of the air out of the lungs and body. Chances are, right now, as you are reading this book, you are relaxed. The contraction and relaxation of your diaphragm are all you need to supply enough gas exchange for your cells to function normally. If, however, you decide to be more active and require more gas exchange, your respirations may become deeper and more rapid. Additional muscle contraction, performed by the *external intercostals muscles* and muscles in the neck and upper chest, raises the ribs to greatly expand the internal volume of the thoracic cavity. This creates an additional vacuum within the lungs, causing you to take in up to six times more air than when relaxed. You can also exhale additional air beyond the expiration of air caused by the relaxation of the diaphragm. *Internal intercostal, abdominal, and lower back muscle* contraction can lower the ribs beyond their relaxed position to force additional air out. By forcing more air in and out, we are able to adjust to higher demands for gas exchange.

Lung Volumes and Capacities

The volumes of air moved by the lungs during various stages of breathing can be measured. These measurements can help us understand what is available to us as we change from relaxed movements to strenuous activities. The following is a list of the volumes and a combination of those quantities to determine the total capacities of air in the lungs (Figure 9.15):

- *Tidal volume* (V_T) is the amount of air moved during relaxed inspiration or expiration when only the diaphragm contracts or relaxes. V_T ranges between 300 and 500 ml (roughly the volume contained in a 470-ml [16-oz.] soft drink bottle.)
- *Inspiratory reserve volume* (IRV) is the quantity of air drawn in beyond relaxed inspiration with maximum exertion by using other muscles to increase thoracic volume as the ribs are raised. This can increase up to six times greater than at rest.

Tidal volume amount of air moved during relaxed inspiration and expiration.

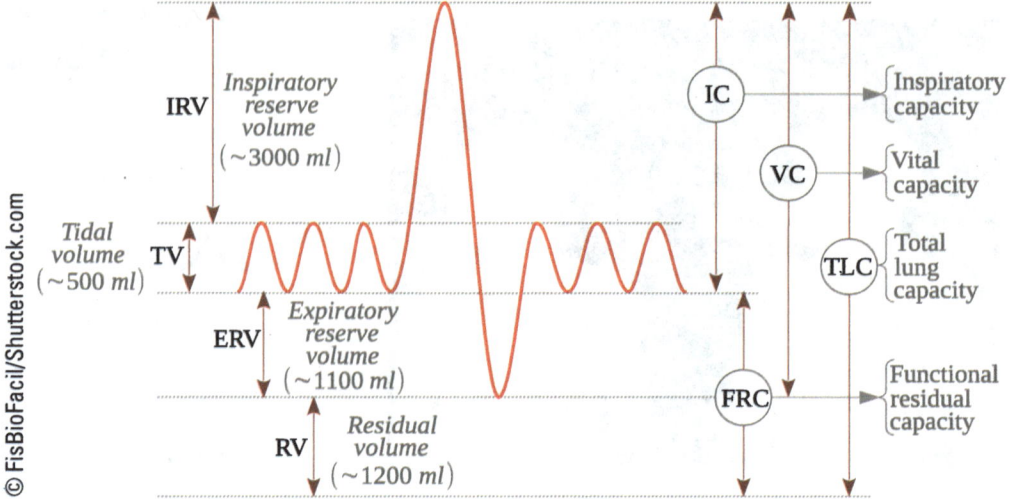

Figure 9.15

Vital capacity the quantity of air the lungs can inspire or expire during maximum exertion.

- *Expiratory reserve volume* (ERV) is the quantity of air exhaled beyond relaxed expiration with maximum exertion. This volume is roughly double the V_T.
- *Vital capacity* (VC) is the quantity of air the lungs can inspire or expire during maximum exertion. It is the sum of the three volumes—one accounting for the volume of air moved during relaxed breathing and the other two occurring by additional force. VC = V_T + IRV + ERV. In a healthy, active male, this can be roughly the volume of two 2-liter bottles.
- *Residual volume* (RV) is the air left in the lungs after forcing out as much as possible. Because the lungs remain attached to the chest wall at all times, there is space within the thoracic cavity that still contains air. Typically this volume is about 1 liter.

Total lung capacity the total amount of air the lungs can hold during maximum exertion.

- *Total lung capacity* (TLC) is the total amount of air the lungs can hold during maximum exertion. It takes into account not only the maximum amount the lungs can move but how much is left in the lungs at the end of the process. It is the sum of the vital capacity and residual volume. TLC = VC + RV.

The measurement of lung volumes is an important factor in determining how well the lungs are functioning. Other factors, such as genetics, height, age, and gender, must be taken into account when determining appropriate lung volumes for normal function. The ability of the lungs to expand and relax is known as *compliance*. High compliance signifies that the lungs are working normally, whereas low compliance is the result of some pathological condition that interferes with air movement.

Comprehension Check-up:

1. To take more air into our lungs, thoracic volume is increased when the ribs are raised by the _____ muscles.
2. The volume of air inhaled or exhaled during relaxed breathing is known as _____.

1. external intercostals 2. tidal volume

> **Is There a Difference in Lung Capacity between Sexes or at Different Ages?**
>
> The vital lung capacities of males tend to be larger than those of females. For males, the average is 70 ml/kg; for females, the average 50 to 60 ml/kg. For example, a male weighing 80 kg (176 lb.) would have a vital capacity of 5600 ml (70 ml/kg × 80 kg), which is about 1.5 gallons. A female weighing 60 kg (132 lb.) would have a vital capacity of about 3300 ml (55 ml/kg × 60 kg), slightly less than a gallon. Vital lung capacity decreases between 200 and 250 ml every 10 years.

> **Emphysema**
>
> Emphysema is a lung disease in which the alveoli break, leaving space within the lung where gases are unable to exchange. Normal healthy lung tissue resembles a foam sponge with alveoli residing side by side. The lung of an individual with emphysema resembles a natural sponge with large spaces inside. Not only is there a loss of gas exchanging ability, but the compliance of the lungs also increases, making the air inside the lungs difficult to be expelled, because the lungs are unable to recoil. As the disease progresses, it becomes necessary to fill as much of each lung as possible, so each inhalation is a forced inspiration. When the ribs are continually raised with each breath, they eventually remain in the upright position causing the individual to have what is referred to as a *barrel chest*. The primary cause for emphysema is irritants in the air, primarily cigarette smoke, meaning that the major cause for this debilitating disease is lifestyle choice.

Gas Exchange

Pulmonary ventilation provides the movement of gases to the lungs, which must then be moved to the cells, or vice versa. It is important to understand that the air contains a high amount of oxygen, approximately 21%, but very little carbon dioxide, 0.04%. During inhalation, a relatively high quantity of oxygen and a very little quantity of carbon dioxide are taken in. As a result, the level of oxygen in the alveoli increases but carbon dioxide levels do not. The high concentration of oxygen in the air sacs allows this gas to diffuse into the blood and into active cells that need oxygen for aerobic metabolism.

The more active a cell is, the more carbon dioxide it produces as a by-product of aerobic metabolism. This carbon dioxide diffuses into the blood and then into the alveoli to increase the concentration of this gas in the air sacs. During exhalation, the excess carbon dioxide is released from its higher concentration in the body to its low concentration in the atmosphere.

Alveolar Exchange of Gases

As a result of pulmonary ventilation, the alveoli contain a high concentration of oxygen and a low level of carbon dioxide. Blood entering the lungs from the right side of the heart has low levels of oxygen and high amounts of carbon dioxide (Figure 9.16). As that blood passes through the capillaries covering the alveoli, the gases are able to diffuse through the alveolar membrane and capillary wall from areas of high concentration to areas of lower concentration. Because the quantity of oxygen is higher in the alveoli than in the blood, oxygen diffuses into the bloodstream. Carbon dioxide is higher in the blood than in the alveoli and diffuses into the alveoli. As a result, the blood exiting the lungs to return to the left side of the heart contains high levels of oxygen and lower amounts of carbon dioxide.

ALVEOLUS GAS EXCHANGE

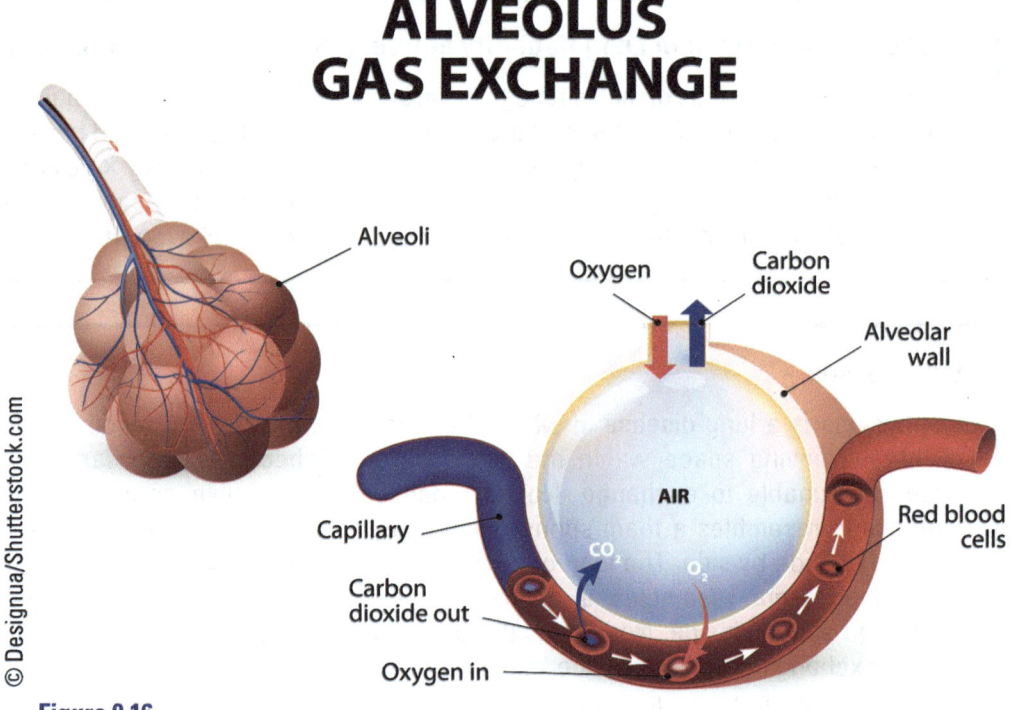

Figure 9.16

Cellular Exchange of Gases

Blood that returns from the lungs to the left side of the heart is pumped to the body. As blood passes through the capillaries in the tissue, gases exchange between the bloodstream and cells (Figure 9.16). If a cell has been active, its need for increased quantities of adenosine triphosphate (ATP) can be met most efficiently by aerobic metabolism. Active cells have been using available oxygen for the production of ATP, so their oxygen content is low. As blood passes through the capillaries, its high concentration of oxygen can diffuse into cells low in oxygen. The high quantity of carbon dioxide in the cell resulting from aerobic metabolism diffuses toward the bloodstream. The result of the exchange can be seen by comparing the color of blood before entering and then again after leaving the tissue. Arterial blood flowing into tissue contains high quantities of oxygen attached to hemoglobin and appears bright red. Venous blood is dark red because it contains low levels of oxygen. It is possible to determine the quantity of oxygen used by cells throughout the body by obtaining arterial and venous blood samples. In some instances this technique may be used in the hospital, but it is not commonly done on most patients because obtaining an arterial sample can sometimes be unpleasant.

Carbon Monoxide—The Undetected Killer

Carbon monoxide is an odorless, colorless gas that results from the incomplete burning of fuels. The attachment of carbon monoxide to hemoglobin is more than 200 times stronger than its attachment to oxygen. Because of the strong attraction of hemoglobin for carbon monoxide, the oxygen-carrying capacity is lost, so even small quantities can have serious effects. Because it causes an inadequate oxygen supply to tissue, if not corrected, carbon monoxide poisoning can be fatal.

> **Comprehension Check-up:**
> 1. The process by which gases exchange between the alveoli and blood or blood and cells from higher concentration to lower concentration is _____.
> 2. The gas that increases in the cell that is producing ATP is _____.
>
> 1. diffusion   2. carbon dioxide

Oxygen Transport

Although it is possible for some oxygen to be transported in the plasma, the bulk of oxygen is carried by red blood cells when oxygen molecules attach to hemoglobin. Hemoglobin is able to hold onto oxygen molecules but can release them when the concentration of oxygen is lower.

Carbon Dioxide Transport

Once carbon dioxide enters the bloodstream most of it attaches to hemoglobin after releasing the oxygen it carried. The carbon dioxide reacts with water in the red blood cell to form carbonic acid, which then ionizes into a hydrogen ion (which lowers blood pH) and a bicarbonate ion (which raises blood pH).

The reversible chemical reaction involved is:

$$CO_2 + H_2O \rightarrow H_2CO_3 \rightarrow H^+ + HCO^-_3.$$

The quantity of hydrogen ions in the blood depends to a large extent on the amount of carbon dioxide in the blood. Bicarbonate ions are excreted or retained by the urinary system and will be discussed in a later chapter. Aerobic metabolism produces carbon dioxide which diffuses into the blood increasing levels of hydrogen ions. In response the individual will breathe more rapidly in order to exhale the additional carbon dioxide in the blood stream, thereby keeping the hydrogen levels and blood pH within homeostatic range.

Someone who has some type of airway obstruction or constriction, such as an asthma attack, is unable to exhale sufficient carbon dioxide. Asthma results when the bronchioles become hypersensitive to substances primarily in the air, although an attack may also be induced by ingested foods or sometimes even exercise. In order to allow the smooth muscle in the bronchiole greater flexibility in changing diameter, these air passages have no cartilages to provide support. If, due to hypersensitivity, the smooth muscle in the bronchiole spasms or its lining becomes inflamed, the resulting increase in resistance to the flow of air can be life-threatening.

As a result of the increasing carbon dioxide remaining in the body, there is a rapid increase in carbonic acid, causing the blood pH to drop. Enzymes in our body have a working pH range. Outside of that range the enzymes are unable to cause chemical reactions at normal rates.

Hyperventilating, on the other hand, may cause an individual to release so much carbon dioxide that the blood may become too alkaline. The breathing rate is normally altered to adjust the amount of carbon dioxide in our blood.

Comprehension Check-up:

1. In blood, oxygen is carried primarily by _____.
2. Airway obstruction results in the retention of too much carbon dioxide, causing the blood pH to _____.

1. hemoglobin 2. decrease

Clinical Skills—Pulmonary Function Testing

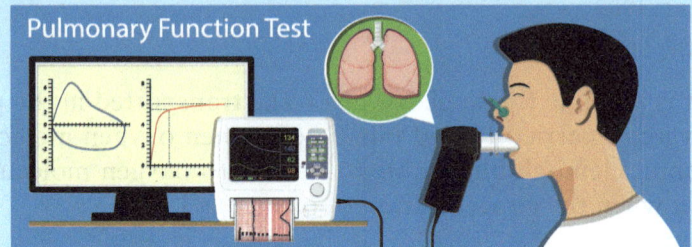

A pulmonary function testing (PFT) is a diagnostic procedure to determine the extent or severity of obstructive or restrictive lung disease. It is also used to check a patient's lung status prior to a major surgery such as chest or abdominal surgery. The testing is done with the patient standing or sitting. Patient is asked to take a deep breath and blow hard and fast through the mouthpiece connected to the PFT machine. During the forceful expiration (called the forced vital capacity or FVC), the patient's lips must be sealed through the mouthpiece to prevent any leaks of exhaled air. A nose clip is also used to prevent exhaling from the nose. The results of the FVC is recorded on the PFT machine shown graphically or numerically, or both.

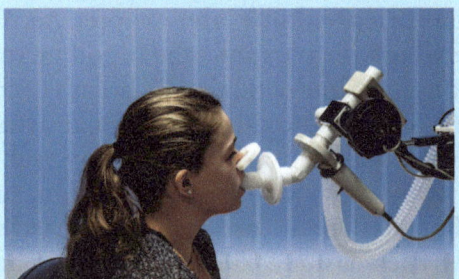
© Goldsithney/Shutterstock.com

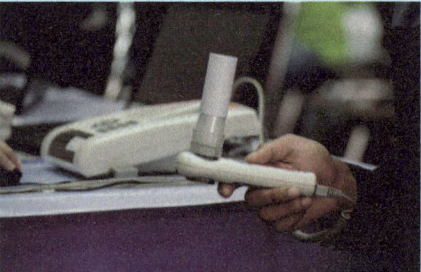
© Chaleewut/Shutterstock.com

Control of Respiration

Changes in lung volume, and thus ventilation, are dependent on the change in thoracic cavity volume. Alterations in the space inside the thoracic cavity are the result of the contraction of the diaphragm and the intercostals muscles, as well as other muscles that can assist in raising or lowering the ribs. Skeletal muscle cannot contract on its own but depends on stimulation from the brain. In the case of involuntary respiration, that stimulation is provided by concentrations of neurons concerned with breathing, known as respiratory centers, which are found in the brain stem (Figure 9.17). For *relaxed breathing*, that is, the minimal exchange of air when the individual is inactive, inspiration requires stimulation of the diaphragm only and expiration needs simply for that

Respiratory centers concentrations of neurons found in the brain stem concerned with breathing.

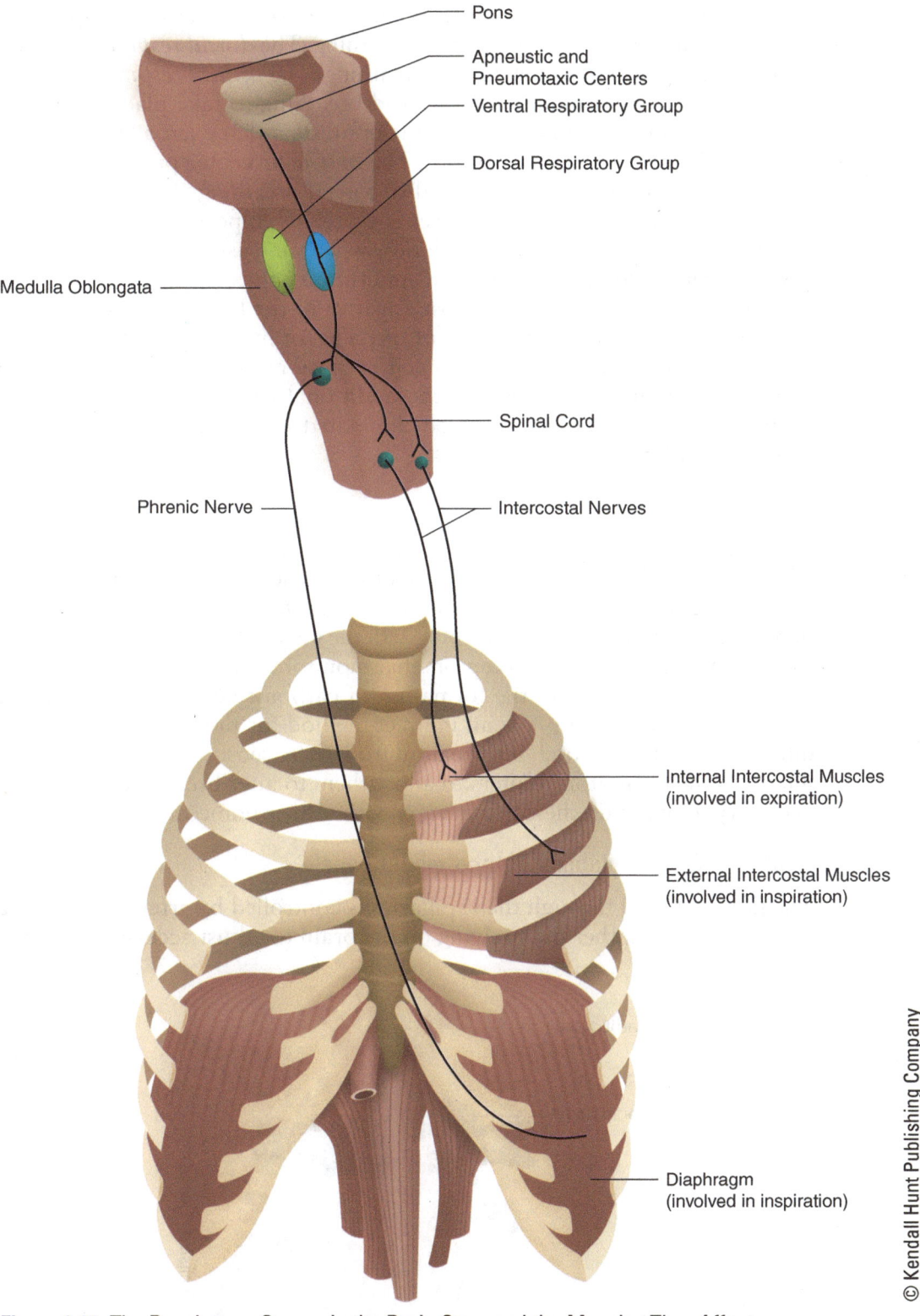

Figure 9.17 The Respiratory Groups in the Brain Stem and the Muscles They Affect.

stimulation to cease and to allow the diaphragm to relax. *Forced breathing*, that is, increased exchange of air because activity requires additional oxygen for increased aerobic metabolism and release of carbon dioxide, requires muscle stimulation in both directions. Forced inspiration requires the contraction of the diaphragm and also of muscles that raise the ribs. Forced expiration involves relaxation of the diaphragm but also stimulation of muscles that lower the ribs beyond their relaxed position. There are two respiratory groups in the brain stem that control these types of breathing. The *dorsal respiratory group* primarily controls relaxed rhythmic breathing. The *ventral respiratory group* adds the necessary stimulation for forced breathing to take place. There are two additional respiratory centers in the brain stem that exert a "fine-tuning" effect on the two respiratory groups. The *apneustic* center prolongs inspiration when transitioning from relaxed to forced breathing. The *pneumotaxic* center inhibits inspiration. This center is especially important during inspiration because it will terminate inhalation when air passages reach the point beyond which further intake of air would rupture alveoli.

The respiratory centers are most sensitive to the level of carbon dioxide in the blood rather than the levels of oxygen. An increase in activity causes an increase in respiration, not so much because more oxygen is needed for aerobic metabolism, but because the additional carbon dioxide produced during aerobic metabolism must be eliminated from the body to maintain blood pH within homeostatic levels.

Chemoreceptors

The *aortic bodies* and *carotid bodies* are special sensory receptors in the aorta where the common carotid artery splits to become the internal and external carotid arteries (Figure 9.18). These *chemoreceptors* detect the oxygen, carbon dioxide, and pH levels of the blood. In the medulla oblongata of the brain stem are also central chemoreceptors that monitor changes in pH of the brain tissue fluid resulting from the formation of carbonic acid in the blood. The detection of changes in pH is an indirect method for monitoring the carbon dioxide levels in the blood. Information from these sensory receptors is transmitted to the respiratory centers to allow them to alter breathing patterns as necessary to maintain these parameters within their homeostatic ranges.

Control of Gas Availability at the Cellular Level

The flow of blood through the capillaries is controlled by arterioles relaxing or constricting. Rather than relying on the brain to adjust accessibility of the

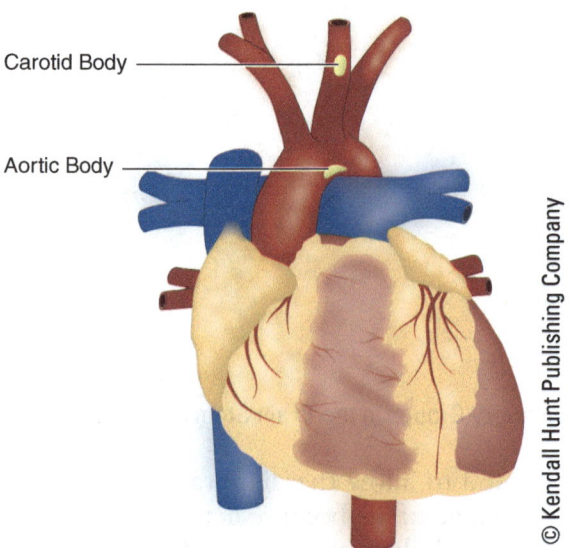

Figure 9.18 Aortic and Carotid Bodies. These receptors detect levels of oxygen, carbon dioxide, and pH levels in the blood.

cells to gas exchange with the blood, the local arterioles respond on their own to tissue levels of oxygen and carbon dioxide and to temperature changes. Decreased oxygen, increased carbon dioxide, and increased temperature are all the result of increased aerobic metabolism. When this condition occurs, the smooth muscle in the arterioles will automatically relax to increase blood flow to these active cells to increase gas exchange.

Respiratory Response during Exercise

Muscle requires additional ATP during exercise. Aerobic metabolism increases the need for oxygen while giving off carbon dioxide and heat. This causes the arterioles controlling blood flow through the local capillaries to relax and increase the amount of blood passing through. The oxygen diffuses out of the capillaries into the muscle fibers while carbon dioxide and heat move into the bloodstream. The bloodstream carries the gases to the alveoli in the lungs while transporting heat to the skin for release into the environment. As blood passes over the alveoli, oxygen diffuses into the capillaries as carbon dioxide passes into the air sacs to be exhaled. The increase in carbon dioxide also increases the production of carbonic acid. Chemoreceptors in the brain stem monitor the lowering of blood pH, which indicates an increase in carbon dioxide. The respiratory centers respond to the input from the chemoreceptors by increasing the respiration rate and depth to cause additional carbon dioxide to be released from the body, thereby increasing the blood pH to maintain homeostatic levels. Even after exercising has ceased, the respiration rate will remain increased until carbon dioxide levels and pH are within homeostatic range.

Whether during exercise or just resting, the respiratory system is at work. It is a powerful homeostatic regulating mechanism, helping to control pH, respiratory gas levels, heart rate, and even blood flow. By understanding its significance and workings, one is better able to more clearly understand the importance of exercise and the avoidance of respiratory disease–causing substances such as smoking or excessive dust. It is also easier to understand why pathological conditions, such as asthma or emphysema, affect the entire body.

Comprehension Check-up:

1. The _____ is the area of the brain that controls respiration.
2. Receptors in the bloodstream that monitor blood pH and oxygen are known as _____.

1. brainstem 2. chemoreceptors

Homeostasis—Holding in Balance

The respiratory system maintains the balance of oxygen and carbon dioxide in conjunction with the brain stem. Stimulation of the muscles of respiration, triggered by the need to retain or exhale carbon dioxide, is a critical parameter not only for the supply of oxygen and removal of carbon dioxide for aerobic metabolism, but also for assistance in the control of blood pH.

The respiratory system also actively participates in the control of blood pressure by producing an enzyme that causes an increase in blood pressure through the constriction of blood vessels and that also results in an increase in plasma volume.

There Is No Substitute for Exercise

Regular, sustained exercise increases lung capacity. Increasing the available area of gas exchange results in increased stamina. Also, the lungs produce an enzyme responsible for an increase in blood pressure. Exercise can improve blood and air flow, decreasing the release of an enzyme that may cause high blood pressure.

Clinical Scenario—Patient with Asthma

A 14-year-old girl was brought by her mother to urgent care at around 9 p.m. The girl was having shortness of breath with a heart rate of 105 beats per minute and a respiratory rate of 30 breaths per minute. The patient's oxygen saturation showed 93%. She was put on oxygen face mask at 5 l per minute. On auscultation, her lungs indicated moderate bilateral wheezing. She was given albuterol by handheld nebulizer. Albuterol is a bronchodilator used to open up the airway to allow breathing. It is classified as a rescue drug for patients having asthmatic attack. After the bronchodilator therapy, the patient's breathing improved with minimal wheezing. Her oxygen saturation increased to 95%. Her respiratory rate was 16, and heart rate was 95.

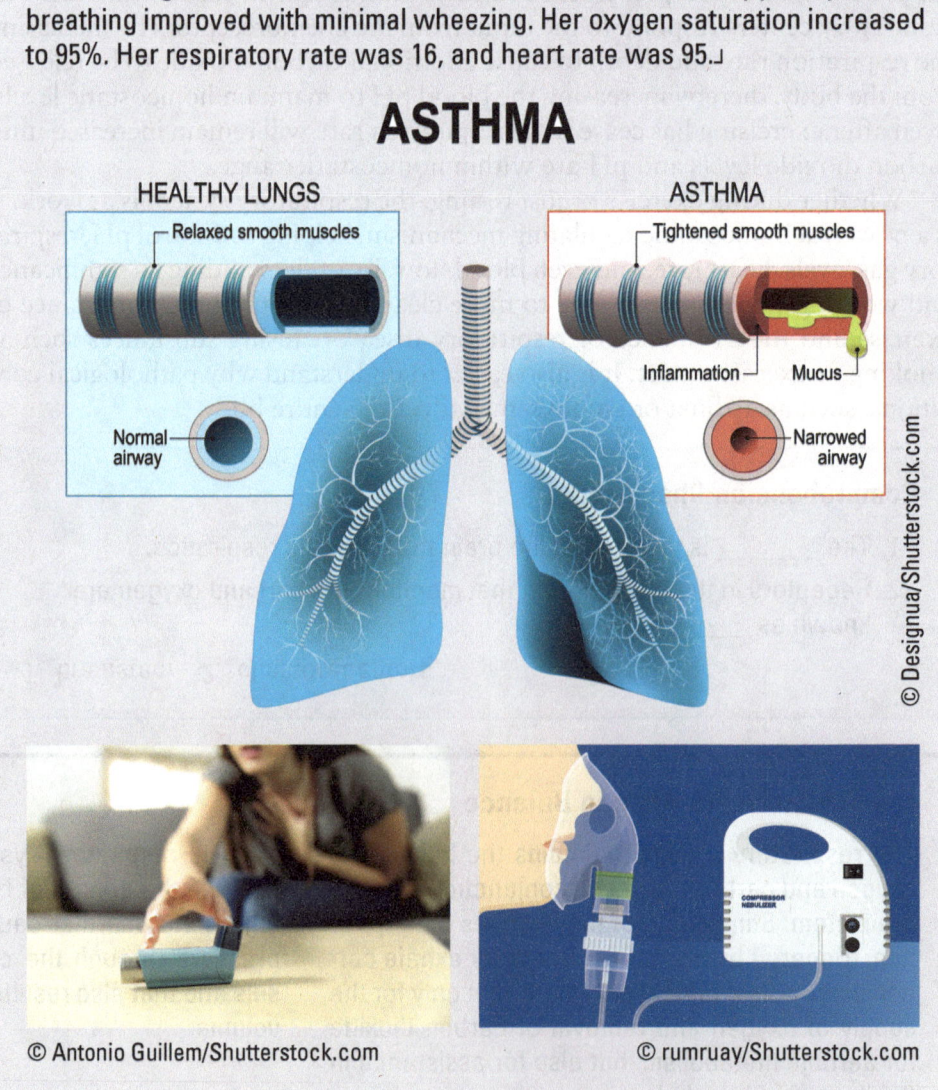

© Antonio Guillem/Shutterstock.com © rumruay/Shutterstock.com © Designua/Shutterstock.com

CLINICAL TERMS TO REVIEW

Hypoxemia, dyspnea, respiratory arrest, COPD
COVID-19 symptoms, CPR, defibrillation
Chest x-ray reading, fractured ribs, pneumonia, and cardiomegaly
Pulmonary function testing
Patient with asthma, wheezing, bronchodilator therapy, oxygen saturation
Inhalation, exhalation, angiotensin-converting enzyme (ACE)
Pulmonary circulation, systemic circulation
Upper respiratory tract—nose, nasal cavity, pharynx
Lower respiratory tract—larynx, trachea, bronchi, bronchioles, alveoli
Paranasal sinuses—frontal, maxillary, sphenoid, ethmoid
Pharynx—nasopharynx, oropharynx, laryngopharynx, uvula
Larynx—thyroid cartilage, vocal cords, arytenoid cartilage, epiglottis
Trachea—C-shaped cartilages, mucus-producing cells
Bronchi—bronchial tree; bronchioles—terminal bronchioles, alveolar duct
Alveoli—type I and type II alveolar cells, macrophages, pulmonary surfactant
Lungs—right, middle and lower lobes; left upper and lower lobes, lingula
Pleura—parietal and visceral pleura
Lung volumes and capacities—tidal volume, vital capacity, total lung capacity, inspiratory reserve volume, expiratory reserve volume, residual volume, functional residual capacity
Lung compliance
Alveolar gas exchange, cellular gas exchange
Oxygen and carbon dioxide transport
Hyperventilation, hypoventilation
Diaphragm, apneustic center, pneumotaxic center
Chemoreceptors—aortic bodies, carotid bodies, pH, carbonic acid

Test Yourself

Choose the best answer to the following multiple choice questions:

1. The type of epithelium that lines most of the respiratory tract is
   a. pseudostratified columnar.
   b. simple cuboidal.
   c. transitional.
   d. stratified squamous.

2. The air we breathe contains approximately _____% oxygen.
   a. 79
   b. 0.04
   c. 12
   d. 21

3. Air spaces within some of the facial bones that resonate the voice are known as
   a. nasal caverns.
   b. paranasal sinuses.
   c. bone deficits.
   d. a pharynx.

4. The respiratory system can affect blood pH by the retention or exhalation of
   a. oxygen.
   b. water vapor.
   c. carbon dioxide.
   d. nitrogen.

5. The sum of all of the air volumes that the respiratory system can move is known as
   a. total lung capacity.
   b. residual volume.
   c. vital capacity.
   d. temporary capacity.

6. The total possible volume the lungs can hold, which includes the air left in the lungs after maximum exhalation, is
   a. total lung capacity.
   b. residual volume.
   c. vital capacity.
   d. temporary capacity.

7. The cartilage flap that covers the closed vocal cords when swallowing is the
   a. uvula.
   b. frenulum.
   c. thyroid cartilage.
   d. epiglottis.

8. The respiratory center that controls relaxed rhythmic breathing is the
   a. superior respiratory group.
   b. dorsal respiratory group.
   c. inferior respiratory group.
   d. ventral respiratory group.

9. The chemical reaction that demonstrates the effect of carbon dioxide on the blood is
   a. $CO_2 + H_2O \rightarrow H_2CO_3$
   b. $CO_2 + ATP \rightarrow CATP + O_2$
   c. $CO_2 + 2\,NaCl \rightarrow CCl_4 + 2\,Na_2O$
   d. $CO_2 + 4\,KOH \rightarrow C(OH)OH_4 + 2\,K_2O$

10. The alveoli are composed of what type of epithelium?
    a. transitional
    b. simple columnar
    c. simple squamous
    d. stratified cuboidal

The Lymphatic System and Immunity

Chapter 10

LEARNING OBJECTIVES

Upon completion of this chapter, you will be able to:

1. Describe the components and functions of the lymphatic system and how it maintains homeostasis.
2. Explain how our bodies defend against foreign organisms and substances, such as bacteria and viruses.
3. Describe the criteria for defense and types of immunity.

CHAPTER OUTLINE

Introduction
The Lymphatic System
Functions of the Lymphatic System
Organization of the Lymphatic System
- Lymphatic Vessels
- Lymph: Its Composition and Formation
- Lymphoid Tissue and Organs
- Cells of the Lymphatic System

Foreign Invaders
- Bacteria
- Viruses

The Lymphatic System and Body Defenses
- Criteria for Defense
- Types of Defenses

INTRODUCTION

The lymphatic system serves to protect the body against bacteria, viruses, and fungi. It also removes toxins and other waste products from the body. It maintains fluid balance by removing excess fluid from the tissues and putting it back into the bloodstream. In other words, the important role of the lymphatic system is protection against infection, removal of toxins and waste products from the body, and maintaining fluid balance. The lymphatic system contains lymph, a clear white or yellowish fluid that contains the lymphocytes, which are white blood cells that fight infection. The lymphatic system closely operates with the immune system, which is the defense system of the body against pathogenic bacteria, viruses, and fungi. Both the lymphatic and immune systems share some organs and functions. Antigens and antibodies fall under the immune system which will be discussed in this chapter. Some of the immune system disorders include lupus, autoimmune disease, and rheumatoid arthritis.

A low-protein diet can weaken one's immune system. The human body manufactures its own proteins during sleep. Without adequate rest and sleep, the body's proteins become low, which affects the ability of the immune system to fight infection. Someone with cancer who is undergoing chemotherapy or taking chemotherapy drugs can also have extremely low immunity.

Therefore, to boost one's immune system, the following would help:

Figure 10.1

THE LYMPHATIC SYSTEM

Our bodies are constantly surrounded by opportunistic organisms and particles seeking entrance wherever possible. We must be constantly vigilant, keenly aware of self-recognition versus identification of foreign intruders, maintaining a strong barrier yet ready to attack invaders at any given instant should that line of defense be breached. The lymphatic system controls and maintains a constantly mobile defense system throughout our body.

FUNCTIONS OF THE LYMPHATIC SYSTEM

Understanding the functions of the lymphatic system involves knowing not only how our body's defenses are designed but also how disease can spread throughout the organism should they, even temporarily, gain the upper hand. There are four basic functions of the lymphatic system:

- Drainage of excess fluid between cells and tissues, known as *interstitial fluid*. As blood passes through most capillaries, some of the fluid filters out of these vessels and flows among the tissue surrounding them. Most of that fluid is reabsorbed by the capillaries. Excess fluid remaining in the tissue is collected by the lymphatic system and is known as lymph.
- Proliferation and distribution of white blood cells known as lymphocytes. They can mark foreign invaders for destruction or control our specific defenses. Lymphocytes divide and are stored in the spleen, lymph nodes, and red bone marrow. These defensive cells are able to move through the lymphatic fluid to the bloodstream.
- Transport of lipids from the digestive tract. Fats and oils, usually in the form of triglycerides, are too large to be absorbed into the capillaries in the intestine like other nutrients. Instead they enter lymphatic vessels that eventually drain into the subclavian veins to transport those lipids to the bloodstream.
- The immune response. The lymphatic system plays a significant role in the control of the body's defenses against trauma, foreign organisms, and abnormal cells.

Lymph excess tissue fluid collected by the lymphatic system.

Lymphocytes white blood cells that can mark foreign organisms for destruction or control our specific defenses.

> **Comprehension Check-up:**
> 1. Lymph collected by the lymphatic system is transported to the _____.
> 2. The lymphatic system also helps with response to infection, which is called _____.
>
> 1. cardiovascular system   2. immunity

ORGANIZATION OF THE LYMPHATIC SYSTEM

The components of the lymphatic system are found almost everywhere (Figure 10.2). The lymphatic system consists of a network of lymphatic vessels, lymphoid organs, and defensive cells.

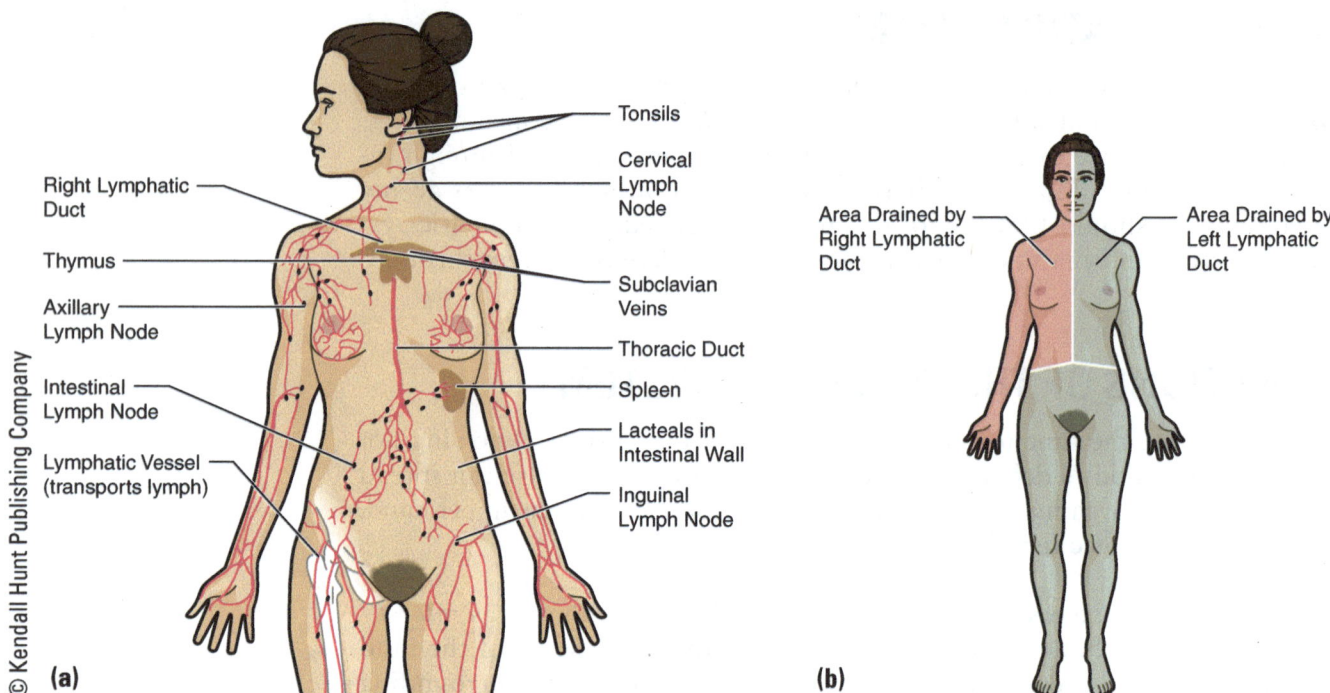

Figure 10.2 Lymphatic System. (a) The left superior body and all areas inferior to the diaphragm drain lymph through the thoracic duct into the left subclavian vein. (b) The right superior body drains lymph through the right lymphatic duct into the right subclavian vein.

Lymphatic Vessels

Lymphatic vessels resemble the venous side of the cardiovascular system in that they drain into vessels of ever-increasing size. Lymphatic capillaries collect excess interstitial fluid which drains into collecting vessels that use one-way valves and the contraction of nearby skeletal muscles to keep lymph moving onward. From the collecting vessels the fluid passes into lymph nodes that contain defensive cells to remove foreign or abnormal cells from the lymph. Once the sterile fluid leaves the lymph node, it flows into lymphatic trunks that drain into lymphatic ducts. Eventually those lymphatic ducts drain into the bloodstream through the subclavian veins.

Lymph Capillaries and Collecting Vessels

Lymphatic capillaries are similar to blood capillaries in that they are composed of a single layer of squamous cells. The cells of the lymphatic capillaries overlap, so fluid can pass between them to enter the vessel. Because they are not delivering fluid to the tissue, these capillaries do not need to have a complete circuit for flow to and from tissue like the bloodstream. Instead, they are closed-ended. Lymphatic capillaries are found throughout almost all vascular tissues, collecting excess fluid that is not reabsorbed by the bloodstream. The lymphatic capillaries drain into collecting vessels with one-way valves similar to those found in veins (Figure 10.3). Once lymph passes the first valve, it cannot backflow into the tissue. Any time these collecting vessels are compressed or the smooth muscle in their walls constricts, lymph is forced through the next valve, keeping the fluid moving onward.

Blocked Lymphatic Vessels—A Source of Edema

Excess interstitial fluid that cannot drain, because of blockage or damage of lymphatic vessels, results in the accumulation of fluid in the tissue known as *edema*. For example, if cancer surgery has resulted in the removal of a fairly large group of lymph nodes, swelling may occur distal to the surgical procedure. The lymphatic system will produce new vessels and nodes to replace those removed, but until that is accomplished, the accumulation of fluid will continue. Severe edema is sometimes seen in tropical climates where a parasitic worm transmitted by a mosquito blocks lymphatic drainage from an extremity. The result is a disease known as *lymphatic filariasis* or *elephantiasis*.

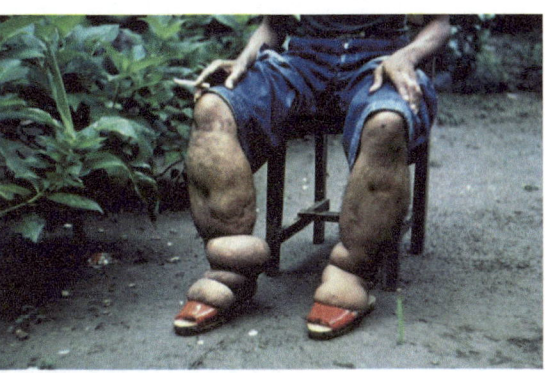

Elephantiasis (lymphatic filariasis) of the lower limb.

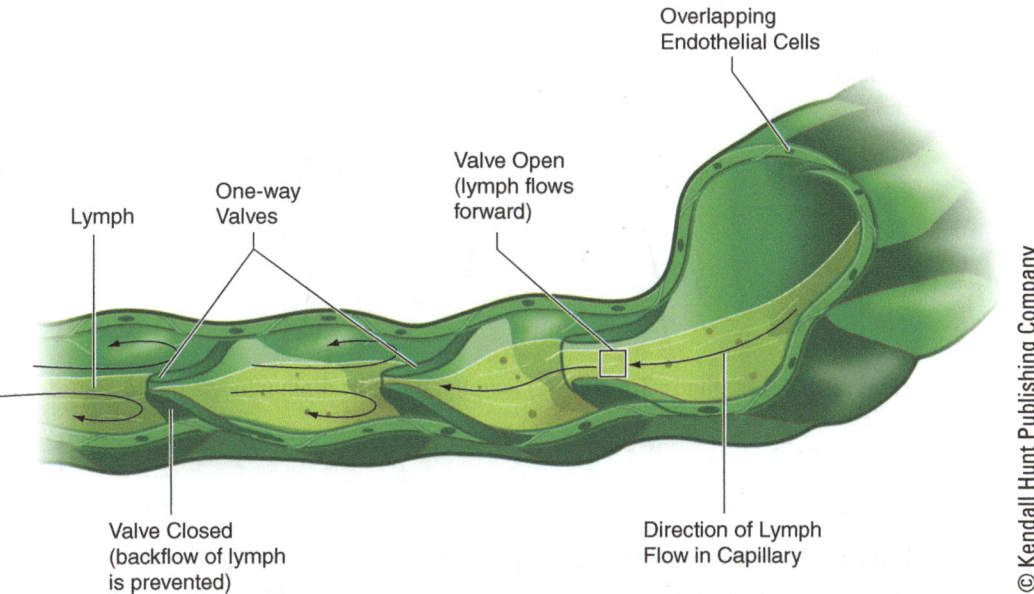

Figure 10.3 Lymphatic Capillary. This vessel is composed of overlapping cells between which extracellular fluid passes as it flows into a collecting vessel. The collecting vessel possesses one-way valves that determine the direction of the flow of lymph.

Lymphatic Trunks

Lymphatic collecting vessels drain regionally into lymphatic trunks. There are five on each side of the body (Figure 10.4):

- Jugular trunks drain the head and neck.
- Subclavian trunks receive lymph from the upper extremities and superficial thoracic wall, including the breasts.
- Bronchomediastinal trunks drain the deep chest.
- Intestinal trunks drain the abdomen.
- Lumbar trunks drain the lower extremities and the wall and organs of the pelvic cavity.

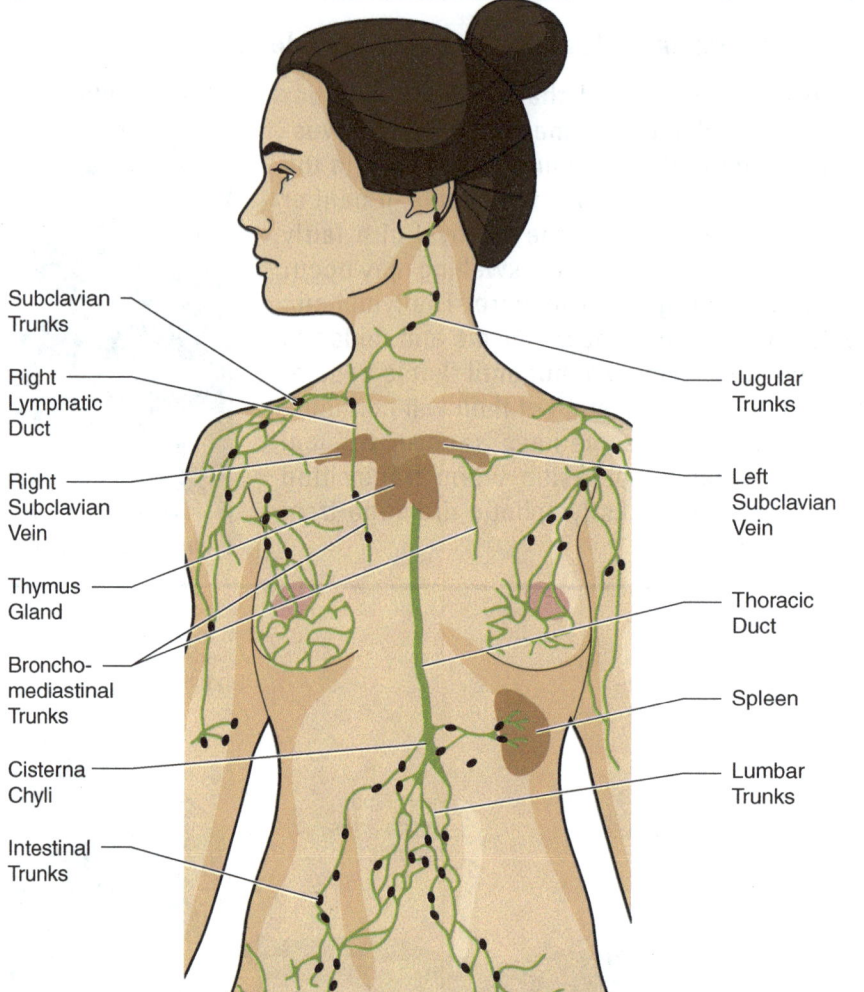

Figure 10.4 Lymphatic Trunks. The lymphatic trunks drain into lymphatic ducts.

Cisterna Chyli

Lymphatic drainage below the diaphragm flows toward a single location in the superior medial abdomen known as the *cisterna chyli*. The drainage from the cisterna chyli passes into the thoracic duct in the chest.

Thoracic Duct

The thoracic duct receives cleansed lymph from below the diaphragm through the cisterna chyli and also from the left upper body, including the left upper extremity and the left side of the head. Lymph from the thoracic duct empties into the left subclavian vein to return the cleansed tissue fluid to the bloodstream.

Right Lymphatic Duct

The right upper body drains lymph into the right lymphatic duct. It receives lymph from the right upper extremity, the right side of the head, and the chest. Drainage from the right lymphatic duct empties into the right subclavian vein.

Thoracic duct major lymphatic vessel that drains lymph from below the diaphragm and from the left upper body into the left subclavian vein to return excess tissue fluid to the bloodstream.

Right lymphatic duct receives lymphatic drainage from the right upper body and returns it to the bloodstream through the right subclavian vein.

> **Comprehension Check-up:**
> 1. Lymphatic drainage through collecting vessels moves in only one direction because _____.
> 2. Lymph reenters the bloodstream through the _____.
>
> 1. they have one-way valves 2. subclavian veins

Lymph: Its Composition and Formation

As blood flows through capillaries, some of the fluid and plasma proteins filter out of the bloodstream into the surrounding tissue as interstitial fluid. Although the formation of interstitial fluid occurs continually, its production increases when an individual is exercising, during some disease processes, or when blood pressure increases. The interstitial fluid flows through the wall of the lymphatic capillaries to become lymph (Figure 10.5). It contains water, proteins (such as albumin and antibodies), respiratory gases, dissolved salts, and lipids. The proteins and lipids, which are relatively large molecules, are able to enter the lymphatic capillaries because of the overlapping cells from which these vessels are constructed. The lymph has almost the same composition as the plasma, although red blood cells are not normally found in the lymph.

It is possible for foreign organisms to invade the tissue. If an injury to the epidermis breaks the barrier, bacteria or other organisms may enter through the damaged tissue. Those organisms have the capacity to flow with the interstitial fluid into the lymphatic capillaries if they are not destroyed at the location of the tissue damage. Foreign organisms may also gain access to the

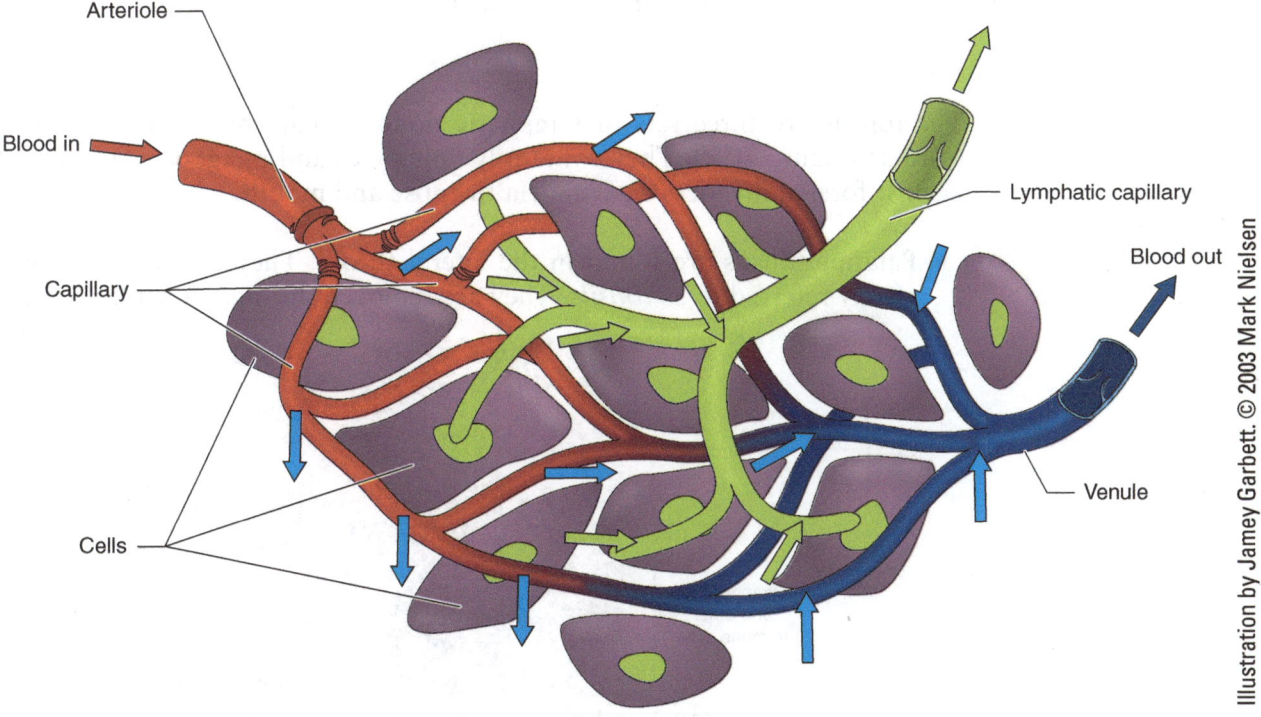

Figure 10.5 Flow of Filtered Fluid from the Capillaries into a Lymphatic Capillary.

body through openings such as the mouth, nose, eyes, anus, urinary tract, or vaginal orifice. Cancer cells may break off from the tumor and enter the lymphatic capillary as well. It is one of the ways cancer cells migrate, that is, metastasize, throughout the body.

Lymphoid Tissue and Organs

Within the lymphatic system are several lymphoid tissues and lymphoid organs that concentrate on intensive defensive action. Lymphoid tissues are concentrations of lymphocytes and phagocytes within mucous membranes, but they are not enclosed within a membrane. Lymphoid organs are also aggregations of defensive cells enclosed within a membrane and are anatomically well defined. These areas are as follows:

- Lymphoid tissue
  - Lymphoid nodules
- Lymphoid organs
  - Tonsils
  - Lymph nodes
  - Thymus gland
  - Spleen

Each is discussed individually.

Lymphoid Nodules

Within the walls of the respiratory, digestive, and urinary tracts are lymphoid nodules. Their location means these areas are constantly exposed to foreign organisms. The nodules are areas where lymphocytes are densely packed but not enclosed by a connective tissue capsule (Figure 10.6). Proliferation of lymphocytes occurs in the center of the nodule.

Tonsils

The tonsils are three sets of lymphatic nodules that encircle the pharynx (throat) (Figure 10.7). They contain lymphocytes and phagocytes to defend against foreign invaders entering via the nose and mouth.

- Palatine tonsils are found in the lateral throat. They are the pair commonly referred to as *tonsils*. They are removed during a tonsillectomy.

> **Lymphoid tissues** tissues associated with the lymphatic system that are responsible for body defenses.

> **Lymphoid organs** organs associated with the lymphatic system that are responsible for body defenses.

> **Lymphoid nodules** cluster of densely-packed lymphocytes and phagocytes not enclosed in a connective tissue capsule.

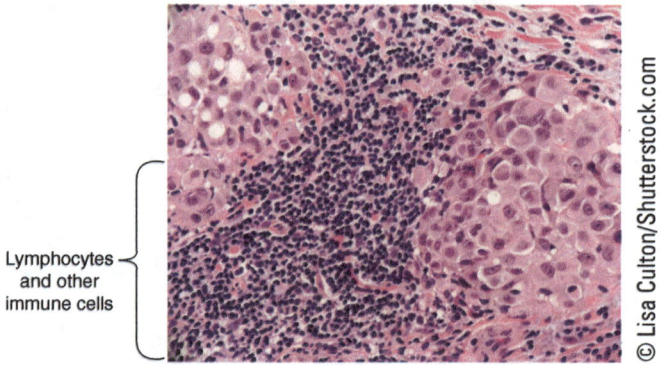

Figure 10.6 **Lymph Nodule.** A lymph nodule is a dense collection of lymphocytes without an enclosing membrane capsule.

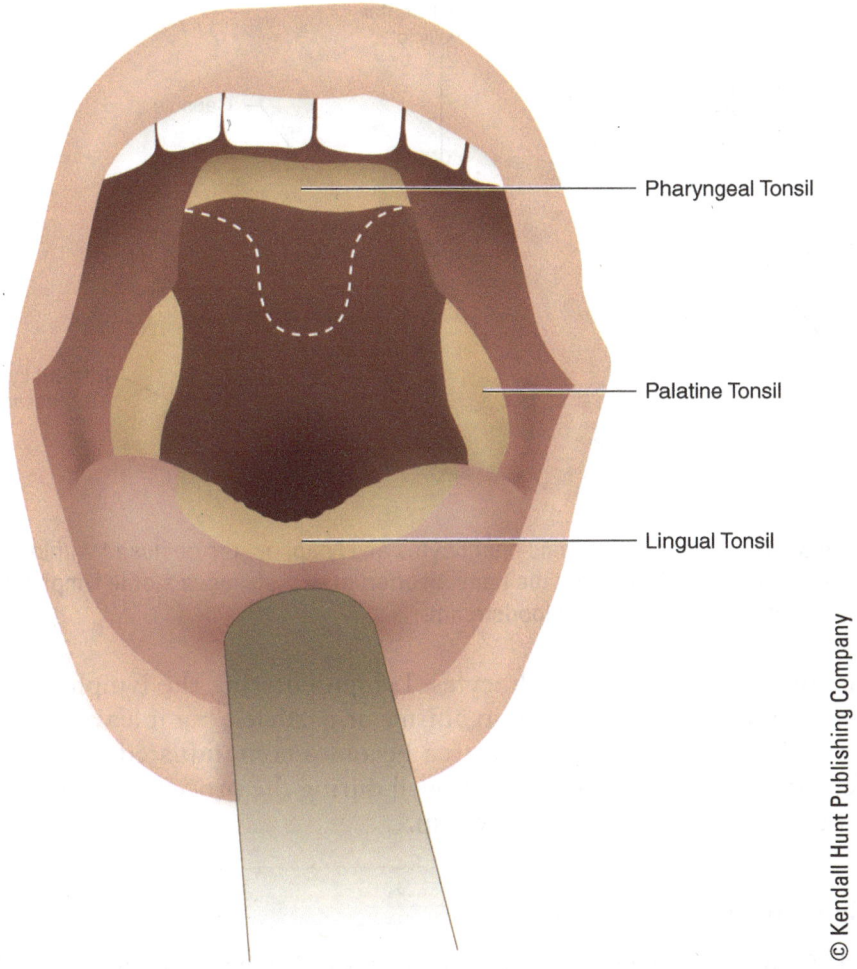

Figure 10.7 Tonsils.

- Pharyngeal tonsils are located in the wall of the pharynx posterior to the nasal passage. They are also referred to as *adenoids*. They may also be removed during a tonsillectomy.
- Lingual tonsils are located on the posterior tongue. They are not removed.

By encircling the pharynx with lymphoid tissues, foreign organisms and particles entering the respiratory tract are minimized. In some cases, the tonsils become so hypersensitive to antigens they become more of a problem than an asset.

Lymph Nodes

Lymph nodes are pea-sized encapsulated structures containing reticular fibers to which defensive cells are attached (Figure 10.8). There are roughly 450 lymph nodes found throughout most of the body, with increased concentrations of them in the groin (inguinal nodes) and axillary areas (axillary nodes), as well as in the neck. The lymph flowing from lymph capillaries into collecting vessels may contain foreign invaders. The collecting vessels drain into lymph nodes. Each lymph node contains lymphocytes, macrophages, and other phagocytes to identify and destroy any foreign organisms or antigens (foreign chemicals, often proteins) that have obtained access to the tissue through injury or opening in the body. The lymph node also contains

> **Lymph nodes** pea-sized structures containing defensive cells that are responsible for removing foreign organisms and substances from lymph drained from local tissue.

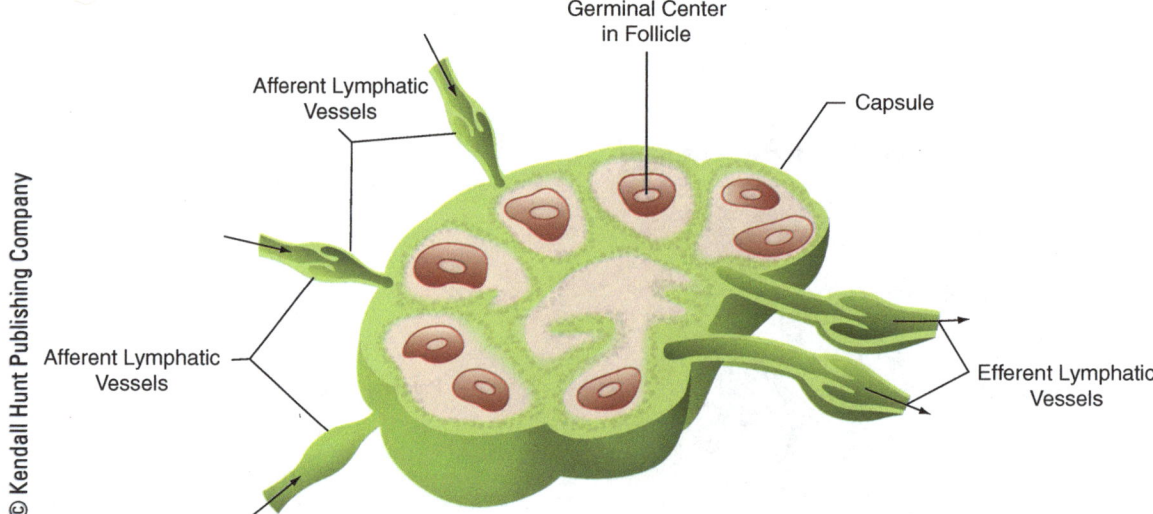

Figure 10.8 Lymph Node. A lymph node contains phagocytes and lymphocytes enclosed within a capsule. Afferent vessels transport lymph into the node as efferent vessels move sterile lymph toward the subclavian veins to return the fluid to the bloodstream.

areas for the mitosis of lymphocytes. Lymph entering the lymph node may be contaminated with harmful organisms or particles, but it leaves the lymph node pathogen-free. If a specific area becomes contaminated by foreign invaders, the local lymph nodes may swell during the process of defense, causing it to become hard, sore, and warm.

> **Comprehension Check-up:**
> 1. Densely packed clusters of lymphocytes not enclosed within a capsule form _____.
> 2. The _____ defend against foreign invaders in the nose and throat.
>
> 1. lymphatic nodules   2. tonsils

Thymus Gland

The thymus gland is located superior to the heart and is wrapped around the trachea (Figure 10.9). At birth the thymus gland is a large organ. It continually enlarges until puberty; after puberty, it begins to shrink and by the age of 40, only a small percentage of the gland remains. The thymus gland produces hormones that cause the maturing of a specific type of lymphocyte known as a *T lymphocyte*. While maturing in the thymus gland, T lymphocytes do not participate in immunity. T lymphocytes are discussed in greater detail later in this chapter.

Spleen

Located in the left upper quadrant of the abdomen (Figure 10.10), the spleen has three functions:

1. It removes foreign organisms from aged, as well as, dead red blood cells from the bloodstream.
2. It is a reservoir of blood.
3. It contains areas for the division and storage of white blood cells and platelets.

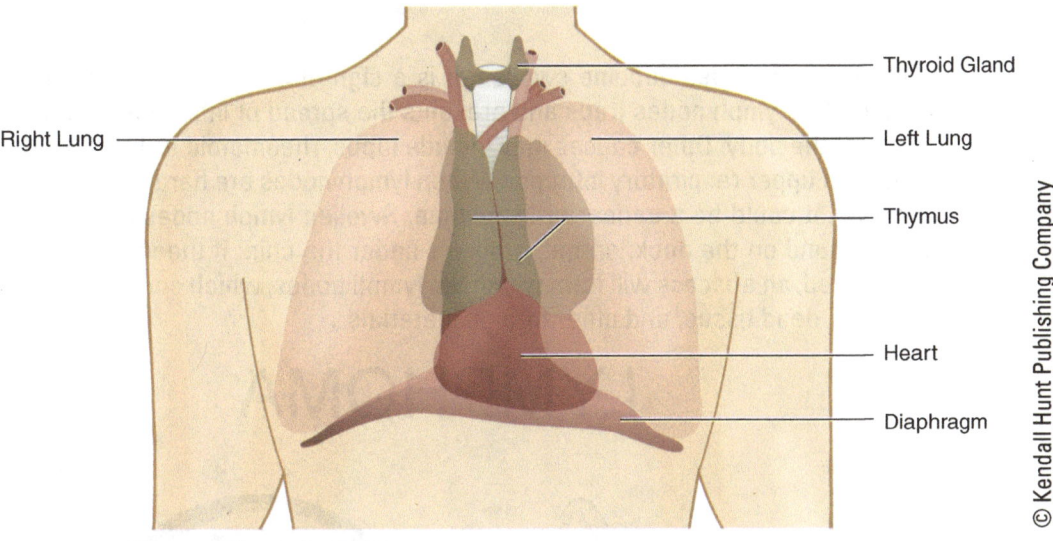

Figure 10.9 Thymus Gland of a Child.

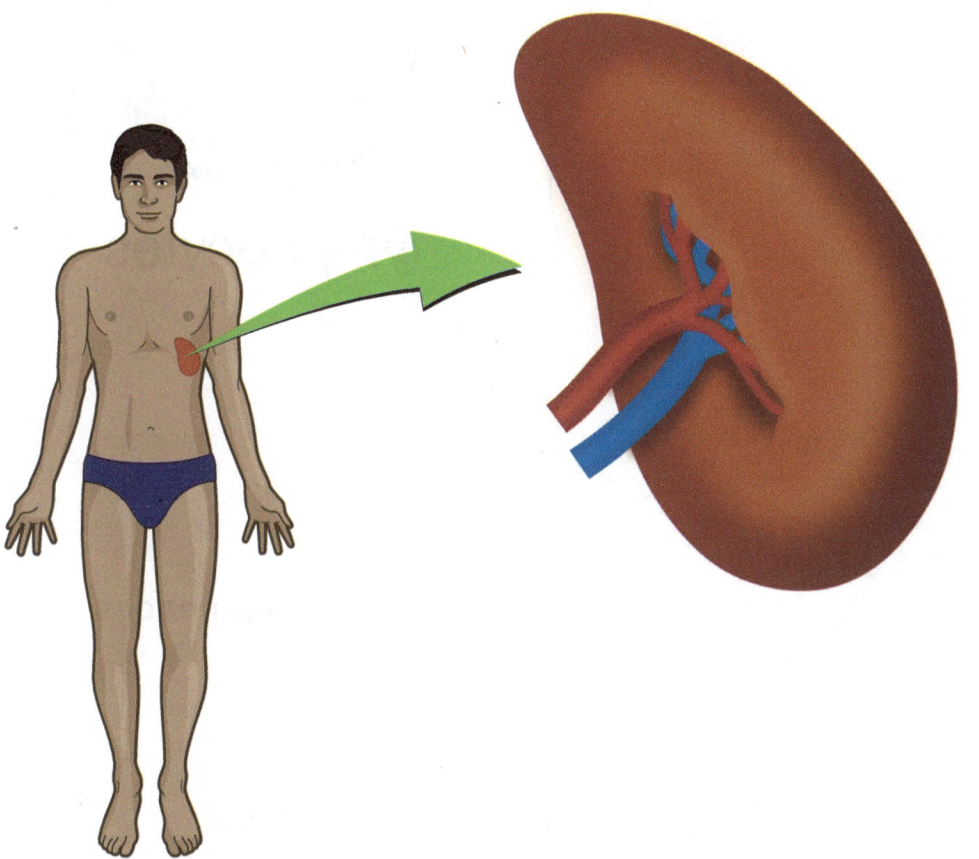

Figure 10.10 Spleen. The spleen filters blood and acts as a center for the germination and storage of white blood cells.

If foreign organisms were somehow able to gain access to the bloodstream, the spleen is able to remove and destroy them. It is normally not necessary to have an excessive number of white blood cells constantly circulating through the cardiovascular system. If, however, a serious infection occurs, the white blood cells stored in the spleen are able to rapidly mobilize to defend the body.

Chapter 10 The Lymphatic System and Immunity

Clinical Notes—Lymph Nodes

When lymph nodes become swollen, it is a sign of a bacterial or viral infection. The lymph nodes traps and prevents the spread of infection to other areas of the body. Other causes may include lupus, rheumatoid arthritis, HIV, or just an upper respiratory infection. When lymph nodes are hard, fixed, and growing, it could be a cancer or lymphoma. Swollen lymph nodes are commonly found on the neck, armpit, groin, or under the chin. If the infection is not treated, an abscess will form within the lymph nodes, which contains pus, bacteria, dead tissue, and other foreign materials.

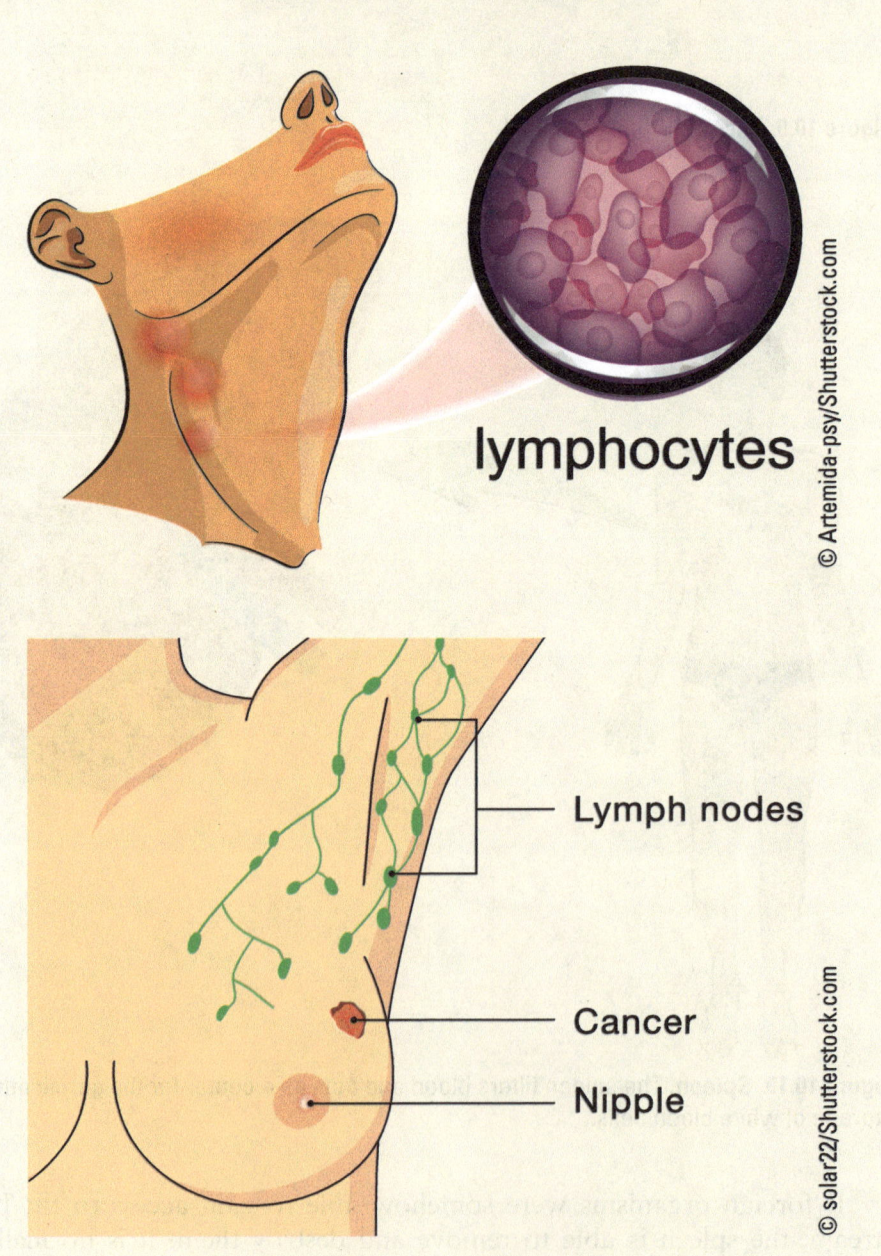

Cells of the Lymphatic System

Three types of lymphocytes associated with the lymphatic system are essential for the normal defense of the body:

- B lymphocytes (B cells)
- T lymphocytes (T cells)
- Natural killer (NK) cells

B Lymphocytes

B lymphocytes (B cells) are so named because they mature in the bursa of fabricus, the site of hemopoietosis in birds. In humans, the B lymphocytes are produced and mature in bone marrow. They provide an essential function in the strategy of defense, but they do not attack foreign invaders. Instead, B cells that differentiate into plasma cells produce antibodies that mark cells for destruction or bind to noncellular antigens to prevent their adverse affect on the body. The processes by which B lymphocytes produce antibodies and their functions are discussed later in this chapter.

> **B lymphocytes (B cells)** white blood cells activated to recognize specific foreign organisms and then produce antibodies that mark those cells for destruction or make toxic substances harmless.

T Lymphocytes

T lymphocytes (T cells) are defensive cells that mature within the thymus gland. There are three types of T cells:

- Cytotoxic cells that destroy virus invaded cells, cancer cells, and transplanted cells
- Helper cells that stimulate the activation of B and T lymphocytes
- Regulatory (suppressor) cells that inhibit activity of B and T lymphocytes

> **T lymphocytes (T cells)** white blood cells that mature within the thymus gland. Some T lymphocytes destroy cells invaded by a specific virus; others control immunity.

The function of cytotoxic cells is discussed in more detail later in this chapter. Helper cells control the activation of the defense system beyond local defenses. It is the helper cells that cause lymphocytes to leave the spleen through the bloodstream to attack antigens. Regulatory cells keep the defense system from going too far and attacking normal body cells.

Natural Killer Cells

Natural killer (NK) cells are a small percentage of lymphocytes. Unlike B and T lymphocytes, NK cells do not have the memory to recognize specific antigens of disease organisms or proteins. Instead, they are able to recognize groups of abnormal cells such as virus-invaded cells or abnormal body cells. Cytotoxic T lymphocytes attack and destroy cells invaded by a specific virus. NK cells rapidly attack any cell invaded by a virus regardless of type. They are our body's best defense against cancer cells, which are usually larger than normal cells.

> **Natural killer (NK) cells** white blood cells that are unable to recognize specific organisms but destroy any virus-invaded cell. They also eliminate abnormal cells from the body.

Comprehension Check-up:

1. Maturing of T lymphocytes occurs in the _____.
2. Antibodies are produced by _____.

1. thymus gland   2. B-lymphocytes

> **Defense Out of Control**
>
> The inability of our bodies to stimulate or limit defense cells may result in devastating consequences. A significant concern of health officials is the human immunodeficiency virus (HIV), which leads to acquired immunodeficiency syndrome (AIDS). This virus infects helper cells, causing those cells to replicate the virus. The processes involved with viral production use up substances within the cell, which causes the infected helper cells to die. The decreased number of helper cells results in an inability to stimulate defenses against certain foreign organisms. Death results when these opportunistic pathogens overwhelm the body. The goal of current drug therapy is to slow the production of HIV within helper cells. This method of treatment does not stop HIV, but it allows the number of helper cells to increase to the point where a reasonable level of immunity control exists.
>
> The opposite condition to AIDS occurs when the regulatory T cells fail to decrease the activity of B and T lymphocytes. Although there is intense research in this area, it is not yet well understood precisely why some normal body tissue comes under attack by our own defenses, resulting in some type of autoimmune disease. It appears that something has altered the self-antigens that identify tissue as belonging to the host. Instead the defense system determines these tissues to be foreign and designated for removal from the body. Some examples of autoimmune diseases are multiple sclerosis, rheumatoid arthritis, and type I diabetes mellitus.

FOREIGN INVADERS

Foreign invaders are any cells or particles with identification different than the host body cell. Most invaders entering the body obtain access through injured skin or through an opening such as the nose and mouth, eyes, anus, or urinary tract orifice. The foreign cells that most commonly cause serious diseases in humans are bacteria, and the most common disease-causing particles are viruses. To better understand the purpose behind our methods of bodily defenses, it is useful to discuss these two pathogens in more detail.

Bacteria

Bacteria are single-celled organisms that fall into a category known as *prokaryotes* (Figure 10.11), cells that have no membrane-bound organelles and no nuclear membrane enclosing its DNA (deoxyribonucleic acid) (Figure 10.12).

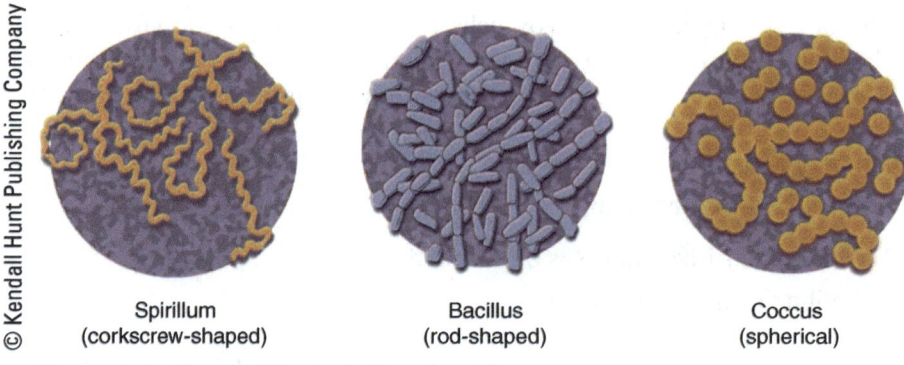

Spirillum (corkscrew-shaped)   Bacillus (rod-shaped)   Coccus (spherical)

Figure 10.11 Types of Bacteria Based on Shape.

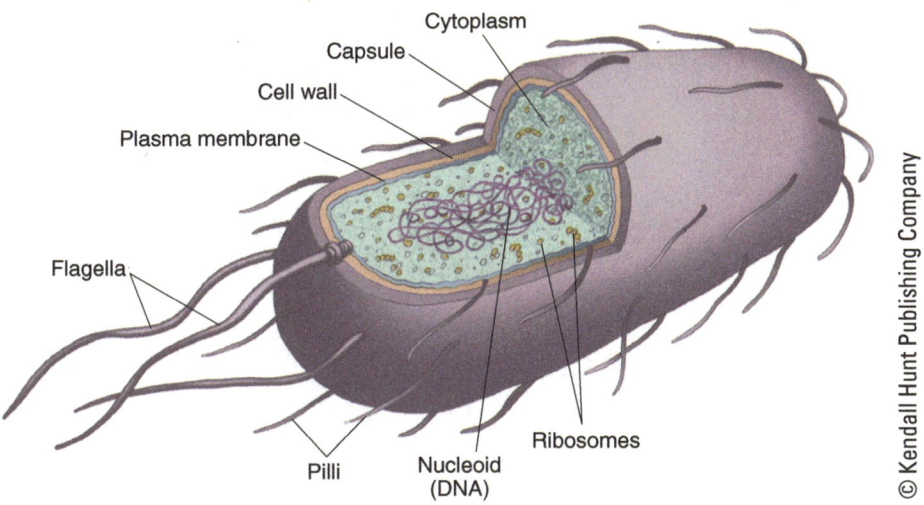

Figure 10.12 Anatomy of a Bacterium.

In terms of defense, the body commonly uses two methods to kill bacteria. First, because bacteria are single cells with no organelles, there is no isolation of their chemical reactions from temperature changes in the environment; therefore, they become temperature sensitive. A relatively slight increase in temperature is sometimes enough to disrupt bacterial chemical processes. Second, the body's defense system can create holes in the bacterial cell membrane. Osmosis draws water into the cell through the holes, causing the bacterium to swell and resulting in rupture of the cell membrane, which kills the cell. This is a typical defense strategy of our body. These defenses are discussed in more detail later in this chapter.

Bacteria have options to defend themselves. They have the unique capability of picking up pieces of DNA from other organisms, even other dead bacteria, and incorporating those genes into their own set of instructions. As a result, bacteria mutate very easily. The rapid division of bacteria combined with the natural tendency of mutations to occur in their genetic code allows them to develop resistance against attack, including antibiotics, especially when there is increased exposure to these drugs.

Viruses

Viruses are composed of a segment of either DNA or RNA within a protein envelope. They do not meet all of the criteria for living organisms. They cannot reproduce themselves and instead rely on living cells for reproduction. The virus attaches to a living cell and then forms a projection similar to a hypodermic needle that pierces the plasma membrane of the cell. The virus then injects its nuclear material along with some key enzymes. Included in the DNA or RNA are instructions for the cell to make more viruses. The cell turns into a virus factory using up its own resources to produce viruses. Eventually the cell will die, releasing viruses into the surrounding tissue (Figure 10.13).

How do our bodies kill a pathogenic particle that was not living to begin with? We cannot. Instead our defense system destroys the virus-infected cell, causing viral reproduction to be terminated.

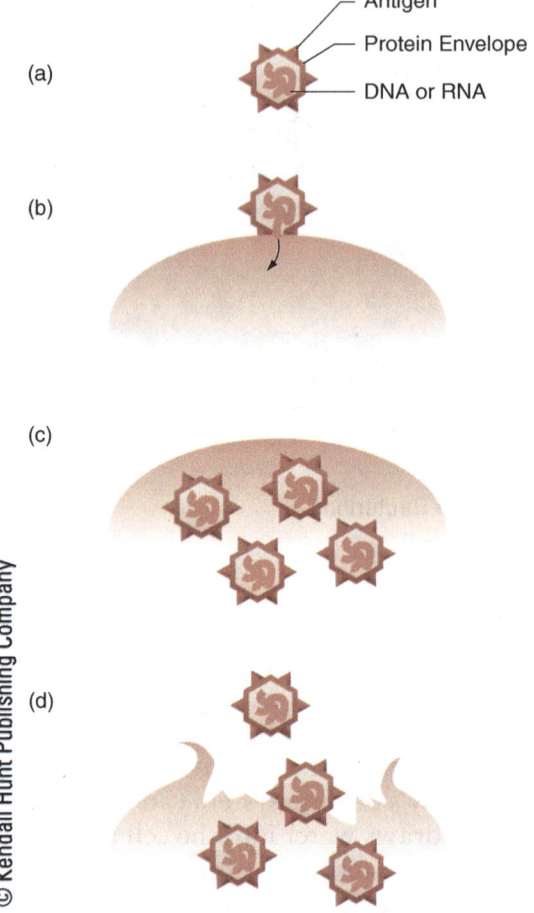

Figure 10.13 Action of a Virus. (a) Structure of a virus. (b) A virus injects its genetic material into a living cell. °(c) The cell produces viruses. (d) When the cell dies and the membrane ruptures, viruses are expelled.

Comprehension Check-up:
1. Two methods to destroy bacteria are _____ and _____.
2. Viruses are composed of _____.

1. increased temperature; create holes in their cell membrane
2. a protein envelope containing a segment of DNA or RNA

THE LYMPHATIC SYSTEM AND BODY DEFENSES

Since the lymphatic system is essentially draining all excess tissue fluid back into the bloodstream, it is of critical importance that lymph entering the cardiovascular system is free of any foreign invaders that may have gained entrance to the body. Without this defense, any external pathogens could potentially have access to the entire body. The lymphatic system specializes in the ability to detect and identify specific foreign antigens that may enter the body and cause them to be destroyed or become ineffective. It also controls the activity of immunity by stimulating or inhibiting lymphocytes.

Criteria for Defense

There are four basic criteria that are essential to maintain defense. The immune system must be able to:

- Recognize "self." This function is provided by the self-antigens on the plasma membrane of our body cells indicating that each cell belongs to us, the host.
- Recognize "non-self." It is critical that our defense system be able to identify foreign invaders. Also included in this group are our own body cells that are behaving in an abnormal manner, such as cancer cells.
- Neutralize "non-self." It is not essential to kill or destroy any cell or material considered non-self. Encapsulating the offending tissue or object can isolate it from the rest of the body.
- Not harm "self." While the defensive process may cause us to feel sick, the overall goal is to keep the damage to our healthy self to a minimum.

If our defenses lose the ability to meet all of these criteria, we become ill. If the immune system cannot recognize "self," our defense system attacks our own tissue. This occurs in autoimmune diseases. The inability to recognize "non-self" could potentially give foreign invaders free reign over our body, with extremely destructive results. In some cases we may settle for some harm to our healthy self in order to neutralize non-self. Some cancer treatments, such as radiation and chemotherapy, involve harmful treatments to self but the potential of successfully eliminating these abnormal cells (non-self) is high.

Types of Defenses

The human body has two types of defenses:

1. Innate immunity is a nonspecific defense to ward off any potential enemy at any given time. The response is virtually the same each time an invasion occurs and there is no memory of the antigen against which the defenses were engaged because the immune system was not required.
2. Adaptive immunity consists of defenses that attack a specific invader. Each time an invasion of a specific antigen occurs, adaptive immunity possesses memory of the antigen and becomes more efficient at controlling the potential threat.

To provide effective defenses, both systems must be fully functional.

Comprehension Check-up:

1. If the body loses the ability of some cells to be recognized as "self," the result is an _____ disease.
2. The type of immunity that requires memory of a specific antigen is _____ immunity.

1. autoimmune 2. adaptive

Innate Immunity

Innate immunity, also known as non-specific immunity, occurs as prevention of invasion or as a means of fighting off any potential attack by foreign invaders that have gained access to the body. There are five types of innate body defenses:

- Physical barriers
- Inflammation
- Fever
- Complement
- Macrophages and other phagocytes

> **Innate immunity** the prevention of invasion or a means of fighting off any potential attack by foreign organisms that have gained access to the body.

Physical Barriers

The skin is our greatest defense against external invasion. Infection below the skin occurs only when that barrier has been damaged or traumatized. Recall from earlier in this chapter that there are also lymphoid nodules around openings to internal organs to maximize our defense in those areas. In some cases, such as the stomach or the vagina, acid kills living organisms that may have gained access. There are other fluids, for example, mucus, oils, wax, saliva, and tears, on the skin or other exposed areas such as the mouth, nasal cavity, and vaginal orifice that also protect against potential invasion.

Inflammation

Inflammation, the response initiated when physical barriers have been compromised, is a four-step process the body uses to defend against invaders and then prepare the damaged tissue for repair (Figure 10.14).

> **Inflammation** the body's response initiated when physical barriers have been compromised to defend against invaders then prepare the damaged tissue for repair.

1. Macrophages are first on the scene to kill any living invaders that may have entered the area. Chemical messages, known as chemotoxins, placed in the bloodstream call in reinforcements, if necessary. The release of the chemical histamine by mast cells and basophilis causes excess fluid to filter from the capillaries, which causes local flooding of tissue.
2. The histamine release increases blood pressure in the capillaries and increases the size of capillary pores, allowing more fluid to pass from the bloodstream into the tissue. This increased blood flow causes swelling, pain, a rise in temperature, and redness in the area.
3. Phagocytes arrive in the damaged area to remove dead tissue so that the wound can be cleaned for repair.
4. The damaged tissue is replaced by mitosis of neighboring cells, or if the wound is too extensive, scar tissue is produced to hold the edges together again.

Fever

Another mechanism of defense is increased body temperature. Fever, body temperature greater than 99.9°F (37.7°C), may have several causes. Exercise, anxiety, or dehydration may cause higher temperatures in healthy individuals. Infections may also result in a fever when a group of chemicals known as *pyrogens* ("pyro" means fire) increase the setting of the hypothalamus, the controller of body temperature. Pyrogens are typically derived from three sources: they may be released by phagocytes, they may be bacterial toxins, or they may be the cytoplasmic contents of destroyed bacteria. Fever increases leukocyte activity and decreases bacterial growth.

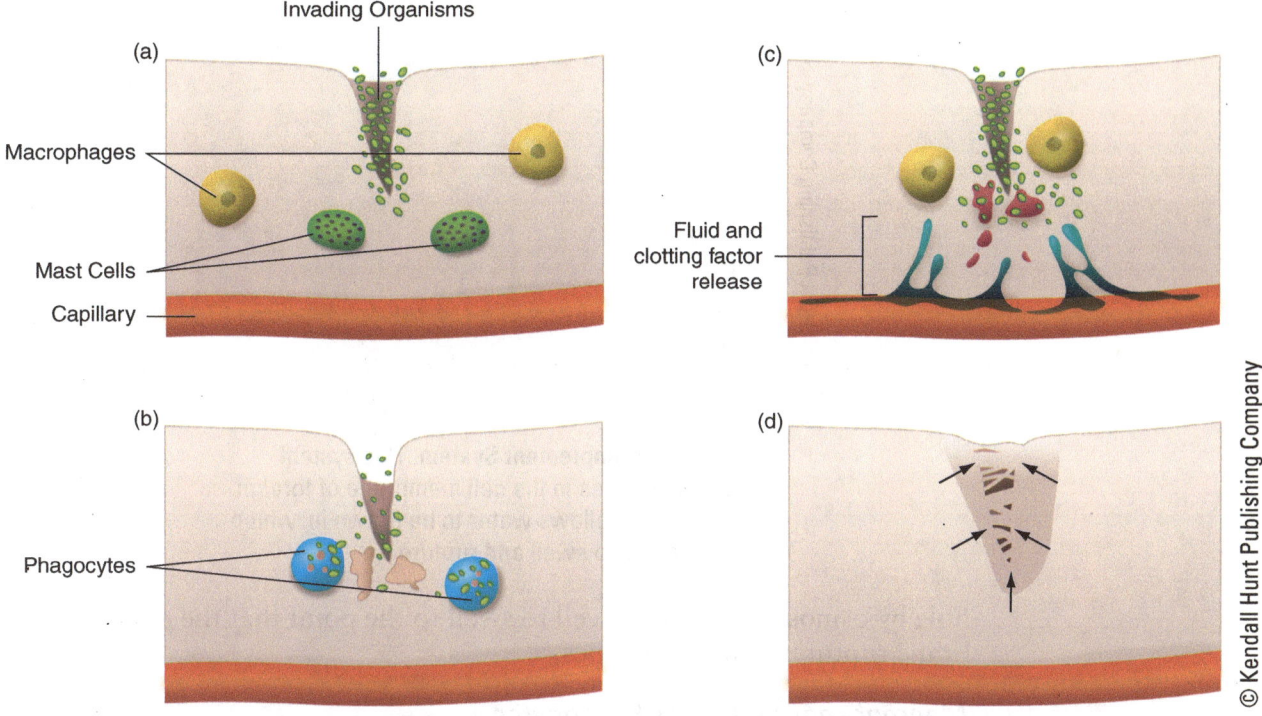

Figure 10.14 Processes of Inflammation. (a) Macrophages arrive to destroy invading organisms, and histamine is released. (b) Histamine causes release of additional fluid and clotting factors from capillaries, resulting in interstitial fluid clotting and trapping of foreign organisms. (c) Phagocytes remove debris. (d) Mitosis of surrounding tissue replaces dead tissue, or it is filled in with scar tissue.

High body temperatures may cause heat exhaustion, heat stroke (which causes permanent brain damage), or even death. In cases of heat exhaustion, the individual sweats excessively, which results in excessive fluid loss. The fluid loss causes low blood pressure, and this leads to loss of consciousness. Heat stroke, on the other hand, is the result of a breakdown of the cooling mechanism of the body. The unconscious individual suffering from heat stroke does not sweat. The lack of sweating results in the internal body temperature increasing even more. This individual is in serious danger of severe brain damage unless the temperature can be reduced.

Fever-reducing drugs such as aspirin and ibuprofen can also relieve additional symptoms of fever such as headache, rapid pulse, hot dry skin, and chills. There is considerable debate among physicians if these drugs should be given unless the individual experiences a fever for an extended period, such as for several days. Some advocate that taking fever-reducing drugs decreases the body's ability to defend against foreign invaders. Others counter that the beneficial effects of this group of drugs far outweighs the benefits of fever reduction, which they consider to be minimal.

Complement System

The *complement system* consists of about 20 inactive proteins found in our bloodstream and tissue fluid. When triggered into its active state by antibodies or other regulating proteins, this system complements or enhances the actions of other defenses. For example, the complement system stimulates the release of histamine to dilate local blood vessels. It may function as a marker that identifies a cell that needs to be destroyed by phagocytes. It may also form holes in the plasma membrane of unwanted living cells, resulting in their death (Figure 10.15). Water enters the cell through the

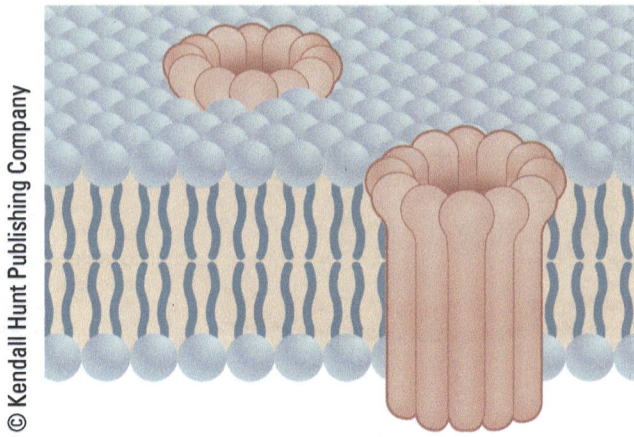

Figure 10.15 Complement System. This system may produce holes in the cell membrane of foreign organisms. This allows water to be drawn in, which causes the cell to swell and rupture.

hole by osmosis, causing the cell to swell to the point that the plasma membrane ruptures and the cell dies.

Macrophages and Other Phagocytes

Macrophages, large phagocytes derived from monocytes in the bloodstream, are very large defensive cells that attack anything considered "non-self." Phagocytes engulf and consume enemies, but they also remove the debris from the battle that has occurred in defense of the body. Both cells are very effective at what they do, but they do not possess memory for identifying specific organisms for destruction. To work most efficiently, these cells often rely on the antibodies produced by B lymphocyte plasma cells to mark foreign organisms for destruction.

> **Comprehension Check-up:**
> 1. The purpose of a fever is to _____ .
> 2. The function of the complement system is to _____ .
>
> 1. kill temperature-sensitive bacteria
> 2. produce holes in the cell membrane of an unwanted cell to cause it to rupture

Adaptive Immunity

> **Adaptive immunity** the ability of the body to identify specific antigens and prevent them from doing harm.

Adaptive immunity, also referred to as specific immunity, recognizes specific pathogens, responds with active defenses against them, and remembers their foreign antigens. There are two types of adaptive immunity: cell-mediated immunity and antibody-mediated immunity, also known as humoral immunity. To understand how this process works, it is useful to describe the activities of each component involved and the result of a loss of control. These areas of adaptive immunity we discuss include:

- Antigens and antibodies
- Types of adaptive immunity

- B lymphocytes
- T lymphocytes
- Hypersensitivity
- Autoimmune diseases

Antigens and Antibodies

An antigen is a protein or other chemical not produced by the host that stimulates an adaptive response. These antigens could be found on cell membranes, such as bacteria, fungi, or transplanted tissue, but they may also be substances such as viruses, pollen, poison ivy oil, animal dander, and food. Antibodies are Y-shaped proteins that attach to specific antigens. Once these antibodies are fixed to a foreign cell's antigens, other defensive cells go into action to destroy the marked cell. Antibodies may also neutralize toxins and agglutinate red blood cells of the wrong type (Figure 10.16).

Antigen a protein in the cell membrane that identifies the type of organism it is.

Antibodies a type of globulin. They act as chemical tags that identify foreign or abnormal cells to be destroyed by body defenses. They may also neutralize toxins produced by foreign organisms.

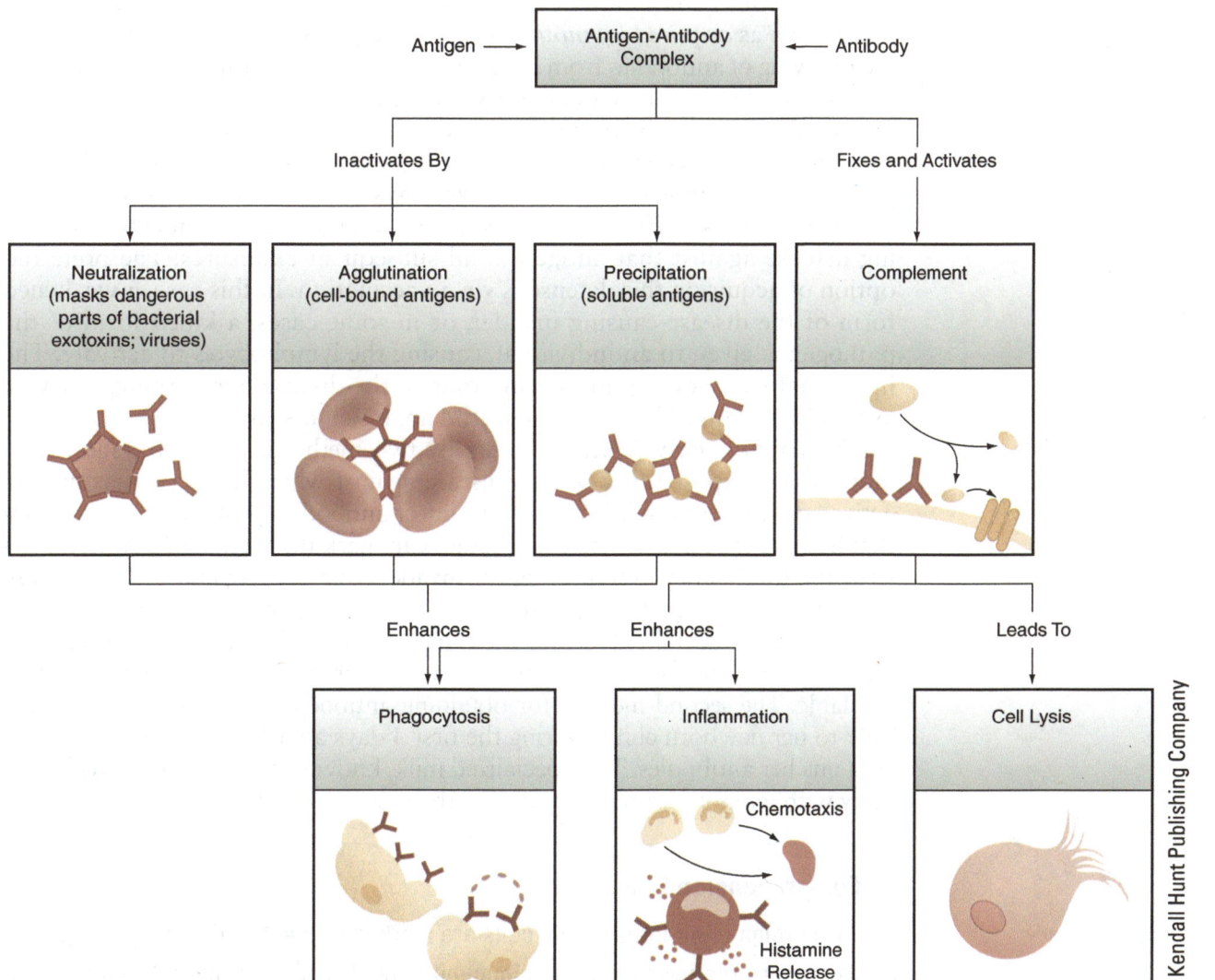

Figure 10.16 Mechanisms of Antibody Action. They may detoxify pathogenic substances, agglutinate the wrong blood type, or precipitate antigens dissolved in the plasma or tissue fluid. They may also activate defensive mechanisms or mark cells for destruction.

Smallpox and Polio Vaccines Changed the World

Some vaccines have had a tremendous effect on the entire world. Most notable is the smallpox vaccine. Smallpox caused high fever, headache, and a red rash that left scars on those who survived infection. Its mortality rate was very high, killing more individuals in the 20th century than all other infectious diseases combined. Through an extensive vaccination program, this dreaded disease has been virtually eliminated from the planet. The last case of smallpox occurred in 1977.

Another vaccine that has had a widespread effect is the polio vaccine. Polio caused flulike symptoms in many cases, but in some infected individuals, damage to nerve endings resulted in paralysis that could lead to death. In the 1950s a polio vaccine was developed that decreased the spread of the disease. Over the next 30 years, polio was virtually eliminated in the United States and most of the world. There are, however, a few countries where outbreaks of polio still occur.

Types of Adaptive Immunity

Lymphocytes are responsible for adaptive immunity and to be able to identify a specific organism, they must be activated. This process begins shortly after birth. It is known as *acquired immunity* because it requires exposure to an antigen or the receiving of antibodies from another source as a temporary means of defense.

Acquired immunity can be in two forms: active or passive. *Active immunity* results in the activation of a lymphocyte's memory as the result of exposure. It is able to recognize a specific antigen if it enters the body again. This process is accomplished in one of two ways. If an individual has a disease, memory of the specific antigen is kept in some of the lymphocytes, providing defense against that antigen in all subsequent exposures. The preferred option of acquiring this defense is via a vaccination. In this case, a weakened form of the disease-causing invader, or in some cases, a killed form of the pathogen is given to an individual, causing the lymphocytes to activate. The individual may develop mild symptoms of the disease but nothing as severe as would be caused by exposure to the unaltered disease.

Passive immunity typically occurs by two methods. In one instance, if an individual is exposed to a serious disease against which he or she has not yet activated lymphocytes, it is possible to inject antibodies from another organism that has had the disease. These antibodies can mark the foreign organism for destruction by phagocytes to prevent the invader from causing illness. This defense is temporary because host lymphocytes have not been activated to produce their own antibodies and the B memory cells have not been formed. Once the injected antibodies have broken down, usually in a few months, the immunity is no longer available. The second method for obtaining antibodies is through the mother's milk to her newborn child. During the first 3 days after birth, the mother's milk contains her antibodies. This specialized milk, known as *colostrum*, provides the baby with temporary immunity during the first few months of life.

Comprehension Check-up:

1. Foreign or abnormal body cells are marked for destruction by _____.
2. A baby receives temporary immunity from his or her mother through the intake of colostrum. This is a form of _____ immunity.

1. antibodies 2. passive

B Lymphocytes

When certain B lymphocytes first become exposed to a specific type of foreign antigen, plasma cells are formed to produce antibodies to mark the pathogen for destruction. Memory cells are formed after this initial exposure to recognize the specific antigen if it enters the body again. If a second exposure occurs, the memory cell divides into another memory cell and a plasma cell. The plasma cell clones into many plasma cells, which actively produce antibodies (Figure 10.17). Those antibodies are released into the blood plasma and attach to the specific organism they were designed to mark. Phagocytes are attracted to those antibodies and move in to destroy the antigen. The plasma cells produce antibodies so aggressively that they do not survive for a very long time; however, their antibodies continue to remain in the plasma for months, long enough to mark disease-causing invaders while the exposure remains. Left behind after the plasma cells die are memory cells. Every time this process occurs against a specific antigen, the number of memory cells increases, making the body more efficient at removing this type of foreign antigen with each exposure. This is anti-body mediated (humoral) immunity.

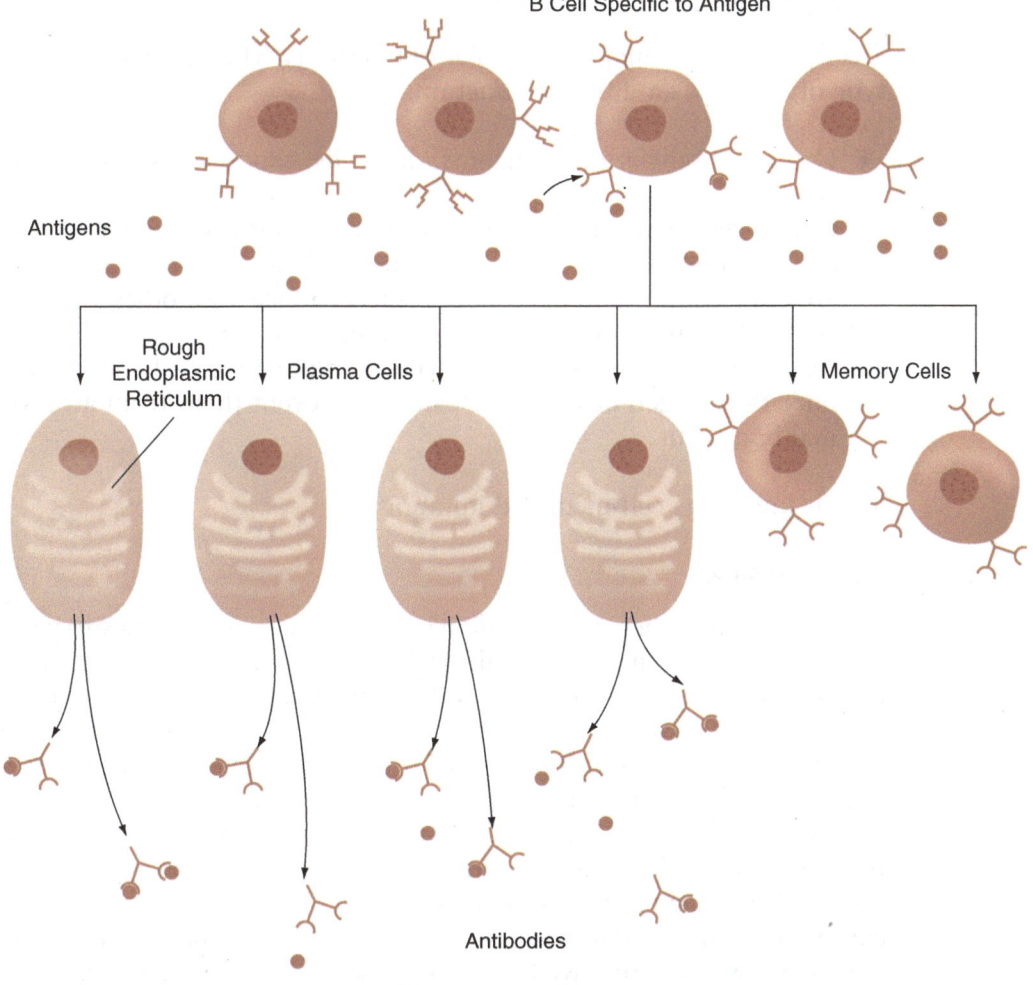

Figure 10.17 B Lymphocyte Action against an Antigen. Once a disease enters the body, the lymphocyte, able to recognize that specific antigen, divides into plasma cells (which produce antibodies until they die a short time later) and memory cells (which remain behind to recognize the pathogen if it enters the body again).

> **Allergies—Caused by Genetic Predisposition and Improved by Hygiene**
>
> There appears to be a strong genetic link between parents with allergies and their children. That is, some individuals have a genetic predisposition to produce an excessive quantity of immunoglobulin to a specific foreign antigen, that results in a hypersensitive immune system.
>
> Another cause for an increase in allergies in industrialized countries is that improved hygiene has decreased exposure to antigens, resulting in the increased sensitivity of the immune system of some individuals to certain allergens. Exposure to foreign antigens during early childhood increases the production of substances that decrease the immune response to prevent hypersensitivity. It was found that large families have fewer children with allergies than those with only one child. The conclusion was that the greater number of children meant the other family members were exposed to more foreign antigens than are small families. Epidemiological studies throughout the general population lend support for this conclusion.
>
> Some studies have linked the use of antibiotics during the first year of life with an increased risk of developing asthma or other allergic reactions. The overuse of antibacterial cleaning products has also been associated with an increased incidence in asthma.

T Lymphocytes

T cells become the major controller of immunity when exposure occurs. Each type of T cell has a different function.

- Cytotoxic cells attack virus-invaded cells, cancer cells, and transplanted cells.
- Helper cells stimulate T and B lymphocytes into action. Cytotoxic T cells are activated to attack virus-invaded cells. The appropriate B lymphocytes are induced to become plasma cells to produce antibodies against foreign cells, while macrophages are stimulated into action against invaders.
- Regulatory (suppressor) cells decrease the activity of the B and T cells once the antigen has been defeated to prevent them from attacking normal body cells.

This is known as cell-mediated immunity.

Hypersensitivity

There are times when an individual's immune system becomes hypersensitive to a specific antigen. This condition is referred to as an *allergy;* the antigen causing the excessive response is known as an *allergen*. The response may be an immediate release of histamine, which causes symptoms such as a burning sensation, swelling and drainage from mucous membranes in the nasal passage, or a rash within minutes of exposure. The intensity of the response varies widely among the general population and even within families. Severe reactions to specific allergens may become very serious and even life-threatening. For example, histamine release may occur over a large portion of the body, causing the individual to break out in hives. This may be seen with food allergies or bee stings. If the swelling involves the throat or air passages, the airway may become seriously restricted. It is important that the individual with severe allergies alert friends and family of the situation and, depending on the severity of the allergy, wear a medical alert bracelet or necklace. In some cases the reaction may be so severe that the individual carries an injectable antihistamine.

> **Common Allergens**
>
> Pollen
> Mold
> Pet dander
> Dust
> Food
>   Examples: Nuts such as peanuts, walnuts, almonds, and cashews
>   Seafood such as shellfish
> Drugs
>   Examples: Penicillin and sulfa drugs
> Insect venom
>   Examples: Bee and wasp stings
> Jewelry
> Cosmetics

There are times when there is a response to a specific allergen hours or even days after the exposure. This delayed hypersensitivity may occur when an individual is exposed to poison ivy or poison oak or some other chemical commonly found in jewelry or cosmetics.

Autoimmune Diseases

If immune cells are inappropriately stimulated, or if regulatory (suppressor) cells are unable to decrease the immune response, the result may be that the defense system attacks normal host body cells. Some common autoimmune diseases include:

- Arthritis, in which the articular cartilage in joints become inflamed, roughened, and swollen
- Multiple sclerosis, which results from the immune system destroying some of the myelin sheath around nerve axons
- Lupus erythramatosis, during which the immune system randomly attacks some organ in the body, often a kidney or the heart

There are many others. Current investigation supports the possibility of an alteration of self-antigens, resulting in the activation of the defenses against the host in several of the autoimmune diseases. The causes for this self-antigen change or inability of the regulatory T cells to limit the activity of our defenses are not well understood.

> **Comprehension Check-up:**
>
> 1. Antibodies are produced by _____.
> 2. Lymphocytes that kill virus-invaded cells are known as _____.
>
> 1. B lymphocytes   2. cytotoxic T lymphocytes

Homeostasis—Holding in Balance

The lymphatic system assists in maintaining the balance of fluid in the tissue. Excess tissue fluid not reabsorbed into the capillaries is collected by the lymphatic system and purified so that it is pathogen-free before being returned to the bloodstream. It is from the interstitial fluid that many organs obtain the fluid to produce the secretions needed to perform various functions throughout the body.

The lymphatic system is also responsible for the defense of the body. Not only does it possess a mobile force of vigilant defensive cells, but lymphocytes also control the action of much of our immunity. While the immune system does not directly maintain homeostasis, it defends other systems so they are able to perform their homeostatic functions.

There Is No Substitute for Exercise

Regular moderate exercise has been shown to improve your defenses through the following processes:

1. Increased depth and rate of respirations help to flush bacteria out in the lungs;
2. Increased heart rate and higher blood pressure increases circulation of antibodies throughout the body speeding up the marking of invaders for destruction;
3. The increase in metabolism raises body temperature which also inhibits bacterial growth; and

Exercise provides a release of stress which results in decreased release of cortisol and other stress related hormones. Chronic stress often decreases immune system activity so a constructive release of stress allows our immune system to be more vigilant. Some oncologists encourage their cancer patients to regularly exercise as a method for increasing activity of natural killer cells which defend against abnormal body cells as are found with cancer. One word of caution, extreme exercise may actually decrease immune system activity.

Clinical Notes—Vaccines, Antibodies, and Antigens

Vaccines contain weak strains of bacteria, viruses, or toxins that are introduced into the body to stimulate the production of antibodies. Antibodies are proteins produced by the body's immune system to neutralize pathogens. Once the body is exposed to these weak strains, the antibodies can identify the specific bacteria, viruses, or foreign substances and then neutralize them to prevent infection. These foreign invaders are referred to as antigens. Antibodies destroy these antigens. Antibodies activate the so-called complement proteins that can rupture the invading cells of bacteria or viruses, or cause clumping of these invading cells, or disable foreign antigens.

An antibody is also known as immunoglobulin (IgG).

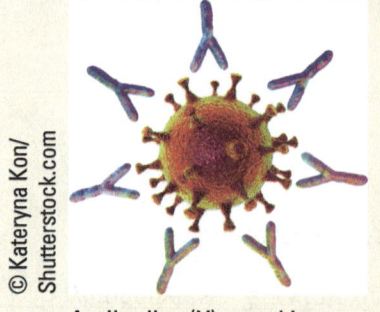
Antibodies (Y) attacking antigen virus

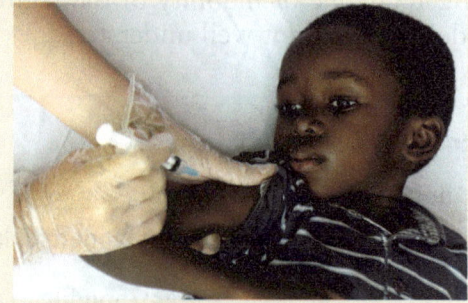
Vaccination

CLINICAL TERMS TO REVIEW

Immune system, lymph, lymph nodes, lymphocytes
Vaccines, antibodies, antigens
Blocked lymphatic vessels, edema
Thoracic duct, right lymphatic duct
Lymphoid tissue, lymphoid nodules, tonsils
Thymus gland, spleen
B lymphocytes, T lymphocytes, natural killer cells
Bacteria, viruses
Innate and adaptive immunity
Inflammation, fever, macrophages, histamine release, phagocytes, complement system
Hypersensitivity, antihistamine, allergens
Autoimmune diseases—arthritis, multiple sclerosis, lupus erythematosus

Test Yourself

Choose the best answer to the following multiple choice questions:

1. Which nutrient is transported by the lymphatic system?
   a. carbohydrates
   b. proteins
   c. lipids
   d. vitamins

2. All of the lymphatic drainage below the diaphragm flows to a common location known as the
   a. cisterna chyli
   b. spleen.
   c. mediastinum.
   d. foramen ovale.

3. The type of lymphocyte that stimulates other T and B lymphocytes into action against a specific foreign antigen is known as a
   a. plasma cell.
   b. cytotoxic cell.
   c. regulator (suppressor) cell.
   d. helper cell.

4. Lingual tonsils are found
   a. in the posterior nasal passage.
   b. in the lateral throat (pharynx).
   c. on the posterior tongue.
   d. on the uvula (soft palate).

5. When the complement system attacks a cell, what causes its death?
   a. When complement forms a hole in the cell membrane the cellular contents leak out.
   b. Complement stops the metabolism of the cell causing it to die.
   c. When complement forms a hole in the cell membrane, water is drawn in by osmosis, causing the cell to swell and the membrane to rupture.
   d. It releases toxins into the cell, causing its death.

6. The processes of defending against foreign antigens, removing the debris after the defensive action has ended, and healing the wound are all involved with
   a. inflammation.
   b. fever.
   c. infection.
   d. the complement system.

7. Receiving someone else's antibodies as a temporary defense against a specific antigen but not activating one's own lymphocytes to recognize the diseasing-causing organism or particle is known as

   a. innate immunity.
   b. complementary immunity.
   c. active immunity.
   d. passive immunity.

8. Before returning to the bloodstream, lymph is filtered and cleaned of all foreign antigens by

   a. lymphatic nodules.
   b. lymph nodes.
   c. the thymus gland.
   d. the spleen.

9. Which type of defensive cell is attacked by HIV?

   a. plasma cell
   b. cytotoxic cell
   c. regulator (suppressor) cell
   d. helper cell

10. The chemical released during inflammation to cause the local area to flood with fluid as a means of isolating the foreign antigen from the host is known as

    a. adrenaline.
    b. histamine.
    c. antidiuretic hormone.
    d. calcitonin.

The Digestive System and Nutrition

Chapter 11

LEARNING OBJECTIVES

Upon completion of this chapter, you will be able to:

1. Describe the components and functions of the digestive system and how it maintains homeostasis.
2. Identify the six basic nutrients and provide examples of each.
3. Trace the path of food through the digestive system and describe where each component of food is digested and absorbed.
4. Describe how the body metabolizes and provides sources of different nutrients.
5. Identify and describe the different types of metabolism.
6. Describe the body temperature, its source, and its regulation.
7. Explain the role of a balanced diet and physical activity in a healthy lifestyle and regulation of body weight.

CHAPTER OUTLINE

Introduction
The Digestive System
- Nutrients and Water

Digestive System Functions
Histologic Structure of the Digestive Tract
- Mucosa
- Submucosa
- Muscularis Externa
- Serosa

Movement of the Digestive Materials
- Peristalsis
- Segmentation

The Movement of Food through the Digestive System
Anatomy of the Digestive System
- Oral Cavity
- The Pharynx
- The Esophagus
- The Stomach
- Small Intestine
- Colon (Large Intestine)
- Rectum and Anus

The Accessory Glands
- The Salivary Glands
- Exocrine Pancreas
- Liver
- Gallbladder

Nutrients and Water
- Carbohydrates
- Proteins
- Lipids
- Vitamins
- Minerals
- Water

CHAPTER OUTLINE (CONTINUED)

The Difference between Food and Nutrients
- Food Groups
- Food Pyramid
- Metabolism
- Types of Metabolic Reactions
- Carbohydrate Metabolism
- Lipid Metabolism
- Protein Metabolism
- Body Temperature

INTRODUCTION

The digestive system is the only system that allows the entry of extrinsic or foreign materials, like food and water, into the human body. By doing so, the digestive system is equipped with chemical substances to guard against entry of harmful substances or pathogenic bacteria into the human body. Harmful substances are expelled by the digestive system, and pathogenic microorganisms are killed by lysosomal enzymes found in the saliva. The acidity of the stomach at a pH of 1.0–2.0 can neutralize pathogenic bacteria when they enter the stomach. The process of digestion with the help of the digestive enzymes is effective at a low pH, which is the reason why the stomach is acidic. The breakdown of food into smaller particles by digestion is essential to proper absorption of nutrients by the small intestine. Digestive juices or enzymes from the pancreas and liver are secreted into the small intestine to help in the digestion and absorption of nutrients.

Common problems of the digestive system include diarrhea, constipation, heartburn or acid reflux, abdominal bloating, and hemorrhoids. Heartburn or acid reflux happens when there is a backflow of gastric acid and stomach contents into the esophagus. The extreme acidity causes a "burning" pain in the esophagus.

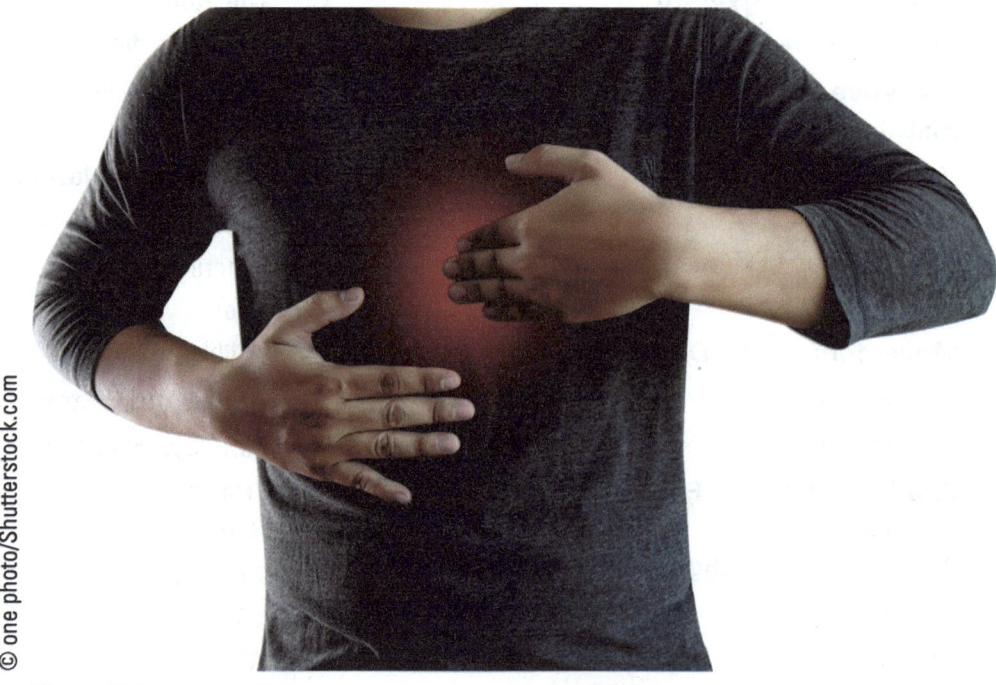

© one photo/Shutterstock.com

Figure 11.1

Figure 11.2

⌐To improve one's digestion and have a healthy lifestyle, eating smaller meals can improve metabolism and prevent overeating which can lead to increased weight. One of the causes of heartburn is eating a big meal, which can stress out the stomach in its digestive function leading to acid reflux, nausea, and vomiting. Eating foods high in fiber is the key to weight loss. Fiber makes one feel "full" and prevents from eating more. Fruits, vegetables, and grains are also recommended. Exercise is the other key component to control one's weight. Diet and Exercise are, therefore, important to healthy life.⌐

THE DIGESTIVE SYSTEM

Just as a single-celled organism is able to absorb gases from the environment into the cytoplasm, nutrients are also able to diffuse through the plasma membrane. Complex organisms, like humans, rely on access to the bloodstream to transport substances throughout the body to every cell. Enzymes in the digestive system break food down into their small subunits. These nutritive compounds pass through the wall of the digestive system into the bloodstream for distribution throughout the body.

Nutrients and Water

Nutrients are chemicals needed by cells to perform their individual functions. They may provide a source of adenosine triphosphate (ATP), or they may be components required to form larger molecules. They may be substances essential to cause cellular action such as muscle contraction or nerve conduction, or they may be necessary ingredients that cannot be manufactured by the body. They may add fluid volume to our plasma and interstitial fluid. There are six nutrients essential to maintaining the chemical processes required by the body for normal function:

- Carbohydrates—sugars
- Lipids—fats and oils

- Proteins—chains of amino acids
- Vitamins—chemicals not usually manufactured by the body that are specifically for enzymes to perform their functions
- Minerals—ions such as sodium, potassium, calcium, and iron, to name a few, that are essential for functions such as nerve conduction, muscle contraction, and transport of oxygen
- Water—the most abundant nutrient in the body; primarily needed to perform all enzymatic reactions and to transport other substances throughout the body

Without all of these nutrients taken in on a regular basis our bodies are not able to continue normal functions.

DIGESTIVE SYSTEM FUNCTIONS

In order for nutrients to reach virtually all of the cells in our body, it is necessary for the digestive system to provide a series of functions:

- Ingestion—Take in food or beverage containing nutrients.
- Digestion—Through the use of muscle contractions and enzymes, break down the food into its simplest components mechanically and chemically.
- Absorption—Move the nutrients into the bloodstream to be transported throughout the body.
- Excretion—Remove unusable substances resulting from the digestive process out of the body.

> Digestion mechanically and chemically, through the use of enzymes, breaking down nutrients to their simplest component.

> Absorption moving nutrients into the bloodstream to be transported throughout the body.

Comprehension Check-up:

1. Chemicals not manufactured by our bodies that are essential for other chemical processes to occur are _____.
2. The movement of nutrients from the digestive system into the bloodstream is known as _____.

1. vitamins 2. absorption

HISTOLOGIC STRUCTURE OF THE DIGESTIVE TRACT

For the digestive tract to function properly, there must be tissue capable of producing enzymes that are used to complete the process of digestion. The lining of the digestive tract must then be able to protect the surrounding tissue from its contents, not only from ingested substances, but also from chemicals produced for the degradation of food. The body must also be protected against invasion of foreign antigens. And, of course, the food needs to be propelled from one end of the digestive tract to the other. These functions are possible because of the four layers of tissue found in the digestive tract (Figure 11.3):

- Mucosa
- Submucosa
- Muscularis externa
- Serosa

Let's consider each layer in detail.

STOMACH WALL

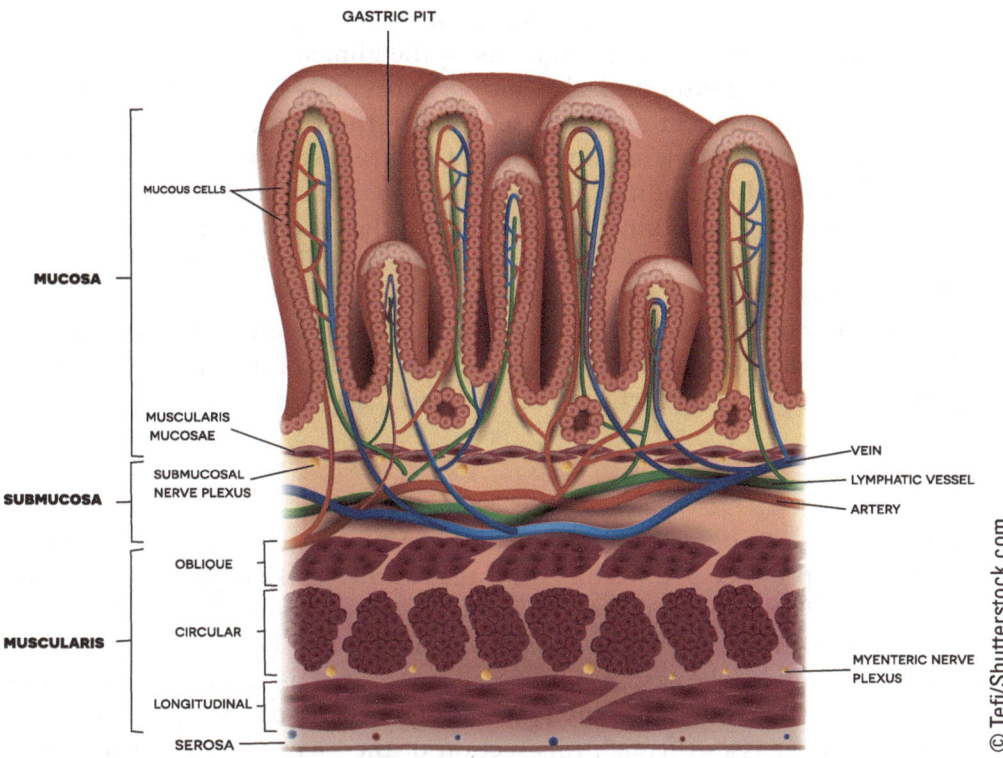

Figure 11.3

Mucosa

The mucosa is the innermost lining of the digestive tract. Recall that mucous membranes line hollow organs and produce a thick protective fluid known as *mucus*. The chemical processes involved with the breakdown of nutrients are also destructive to our own digestive system tissue. The mucus that coats the lining of the digestive tract prevents this self-destruction. The bulk of the digestive system, other than the oral cavity, upper esophagus, and anus, is lined with simple columnar epithelium. Included among these columnar cells are mucus-producing cells. The oral cavity, upper esophagus, and anus are lined by stratified squamous epithelium similar in construction to the skin.

> **Mucosa** the innermost lining of the digestive tract.

Submucosa

The *submucosa* is a layer of loose connective tissue surrounding the mucosa. The submucosa contains blood vessels for absorption of nutrients and nerves that control digestive function. Within this layer are lymphatic nodules; these provide protection against foreign antigens as needed.

Muscularis Externa

The propulsion of food through the digestive tract and the mixing of enzymes to break nutrients down into their simplest components are functions of the two layers of smooth muscle wrapped around the submucosa to form the *muscularis externa*. The circular layer of smooth muscle spirals around the alimentary canal. Its contraction decreases the diameter of the tube. This reduction in the diameter occurs in waves, forcing food farther down the length of the digestive tract. There is also a longitudinal layer of smooth muscle. Its contraction causes shortening of the tube. The combined contraction of these two layers mixes the contents of the alimentary canal by a process known as *segmentation* and also moves substances onward by peristalsis. Both types of movement are discussed later in this chapter.

Between the submucosa and the muscularis externa, and also between the circular and longitudinal layers of the muscularis externa, are two networks of nerves: the submucosal plexus and the myenteric plexus. The submucosal plexus controls secretions produced within the digestive system. The myenteric plexus is mostly autonomic nerves that can alter the speed of the muscular contraction to increase or decrease the movement of contents through the digestive system. Parasympathetic nerves increase the speed of contraction, and sympathetic nerves decrease gastric motility.

Serosa

The *serosa* is the outer covering of the digestive organs. Recall that serous membranes produce a waterlike fluid that allows these organs to slide over each other. The peritoneum is the lining of the abdominal wall. It forms a continuous membrane that covers the viscera as well. The small intestine not only is covered by the serosa but is suspended by it as well. The small intestine hangs in the abdominal cavity at the end of a tissue sheet, so it is free to move as food passes through this section of the digestive tract. The area of the tissue sheet in contact with the intestine is the serosa. The section that suspends the intestine is the mesentery. Within the mesentery are arteries and veins that carry blood to and from the small intestine.

> **Mesentery** serous membrane of the peritoneum in the abdomen that suspends the small intestine.

Comprehension Check-up:
1. Lymphatic nodules are found in the _____ layer of the digestive system.
2. Autonomic nerves that can alter the rate of smooth muscle contraction are found in the _____.

1. submucosa 2. myenteric plexus or between layers of smooth muscle

MOVEMENT OF THE DIGESTIVE MATERIALS

The contents of the digestive system pass through the alimentary canal via peristalsis and segmentation. Peristalsis occurs throughout the alimentary canal. Segmentation occurs in the small intestine.

Peristalsis

Peristalsis is the propulsive movement that results from the progressive wavelike contractions of the smooth muscle in the muscularis externa; this movement forces the contents onward (Figure 11.4a). Say, for example, you wrapped your fingers around a very flexible hose and then stuffed some soft

> **Peristalsis** the propulsive movement that results from the progressive wavelike contraction of the smooth muscle from the esophagus to the rectum, forcing its contents onward.

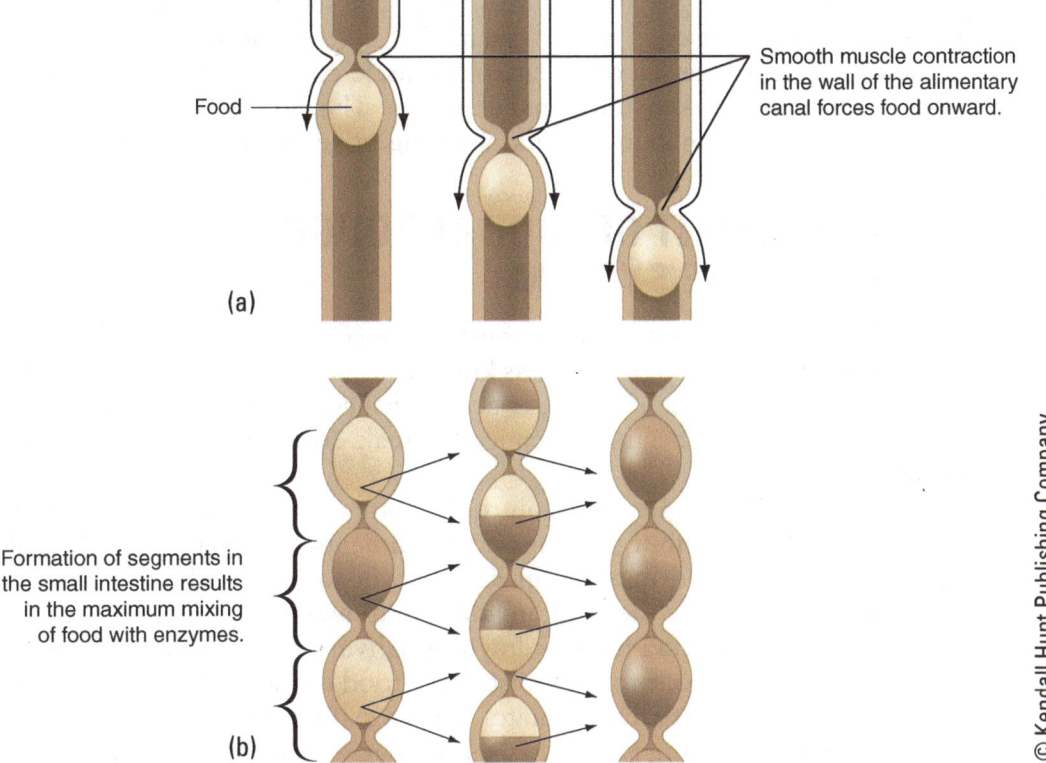

Figure 11.4 Movement through the Digestive System. (a) Peristalsis, propulsive movement. (b) Segmentation, mixing movement.

food into the end of the tube. As you squeezed on the hose, some of the food would be forced downward. If you placed your other hand just below the first hand holding the tube and squeezed, the food would be forced farther down since your hand is preventing it from moving up. If you move the hand originally holding the top of the hose to be below the second hand and squeeze, the contents again move down. By repeating these steps hand-under-hand you can push the contents the entire length of the hose. That is similar to the process of peristalsis. Pacemaker cells in the smooth muscle cause these contractions. Each organ in the digestive system has its own pacemaker cells that determine its rate of contraction.

The rhythmic rate of digestive tract smooth muscle can be altered by the autonomic nerves found in the myenteric plexus. If the sympathetic nervous system of an individual is stimulated and is in "fight-or-flight" mode, it is more critical that blood flow to muscle than to the digestive system to absorb nutrients. In a life-or-death situation, it is exceedingly more important to find a way to ensure survival than it is to put so much effort into digestion. The response of the digestive system during fight-or-flight mode is to decrease the speed of the smooth muscle contractions. Once the crisis has passed, the parasympathetic nervous system, the "rest and digest system," calms the body and speeds up the action of the digestive system.

Segmentation

The small intestine contains a combination of food and enzymes to break nutrients down into the smallest components through a process called segmentation. In segmentation the small intestine forms compartments for

> **Segmentation** the formation of compartments in the small intestine for intense mixing and digestion.

intense mixing and digestion (Figure 11.4b). The segments are brief contractions of circular muscle, visually resembling a chain of sausage links. The circular muscle relaxes as other smooth muscle contracts, as if the sausage links were being broken into smaller pieces. The process of segmentation continues to achieve maximum mixing of digestive enzymes with food.

THE MOVEMENT OF FOOD THROUGH THE DIGESTIVE SYSTEM

To obtain nutrients from food, we must first mechanically break food down into smaller pieces through a process known as masfication, or chewing combined with saliva. Once chewing is accomplished, an enzyme in the saliva, salivary amylase, can begin to chemically break down some of the chemical bonds in starch.

Let's say, for example, that you took a bite of a hamburger. Your teeth tear and grind the hamburger, bun, and condiments into smaller pieces while mixing in saliva. Food must be in liquid form to be tasted, so it is saliva that allows you to taste the food, in addition to assisting in swallowing. The starch in the bun is being broken down by salivary amylase to less complex sugars as you chew. Your tongue pushes the food to the back of your throat, and your brain sequentially stimulates pharyngeal muscles to force food into the esophagus. Breathing is inhibited as you begin to swallow while your vocal cords close and are covered by the epiglottis to ensure that food does not enter your trachea. The esophagus transports the food, saliva, and salivary amylase to your stomach, where it will be churned until it is the proper consistency for further digestion. While in the stomach, hydrochloric acid will kill many living organisms that may be in the food; however, some resistant microbes escape the antimicrobial affect of the stomach acid. Hydrochloric acid also breaks down cell membranes and connective tissue.

The stomach secretes an enzyme that begins breaking down protein. Lipids are broken up into small droplets. As the food passes out of the stomach into the small intestine, pancreatic enzymes are added to digest most of the remaining nutrients. Bile, a secretion produced by the liver, coats fats with negative charges so that they cannot recombine into large globules. As a result, lipids are easier to digest. As the food continues through the small intestine, enzymes are mixed with the nutrients and additional enzymes are secreted to complete the breakdown of food. The nutrients are absorbed into the bloodstream as the unusable material passes into the colon. The large intestine reabsorbs water and, in conjunction with bacteria, breaks down waste, which is later expelled from the body.

> **Mastication** chewing, resulting in mechanically breaking food into smaller pieces.

> **Salivary amylase** an enzyme in the saliva that breaks some of the chemical bonds in the complex carbohydrate, starch.

> **Bile** secretion produced by the liver that coats fats with negative charges so they cannot recombine into large globules, allowing for easier digestion of lipids.

> **Comprehension Check-up:**
> 1. When the smooth muscle of the small intestine forms compartments for intense mixing and digestion of nutrients, this type of movement is known as _____.
> 2. The physiological term for chewing is _____.
>
> 1. segmentation 2. mastication

ANATOMY OF THE DIGESTIVE SYSTEM

The digestive system can be divided into two sections: the digestive tract and accessory organs. The digestive tract is a series of tubelike organs, referred to as the alimentary canal, that begins at the mouth and ends at the anus (approximately 30 feet in length) (Figure 11.5). The *accessory organs* produce chemicals outside the alimentary canal that are essential to digestion.

The digestive tract consists of the following:

- Oral cavity—The mouth contains teeth and the tongue to mechanically break down food into smaller pieces through mastication. The oral cavity is lined with stratified squamous epithelium. Salivary glands secrete saliva, which contains salivary amylase, into the oral cavity as well.
- Pharynx—This common passage for air and food is also known as the *throat*. It covers and protects the vocal cords during swallowing to force food into the esophagus rather than into the trachea. It is lined with stratified squamous epithelium.
- Esophagus—This muscular tube propels food by peristalsis through the thoracic cavity to the stomach. The upper third is lined with stratified squamous epithelium, but the lining converts to simple columnar epithelium by the lower third of the tube.

> **Digestive tract** a series of tubelike organs, which begins at the mouth and ends at the anus, within which nutrients are broken down to their smallest unit and through which those units are absorbed into the bloodstream for transport throughout the body.

> **Alimentary canal** see Digestive tract.

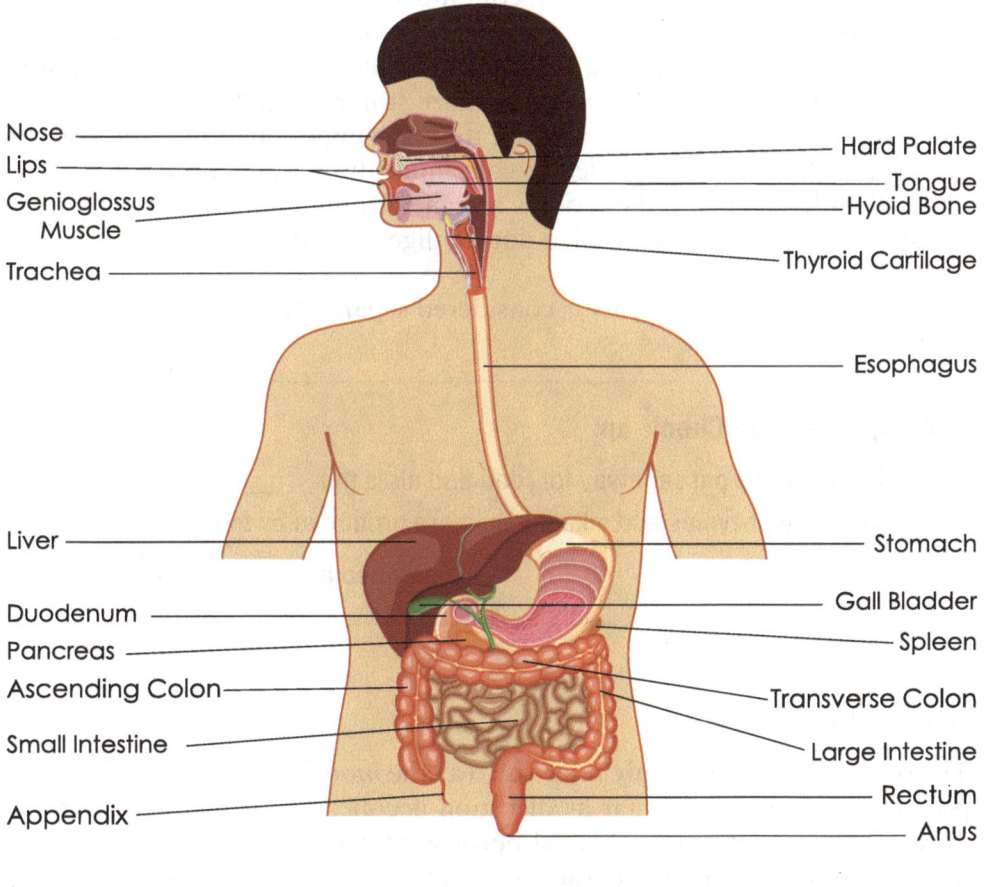

Digestive System

Figure 11.5

- Stomach—The stomach, located in the upper abdomen, primarily stores food until passing it to the intestines. Hydrochloric acid kills foreign organisms in the food as a form of defense. The stomach is lined with simple columnar epithelium and mucus-producing cells to protect against acid and enzymes.
- Small intestine—The small intestine, located mostly in the lower abdominal cavity, mixes enzymes with food to chemically digest nutrients to their simplest components, then absorbs substances through its simple columnar epithelium into the bloodstream.
- Colon (large intestine)—The colon is shaped somewhat like an inverted "U," beginning in the right lower quadrant of the abdomen and passing through all four abdominal sections and ending in the left lower quadrant. It consists of the cecum, ascending, transverse, descending, and sigmoid colon, and it contains bacteria to break down waste. It is lined with simple columnar epithelium.
- Rectum and anus—Found in the center of the pelvic cavity, the rectum stores feces until the anus relaxes to allow waste to be eliminated from the body. It is lined by simple columnar epithelium but becomes stratified squamous epithelium toward the distal end.

The accessory glands are as follows:

- Salivary glands—Three sets of salivary glands are found in the oral cavity. They produce saliva, which contains salivary amylase.
- Exocrine pancreas—The pancreas is located posterior and inferior to the stomach in the left upper quadrant of the abdomen. It produces digestive enzymes and alkaline fluid to neutralize stomach acid.
- Liver—The liver occupies most of the right upper quadrant of the abdomen. It produces bile needed to coat lipids during digestion.
- Gallbladder—The gallbladder is attached to the underside of the liver. It stores excess bile until needed during digestion of fats.

Each of these components is considered separately.

> **Comprehension Check-up:**
> 1. The common passageway for food and air is the _____.
> 2. Digestive enzymes and alkaline fluid are produced by the _____.
>
> 1. pharynx 2. exocrine pancreas

Oral Cavity

The oral cavity is commonly referred to as the *mouth* (Figure 11.6). The purpose of the oral cavity as far as digestion is concerned is ingestion of food. Because the primary physiological purpose of eating is to provide access of nutrients to the bloodstream, the mouth's function is to prepare the entering

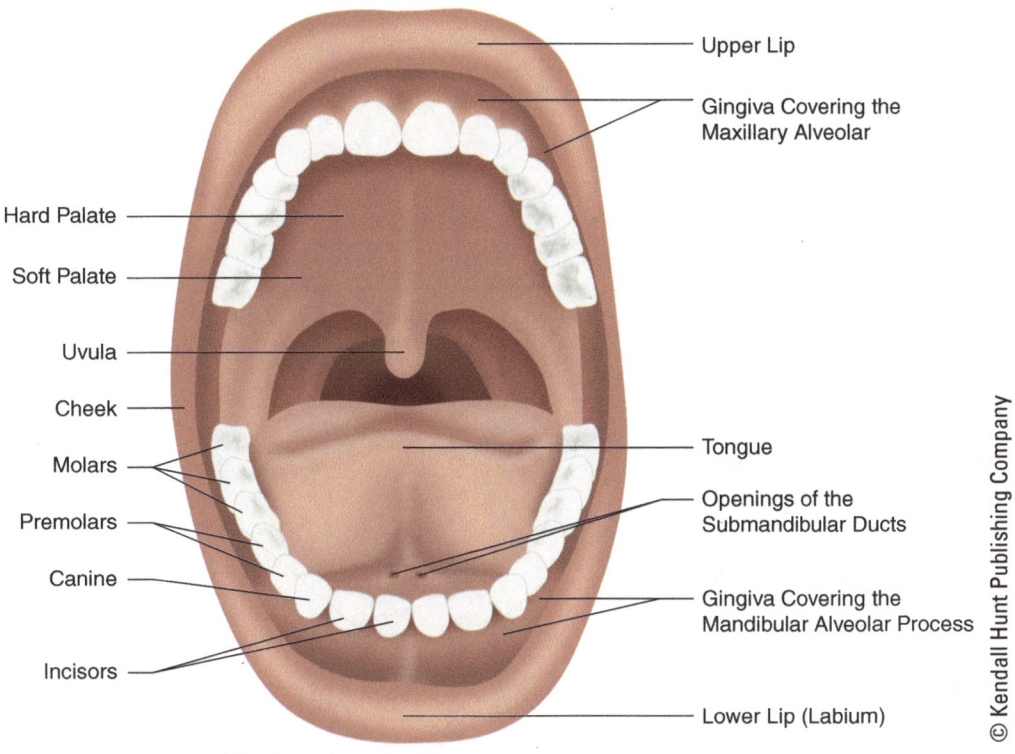

Figure 11.6 The Oral Cavity—Anterior View.

food for digestion. Nutrients are broken down to their simplest components through chemical action, but the smaller each piece of food is, the faster the process can be accomplished.

Teeth in the oral cavity tear and break food into smaller, more easily digested pieces. Humans have two sets of teeth (Figure 11.7). The first teeth appear around 6 months of age and are known as *baby teeth, milk teeth,* or *deciduous teeth*. The term *deciduous* has the same meaning as in *deciduous trees*—that is, trees that shed their leaves in the fall. Deciduous teeth are similarly shed and then replaced by larger permanent teeth. Humans have 32 permanent teeth consisting of a root that is attached to bone and crown that comes in contact with food. The tooth is composed of dentin, which forms the main body of the tooth (Figure 11.8). Enamel, the hardest substance in the body, covers the dentin, forming the crown.

The tongue manipulates food within the oral cavity and may smash soft food against the ridges on the hard palate, commonly known as the *roof of the mouth*. If this process cannot be accomplished, the tongue will push food between the teeth and crush, grind, and tear it, a process called *mastication*. The tongue also forms the food into a ball and moves it to the back of the throat for swallowing.

Salivary glands produce saliva, which moistens and lubricates food. Although the digestion of carbohydrates begins in the mouth (due to salivary amylase), most of the actual breakdown occurs later, simply because food does not spend enough time in the oral cavity for much of the breakdown of starch to take place.

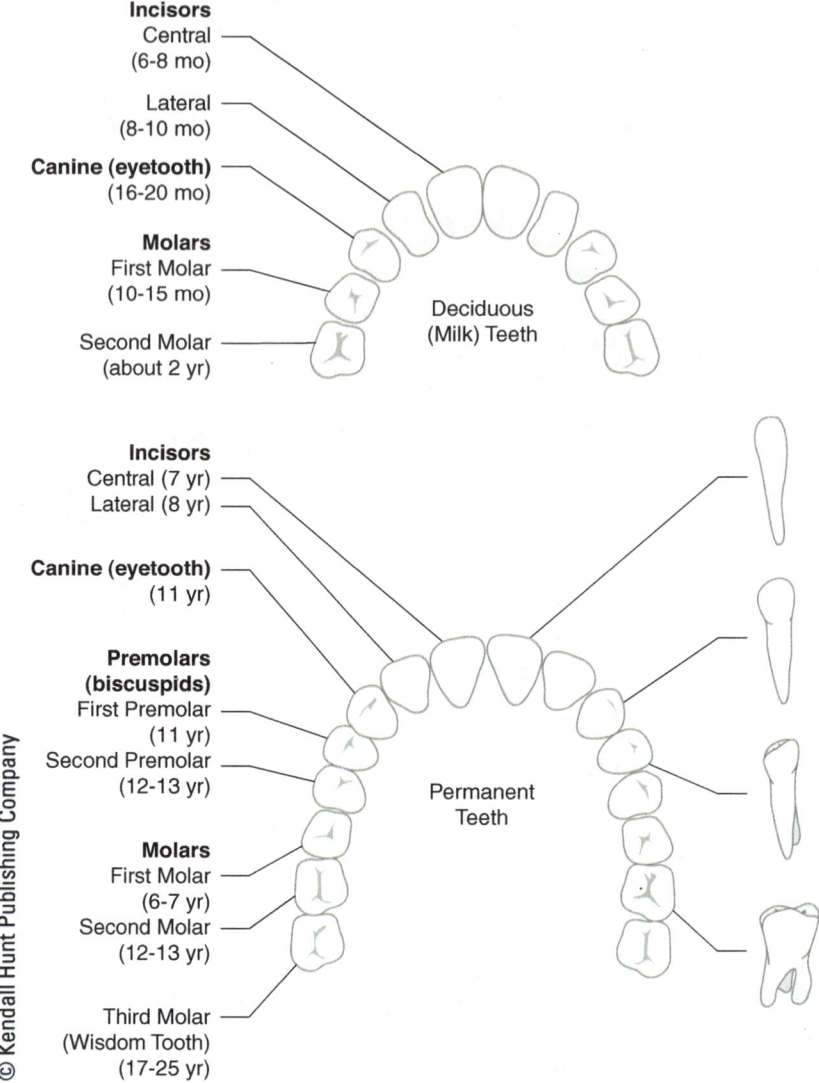

Figure 11.7 Deciduous and Permanent Human Teeth.

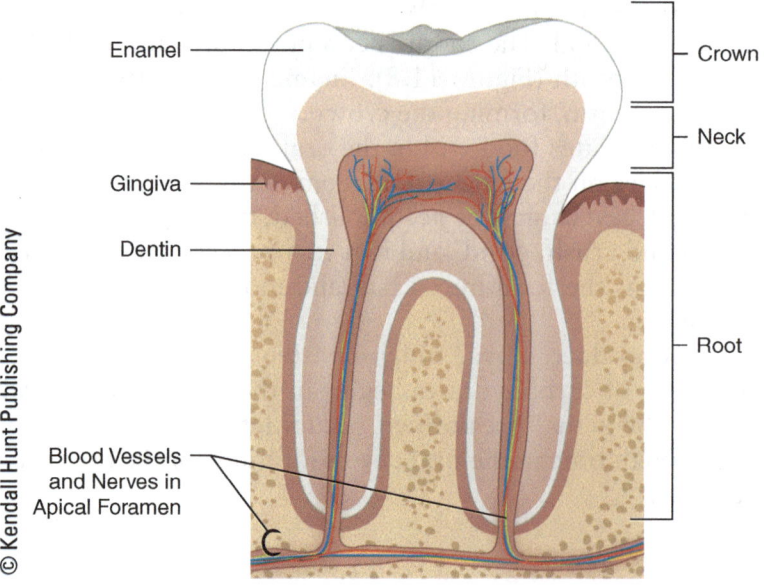

Figure 11.8 The Structures of the Tooth.

The Pharynx

Recall that the pharynx is a common passageway for food and air (Figure 11.9). Highly coordinated contraction of the skeletal muscle in the pharynx allows the complex process of swallowing to take place. Three sets of *pharyngeal constrictor muscles* contract in sequence to force food down the laryngopharynx into the esophagus (Figure 11.10). At the same time, other pharyngeal

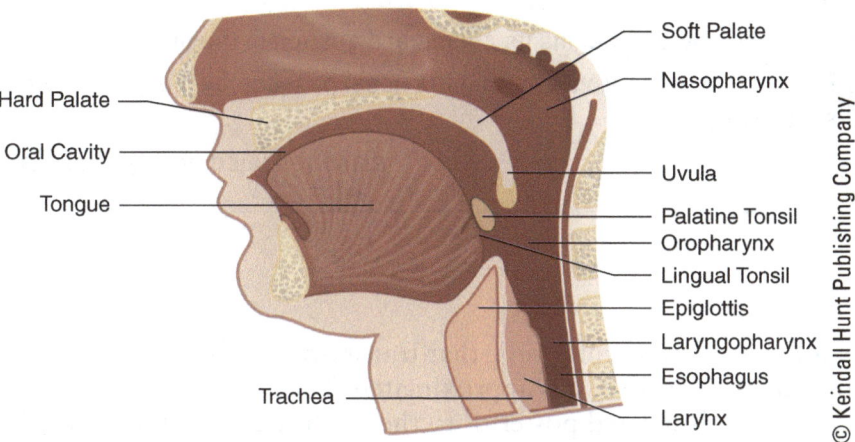

Figure 11.9 The Pharynx. The pharynx is a common passage for food and air. The nasopharynx is found in the posterior nasal cavity, the oropharynx is located in the posterior oral cavity, and the laryngopharynx is superior to the division of the trachea and esophagus.

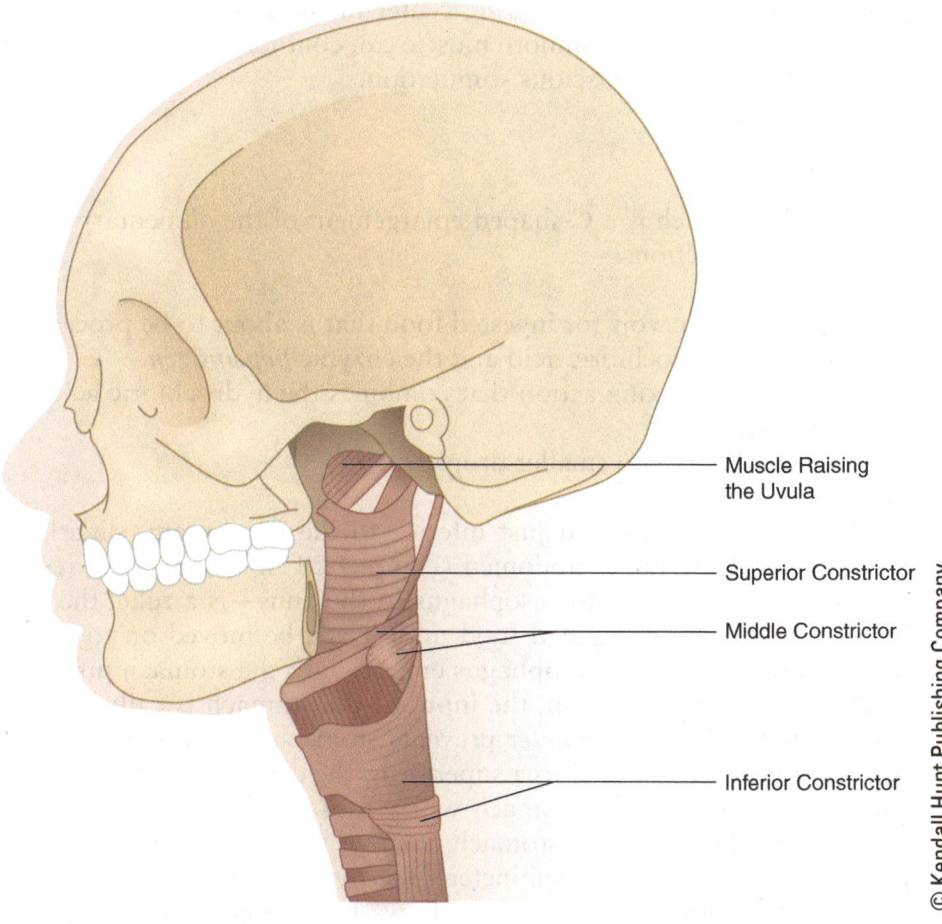

Figure 11.10 Pharyngeal Constrictor Muscles.

muscles raise the soft palate, or *uvula,* to block the nasopharynx so that food does not enter the nasal passage. The loss of stimulation or coordination of the pharyngeal constrictor muscles sometimes occurs as the result of a stroke, making it difficult to swallow.

> **Comprehension Check-up:**
> 1. The only food that begins its chemical digestion in the mouth is _____.
> 2. Food is prevented from entering the nasal passage by the _____ when swallowing.
>
> 1. starch   2. soft palate or uvula

The Esophagus

Esophagus a muscular tube that transports food from the pharynx to the stomach by peristalsis.

The esophagus is a muscular tube that transports food from the pharynx to the stomach by peristalsis. It is approximately 25 centimeters (10 inches) long in the adult and is located posterior to the trachea. In the upper third of the esophagus the muscularis externa contains skeletal muscle; the middle third transitions from skeletal to smooth muscle, and the lower third is wrapped with smooth muscle. Skeletal muscle requires stimulation by the brain, so the process of swallowing begins with conscious control in the oral cavity, although the contraction of skeletal muscle in the esophagus is an automatic process regulated by a swallowing center in the brainstem. Once reaching the lower esophagus, the smooth muscle can continue the movement process without the need for conscious stimulation.

The Stomach

Stomach C-shaped enlargement of the alimentary canal primarily for storage of food as the process of digestion begins.

The stomach, which is a C-shaped enlargement of the alimentary canal, has four primary functions:

1. It acts as a reservoir for ingested food that is about to be processed.
2. It secretes hydrochloric acid and the enzyme *pepsinogen.*
3. It provides a mixing action that combines the hydrochloric acid and enzymes with food.
4. It breaks lipids into smaller droplets.

The stomach is located just inferior to the diaphragm, slightly left of the midline of the upper abdomen (Figure 11.11). Whereas the rest of the alimentary canal—from the esophagus to the anus—is a tube, the stomach balloons out to hold ingested food until it can be moved on to the rest of the digestive system. The esophagus empties into the stomach just posterior to the heart. For this reason, the input of the stomach is called the *cardia.* The *gastroesophageal sphincter* prevents stomach contents from reentering the esophagus. There is an area superior to the cardia known as the *fundus.* The working area of the stomach where most of the churning takes place is known as the *body.* The stomach narrows to form an exit known as the *pylorus,* which also has a sphincter—the *pyloric sphincter.* It is the relaxation of the pyloric sphincter that allows food to continue onward to the

STOMACH
internal structure

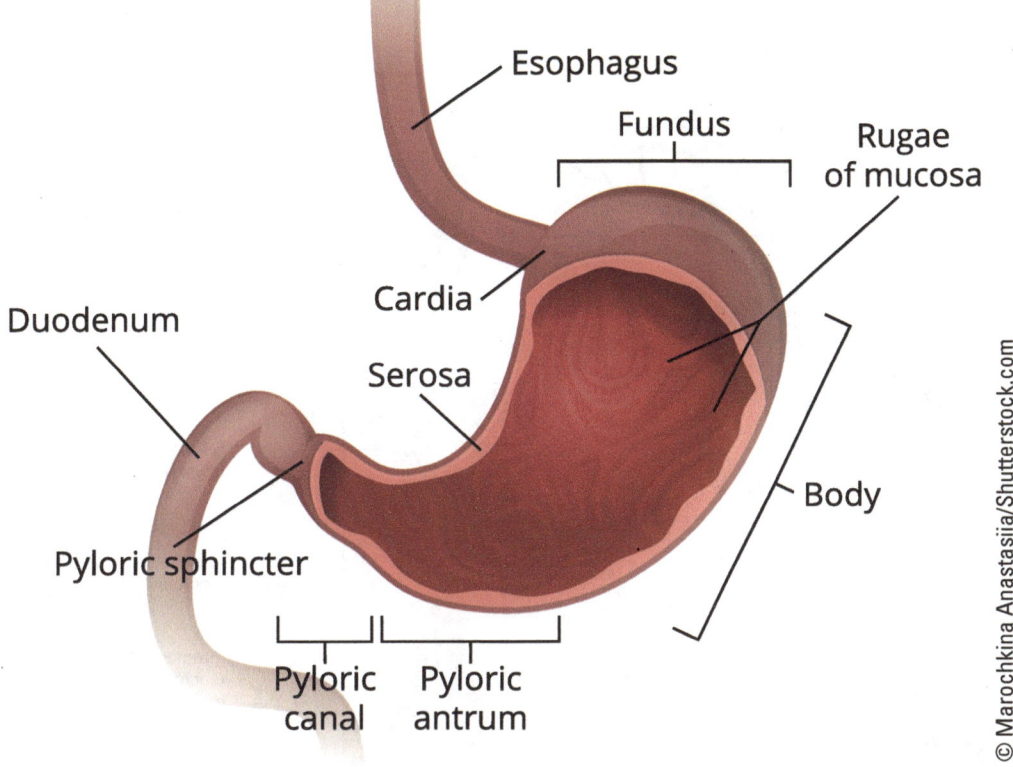

Figure 11.11

small intestine for processing. Because simple columnar epithelium does not stretch, the stomach mucosa is oversized to allow room for increased bulk when eating. When the stomach is empty, the mucosa forms wrinkles known as *rugae* (gastric folds).

The mucosa of the stomach forms secretory glands known as *gastric pits* (Figure 11.12). In the gastric pits are two types of cells that produce chemicals essential for digestion: parietal cells and chief cells. *Parietal cells* produce hydrochloric acid, which kills living organisms, denatures (unravels) proteins, and breaks down cell walls and connective tissue. *Chief cells* make pepsinogen, which is an inactive enzyme. Both of these chemicals are secreted out of the gastric pit into the stomach. Hydrochloric acid activates the inactive enzyme pepsinogen to its active form, pepsin. Pepsin breaks some of the amino acid bonds in protein, causing protein to be digested into shorter chains of amino acids. The upper lining of the gastric pits and the wall of the stomach are abundant with mucus-producing cells. This mucus coating is essential for the protection of the stomach lining against its own secretions.

The muscularis externa of the stomach contains three layers of smooth muscle. Along with the longitudinal and circular layers typically found in the rest of the alimentary canal, the stomach has an inner muscular layer known as the oblique layer.

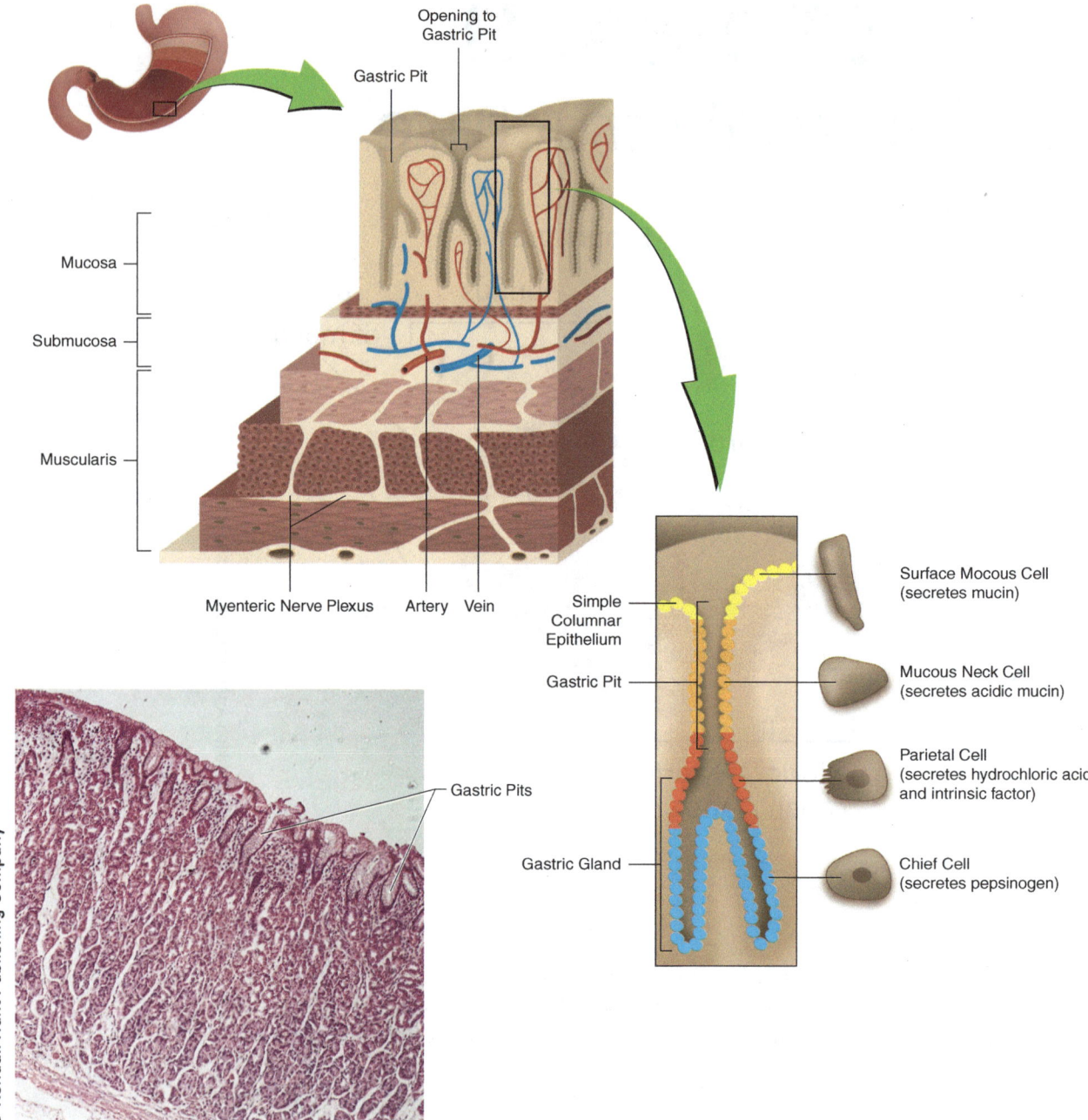

Figure 11.12 Gastric pits contain parietal cells. These cells produce hydrochloric acid and chief cells, which secrete pepsinogen. The pepsinogen is activated to pepsin as the first step in digesting proteins. Mucous cells produce mucus to protect the stomach lining against self-digestion.

Chyme food in the process of being digested.

The stomach churns the food for about 2 to 6 hours, mixing acid and enzymes and breaking up lipids into smaller droplets. Lipids tend to decrease the rate of digestion more than any other nutrient. As a result of the extended time required by lipids for stomach emptying, fats tend to make our stomach feel full longer than other nutrients. When the food being digested (chyme) is at the satisfactory consistency, it passes through the pyloric sphincter into the small intestine.

The regulation of the action of the stomach occurs in three phases. Anticipation of food entering the stomach results from the sight, smell, or even expected mealtime. This can trigger the stomach to start producing acid and enzymes. Because the stimulation begins in the brain and arrives via the parasympathetic nervous system, this first period is called the *cephalic phase*. Once food arrives, its presence causes the continued secretion of acid and enzymes. It is now in the *gastric phase*. As the chyme begins to pass into the small intestine, the stomach enters the *intestinal phase*. If the small intestine requires additional time to digest and absorb nutrients, the emptying of the stomach can be delayed by an inhibitory hormone, gastric inhibitory peptide, which is produced by the duodenum. Stomach emptying can also be affected by the quantity it contains. The more distended the stomach becomes, the more rapidly it will empty. The stomach also acts as an endocrine gland by producing the hormone gastrin, which stimulates its own parietal cells to produce hydrochloric acid when food is present.

Comprehension Check-up:

1. Cells in the stomach mucosa that produce hydrochloric acid are known as _____ cells.
2. The exit to the stomach is known as the _____.

1. parietal 2. pylorus

Clinical Notes—Gastroesophageal Reflux Disease (GERD)

The gastroesophageal sphincter at the junction of the esophagus and stomach prevents acidic stomach contents from being forced back into the esophagus during contraction of the stomach. If some of the stomach contents are allowed to reenter the esophagus through the sphincter, which occurs because of overeating, smoking, or obesity, the esophagus is exposed to stomach acid. The stomach produces large quantities of protective mucus to coat itself against the acid. The esophagus does not produce such high quantities of mucus, so the acid can burn esophageal tissue, causing pain and resulting in gastroesophageal reflux disease (GERD). Because this area is located posterior to the heart, the acid reflux is referred to as heartburn. If this condition persists, significant damage can occur to the esophageal mucosa resulting to erosion or scarring of the tissues. This decreases the ability of the esophagus to stretch when swallowing.

Gastroesophageal reflux disease

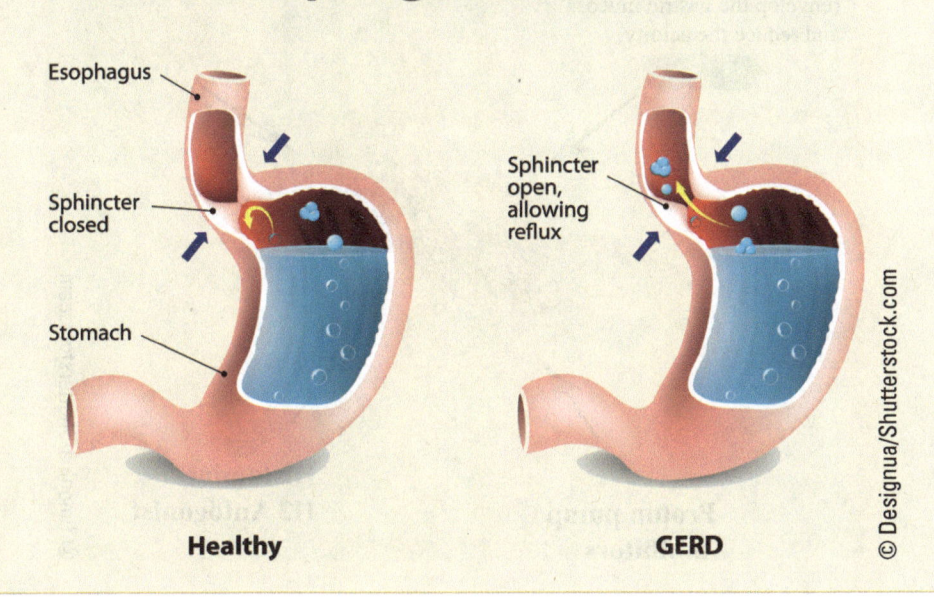

Clinical Notes—Peptic Ulcer

Peptic ulcers were once thought to be the result of stress, smoking, diet, or alcohol. Although these factors may contribute to increased irritation, it is now understood that most peptic ulcers are due to a bacterial infection. Hydrochloric acid kills most living organisms that enter the stomach, but not all. The bacterium, Helicobacter pylori (H. pylori), thrives in an acidic environment. It colonizes in the stomach and interferes with the production of mucus. Hydrochloric acid then damages the stomach tissue and causes an ulcer. Treatment of peptic ulcers includes use of antibiotics to kill the invading bacteria.

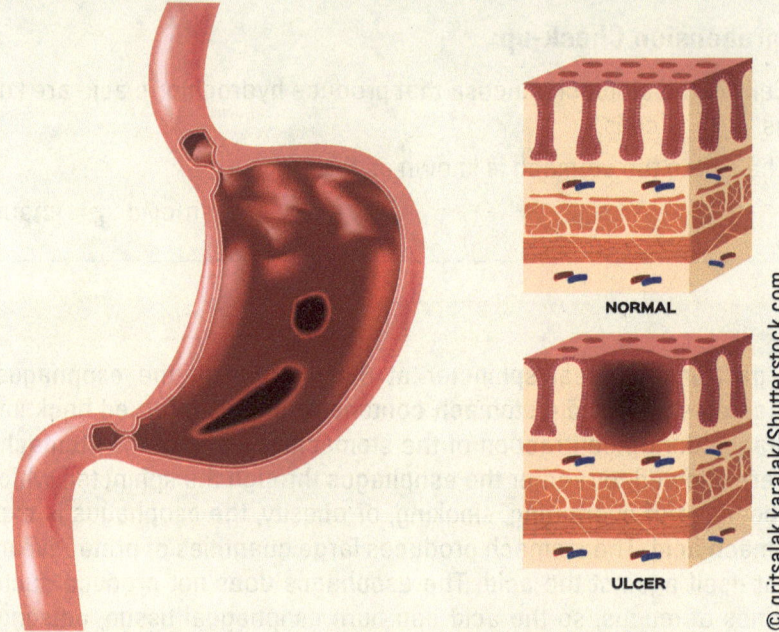

Treatment of Helicobacter pylori

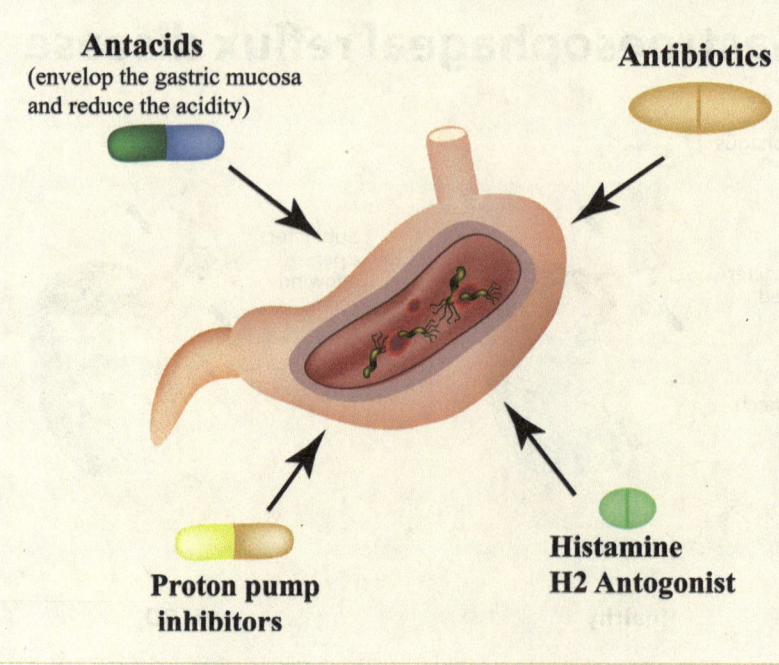

Antacids (envelop the gastric mucosa and reduce the acidity)

Antibiotics

Proton pump inhibitors

Histamine H2 Antogonist

Small Intestine

The small intestine mixes digestive enzymes with the chyme to break nutrients into their simplest components and then absorbs them into the bloodstream. The small intestine is approximately 30 feet in length when the smooth muscle in its muscularis externa is relaxed. To understand the sections of the small intestine, it is beneficial to describe the histology.

The mucosa lining the small intestine is wrinkled into projections known as *circular folds* (Figure 11.13). These projections increase contact of the intestinal lining with chyme. They are also able to assist with the mixing process as enzymes break down nutrients. The circular folds are covered with projections called *intestinal villi*. It is through the villi that nutrients pass from the intestine into the bloodstream. The villi are covered by simple columnar epithelium and contain a network of capillaries ready to carry nutrients to the body. Deep inside each villus is a lymphatic vessel that transports

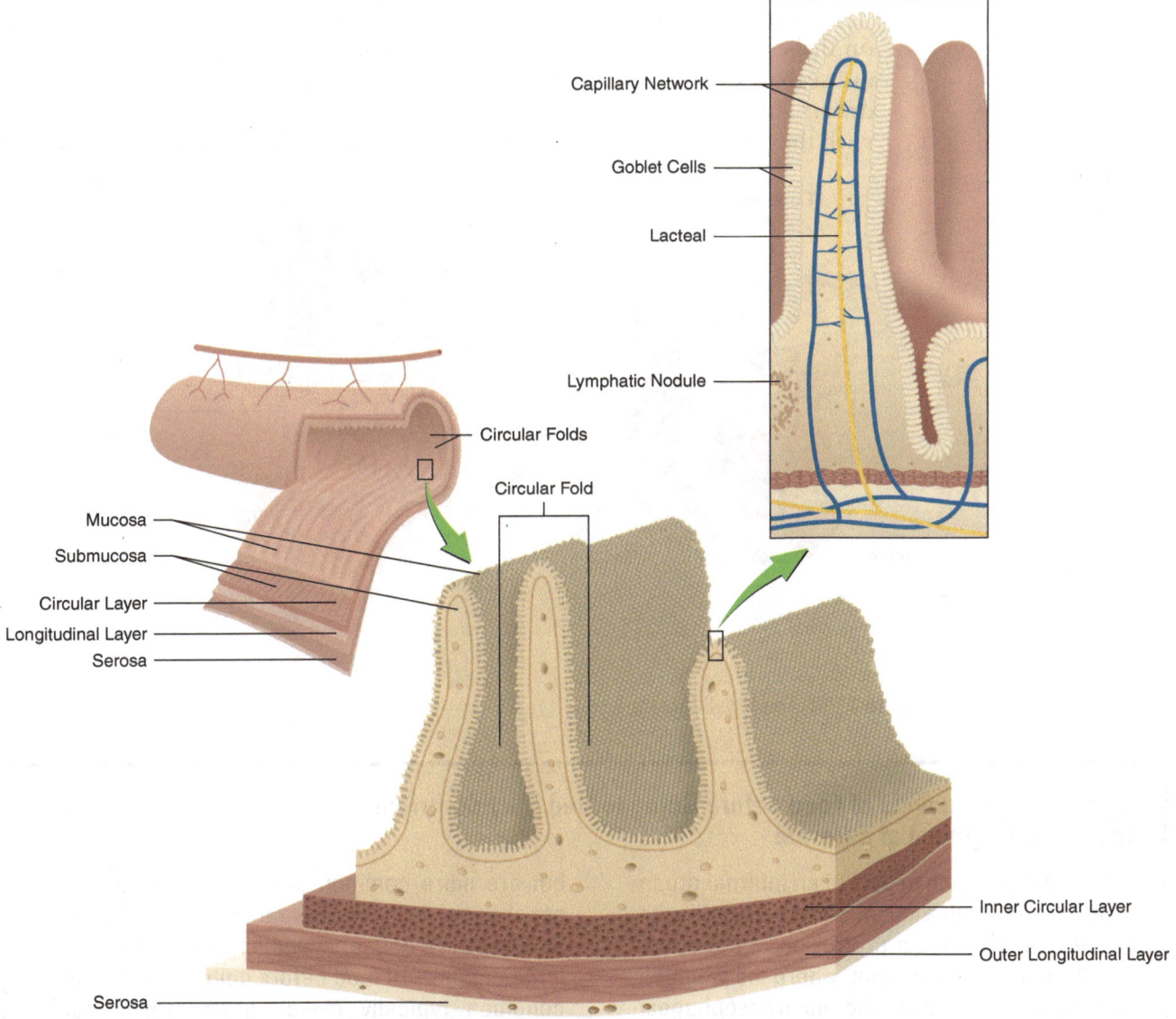

Figure 11.13 Circular Folds and Villi in the Small Intestine. These are designed to increase contact with food being digested. Nutrients pass through the cells of each villus to be absorbed into the bloodstream or, in the case of lipids, pass into lymphatic vessels (lacteals) to be transported to the bloodstream by the lymphatic system.

triglycerides to the bloodstream. The simple columnar cells have stiff hairlike projections on their apex known as *microvilli* or *brush border* that increase absorption of nutrients because they increase contact with the chyme as it passes by. The microvilli also produce enzymes that complete the final breakdown of nutrients to their smallest component.

The small intestine can be divided into three sections (Figure 11.14).

> **Duodenum** the shortest section of the small intestine. It receives acidic chyme from the stomach, neutralizing alkaline fluid and digestive enzymes from the pancreas, and lipid-coating bile from the liver and gallbladder.

- Attached to the distal end of the stomach, the duodenum is only about a foot long, yet is exceedingly active in the process of digestion. It receives chyme from the stomach, pancreatic juices containing enzymes to break down each nutrient, and fluid high in sodium bicarbonate, which is a base that neutralizes stomach acid. Also entering the duodenum is bile from the liver and gallbladder. *Bile* is an emulsifier; that is, bile molecules coat droplets of lipids with negative charges so that they cannot recombine to form large fat globules. Because these droplets now have charges on them, bile allows lipids to move freely in water.

A common emulsifier used every day is soap. Soap molecules coat oils on dishes, thereby preventing fats from sticking to them. Because these oils are surrounded by negative charges, they float freely in water and down the drain. The emulsification of fats in the digestive system allows them to mix with the water rather than separating into polar and nonpolar substances.

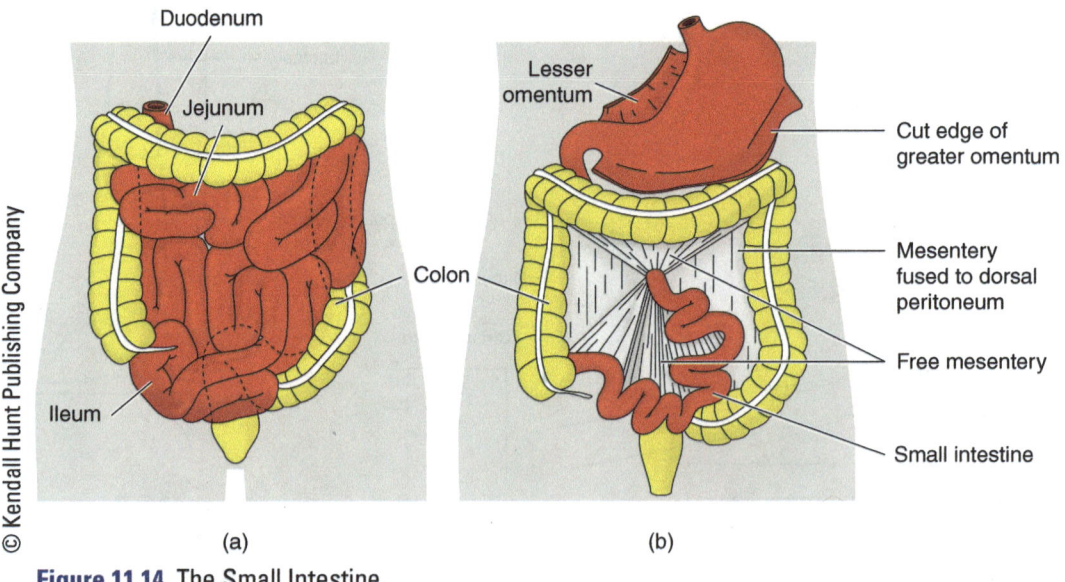

Figure 11.14 The Small Intestine.

Eating Large Amounts of Food before Going to Bed May Contribute to the Development of a Hernia

A hernia occurs when part of an internal organ protrudes through a muscular wall. People with a hiatal hernia have a protrusion of the stomach through the diaphragm, where the stomach and esophagus meet, the gastroesophageal junction. Hiatal hernias may occur in anyone but are more common in people with gastroesophageal reflux disease and among individuals who are obese, especially those who eat large amounts of food before going to bed. This condition typically makes gastroesophageal reflux worse.

The duodenum also produces hormones to delay stomach emptying. These hormones also cause the pancreas to secrete digestive enzymes or alkaline fluid and cause the gallbladder to contract and consequently increase the amount of bile entering the small intestine.

- The jejunum is the middle section and is about two-fifths the total length of the small intestine. Viewed under a microscope, the jejunum has long, branched circular folds (Figure 11.15). Primarily the jejunum mixes enzymes with the chyme to break each nutrient into its simplest component.
- The ileum is the final section of the small intestine, and it comprises roughly three-fifths its total length. Histologically, the ileum has shorter circular folds than the jejunum (Figure 11.16). It absorbs nutrients into the bloodstream or lymphatic vessels.

Single sugars and amino acids are pumped into the simple columnar epithelium on the villi to diffuse out the other side and into capillaries. Triglycerides coated with bile are disassembled so that they can pass through the apex of the epithelium. Inside the simple colunmnar cells of the mucosa, the triglycerides are reassembled and then coated with charged proteins.

> **Jejunum** the middle section of the small intestine. Primarily the jejunum mixes enzymes with the chyme to break each nutrient into its simplest component.

> **Ileum** the final section of the small intestine. It absorbs nutrients into the bloodstream or lymphatic vessels.

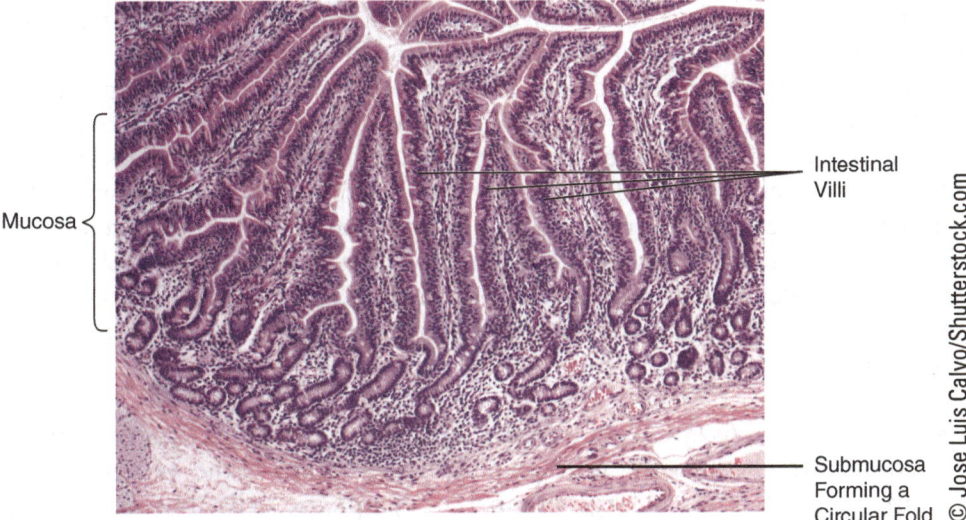

Figure 11.15 Histology of the Jejunum.

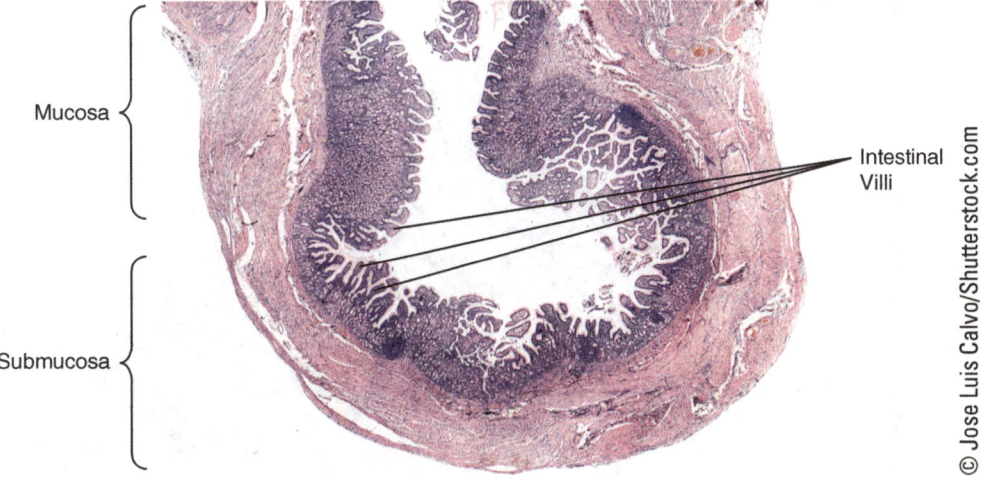

Figure 11.16 Histology of the Ileum.

These coated triglycerides are too large to diffuse into the bloodstream directly. Instead they diffuse into the lymphatic vessels (lacteals), where they eventually enter the bloodstream through the left subclavian vein.

The rhythmic contraction of smooth muscle in the small intestine is controlled by pacemaker cells. The rate can be decreased by the sympathetic nervous system or increased by the parasympathetic nervous system. The average time for food to be in the small intestine is 6 to 8 hours.

Colon (Large Intestine)

By the time the chyme has reached the colon, the desired nutrients have been removed. Except for reabsorbing some of the water in the chyme, the remaining purpose of the colon is to prepare the waste for excretion from the body. The breakdown of waste is actually accomplished in the large intestine by numerous microorganisms. As chyme passes through the 5 feet of colon, more and more water is removed and the waste thickens to a dense mass.

Looking at the histology of the large intestine you can see that there are no longer villi present (Figure 11.17). Abundant mucus-producing cells lubricate the movement of fecal material. The gross anatomy of the colon is also different from that of the small intestine. The large intestine forms pouches known as *haustra* that can constrict to move dense fecal material. Each haustrum distends as it fills with fecal material, thereby stimulating the surrounding muscles to contract and push the contents to the next haustrum by segmental peristaltic contractions. Peristalsis requires powerful muscle contraction to move feces. On the outer surface of the colon are bands of smooth muscle to assist in the process of excreting waste. The stimulation of this movement is transmitted as a reflex by the autonomic nervous system and is affected by the distension of the stomach and duodenum. The time in the colon is approximately 24 hours.

The colon can be divided into five sections (Figure 11.18). The ileum attaches to the large intestine in the right lower quadrant of the abdomen. The sections of the colon are as follows:

- **Cecum**—This is a closed-ended tube where the ileum attaches to the colon. Attached on the inferior end of the cecum is the appendix. The appendix contains a large number of lymphatic cells, but its purpose has not been conclusively defined. Inflammation of the appendix, that is, appendicitis, can cause serious infection if it is not treated. Treatment is surgical removal of the appendix.

> **Colon** the colon is an organ containing bacteria that is needed to break down wastes from digestion. Water is also reabsorbed into the bloodstream through the colon. Also called the *large intestine*.

> **Cecum** the closed-ended section of the colon into which the remaining waste from the small intestine passes. Attached to the inferior end of the cecum is the appendix.

> **Appendix** wormlike appendage attached to the interior end of the cecum. It contains a large quantity of lymphatic cells, but its purpose has not been conclusively defined.

Figure 11.17 Histology of the Colon.

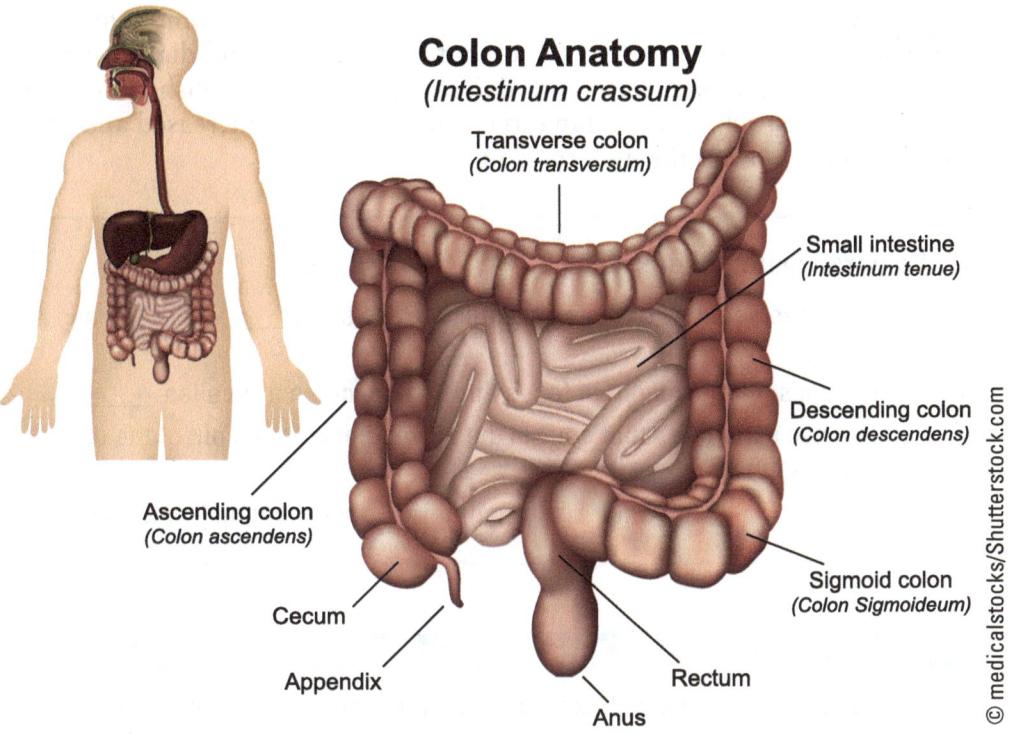

Figure 11.18

- Ascending colon—From the cecum the large intestine passes cranially toward the liver as the ascending colon.
- Transverse colon—Once reaching the right upper quadrant of the abdomen, the large intestine bends and passes across the anterior upper abdomen as the transverse colon.
- Descending colon—In the left upper quadrant of the abdomen, the large intestine bends toward the caudal abdomen as the descending colon.
- Sigmoid colon—In the left lower quadrant of the abdomen, the large intestine forms an S-shaped section resembling the Greek letter sigma (Σ). It is referred to as the *sigmoid colon*.

Chyme entering the cecum is thickened but still liquefied. By the time it reaches the sigmoid colon, most of the water has been extracted and it is a dense soft solid. Throughout the digestive process, water has been added to the chyme in the form of saliva, gastric juices, pancreatic juices, bile, and intestinal secretions. The total quantity added throughout a 24-hour period is about 6500 ml. If that fluid was not reabsorbed in the distal end of the alimentary canal, an individual could become dehydrated very rapidly.

Rectum and Anus

The rectum is the distal end of the digestive system; its function is to store feces until release. Two sphincters surround the distal end of the alimentary canal as it forms the anus, the opening through which waste passes out of the body. The internal anal sphincter is composed of smooth muscle, which can hold for extended periods without fatigue. The external anal sphincter is skeletal muscle and under conscious control.

Sigmoid colon the S-shaped distal end of the colon found in the left lower quadrant of the abdomen.

Rectum the organ at the distal end of the alimentary canal for storage of feces until released.

Anus composed of two sphincters, one smooth muscle and the other skeletal muscle, on the distal end of the digestive system to control the release of waste.

| Defecation release of feces. |

The release of feces is known as **defecation**. As more and more fecal matter enters the rectum, pressure sensors inform the brain and digestive system of the need to eliminate waste. For defecation to be accomplished, both anal sphincters must relax. A combination of peristalsis and increased abdominal pressure pushes feces through the anus.

Comprehension Check-up:

1. The three sections of the small intestine are the _____, _____, and _____.
2. Pouches for movement of feces through the colon are called _____.
3. The two sphincters that control the movement of feces out of the rectum form the _____.

1. duodenum, jejunum, ileum   2. haustra   3. anus

Clinical Skills—Patient Assessment in Appendicitis

The appendix, about 4 inches long, is located at the junction of the small and large intestines. It is in the lower right quadrant of the abdomen. It becomes inflamed when there is an obstruction in the appendix lumen and is commonly referred to as appendicitis. A person with appendicitis would show abdominal pain that radiates to the right lower quadrant (RLQ), nausea, fever, anorexia, and vomiting. Diarrhea or constipation may also be observed. There is elevation of the white blood cells (WBC) above 10,000 per cubic millimeter and also an increase of neutrophils, in the complete blood count (CBC) lab test. Imaging such as abdominal ultrasound may be indicated if there is uncertainty in the diagnosis. Emergency surgical removal of the appendix and use of antibiotics are the most common treatments.

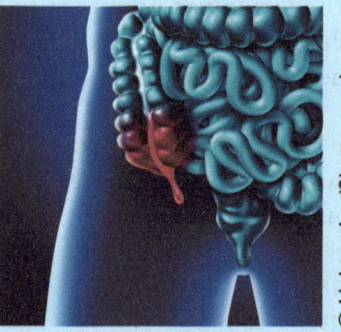

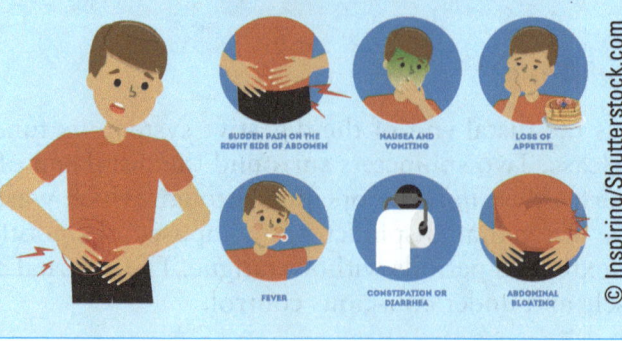

THE ACCESSORY GLANDS

Accessory glands are organs outside the digestive tract that produce essential substances for digestion.

The Salivary Glands

The purpose of the salivary glands is to produce saliva. Three sets of salivary glands are found in the oral cavity (Figure 11.19):

- Parotid glands—found in the cheek anterior to the ear
- Sublingual glands—located under the tongue
- Submandibular glands—located on the medial inferior side of the mandible

The parotid glands secrete above the tongue, and the other two release saliva inferior to the tongue. The presence of food in the mouth stimulates production of saliva; however, the mere anticipation of food can also cause its production. Saliva lubricates and moistens food. Without sufficient moisture in our food, we could not swallow.

> **Salivary glands** three pairs of glands lateral or inferior to the oral cavity that produce saliva and secrete it into the mouth when food is present.

Exocrine Pancreas

The exocrine pancreas is a triangular organ located inferior and posterior to the stomach and lying across the upper abdomen between the duodenum and spleen (Figure 11.20). Recall that the pancreas is both an exocrine and an endocrine gland. We have already discussed the endocrine portion of the pancreas (Chapter 7). Within the pancreas are pancreatic islets that produce hormones to regulate the level of glucose in the bloodstream.

> **Exocrine pancreas** located inferior and posterior to the stomach, the exocrine pancreas produces alkaline fluid to neutralize stomach acid and enzymes to break nutrients into simpler components.

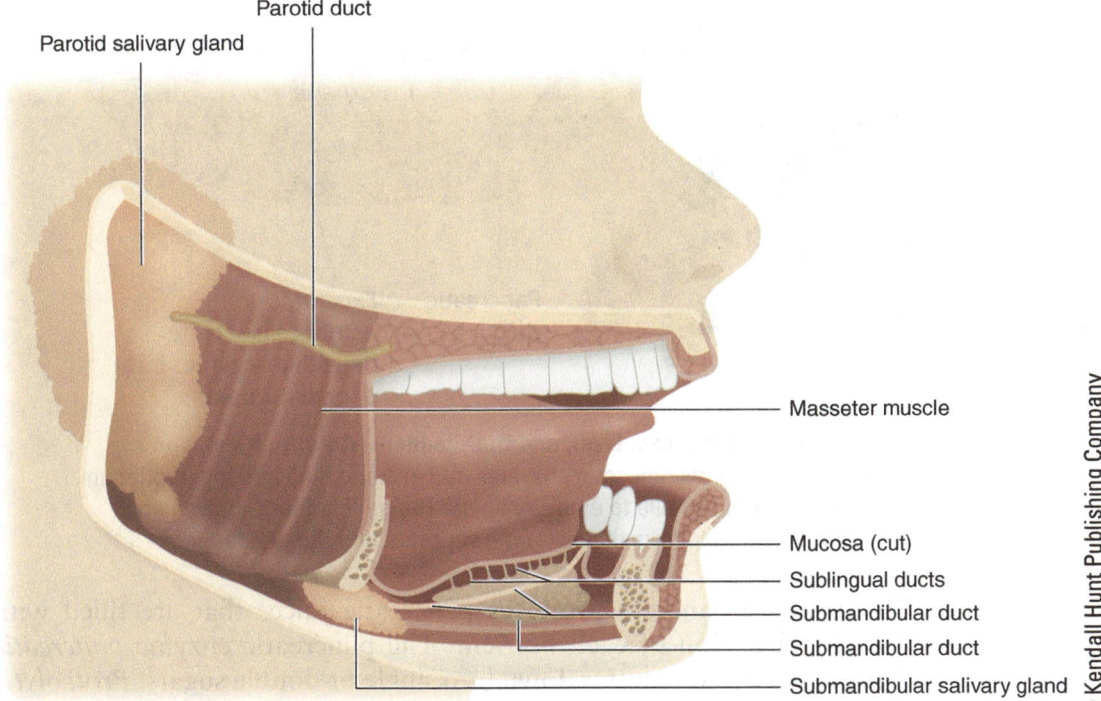

Figure 11.19 The Salivary Glands. The parotid glands are found in the cheek, the sublingual glands are inferior to the tongue, and the submandibular gland is located on the inferior medial side of the mandible.

Liver, Gallbladder, Pancreas and Bile Passage

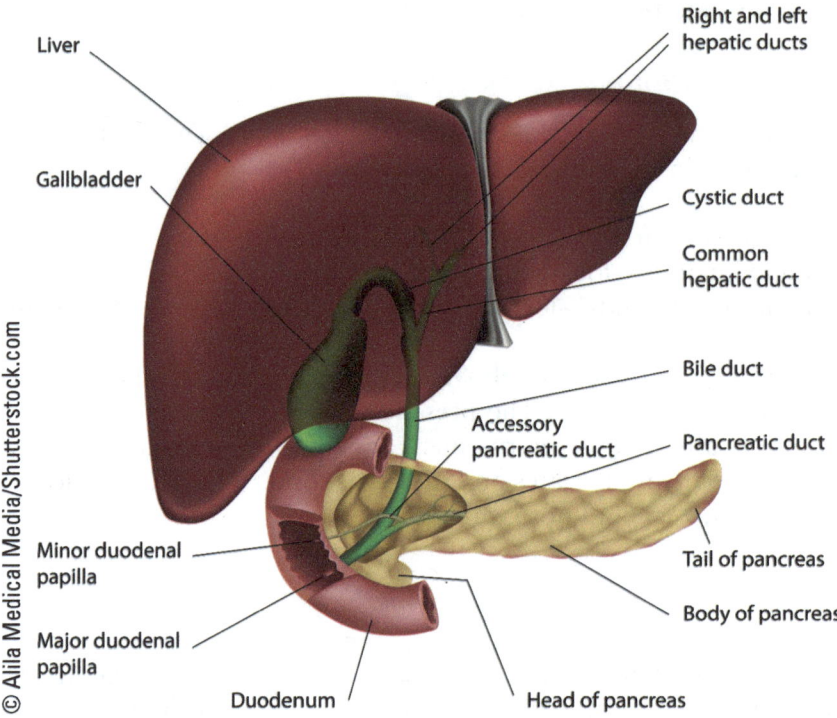

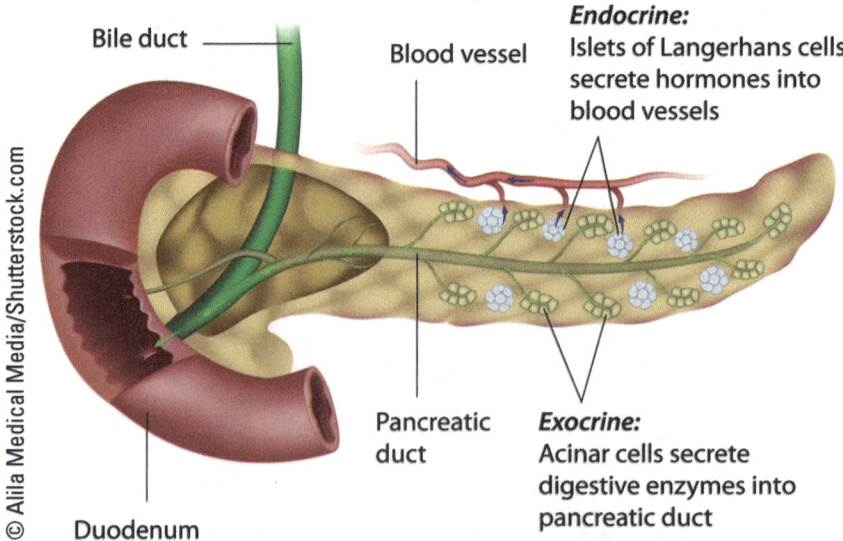

Figure 11.20 Pancreas, Liver, and Gallbladder—Anterior View. The pancreas secretes digestive enzymes and alkaline fluid into the duodenum. The liver produces bile to emulsify lipids. The gallbladder receives and stores excess bile from the liver until secreted into the duodenum.

The exocrine pancreas produces pancreatic juices that are filled with enzymes for digestion of each nutrient. The pancreatic enzyme *pancreatic amylase* breaks carbohydrates down into single or double sugars. *Proteolytic (protein-breaking) enzymes* break proteins down into single or double amino acids. Triglycerides are disassembled by the pancreatic enzyme *pancreatic lipase* into monoglycerides and free fatty acids. This allows them to pass through the cells of the villi to enter lymphatic vessels.

The exocrine pancreas also produces fluid high in sodium bicarbonate ions (alkaline fluid) to neutralize stomach acid. Once the chyme leaves the stomach, the acid is no longer necessary. The secretion of pancreatic enzymes and alkaline fluid are each under the control of hormones produced by the duodenum.

Liver

The liver is the largest internal organ in the body. It is composed of four lobes and is located in the right upper quadrant of the abdomen (Figure 11.21). The liver performs numerous types of chemical reactions that not only contribute to the digestive process but also to the maintenance of homeostasis. Within the liver are *hepatocytes,* liver cells that contribute to the digestive system by:

- Processing some substances from one form to another before sending them to the body. For example, it may form lipids into phospholipids to be used to make cell membranes. It may also convert one nutrient into another type, such as it may process amino acids so that they can be converted into ATP, lipids, or glucose.
- Removing and storing some substances, such as extra glucose, by converting them into glycogen.
- Detoxifying potentially poisonous substances. For example, a by-product of the use of amino acids for ATP production is toxic ammonia. The liver converts ammonia to less toxic urea, which can be excreted by the kidneys.
- Producing bile to emulsify lipids. Contained in the bile is bilirubin from the breakdown of dead red blood cells. Bilirubin is responsible for the brown color of feces. Bile passes out of the liver through the hepatic duct, which drains through the common bile duct into the duodenum.

> **Liver** the largest internal organ in the body. It has four lobes and is located in the right upper quadrant of the abdomen. It receives all venous blood from the abdominal viscera, allowing it to alter or store nutrients from the bloodstream before releasing those substances into the vena cava for transport throughout the body. It produces bile, which is secreted into the duodenum for the emulsification of fat.

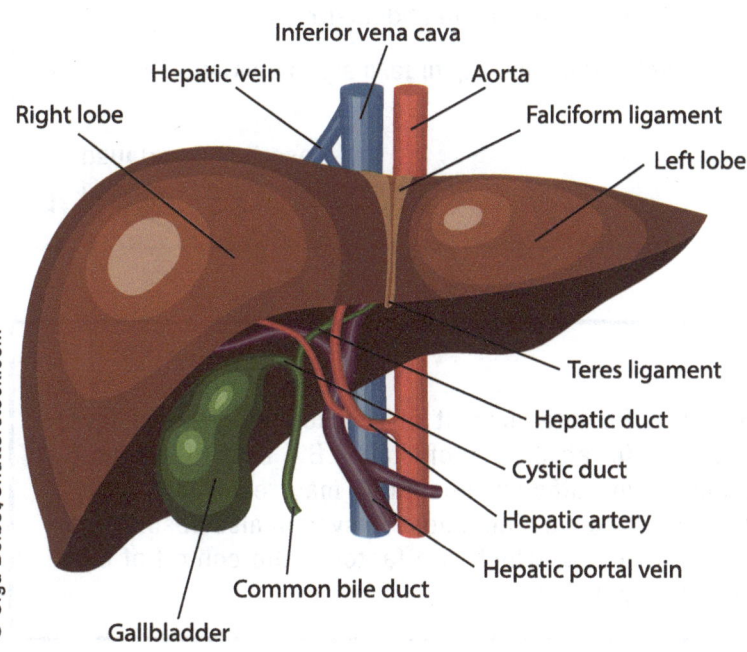
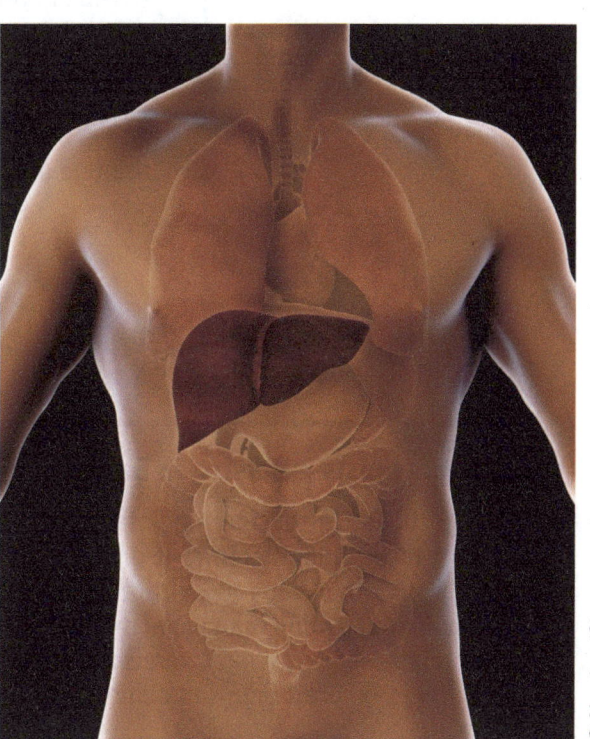

Figure 11.21 Lobes of the Liver.

There are some nondigestive functions the liver performs as well, such as:

- Producing clotting factors essential for the clotting of blood
- Producing albumin, which stays in the plasma to draw interstitial fluid back into the bloodstream by osmosis, and producing globulins, which are antibodies or molecules that transport some substances in the plasma
- Increasing blood glucose, which results from performing nutrient-altering chemical reactions
- Converting excess glucose into lipids for storage in adipose tissue

All venous blood from the abdominal organs flows to the liver through the hepatic portal system; therefore, the liver is able to provide the aforementioned functions before allowing the blood to have access to the rest of the cardiovascular system and the body.

Gallbladder

> **Gallbladder** this organ is attached to the underside of the liver. It stores excess bile for release into the duodenum as needed.

The gallbladder is attached to the underside of the liver (Figures 11.20 and 11.21). If some of the bile formed by the liver is not needed, it can be stored in the gallbladder by passing from the hepatic duct into the cystic duct into the gallbladder. If additional bile is required when a high-fat meal is consumed, the duodenum can create a hormone that causes contraction of the gallbladder to secrete additional bile into the duodenum.

Comprehension Check-up:

1. The salivary gland found in the cheek is the _____ gland.
2. The enzyme produced by the exocrine pancreas that continues the breakdown carbohydrates into single and double sugars is _____.
3. The liver has _____ lobes.
4. Where is the gallbladder located in the abdomen?

1. parotid
2. pancreatic amylase
3. 4
4. On the underside of the liver in the right upper quandrant of the abdomen

---

Homeostasis—Holding in Balance

The digestive system maintains the nutrient level in the bloodstream. It provides a process to break down food to extract nutrients in their simplest units and then move those units into our bloodstream to increase levels available to cells. It stores or preprocesses some of the nutrients for later use. It also produces intrinsic factor for absorption of vitamin B_{12}. Lack of vitamin B_{12} in sufficient quantities may result in a form of anemia. The digestive system also absorbs vitamin D, which is a factor in the control of blood calcium.

> **There Is No Substitute for Exercise**
>
> Even moderate exercise can aid in the movement of food through the digestive system. Increased blood supply throughout the body as the result of increased activity allows for more rapid absorption of nutrients. Exercise decreases visceral fat, that is, excess fat around visceral organs. Excess visceral fat has been found to be a factor contributing to insulin resistance, cardiovascular disease, and stroke.

If we define nutrients as substances we need to perform chemical and electrical processes within our body, do we all need the same nutrients? What happens if those nutrients are in short supply? In our brain, the hypothalamus makes us feel hungry when our nutrient levels are low. Why do some people prefer sweets over meat and potatoes? Why might a pregnant woman crave unusual combinations of food? The study of the substances ingested and their effect on the function of the body is known as nutrition.

The chemical processes that occur in the body, especially concerning the conversion of available energy into adenosine triphosphate (ATP), are known as *metabolism*. Virtually every living cell in the human body performs a form of metabolism, although some very active cells, such as muscle fibers or neurons, require immensely higher levels of ATP than passive adipocytes or bone cells. In this chapter we first consider the chemicals we must ingest as nutrients and then discuss the processes by which energy is extracted from nutrients and used to produce ATP.

> **Nutrition** the study of the substances ingested and their effect on the function of the body.

NUTRIENTS AND WATER

Nutrients are chemicals from which we obtain the essential ingredients to perform all of the chemical and electrical processes the body requires to maintain homeostasis. Nutrients provide the following functions:

- They are a source of energy. The quantity of energy available from a nutrient source is measured in terms of calories. Lipids are the greatest source of energy, containing 9 calories per gram while carbohydrates and proteins each have 4 calories per gram.
- They are the components, such as amino acids, necessary to produce molecules involved in chemical reactions, synthesis of enzymes, and the building of cellular structures, such as muscle or cellular membrane proteins.
- They are cellular signaling molecules that initiate the functions of each cell.

> **Nutrients** chemicals from which we obtain the essential ingredients to perform all of the chemical and electrical processes the body requires to maintain homeostasis.

Six substances are essential to the normal functioning of the human body:

- Carbohydrates
- Proteins
- Lipids
- Vitamins
- Minerals
- Water

Each of these components must be taken in from external sources on a regular basis.

Carbohydrates

> **Carbohydrates** sugars.

Carbohydrates are sugars. Recall from Chapter 1 that carbohydrates can be found in three forms: (1) single sugars—monosaccharides, (2) two single sugars bonded together—disaccharides, or (3) many single sugars chained together—polysaccharides. Single sugars can be absorbed through the wall of the jejunum and ileum into the bloodstream without any alteration. Disaccharides and polysaccharides must be broken apart by enzymes into single sugars in order to pass through the columnar cells of the small intestine and be absorbed into the bloodstream.

Proteins

> **Proteins** chains of amino acids.

Proteins are chains of amino acids. Proteins are too large to be absorbed directly into the bloodstream from the small intestine. Enzymes break the protein chains into individual amino acids so that they can pass through the intestinal wall and diffuse into the bloodstream. Cells can bond the amino acids together to form appropriate proteins to meet their needs. For example, cells can form amino acid chains into membrane proteins such as ion channels or receptor sites then insert them into its cell membrane. Outside the cell, collagen fibers, made of protein, may be knit together in the matrix around cells such as osteoblasts or fibroblasts to form additional bone or scar tissue. The human body uses 20 amino acids to form proteins. Of these, nine are considered *essential amino acids* because the body cannot synthesize them and they must be taken in from other sources. The remaining amino acids are designated as *nonessential*, not because they are not important, but because they can be synthesized from the essential amino acids when needed. Proteins containing the nine essential amino acids are referred to as *complete proteins;* these include milk, cheese, eggs, meat, fish, and poultry. *Incomplete proteins* do not possess all the essential amino acids. Some incomplete proteins are legumes (beans or peas), grains, and leafy green vegetables.

Lipids

> **Lipids** fats and oils.

Lipids, such as fats and oils, are handled differently than carbohydrates and proteins. In Chapter 1 we discussed the fact that water is polar, possessing an electrical charge, and oils are nonpolar, or uncharged. Because of this electrical difference, oil and water do not mix. Carbohydrates and proteins are polar and can dissolve in water and pass directly into the bloodstream. Lipids, because they cannot mix with water, require another form of transport throughout the body. Most lipids in the digestive system are in the form of triglycerides and have been coated with bile, charged particles that emulsify fats (see Chapter 11). Triglycerides are too large to pass through the intestinal mucosa, so they are partially disassembled by pancreatic lipase, an enzyme that allows the parts to pass through the lining of the jejunum and ileum. Inside the cells of the mucosa the triglycerides are reassembled and then coated with charged proteins. These coated molecules are too large to diffuse into capillaries as single sugars and amino acids do when they exit the other side of the mucosa, so they pass instead into lymphatic vessels. They circulate through the lymphatic system and into the left subclavian vein to enter the bloodstream indirectly.

> **Comprehension Check-up:**
>
> 1. The six types of nutrients are _____ , _____ , _____ , _____ , _____ , and _____ .
> 2. Most lipids in the digestive system are in the form of _____ .
>
> 1. carbohydrates, proteins, lipids, vitamins, minerals, water
> 2. triglycerides

Vitamins

Vitamins are cofactors of enzymes that cannot be manufactured in the human body in sufficient quantity, except for vitamin D and must be ingested on a regular basis in order for homeostasis to be maintained. They are essential to the performance of various chemical processes throughout the body. They can be found in food and vitamin supplements. These chemicals must be taken in regularly for homeostasis to be maintained. Vitamins can be divided into two general categories: water-soluble (polar) and lipid-soluble (nonpolar) (Table 11.1). Water-soluble vitamins are the B complex vitamins and vitamin C. The level of the water-soluble vitamins can be controlled by the kidneys. Excess water-soluble vitamins are normally excreted in the urine. Lipid-soluble vitamins include vitamins A, D, E, and K and are not as easily regulated. They are stored and maintained by the digestive system, primarily the liver. Because they are not controlled by the kidneys, it is easier to overdose on lipid-soluble vitamins than on those that are water-soluble.

> **Vitamins** substances that cannot be manufactured in the body in sufficient quantity yet are essential to the performance of various chemical processes throughout the body.

Minerals

Minerals, metals and nonmetals, required by the body are in the form of ions (Table 11.2). They can be absorbed directly into the bloodstream from the small intestine, where they remain in ionized form. We have previously discussed the importance of sodium and potassium ions for depolarization and repolarization of cells. Calcium was also mentioned as being essential for activation of the thin filament in each sarcomere in muscle. Phosphorus is ionized to phosphate and used to make ATP. Calcium and phosphorus are major components in the matrix of our bones. Like vitamins, minerals also serve as cofactors in the function of many enzymes.

> **Minerals** metals and nonmetals required by the body primarily in the form of ions to perform chemical reactions or create action potentials.

Water

Water is found in most areas of the body and is used for the transport of substances to and from each cell. It is also necessary for any metabolic reaction within the cell and is a major component of saliva and secretions from the stomach, pancreas, liver, and small intestine. The average adult typically takes in about 2,500 ml of water in food and beverages. Additional water is added to the digestive tract from saliva (1,500 ml), from the stomach (2,000 ml), from the pancreas (1,500 ml), from the liver (500 ml), and from the small intestine (1,500 ml). All of the secretions from the digestive organs obtain their water from storage in body tissue. In a 24-hour period,

Table 11.1 Vitamins

Vitamin	Significance	Sources	Daily Requirement	Effects of Deficiency	Effects of Excess
Fat-Soluble Vitamins					
A	Maintains epithelia; required for synthesis of visual pigments	Leafy green and yellow vegetables	1 mg	Retarded growth, night blindness, deterioration of epithelial membranes	Liver damage, skin peeling, central nervous system effects of nausea, anorexia
D	Required for normal bone growth, calcium and phosphorus absorption at gut, and retention at kidneys	Synthesized in skin exposed to sunlight	None[a]	Rickets, skeletal deterioration	Calcium deposits in many tissues disrupting functions
E (tocopherols)	Prevents breakdown of vitamin A and fatty acids	Meat, milk, vegetables	12 mg	Anemia; other problems suspected	None reported
K	Essential for liver synthesis of prothrombin and other clotting factors	Vegetables; production by intestinal bacteria	0.7–0.14 mg	Bleeding disorders	Liver dysfunction, jaundice
Water-Soluble Vitamins					
B_1 (thiamine)	Coenzyme in decarboxylation reactions	Milk, meat, bread	1.9 mg	Muscle weakness, central nervous system and cardiovascular problems, including heart disease; called *beriberi*	Hypotension
B_2 (riboflavin)	Part of FMN and FAD	Milk, meat	1.5 mg	Epithelial and mucosal deterioration	Itching, tingling sensations
Niacin (nicotinic acid)	Part of NAD	Meat, bread, potatoes	14.6 mg	Central nervous system, gastrointestinal, epithelial, and mucosal deterioration; called *pellagra*	Itching, burning sensations, vasodilation, death after large dose
B_6 (pyridoxine)	Coenzyme in amino acid and lipid metabolism	Meat	1.42 mg	Retarded growth, anemia, convulsions, epithelial changes	Central nervous system alterations, perhaps fatal
Folacin (folic acid)	Coenzyme in amino acid and nucleic acid metabolism	Vegetables, cereal, bread	0.1 mg	Retarded growth, anemia, gastrointestinal disorders	Few noted except in massive doses
B_{12} (cobalamin)	Coenzyme in nucleic acid metabolism	Milk, meat	4.5 mg	Impaired iron absorption causing *pernicious anemia*	Polycythemia

(continued)

Table 11.1 Vitamins (*Continued*)

Vitamin	Significance	Sources	Daily Requirement	Effects of Deficiency	Effects of Excess
Biotin	Coenzyme in decarboxylation reactions	Eggs, meat, vegetables	0.1–0.2 mg	Fatigue, muscular pain, nausea, dermatitis	None reported
Pantothenic acid	Part of acetyl-CoA	Milk, meat	4.7 mg	Retarded growth, central nervous system disturbances	None reported
C (ascorbic acid)	Coenzyme; delivers hydrogen ions	Citrus fruits	60 mg	Epithelial and mucosal deterioration; called *scurvy*	Kidney stones

[a]Unless there is poor exposure to sunlight for extended periods; alternative sources are provided in fortified milk
FMN = Flavin mononucleotide
NAD = Nicotinamide adenine dinucleotide
FAD = Flavin adenine dinucleotide

Table 11.2 Minerals

Mineral	Significance	Total Body Content	Primary Route	Recommended Daily Intake
Bulk Minerals				
Sodium	Major cation in body fluids; essential for normal membrane function	110 g, primarily in body fluids	Urine, sweat, feces	1.1–3.3 g
Potassium	Major cation in cytoplasm; essential for normal membrane function	140 g, primarily in cytoplasm	Urine	1.9–5.6 g
Chloride	Major anion in body fluids	89 g, primarily in body fluids	Urine, sweat	1.7–5.1 g
Calcium	Essential for normal muscle and nerve function, structural support of bones	1.36 kg, primarily in skeleton	Urine, feces	0.8–1.2 g
Phosphorus	As phosphate in high-energy compounds, nucleic acids, and structural support of bones	744 g, primarily in skeleton	Urine, feces	0.8–1.2 g
Magnesium	Cofactor of enzymes; required for normal membrane functions	29 g, 17 g in skeleton and the rest in cytoplasm and body fluids	Urine	0.3–0.4 g
Trace Minerals				
Iron	Component of hemoglobin, myoglobin, and cytochromes	3.9 g, 1.6 stored (ferritin or hemosiderin)	Urine (traces)	10–18 mg
Zinc	Cofactor of enzymes systems, notably carbonic anhydrase	2g	Urine, hair (traces)	15 mg
Copper	Required for hemoglobin synthesis, as cofactor	127 mg	Urine, feces (traces)	2–3 mg
Manganese	Cofactor for some enzymes	11 mg	Feces, urine (traces)	2.5–5 mg

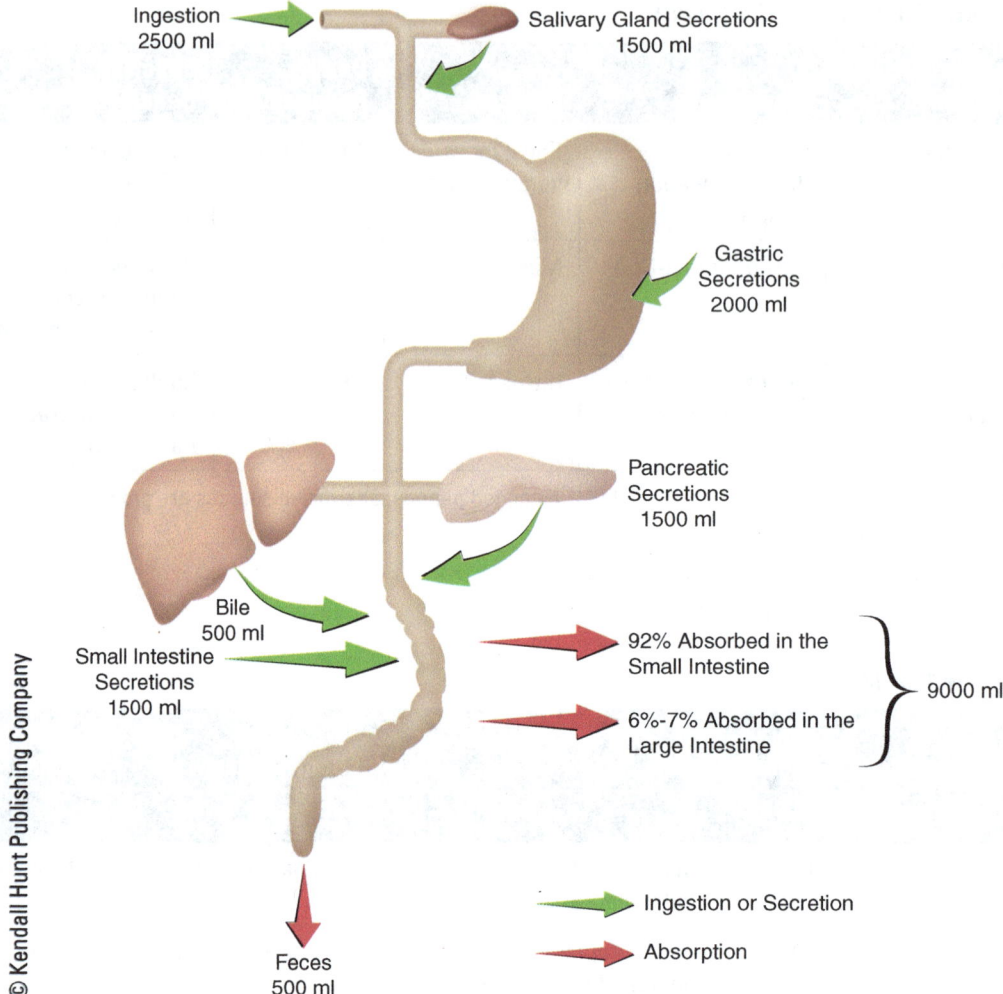

Figure 11.22 Movement of Fluid in and out of the Digestive System.

approximately 9,000 to 9,500 ml of water passes through the digestive system (Figure 11.22). The intestines reabsorb all but about 500 ml of this water back into the bloodstream. If a significant amount of the water in the digestive system is not reabsorbed, as occurs with diarrhea, dehydration may occur.

The kidneys, along with input from the brain, can then determine how much of the water in the bloodstream is essential to maintain homeostasis. The excess is excreted in the urine. The 500 ml remaining in our feces simply provides some softness. Water is also lost from the body through sweating or exhaled during respiration.

Comprehension Check-up:
1. The two general categories of vitamins are _____ and _____.
2. Sodium, potassium, and calcium are _____, a type of nutrient.

1. water soluble, lipid soluble   2. minerals

> **Diarrhea and Its Effect on the Body**
>
> Almost everyone has experienced diarrhea at some point. Typically the cause is some type of irritation to the intestinal mucosa, as is commonly caused by bacteria or some type of virus. The response of the digestive system is to empty its contents as quickly as possible. When gastric motility, that is, the speed of the digestive process, increases substantially, there is little time to reabsorb water and electrolytes and water is lost from the body. This excessive loss may lead to serious health concerns, especially in children. The digestive organs obtain water and nutrients that are essential to the production of their secretions from the bloodstream and from internal tissue fluid. Typically 9000 ml of water is reabsorbed by the intestine back into the cardiovascular system. If, however, a large volume of that normally reabsorbed water is lost as a result of diarrhea, the cardiovascular system may have a significant loss of plasma volume, making it difficult for the heart to maintain blood pressure. Most systems in the body can be affected by inadequate blood flow. Diarrhea is of particular concern in small children because they do not have the large reservoir of fluid in their tissue typically found in adults. As a result, diarrhea can become serious much more quickly in the small child.

THE DIFFERENCE BETWEEN FOOD AND NUTRIENTS

Not everything we eat as food is nutritious. There are some substances in our food for which we do not possess enzymes to break their chemical bonds to reduce them to their simplest components, for example, cellulose. Commonly known as *plant fiber,* cellulose is composed of chains of glucose. Although it would seem to be a great source of nutrition, animals do not produce the enzymes necessary to break these bonds. Instead, fiber primarily found in fruits and vegetables adds bulk to our digestive system and helps maintain intestinal health by keeping the intestinal wall clean. It can also bind cholesterol to assist in removing excess levels from the body.

Food Groups

The food we eat can be divided into five distinct groups depending on their source and content: fruits, vegetables, grains, milk products, and meats (Table 11.3).

Food Pyramid

The U.S. Department of Agriculture developed a food pyramid to assist individuals in determining the quantity of each food group to take in each day (Figure 11.23).

Each color in Figure 11.23 represents a different food group and the quantity to be consumed:

- ▲ Grains such as rice, bread, cereals, and pasta— 6 to 11 servings per day
- ▲ Vegetables—3 to 5 servings per day
- ▲ Fruits—2 to 4 servings per day
- ▲ Fats, oils, and sweets—occasional use
- ▲ Dairy products, including cheese and yogurt— 2 to 3 servings in 2 days
- ▲ Meat, fish, eggs, nuts, and beans or legumes—2 to 3 servings per day

Table 11.3 Basic Food Groups, Their Sources, and Nutrient Content

Group	Example Foods	By All in Group	By Only Some in Group
Fruits	Apples, bananas, dates, oranges, tomatoes	Carbohydrate Water	Vitamins: A, C, folic acid Minerals: iron, potassium Fiber
Vegetables	Broccoli, cabbage, green beans, lettuce, potatoes	Carbohydrate Water	Vitamins: A, C, E, K, and B vitamins except B_{12} Minerals: calcium, magnesium, iodine, manganese, phosphorus Fiber
Grain products (preferably whole grain; otherwise, enriched or fortified)	Breads, rolls, bagels, cereals (dry and cooked); pasta, rice, other grains; tortillas, pancakes, waffles; crackers; popcorn	Carbohydrate Protein Vitamins: thiamin (B_1), niacin	Water Fiber Minerals: iron, magnesium, selenium
Milk products	Milk, yogurt; cheese; ice cream, ice milk, frozen yogurt	Protein Fat Vitamins: riboflavin, B_{12} Minerals: calcium, phosphorus Water	Carbohydrate Vitamins: A, D
Meats and meat alternates	Meat, fish, poultry; eggs; seeds; nuts, nut butters; soybeans, tofu; other legumes (peas and beans)	Protein Vitamins: niacin, B_6 Minerals: iron, zinc	Carbohydrate Fat Vitamins: B_{12}, thiamin (B_1) Water Fiber

Source: Christian, Janet, and Janet Greger. *Nutrition for Living*, 3rd ed. San Francisco: Benjamin Cummings, 1991.
Images © Shutterstock, Inc.

Figure 11.23 Food Pyramid.
(From: http://www.righthealth.com, Food Pyramid.)

Grains are to be consumed in the largest quantities because they are composed primarily of complex carbohydrates (polysaccharides as well as protein). Chains of sugars take more time to digest than disaccharides (two single sugars bonded together), which are typically found in sweets. As a result, complex carbohydrates provide a source for energy over a much longer period than do disaccharides.

Fruits and vegetables also contain carbohydrates. They are an excellent source of many of the vitamins and minerals needed for normal body function. Undigestible plant fiber helps clean the walls of the intestine and also provides bulk to keep the movement of food through the digestive system efficient.

Dairy products contain protein as well as vitamins and minerals. They are, however, also high in fat, which is why they are recommended to be consumed in lower quantities in adults than the previously mentioned food groups.

Meats, fish, eggs, nuts, beans, and legumes are an excellent source of protein. They may also contain vitamins, minerals, and carbohydrates. The proteins in this food group are broken down into amino acids for reassembly into needed proteins throughout the body.

For maximum health and wellness, it is recommended that fats, oils, and sweets be limited to only occasional servings. We can obtain the lipids and sugars we need from the other groups. It is not necessary to take them in separately. Although fat is an excellent energy source, the consumption of excessive fat, especially for individuals with low activity levels, causes the body to store more than is used. There are two types of fats found in our diet: saturated and unsaturated. Saturated fats have been linked to a buildup of plaque in blood vessels, resulting in heart disease and stroke. Polyunsaturated fats may actually decrease the inflammation of blood vessels that leads to the buildup of plaque. Sweets (disaccharides) are absorbed into the bloodstream too fast, causing a rapid rise in blood glucose followed by a large increase in insulin. The purpose of insulin is to lower blood sugar back to homeostatic levels, but it also causes additional storage of other nutrients and increases the content of adipose tissue. The effects of increased insulin levels are discussed in greater detail later in this chapter.

The concept of the food pyramid is that appropriate consumption of each food group, combined with daily physical activity, provides the level of nutrients needed to perform all of the necessary chemical reactions and processes in the body while helping the individual maintain his or her current weight and increase overall health. It is important to understand that physical activity is an essential factor in maintaining health and weight. There is no substitute for exercise.

The hypothalamus controls our food intake. It contains both a hunger center, which increases our desire to eat, and a satiety center, which determines our nutrient level is sufficient and decreases our hunger. If we have low levels of a specific nutrient we may have cravings for foods containing that substance. This often affects women during pregnancy when cravings for specific foods become very strong. We do not just eat because our hypothalamus stimulates our hunger center. We may consume food because we are bored, tired, or stressed. Perhaps you have heard the term, "comfort foods." There may also be foods we associate with unpleasant experiences which we refuse to eat. We may even be conditioned to like certain foods, such as sweets or meat, over others, like fruits and vegetables.

It is a concern that there is an increase in obesity in the United States, especially in children. Food portions are becoming larger and physical activity

---

Advantages of Physical Activity

Studies have shown that obese individuals typically do not overeat; they underexercise. There are numerous advantages to getting off the couch and taking a walk.

Some of the benefits of exercise are that it:

- Increases metabolism, which uses excess fat, resulting in weight loss or allowing the individual to stay trim.
- Improves circulation as blood flow increases to muscles performing their work.
- Increases assistance of venous return of blood to the heart.
- Increases stroke volume because the heart pumps higher volumes of blood during exercise, so during relaxation the larger stroke volume allows the heart more time to rest while it maintains normal cardiac output.
- Improves movement of extracellular fluid out of the tissue to decrease potential swelling.
- Strengthens bones.
- Keeps joints lubricated and moving freely.
- Relieves stress, which allows your defensive system to work more efficiently.
- Improves muscle tone, which reduces the likelihood of injury.
- Releases neurotransmitters in the brain that improve mental health.
- Allows you to think more clearly.
- Increases stamina.
- Improves the body's ability to maintain homeostasis.

None of these advantages can be achieved only by diet. If you provide the body with the appropriate level of nutrients combined with physical activity you increase probability of improved health and well-being.

is decreasing. It is an issue that needs to be addressed today because the patterns learned as children are very likely to continue throughout life. Perhaps the best alternative is to not only provide appropriate diets for children but also to encourage them to be physically active.

> **Comprehension Check-up:**
> 1. _____ and _____ are two food groups that are a good source for vitamins, minerals, and fiber.
> 2. The food group that is an outstanding source of protein is _____ .
>
> 1. Fruits, vegetables
> 2. meats, fish, eggs, nuts, beans, and legumes

Metabolism

Metabolism is the series of chemical reactions that occur in the body that result in the conversion of energy in the form of nutrients into a form our cells can use to perform whatever work is needed at the time. For example, energy is needed to cause muscle contraction, to move nutrients through the digestive mucosa, for the kidneys to separate excess substances in the plasma and excrete them as urine, for the maintenance of ion balance to allow nerve impulse conduction, and for contraction of the heart to pump blood. Recall from Chapter 1 that all energy on the surface of the Earth comes from the sun. Plants capture the solar energy by a process known as photosynthesis. During photosynthesis the plant uses the energy from the sun to cause the single electron around hydrogen atoms to spin faster. Once the electron possesses the additional energy, it continues to retain that energy until a later time when the energy can be released and used by the plant. The hydrogen atoms come from water in the plant. The plant then attaches those energized hydrogen atoms to carbon dioxide molecules to form glucose ($C_6H_{12}O_6$) and oxygen. The chemical equation for photosynthesis is:

$$6\ H_2O + 6\ CO_2 + \text{solar energy} \rightarrow C_6H_{12}O_6 + 6\ O_2$$

We, as animals, cannot use light as a source of energy in our bodies. We eat plants or we eat animals that eat plants and obtain energy-rich molecules from them for metabolism. We can reverse the process by breaking glucose down, when oxygen is available, to water and carbon dioxide, and we can also release the energy contained in the glucose molecule.

$$C_6H_{12}O_6 + 6\ O_2 \rightarrow 6\ H_2O + 6\ CO_2 + \text{energy}$$

It is possible to break down glucose without oxygen to release a small amount of energy, but the process is temporary and not available to all cells.

Animals store the released energy from the breakdown of nutrients in the high-energy bonds of the ATP molecule. The energy is then transferred by ATP to the appropriate locations in the cell for work to be accomplished. This process that occurs within the cell is known as *cellular respiration*.

> **Metabolism** the series of chemical reactions that occur in the body that result in the conversion of energy in nutrients into a form our cells can use to perform work.

Types of Metabolic Reactions

Throughout the process of metabolism, two types of chemical reactions occur. The combining of small molecules that results in the synthesis of larger ones is referred to as anabolism. The breakdown of large molecules into smaller ones is known as catabolism. For example, in Chapter 1 we discussed that ATP (Adenosine-Pi~Pi~Pi where ~ represents a high energy bond) was constructed by adding a third phosphate molecule to adenosine diphosphate (ADP; 2 phosphates) but energy was required to add the third phosphate onto ADP. Later, the third phosphate can be broken off ATP to release that energy for use elsewhere. Adding the third phosphate to ADP builds the larger molecule ATP by anabolism. To obtain the energy necessary to add that third phosphate, through catabolism we must break glucose down into smaller molecules to cause its energy to be released.

The general purposes of metabolism occur in the following order:

1. Remove the energized hydrogen atoms containing additional energy from their source.
2. Cause the hydrogen electron to release its additional energy.
3. Use that energy to produce ATP in order to transfer the energy to the appropriate location in the cell.

Energy is available from carbohydrates, lipids, and proteins. The process of removing this energy from nutrients is not very efficient. Only 25% of the available energy actually is used for work such as causing muscle contraction or the pumping of ions. Seventy-five percent of the energy in our nutrients becomes heat, raising the temperature of the body. The processes involved with extracting energy from these nutrients are addressed individually.

> **Anabolism** the combining of small molecules that results in the synthesis of larger ones.

> **Catabolism** the breakdown of large molecules into smaller ones.

> **Comprehension Check-up:**
> 1. The source of energy on the surface of the earth is the _____ .
> 2. Of the energy in nutrients, ___% of that energy actually is used to perform work and the remaining _____% turns into heat.
>
> ans. 1. sun 2. 25, 75

Carbohydrate Metabolism

Glucose is a major source of available energy. The catabolism of glucose to extract excess energy to make ATP occurs in two pathways. The first series of chemical reactions does not require oxygen to complete the process and is referred to as anaerobic metabolism. Essentially, only carbohydrates are catabolized during anaerobic metabolism. The second pathway is oxygen dependent and is known as aerobic metabolism. Lipids and proteins can also be catabolized during aerobic metabolism. Anaerobic metabolism involves only a few chemical reactions to complete its process, so it is a rapid source of ATP but its yield per glucose is small. Aerobic metabolism, on the other hand, involves many chemical reactions, but the yield of ATP per glucose is much greater. The process involving oxygen during the production of ATP is referred to as *oxidative phosphorylation*. During skeletal muscle

> **Anaerobic metabolism** the production of a small amount of ATP per glucose molecule without the use of oxygen.

> **Aerobic metabolism** a series of oxygen-dependent chemical reactions by which a relatively large amount of ATP is produced.

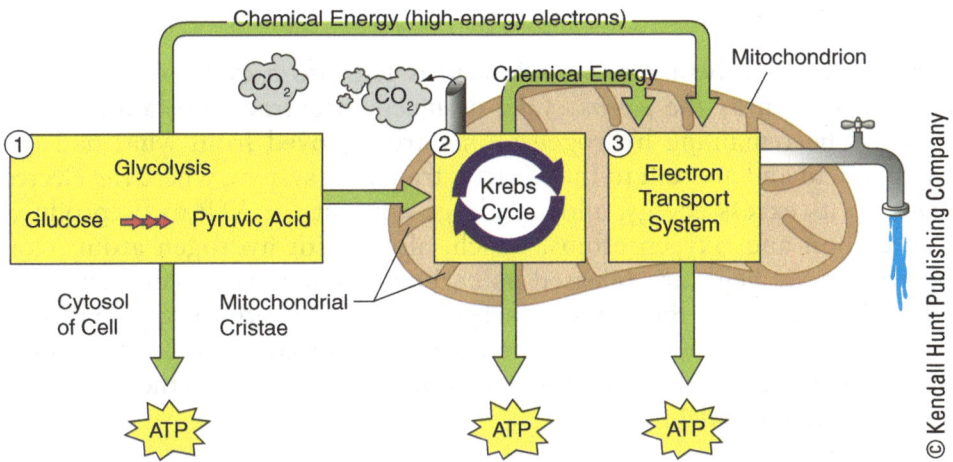

Figure 11.24 Glucose Metabolism. (1) In the cytsol, glycolysis breaks glucose into 2 pyruvic acids and produces a small amount of ATP. (2) Pyruvic acid enters the Krebs cycle in the mitochondrion, where hydrogen atoms with their high-energy electrons are removed. In the process, a small amount of ATP and carbon dioxide is produced. (3) The chemical energy from glycolysis and the Krebs cycle is transferred to the electron transport system, where the energy contained in the high-energy hydrogen electron is used to form ATP. Hydrogen atoms that have given up their excess energy are bonded with oxygen to form water.

contractions, we often use anaerobic metabolism as a quick source of ATP production and then later, when oxygen is available, we continue to aerobic metabolism for increased ATP production.

Anaerobic Metabolism

The catabolism of glucose is accomplished by a series of chemical reactions that breaks glucose in half and rearranges it into two molecules known as *pyruvic acid*. In the process of breaking glucose apart, called *glycolysis*, a small amount of ATP is produced (Figure 11.24-1). In most cases, the pyruvic acid moves on to aerobic metabolism. If oxygen is in short supply, however, it becomes essential to do something to remove the pyruvic acid; otherwise, the catabolism of glucose will back up and come to a halt. The production of pyruvic acid is somewhat like an object being made and then placed on a conveyor belt. At the end of the belt someone must remove the product, or it will stack up so high that work cannot continue. In the same way, pyruvic acid needs to be moved out of the way and a coenzyme, NAD, which is a carrier of energized electrons during the process, needs to be regenerated so that the production of ATP through anaerobic metabolism can continue. Alternatively, muscle can convert pyruvic acid into lactic acid. The second phase of anaerobic metabolism results in the production of lactic acid which regenerates NAD, allowing the breakdown of glucose (glycolysis) to continue. Lactic acid is somewhat like taking out a loan until oxygen is available, sometimes referred to as *oxygen debt*. This allows muscle to continue working even though the bloodstream is unable to transport sufficient oxygen at the moment to keep up with the demand.

If there is excess pyruvic acid and the demand for ATP has decreased, pyruvic acid can be converted back into glucose as an anabolic reaction. This process occurs in the liver and is known as *gluconeogenesis* (the creation of new glucose).

Aerobic Metabolism

When oxygen is available, pyruvic acid is fed into a cycle of chemical reactions, known as the *Krebs cycle,* designed to extract substantially more energy. The remaining hydrogen atoms are removed from what had been pyruvic acid and are sent to the electron transport system, where the electron gives up its excess energy, allowing large amounts of ATP to be produced. The carbon and oxygen atoms, which, along with hydrogen atoms, composed pyruvic acid, are released as carbon dioxide. The hydrogen atoms, which have given up their excess energy, are bonded with oxygen to produce water. The carbon dioxide and water are transported to the lungs from the bloodstream so that they can be exhaled out of the body into the atmosphere (Figure 11.24-2 and 11.24-3).

Lipid Metabolism

Fats are carbon chains surrounded by hydrogen atoms. Those hydrogen atoms contain excess energy, making lipids an excellent option for storing large amounts of energy. In fact, there is more than twice the amount of energy in fat than in either carbohydrates or protein, so the resulting energy from fat allows us to be active for a longer period than that provided by the other two nutrients. When oxygen is available, the liver can catabolize fatty acids by a process known as *lipolysis* and, like pyruvic acid, send them into the Krebs cycle and electron transport system to produce a great deal of ATP (Figure 11.25). The

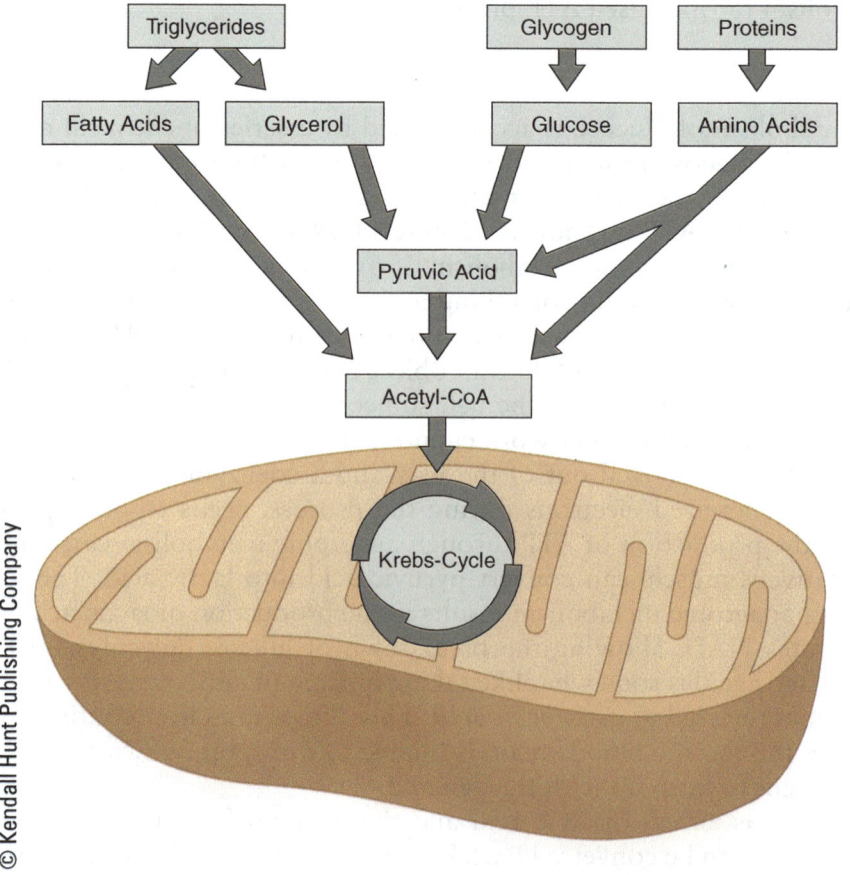

Figure 11.25 Carbohydrates, Lipids, and Proteins. These can be sources of ATP through aerobic metabolism in mitochondria.

process of lipolysis produces some by-products known as *ketone bodies* that have a fruity odor. If a diabetic individual has taken too much insulin for the amount of food eaten and is unconscious, often the person's breath will have a fruity odor, indicating that the body is rapidly catabolizing lipids because their available glucose is too low.

If there are excess carbohydrates, they can be converted into fatty acids and stored in liver or adipose tissue. The creation of lipids by the body is known as *lipogenesis*.

Protein Metabolism

Proteins are chains of amino acids. Amino acids may be used as a source for ATP as well, although they contain a nitrogen atom, which is unusable by the Krebs cycle. As a result, during amino acid catabolism, the nitrogen atom is removed and discarded as urea. Urea is eliminated in the urine. Once urea has been exposed to the air, it incorporates another hydrogen atom, converting it to ammonia. It is the excretion of urea that causes a diaper pail to smell of ammonia.

Excess pyruvic acid can also be converted into amino acids when a nitrogen atom from some of the urea remaining in the body is added to the molecule. Those amino acids can then be bonded together in a specific order inside cells to form protein through protein synthesis.

Comprehension Check-up:

1. The primary nutrient that has some energy extracted to make ATP without the use of oxygen is _____.
2. The nutrient that contains more energy than the other two combined is _____.
3. To use amino acids for an energy source, our cells must first remove _____.

1. carbohydrates or glucose   2. lipids   3. nitrogen

Body Temperature

Body temperature is regulated at an average of 98.6°F (37°C). The heat for maintaining our temperature comes from metabolism. As we convert the energy in nutrients into ATP, 50% of that energy escapes as heat. When transferring the energy in ATP to its intended process, another 25% of that energy becomes heat as well. That heat energy is available, somewhat like a constantly running furnace, to increase body temperature.

Regulation of Body Temperature

Our body temperature is derived from metabolism. We can regulate our temperature by (1) retaining or releasing heat into the environment through our skin and (2) altering the rate of metabolism to adjust the generation of heat.

The hypothalamus controls body temperature to maintain it within homeostatic range. If we become too warm—for example, if we are exercising—the hypothalamus causes an increase in blood flow to the skin and increases sweating to allow evaporation to draw additional heat from the skin and cool the body. If we become cold, the hypothalamus causes a decrease in blood flow to the skin to minimize heat loss. We also increase the need for more ATP by shivering, a rapid contraction of muscle that produces no coordinated movement but uses up ATP, which results in an increase in metabolism to generate more heat. Another method for increased heat production is to stimulate our sympathetic nervous system, our "fight-or-flight" system. The process of getting our body ready for action increases the production of ATP, which can also warm us when we are cold.

Thermal receptors in our skin can inform the brain about the temperature of our environment. For example, stepping into a sauna causes the sensing of heat. This results in the appropriate response by our temperature-regulating system. If we jump into a cold swimming pool, the sensing of the temperature of the water causes responses to cold.

Regulation of Body Weight

> **Calorie** a measure of the amount of energy in a substance, usually determined in food.

Scientists determined a measure of heat energy in terms of calories. A calorie is the amount of energy required to raise 1 g of water 1°C. The levels of calories in food, and also those required by our body, have been determined and are in the millions. When dealing with relatively large numbers, it is easier to discuss heat energy in terms of kilocalories (1000 calories), also written as Calories. Scientists then determined the minimum level of Calories (kilocalories) required by a resting, alert individual to perform essential activities, such as breathing and circulating blood. They called this the basal metabolic rate (BMR). Any activity performed during the day requires additional energy above BMR.

> **Basal metabolic rate (BMR)** the minimum level of Calories (kilocalories) required by a resting, alert individual to perform essential activities such as breathing and circulating blood.

If the Calories taken in equals the energy used, then weight remains unchanged. When more energy is taken in than Calories used, the body stores the excess primarily as fat because it is the most efficient means to do so. As a result, there is weight gain. The reverse is also true; when the individual uses more energy than is taken in, the body draws from its reserves and weight is lost.

For the body to perform all of the functions necessary to maintain homeostasis and well-being throughout life, it is important that it receive the essential nutrients (9 amino acids, most vitamins, and a few fatty acids) on a regular basis; most nutrients can be synthesized in the body from other nutrients. Each system is more capable of performing its tasks when it possesses all of the chemicals it requires than when those chemicals are in short supply. As a result, the interaction of each system improves because they are more capable of meeting the other's needs. Not only are our defenses more capable of reacting to invasion, but repair of tissue damage is also more rapid. Exercise relieves stress, improves circulation, strengthens bones, and provides numerous health benefits. Our bodies become trim and resilient, so injury is minimized. It has even been found that running, for example, can be as effective an antidepressant as drugs. Adequate nutrition combined with physical activity improves the quality of life because it provides the body with what it needs and creates the opportunity for our body to function in the way it was designed.

> **Comprehension Check-up:**
> 1. The control center for body temperature is the _____.
> 2. The minimum amount of energy in Calories that a resting, alert individual needs to perform minimal activities is known as _____.
>
> 1. hypothalamus   2. basal metabolic rate

Body Mass Index versus Percentage of Fat Content

A rough estimate of the condition of the body can be made by comparing weight to height. This is known as **body mass index (BMI)**. BMI is calculated by the following equation:

$$BMI = \frac{Wt(kg)}{(Height\ [meters])^2} \text{ or } \frac{Wt\ (lb.) \times 703}{(Height\ [inches])^2}$$

Normal BMI is between 18 and 25. A person with a BMI of less than 16.5 is considered severely underweight. A person with a BMI between 25 and 30 is overweight, and someone with a BMI greater than 30 is defined as obese.

For example, if a 5'5" woman weighed 135 pounds, her BMI would be calculated as follows:

$$BMI = \frac{135 \times 703}{(65)^2} = \frac{94,905}{4225} = 22.46$$

The BMI for this woman is within normal range.

There are times when BMI can be misleading. For example, a professional football player may have a high quantity of muscle mass even though he is relatively lean, causing his BMI to indicate an unhealthy weight. Conversely, an inactive person may not be overweight but his proportion of fat to muscle may be at an unhealthy level even though his BMI would calculate him to be within normal range.

Although BMI is a method for calculating the general relationship between weight and height, it is not taking into account the ratio of fat to muscle. It is possible, through measurement of the thickness of specific areas of the pinched skin, to determine the fat content of the body, giving a more accurate health assessment than BMI. Normal body fat content should be around 8% to 22% for men and 22% to 33% for women. There is a slight normal increase with age but not more than a few percentage points every 20 years. An individual who is greater than 20% over the normal fat content for age and gender is considered obese.

> **Body mass index (BMI)** comparison of body weight to height as a rough estimate of the condition of the body.

Clinical Notes—Weight Control

LMany factors are involved in weight control such as genetics, hormone levels, and disease processes. For the average person, there are four major factors that affect weight:

- Do not allow your body to go into starvation by skipping meals in order to lose weight. Because there is no food consumed, the body slows down the metabolism to conserve the remaining food supplies. Skipping a meal will typically result to gaining weight about 6-8 lbs. Eating smaller meals every 3 hours will have better metabolism. Stay away from foods with high sugar content.

- Avoid foods that stimulate an increase in insulin level. It causes the insulin to store excess glucose as fats. Sucrose or table sugar causes the insulin to store more glucose as fats. When more glucose is stored, it results to low blood sugar causing the person to be irritable, confused, and fatigued. It is important not to totally deprive the body of carbohydrates because it can lead to hypoglycemia, an abnormally low blood glucose.

- Follow the suggested amounts for each food group in the food pyramid to obtain the nutrients your body requires. Do everything in moderation only.

(continued)

- Get regular exercise, which increases your basal metabolic rate. Much of your use of nutrients occurs after the exercise is over, as you replenish the energy used during the activity. Vigorous exercise 6 days a week is optimal. Three days should involve aerobic exercise to help increase blood flow and allow your heart to work more efficiently.

CLINICAL TERMS TO REVIEW

Pathogenic microorganism, stomach acidity, lysosomal enzymes
Acid reflux, abdominal bloating, diarrhea
Gastroesophageal Reflux Disease (GERD)
Helicobacter pylori, peptic ulcer, appendicitis, WBC, neutrophils
Weight control
Carbohydrates, lipids, proteins, vitamins, minerals.

Water, digestion, absorption
Mucosa, submucosa, serosa, muscularis externa
Peristalsis, segmentation
Mastication, amylase, bile
Esophagus, stomach, hydrochloric acid
Small intestine, duodenum, jejunum, ileum, colon, cecum, sigmoid colon, rectum, anus
Salivary glands—parotid, sublingual, submandibular

Pancreatic enzymes, proteolytic enzymes
Liver, detoxification, clotting, albumin, adipose tissue, gallbladder
Anabolism, catabolism
Aerobic and anaerobic metabolism
Lipolysis, lipogenesis, body temperature, hypothalamus
Basal metabolic rate (BMR), body mass index (BMI)
Food pyramid

Test Yourself

Choose the best answer to the following multiple choice questions:

1. When the stomach is empty, the muscoa forms folds known as
   a. rugae.
   b. peptic folds.
   c. gastronomic inflections.
   d. frenula.

2. The breakdown of nutrients into their individual components is known as
   a. absorption.
   b. reconstitution.
   c. induction.
   d. digestion.

3. The enzyme pepsinogen is produced in the gastric pits of the stomach by
   a. goblet cells.
   b. parietal cells.
   c. chief cells.
   d. squaw cells.

4. Excess bile is stored in the
   a. pancreas.
   b. gallbladder.
   c. rectum.
   d. spleen.

5. Baby teeth that are replaced as a child ages are also known as
   a. a temporary plate.
   b. semipermanent teeth.
   c. false teeth.
   d. milk teeth or deciduous teeth.

6. The hardest substance in the body is
   a. bone.
   b. tooth enamel.
   c. fibrocartilage.
   d. a stubborn mind.

7. The period of stomach activity that occurs as the brain stimulates the stomach to prepare to receive food is known as the
   a. cephalic phase.
   b. intestinal phase.
   c. caudal phase.
   d. gastric phase.

8. There are projections on mucosa of the small intestine that increase contact with digesting food. Which of the following is not a type of projection?
   a. villi
   b. circular folds
   c. flagella
   d. microvilli

9. Fats are emulsified, that is, coated with negative molecules, to prevent them from reforming back into globules and to allow them to move freely in water. In humans, fat is emulsified by
   a. enzymes.
   b. hormones.
   c. bile.
   d. hydrochloric acid.

10. The appendix is attached to the
    a. ileum.
    b. cecum.
    c. sigmoid colon.
    d. rectum.

11. Although dairy products contain protein, vitamins, and minerals, they should not be eaten in large quantities because they also contain
    a. contaminants.
    b. too many sugars.
    c. toxins.
    d. fat.

12. Why is it recommended that we take in few fats, oils, and sweets?
    a. They are toxic to us.
    b. We take in sufficient quantities in other food groups.
    c. They will give us too much energy.
    d. We cannot break them down.

13. The process by which plants convert solar energy into glucose and oxygen is known as
    a. photosynthesis.
    b. parthenogenesis.
    c. solar metabolism.
    d. light chain reaction.

14. Lipid-soluble vitamins are vitamins
    a. A, B, C, and D.
    b. A, D, E, and K.
    c. D, E, and K.
    d. B and C.

15. When we are cold, we shiver. Why?
    a. It helps shake nutrients through our digestive system faster.
    b. We release so much adrenaline when cold that our fight-or-flight system makes us jittery.
    c. It is rapid muscle contraction that increases the need for ATP to speed up metabolism and generate heat.
    d. The cold makes us so weak that we shake while trying to remain upright.

16. In the small intestine, lipids are absorbed
    a. into the urinary system before being recycled into the bloodstream.
    b. directly into capillaries within the intestinal lining.
    c. only in the colon, where bacteria can break them into small enough droplets to be reabsorbed.
    d. into lymphatic vessels and passed through the lymphatic system into the bloodstream.

17. The average adult takes in about _____ ml of water in food a beverages per day.
    a. 1,000
    b. 350
    c. 6,000
    d. 2,500

18. During anaerobic metabolism, glucose is converted into pyruvic acid to produce 2 ATP molecules. If oxygen is still not available, muscle will temporarily convert pyruvic acid into _____ to allow the conversion to continue. Excessive levels of this product causes muscle soreness.
    a. lactic acid
    b. uric acid
    c. hydrochloric acid
    d. nicotinic acid

19. When cold, the body can increase the generation of heat through metabolism and can
    a. increase sweating.
    b. increase absorption of nutrients.
    c. decrease loss of heat to the environment.
    d. decrease heart rate so that less heat is transported through the body.

20. A graphic method for demonstrating the quantities of food groups we should take in daily is known as the
    a. Venn diagram.
    b. Food Pyramid.
    c. Nutrient Circle.
    d. Square Meal Chart.

Chapter 12

The Urinary System

LEARNING OBJECTIVES

Upon completion of this chapter, you will be able to:

1. Describe the components and functions of the urinary system and how it maintains homeostasis.
2. Trace the flow of fluids through the urinary pathway.
3. Describe the contents and characteristics of normal urine.

CHAPTER OUTLINE

Introduction
Functions of the Urinary System
Organization of the Urinary System
The Flow of Fluid through the Urinary Pathway
The Kidneys
- External Anatomy
- Internal Anatomy
- Renal Blood Supply
- Nephron
- Role of the Kidney in Maintaining Water, Electrolytes (Ions), and pH Balance
- Functions of the Nephron
- Glomerular Filtration
- Tubular Reabsorption
- Tubular Secretion
- Autoregulation
- Hormonal Regulation
- Neural Regulation

Urine
- Micturition

INTRODUCTION

The urinary system is also referred to as the renal system. The human body needs to perform the process of elimination or excretion. The urinary system primarily serves this function by eliminating waste products and producing urine. Urine production and micturition are carried out by the kidneys, urinary bladder, urethra, and ureters. The kidneys control water, electrolytes, and blood pH. The kidneys can identify which chemicals or substances dissolved in the blood are needed by the body and so they are reabsorbed back into the system. Any unwanted chemicals or substances are excreted into the urine. This process is called filtration by the kidneys, a particularly important function to keep the body's fluid in balance. The kidneys depend on the hydration of the human body in order to do this function. It is, therefore, important to keep oneself adequately hydrated by drinking water. With inadequate water or hydration, the bladder and kidneys may get damaged because of low-volume urine containing high concentration of toxins. The kidneys need diluted urine to do the function of filtration.

One of the most common problems of the urinary system is urinary tract infection (UTI). This is more common in women but can also occur in men. The kidneys, bladder, ureter, or urethra may become infected by pathogenic bacteria. The symptoms include burning sensation when urinating, frequent urge to urinate, and not being able to empty the bladder. It is therefore important to keep oneself hydrated by drinking lots of water and fluids to help eliminate or flush out the bacteria. Minimize salt and caffeine. Seek medical attention when UTI is suspected.

Urinary Tract Infection

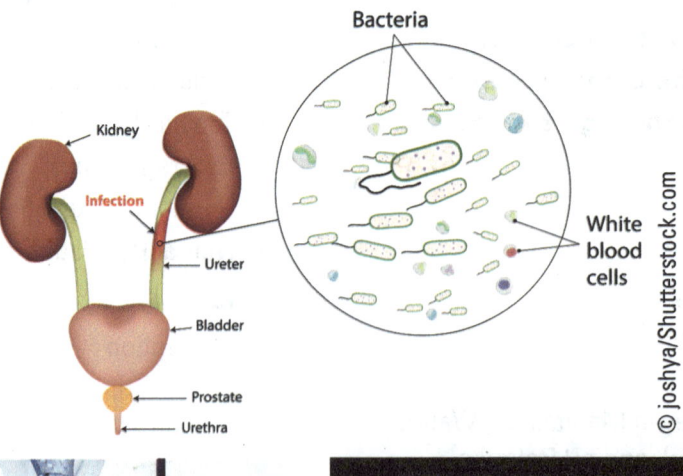

Figure 12.1

It is important that the contents of the plasma be regulated as it is pumped throughout the body. Ions, for example, need to be kept within homeostatic range for cells to function properly. Wastes from cellular activity that diffuse into the bloodstream need to be eliminated from the body. The urinary system is responsible for regulating many of the substances in the plasma within homeostatic range so that other organ systems can perform their functions.

FUNCTIONS OF THE URINARY SYSTEM

The urinary system performs many functions involved with maintaining homeostasis of the plasma. Other body systems also alter the contents of the blood and instruct the kidneys to alter plasma volume. The major functions of the urinary system are:

- Regulation of ion levels in the bloodstream
- Regulation of blood volume
- Assistance in the regulation of blood pressure
- Participation as a major component in the regulation of blood pH
- Production of a hormone that increases the production of red blood cells
- Elimination of wastes

ORGANIZATION OF THE URINARY SYSTEM

The urinary system is located in the abdominopelvic cavity (Figure 12.2). The kidneys function to keep the plasma within homeostatic range while the other organs in this system eliminate substances discarded by the kidneys. The organs of the urinary system are as follows:

- Kidneys—The kidneys are reddish brown bean-shaped organs located in the upper abdomen posterior to the peritoneum, that is, *retroperitoneal*. They are held in place by fibrous connective tissue and are accompanied by adipose tissue to provide protective padding. Superficial to each kidney are a pair of floating ribs that provide protection on the posterior side.
- Ureters—Ureters are muscular retroperitoneal tubes running down the posterior wall of the abdomen that drain urine from the kidneys to the urinary bladder by smooth muscle peristalsis. Urine can be transported to the urinary bladder regardless of body position.
- Urinary bladder—A muscular hollow organ found in the pelvic cavity, the urinary bladder is lined with transitional epithelium. It stores urine (400 to 620 ml) until it is voided from the body. For urination to occur, smooth muscle found within its walls contracts to push urine out of the body.
- Urethra—The urethra is a tube for the passage of urine from the urinary bladder to the exterior of the body. The length of the urethra in females is approximately 1 inch (2.5 to 3 cm). In males the additional length of the urethra through the penis makes the average length of the urethra about 7 or 8 inches (18 to 20 cm). It is the short distance of the female urethra as well as the urethra's close proximity to the anus that allows bacteria to migrate from the external orifice to the urinary bladder, resulting in urinary bladder infections being much more common in women than in men.

> **Kidneys** located retroperitoneally in the upper posterior abdomen, the kidney functions primarily to remove excess or unwanted substances from the plasma. The kidneys also control red blood cell production and assist in the management of blood pressure.

> **Ureters** muscular tubes that transport urine from the kidneys to the urinary bladder.

> **Urinary bladder** a muscular hollow organ for the storage of urine located in the pelvic cavity.

> **Urethra** a tube for the passage of urine from the urinary bladder to the exterior.

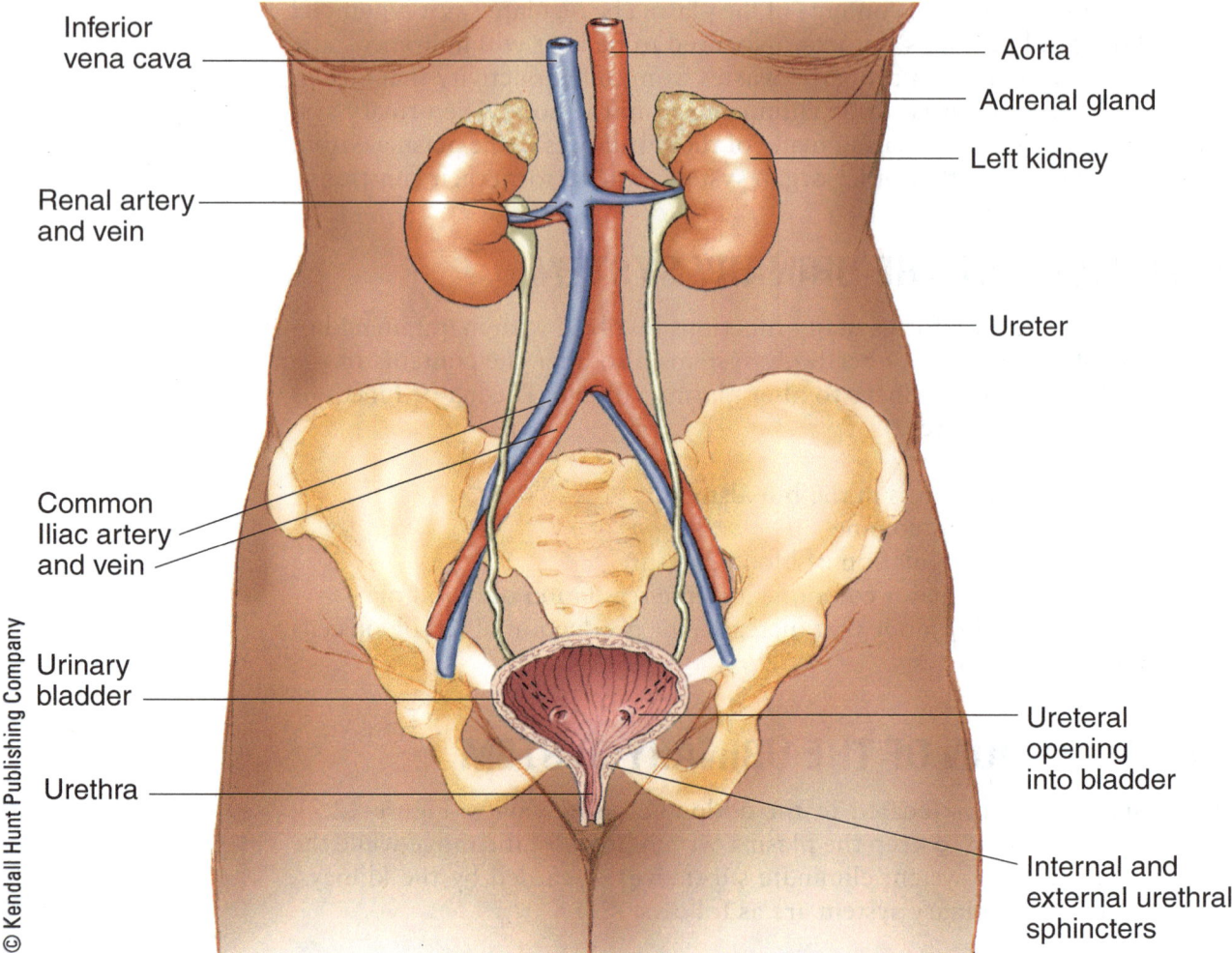

Figure 12.2 Urinary System.

THE FLOW OF FLUID THROUGH THE URINARY PATHWAY

Urine is formed in the kidneys. From there it enters the ureters, which move the urine from the upper posterior abdomen caudally through the posterior abdomen to the pelvic cavity, where it enters the urinary bladder. The urine remains in the urinary bladder until enough is present to distend the walls sufficiently to cause stretch receptors to signal the brain that it is time to drain the fluid through the urethra out of the body.

Comprehension Check-up:
1. The process which helps urine move through the ureters is _____.
2. Urine is stored in the _____ until released.

1. peristalsis   2. urinary bladder

THE KIDNEYS

Most of the urinary system's functions are performed by the kidneys. The rest of the system focuses on the elimination of urine. Our discussion begins with the gross anatomy of the kidney before describing renal histology. To understand the physiological processes of the kidneys, it is necessary to discuss renal histology.

External Anatomy

The kidneys are enclosed by a connective tissue membrane called the *renal capsule.* Connective tissue attaches each kidney firmly to the posterior abdominal wall. On the medial side of each kidney is an indentation known as the *hilus,* where the ureters and renal blood vessels attach.

Internal Anatomy

Looking at the internal anatomy of the kidney we can see that there are two distinct layers (Figure 12.3). The outer layer is the *cortex,* and the inner layer is the *medulla.* It is helpful to discuss the purpose of each layer. Blood plasma is transferred from the bloodstream into tubules, which together form a nephron, the blood-filtering urine-producing unit of the kidney where substances that need to be retained by the body can be reabsorbed back into the bloodstream. Excess or nonnutritive chemicals remain in the nephron to be emptied into the ureters and transported to the urinary bladder.

The medulla consists of two primary structures: renal pyramids and renal columns. The *renal pyramids,* as the name implies, are pyramid-shaped structures. The apex of each renal pyramid faces the hilus, with the base of

> **Nephron** the tubelike structure within the kidney that performs renal functions.

KIDNEY ANATOMY

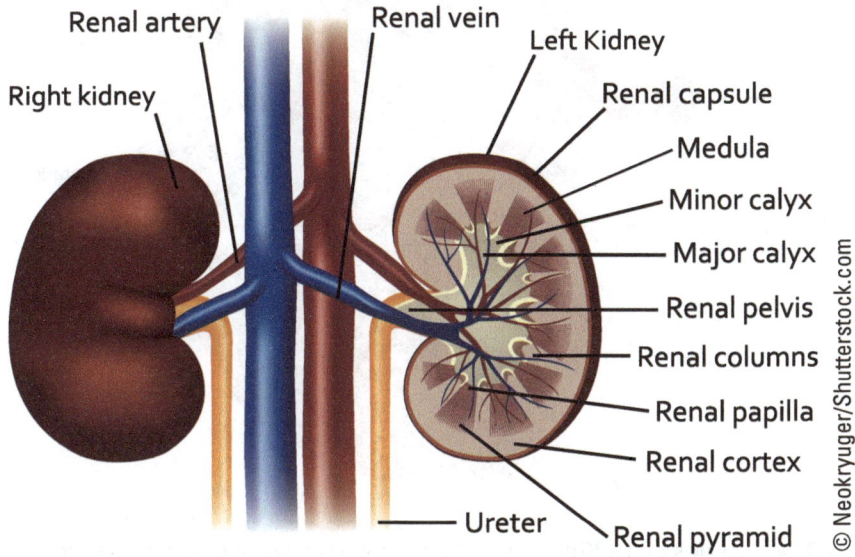

Figure 12.3

the pyramid oriented toward the cortex. The U-shaped nephron loop extends into the pyramid to allow water and salt to be reabsorbed back into the bloodstream. Also passing through the pyramids are collecting ducts that drain urine from the distal end of the nephron. The area between each pyramid forms *renal columns*. Blood vessels pass through the columns toward the cortex.

As urine is formed and passes through collecting ducts in the renal pyramids, those ducts drain through the apex of the pyramids into large drainage tubes known as *minor calyces*. Each minor calyx empties into larger *major calyces* that collect all of the urine and transport it to the ureter.

Renal Blood Supply

Blood enters the kidneys through renal arteries (Figure 12.4). Once inside the kidney, the renal artery branches through the renal columns as *interlobar arteries*. When the interlobar arteries reach the cortex, they branch over the base of the renal pyramids as *arcuate arteries*. Arcuate arteries form smaller branches, called *interlobular arteries*, which radiate through the cortex toward the surface. Off of those radiating branches are afferent arterioles that control blood flow into tufts of capillaries known as glomeruli. Imagine taking a sprinkler hose and folding it into a giant pompom. Now reduce that arrangement to microscopic size. That's the general shape of each glomerulus. The interlobular arteries are surrounded by glomeruli. Blood enters the glomerulus, where its plasma (minus substances composed of large molecules, such as proteins or lipids) filters through so that needed substances can be retained and wastes and unusable chemicals excreted.

> Glomeruli a tuft of capillaries through which plasma minus protein filters into the nephron.

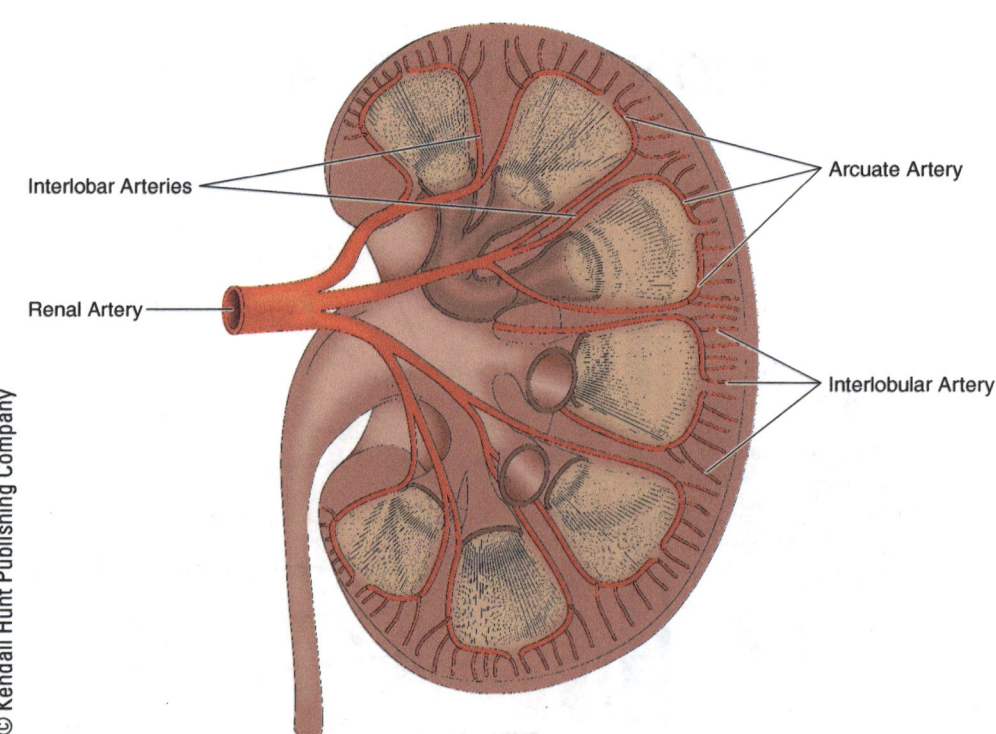

Figure 12.4 **Arteries of the Kidney.** Interlobar arteries pass between renal pyramids in the renal columns. Arcuate arteries arch over the base of each pyramid. Interlobular arteries branch off the arcuate arteries to radiate toward the surface of the kidney.

After the blood vessels form into the glomeruli, an efferent arteriole forms and the vessels continue on as capillaries wrapped around the nephron. These are known as peritubular capillaries and vasa recta. Substances to be retained from the tubular fluid are reabsorbed by these capillaries; unwanted substances are secreted from the capillaries to the tubules. Peritubular capillaries and vasa recta drain into veins taking the appropriate level of nutrients and chemicals back to the rest of the body through the renal veins.

> **Peritubular capillaries and vasa recta** capillaries and veins wrapped around the nephron that reabsorb substances transported through the wall of the nephron to be returned to the bloodstream.

> **Comprehension Check-up:**
> 1. The area on the medial side of the kidney where the ureter and renal blood vessels enter and exit is known as the _____.
> 2. Large drainage tubes within the kidneys for urine are known as _____.
>
> 1. hilus 2. calyces

Nephron

Nephrons are the functional unit of the kidney (Figure 12.5). The nephron begins with the glomerulus. The glomerulus is surrounded by a bulblike structure called the *glomerular capsule* (also known as Bowman's capsule). Recall that the glomerulus is similar to a sprinkler hose formed into a pompom. The glomerular capsule envelopes the glomerulus to capture all of the fluid filtering from it and forming the renal tubule.

The glomerular capsule forms a twisted tube first known as the *proximal convoluted tubule*. It then narrows and straightens into a U-shaped tube known as the *nephron loop* (also known as the loop of Henle). The side of the nephron loop that is closest to the proximal convoluted tubule is the *descending limb* and the other side is the *ascending limb*. At the upper end of the ascending limb the tube becomes widened and twisted again; this is called the *distal convoluted tubule*. It drains into a collecting duct, which receives urine from many nephrons surrounding it.

Peritubular capillaries, surrounding the proximal and distal convoluted tubules, and vasa recta, found with the nephron loops and collecting ducts, wrap around the tubular nephron to reabsorb substances released from the renal tubules and to secrete unwanted substances to the tubules. There is a small space surrounding each nephron known as the *peritubular space*. It becomes a reservoir for substances that are transported through the wall of the nephron to be returned to the bloodstream through the peritubular capillaries and vasa recta.

Role of the Kidney in Maintaining Water, Electrolytes (Ions), and pH Balance

The kidneys continually return most, if not all, of each nutrient in the plasma back to the bloodstream while removing the excess to be excreted from the body. They also maintain, under the direction of a hormone, aldosterone, the balance between the electrolytes sodium and potassium. There are times when additional water or sodium needs to be retained to keep the composition or volume of the plasma constant.

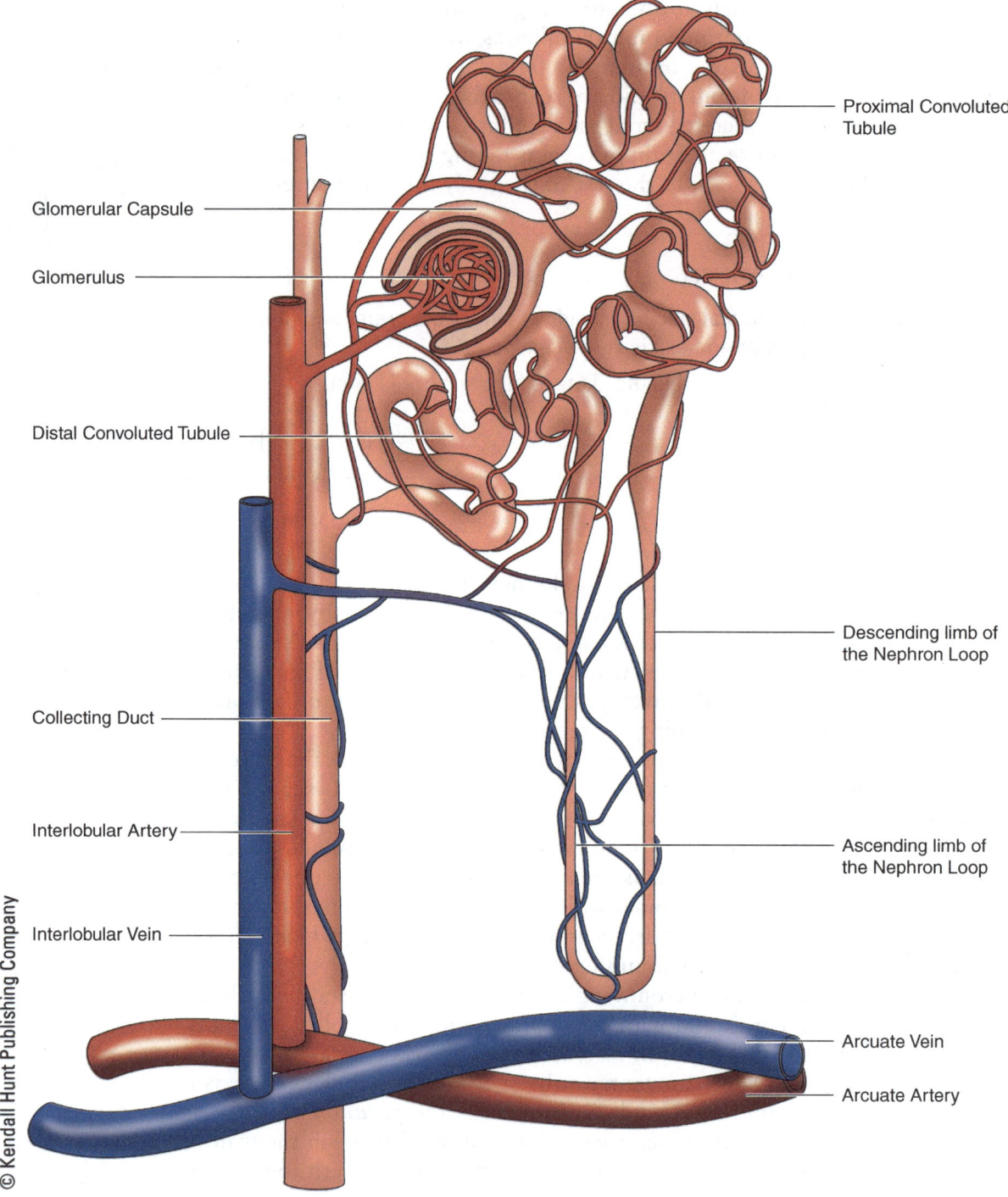

Figure 12.5 Nephron. The nephron is the function unit of the kidney.

The hypothalamus monitors the concentration of the plasma. If the plasma becomes too concentrated, cells will shrink as water is drawn out of them and put back into the bloodstream. If the plasma becomes too diluted, water will be drawn into the cells, making them swell. If the hypothalamus determines the osmotic pressure of the plasma is above homeostatic levels, it produces antidiuretic hormone (ADH) to cause the kidneys to retain additional water to dilute the plasma. If the osmotic pressure it too low, ADH is not produced, so excess water is not retained in the plasma and it passes out of the body as urine.

The adrenal cortex produces aldosterone to cause additional sodium ions to be retained by the kidneys. Retaining additional sodium also causes the kidneys to keep additional water. The overall response to aldosterone is to increase plasma volume. The increase in plasma volume in the cardiovascular system results in an increase in blood pressure as additional fluid is pushed through blood vessels.

There are numerous potential causes for the pH of our blood to be altered, yet, in most cases, we are able to compensate for these changes. Some causes include the beverages we drink, acids resulting from metabolism, breathing rate influencing our production of carbonic acid from carbon dioxide we retain, and digestive problems such as vomiting, which is loss of stomach acid, or diarrhea, resulting in a loss of base. The kidneys provide a major source of pH control by excreting additional acid or by actively retaining or excreting base to neutralize acid or base imbalance. The maintenance of acid-base balance is important for allowing chemical reactions to occur in the body; some enzymes function only within a narrow pH range.

> **Comprehension Check-up:**
> 1. Surrounding the nephron are blood vessels known as _____ and _____.
> 2. _____ is the hormone that causes additional sodium retention by the kidneys.
>
> 1. peritubular capillaries, vasa recta 2. Aldosterone

Clinical Notes—Kidney Stones

Kidney stones can form in the kidneys as a result of low-volume urine. When a person is dehydrated, the salts in the urine are not diluted by water causing kidney stones to form. There are risk factors that can cause kidney stones to form, and these include obesity, type 2 diabetes, family history of kidney stones, gout, Crohn's disease, eating foods rich in animal proteins, and taking calcium supplement. It is common in males. Kidney stones come in four types: calcium, uric acid, cysteine, and struvite stones. The most common type of kidney stones are calcium oxalate and calcium phosphate, which account for 75%–80% of the kidney stones.

A sharp or stabbing pain occurs in the lower back when a kidney stone leaves the kidney and get stuck in the ureter. The urine flows back to the affected kidney instead of moving into the bladder. Other symptoms include pain during urination, presence of blood in the urine, and foul-smelling urine. The presence of fever along with the kidney stones is a sign of infection and needs medical attention immediately. Sometimes small-size kidney stones can pass along during urination.

(continued)

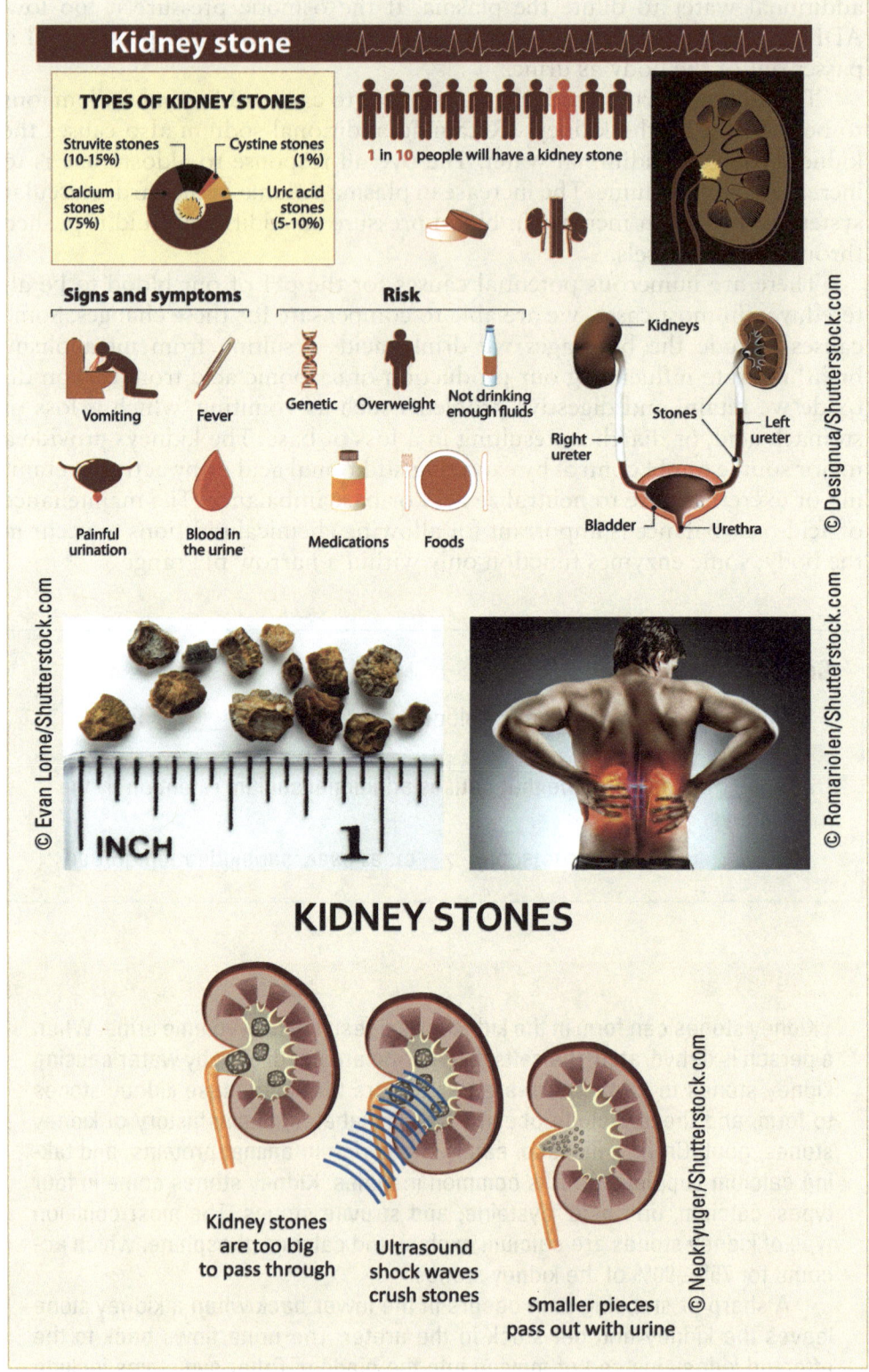

Functions of the Nephron

The primary function of the nephron is to maintain the concentration of various substances in the plasma within homeostatic levels and to remove wastes. To do so, plasma minus large molecular substances filters from the glomerulus into the renal tubules, where chemicals needed by the body that are in the filtrate are returned to the bloodstream. The excess continues through the renal tubules, where additional water and ions, primarily sodium, can be reabsorbed if needed. The remaining unnecessary substances pass out of the kidney as urine. Three general processes occur in the nephron:

- Glomerular filtration
- Tubular reabsorption
- Tubular secretion

We will discuss each in detail.

Glomerular Filtration

The glomerulus contains many holes, called *fenestrations*, across the simple squamous cells in this specialized tuft of capillaries (Figure 12.6). As blood

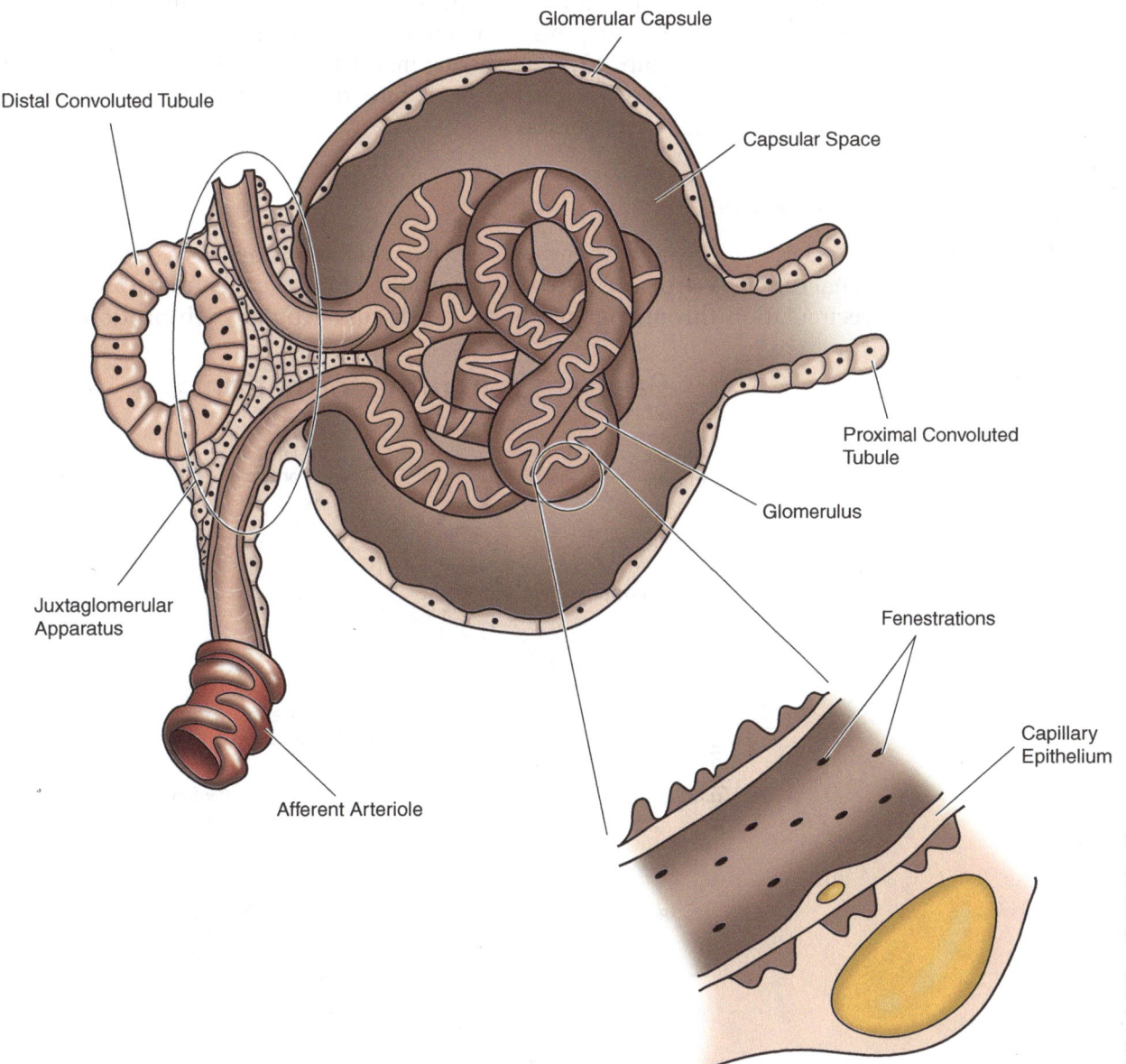

Figure 12.6 Glomerulus. The glomerulus is a tuft of capillaries surrounded by a glomerular capsule, which catches fluid filtering out the fenestrations in the glomerulus. The blood pressure inside the glomerulus is controlled by the afferent arteriole. The distal end of the nephron comes in contact with the glomerulus to assist in the control of the glomerular filtration rate.

passes through the glomerulus, the liquid of the plasma passes through the fenestrations. Proteins and other large molecular substances are prevented from passing through the fenestrations because they are too large. The result is that plasma minus protein filters out of the glomerulus into the glomerular capsule.

Going back to the analogy of the sprinkler hose, the higher the water pressure, the faster the water squirts through its holes. The same process is also true for the glomerulus. The blood pressure inside the glomerulus affects how fast fluid can filter through the fenestrations. Arterioles can automatically control the pressure in each glomerulus much like a faucet controls the pressure of water inside a sprinkler hose. Arterial blood pressure can affect glomerular filtration rate (GFR), the speed at which filtrate passes out of the glomerulus. The higher the arterial pressure, the greater the filtration rate. GFR is often used as a measure of kidney function and is normally about 125 ml per minute for both kidneys. That results in the production of 7.5 liters per hour or 180 liters of filtrate per day. Considering that 1 liter is roughly equal to 1 quart, this means that we produce roughly 45 gallons of filtrate per day (Figure 12.7), yet a vast majority of the filtrate is reabsorbed. The excess that does not diffuse back into the bloodstream and becomes urine is about 1.2 liters per day.

> **Glomerular filtration rate (GFR)** the rate at which the filtrate from the glomerulus passes through the nephron.

> **Tubular reabsorption** the process by which substances in the nephron are transferred back into the bloodstream and reabsorbed for recirculation throughout the body.

Tubular Reabsorption

Tubular reabsorption is the process by which substances in the tube—the renal tubules—are transferred back into the bloodstream and reabsorbed for recirculation throughout the body. To discuss tubular reabsorption, it will

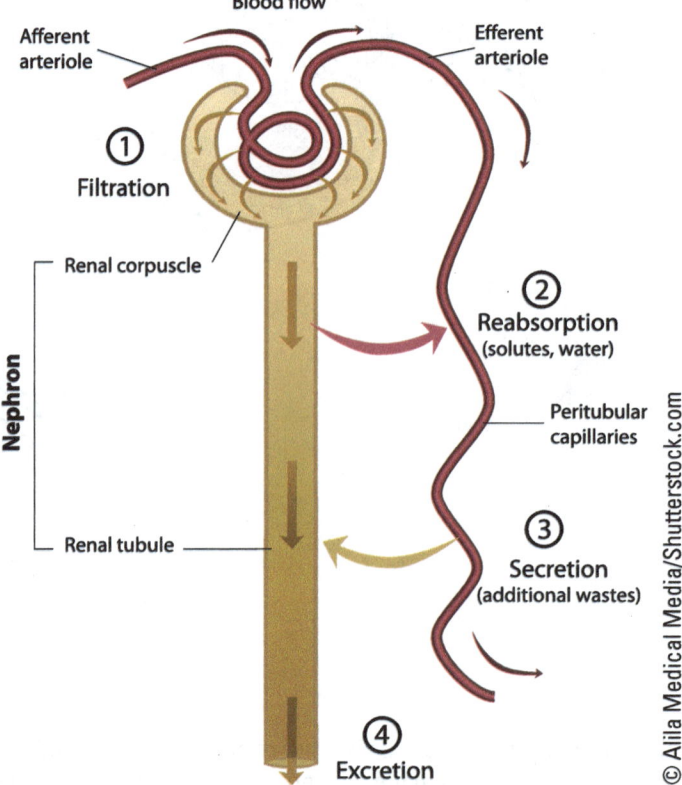

Figure 12.7

be easier if we simplify the nephron into a U-shaped tube. At the top of the U on one side is the glomerulus. Filtrate exits the glomerulus and descends through one side of the U (glomerular capsule, proximal convoluted tubule, and descending limb of the nephron loop), referred to as the *descending side*. Filtrate travels up the ascending side (ascending limb of the nephron loop and distal convoluted tubule), where it empties into the collecting duct at the top of the other side of the U.

Once the filtrate has been collected by the glomerular capsule, it passes into the descending side of the renal tubule. As the filtrate descends, cells forming the walls of the proximal convoluted tubule pump substances in the tubular fluid that are needed by the body into the peritubular space. If there are more molecules than can be transported through the renal tubules, the excess remains inside.

As the concentration of various substances becomes greater in the peritubular space than inside the renal tubule, water is drawn out of the renal tubule by osmosis. By the time the tubular fluid reaches the distal end of the U, virtually all that remains inside the renal tubule are excess or nonnutritive substances, such as food additives or drugs and metabolic wastes, for example: urea, ammonia, and bilirubin.

Recall that capillaries, after forming the glomerulus, wrap around the renal tubule as peritubular capillaries and vasa recta. As substances accumulate in the peritubular space, their concentration becomes greater than in the peritubular capillaries. Remember that a substantial portion of the plasma and its chemicals have passed into the renal tubule, so the level in the peritubular capillaries and vasa recta is reduced. The substances in the peritubular space diffuse from areas of higher concentration to areas of lower concentration into the peritubular capillaries and are reabsorbed back into the bloodstream.

The ascending limb of the nephron loop and the distal convoluted tubule are impermeable unless hormones cause substances to be moved through its walls. As the filtrate ascends, if additional sodium is needed to be reabsorbed, the hormone aldosterone, produced by the adrenal gland, will cause some of the sodium in the excess to be pumped into the peritubular space. If additional water is needed by the body because the plasma has become too concentrated, ADH is released from the posterior pituitary gland to allow water to be reabsorbed. Tubular fluid still in the nephron when it reaches the collecting duct becomes urine. As the urine passes through the collecting duct, if ADH is released, additional water is drawn out until it reaches the apex of the pyramid and empties into a calyx.

Tubular Secretion

Some substances are unable to filter through the glomerulus but are not wanted by the body. Examples of those substances are foreign compounds such as environmental pollutants (such as pesticides) and also many drugs (such as penicillin or nonsteroidal anti-inflammatory drugs). It is still useful to remove these substances from the bloodstream. There is an alternative: as blood passes through the peritubular capillaries, those undesired substances can be moved out of the capillaries and into the proximal and distal convoluted tubules so that they can be secreted into the developing urine. There is no attempt to control the level, only to remove it from the body. This process of moving previously unfiltered unwanted substances into the tubular section of the nephron for excretion in the urine is defined as **tubular secretion**.

> **Tubular secretion** the process of moving previously unfiltered, unwanted substances into the distal end of the nephron for excretion in the urine.

> **Comprehension Check-up:**
>
> 1. The fluid filtering through the wall of the glomerulus into the glomerular capsule is composed of _____.
> 2. Tubular reabsorption refers to _____.
> 3. Tubular secretion refers to _____.
>
> 1. plasma minus proteins
> 2. movement of a substance from the renal tubule into the blood stream
> 3. movement of a substance from the bloodstream into the renal tubule
> Making Artificial Kidneys with Cellophane

Autoregulation

The nephron can alter the blood pressure and flow into the glomerulus by a process known as *autoregulation*. As an analogy, say I am using a garden hose. The best way to determine whether the water flow is adequate is to look at the nozzle end. If the flow of water out the end of the hose is sufficient, then the whole water system must be working properly. If the output is not what I expected, and assuming there are no kinks or blockages in my hose, I have two places to look to determine the cause of inadequate water flow. I can open the faucet more to increase flow, and if that still doesn't work, then perhaps there is a problem with the water system coming from the water company or, if I have a well, the pump.

The nephron works in a similar manner. The distal end of the renal tubule passes next to the glomerulus to form the *juxtaglomerular apparatus* (*juxta* meaning "near" or "next to") that senses both the blood pressure going into the glomerulus (similar to detecting the pressure inside the garden hose) and the flow rate of tubular fluid through the renal tubule. If the nephron flow is too low, the juxtaglomerular apparatus will cause the arterioles controlling blood flow through the glomerulus to increase blood pressure, in the same way opening the faucet increases water pressure inside the garden hose. If that restores the desired filtration rate and flow through the nephron, no further action is needed. If, however, that process does not result in an improvement of flow, then the kidneys indicate that the blood pressure may be too low by producing the enzyme renin, which converts angiotensinogen, an inactive chemical in the plasma, to its active form, angiotensin I. Angiotensin I in the bloodstream acts as a signal that the blood pressure may be too low. If the blood pressure is in fact too low, the lungs produce angiotensin-converting enzyme (ACE), which facilitates the conversion of angiotensin I to angiotensin II and results in the constriction of blood vessels and the release of aldosterone, raising the blood pressure. If the blood pressure is within normal homeostatic range, then ACE is not released, so the sequence causing in the release of aldosterone is not activated and blood pressure is not increased. This second process is discussed more in the next section.

```
                    Renin              ACE
                      |                 |
Angiotensinogen → Angiotensin I → Angiotensin II → Aldosterone
                   → Increased blood pressure
```

Renin a hormone produced by the kidneys to determine whether the blood pressure is too low.

Angiotensin-converting enzyme (ACE) after renin has been released by the kidneys to determine whether the blood pressure is too low, the lungs respond, if the blood pressure is in fact low, by producing this enzyme, which results in the secretion of aldosterone to increase in blood volume, which increases blood pressure.

Hormonal Regulation

If an individual sweats from activity, eats very salty food, or has a case of diarrhea, there are some significant changes in the sodium chloride or water content in the plasma that require regulation. Recall that the ascending renal tubule will allow substances to pass through its wall only when specific hormones are present. The two hormones that affect the ascending side of the renal tubule are aldosterone and antidiuretic hormone.

Aldosterone is produced by the adrenal cortex and causes additional sodium ions to be pumped out of the excess in the ascending renal tubule into the peritubular space, to retain more sodium. As the extra sodium accumulates in the peritubular space, it increases the concentration of the fluid in that area. The increased sodium concentration draws additional water out of the descending renal tubule into the peritubular space by osmosis. The end result is that retaining additional sodium includes the retention of additional water as well. As this additional sodium and water are reabsorbed into the bloodstream, they increase the volume of plasma in the cardiovascular system. The additional plasma volume entering the bloodstream results in an increase in blood pressure. The previous section discussed the process by which the kidneys produce renin if the blood pressure is too low. If the blood pressure is low, the secretion of ACE results in the constriction of blood vessels to increase blood pressure and causes the adrenal gland to secrete aldosterone into the bloodstream to increase plasma volume. Because more fluid is in the cardiovascular system, each heart contraction results in higher blood pressure.

Antidiuretic hormone, secreted by the posterior pituitary gland, causes the retention of additional water from the distal convoluted tubules and collecting ducts. Sodium is not included in this process, so the result is to dilute the plasma when the body has started to dehydrate and the plasma begins to become too concentrated.

The kidneys also monitor the oxygen content of the blood. If the ability to transport oxygen decreases, the hormone erythropoietin is produced by the juxtaglomerular apparatus to stimulate bone marrow to make additional red blood cells to carry available oxygen.

Neural Regulation

The kidneys receive about 22% of the blood pumped out of the heart; therefore, there is a substantial quantity of blood passing through the kidneys at any given time. If we have a stressor and the sympathetic nervous system causes us to go into "fight-or-flight" mode, it is essential that we provide our skeletal muscles with the maximum amount of blood flow. Neurons from the sympathetic nervous system innervate the kidneys to decrease renal blood flow during critical situations in order to divert as much blood as possible to muscle.

> **Aldosterone** a hormone produced by the adrenal cortex that causes the kidneys to retain additional sodium and water, resulting in an increase in plasma volume.

> **Antidiuretic hormone** a hormone produced by the posterior pituitary gland that causes the kidneys to retain additional water to dilute plasma that has become too concentrated.

> **Erythropoietin** a hormone produced by the kidneys in response to decreased oxygen-carrying capacity of the blood. It causes an increase in red blood cell production.

Comprehension Check-up:

1. The _____ is a sensor that determines the flow rate of tubular fluid through the nephron tubule.
2. _____ is the enzyme produced by the kidneys that may result in the increase of blood pressure.

1. juxtaglomerular apparatus 2. Renin

Drugs and Their Detectable Trail

We take drugs to temporarily remove a symptom or to achieve some desired effect. The drug level in the bloodstream must be high enough for the results to be noticeable. Often, once a drug is absorbed into the bloodstream, it is gradually removed by tubular secretion. When the plasma level decreases below a specific point, the therapeutic effect is lost. Typically it is roughly 4 hours. For this reason, many drugs are prescribed to be taken every 4 hours. However, just because the therapeutic effect is gone, this does not mean the drug is completely removed from our plasma. In fact, the drug may be secreted over the next several weeks. This allows urine drug screening to be an effective source to determine an individual's drug use.

URINE

Urine is more than just a yellow-colored liquid. It contains ions such as sodium, chlorine, and potassium, as well as suspended solids, known as *sediments*, such as cells, mineral crystals, mucus threads, and even sometimes bacteria. The pH of urine is normally between 4.6 and 8.0. Examining the contents of the urine can provide clues about internal body processes by quantifying the by-products excreted. As a result, urinalysis is a diagnostic tool used to identify abnormal processes occurring within the body. Deviations from normal ranges provide clues about metabolic problems in the body without requiring any invasive techniques (Table 12.1).

Because urine is excess, its contents are influenced by the foods and beverages we ingest. We may lose fluid elsewhere, such as through sweating or diarrhea, which causes the urine to become more concentrated. Acids produced through metabolism and excreted can change the pH of our urine. Even changes in breathing rate can result in a change in urine pH as excess acids or bases are excreted to maintain plasma pH.

> **Urinalysis** analysis of urine contents as a diagnostic tool for determining abnormal processes occurring within the body by using deviations from normal ranges as clues to metabolic problems.

Micturition

Urination is technically known as micturition. Once the volume of urine in the urinary bladder exceeds 200 ml, on average, stretch receptors in its walls send impulses to the brain, indicating the need to eliminate its contents. When the decision is made to urinate, nerves from the parasympathetic nervous system stimulate smooth muscle in the urinary bladder to contract. Two sphincters—*internal urethral sphincter* (a smooth muscle), and *external*

> **Micturition** urination.

Table 12.1 Abnormal Constituents in Urinalysis

Substance Found	Potential Cause for Its Presence
Glucose	Pathological: diabetes mellitus Nonpathological: excessive sugar intake
Blood	Pathological: bleeding in the urinary tract as a result of bacterial infection or kidney stone; chronic sympathetic stimulation
Protein	Pathological: kidney disease—glomerulonephritis; hypertension Nonpathological: excessive exercise, pregnancy
Pus	Pathological: urinary tract infection
Bilirubin	Pathological: if excessive, liver malfunction
Ketones	Nonpathological: excessive breakdown of lipids

urethral sphincter (a skeletal muscle)—must both relax for urine to exit the urinary bladder. As those sphincters relax, smooth muscle in the wall of the urinary bladder constricts in what is known as the **micturition reflex** to force urine into the urethra and out of the body.

> **Micturition reflex** as sphincters which normally hold urine in the urinary bladder relax, smooth muscle in the bladder constricts to force urine into the urethra and out of the body.

Comprehension Check-up:

1. The technical term for urination is _____.
2. Suspended solids found in the urine are called _____.

1. micturition 2. sediments

Clinical Skills—Urinalysis by Dipstick

Doing urine analysis or urinalysis (UA) by using Dipstick is a simple procedure to check for kidney and urinary bladder problems, and also for presence of glucose or sugar in the urine. Dipsticks have several colored chemical squares that correspond to urine pH, specific gravity, leukocytes or WBC, blood, protein, glucose, nitrites, ketones, urobilinogen, and bilirubin. The color of the urine must be observed also.

1. Wash your hands, dry them, and wear gloves. Transfer the urine sample from the container into a test tube. Note down urine color.
2. Dip the test strip into the test tube with the urine sample. Make sure that all the test squares are completely submerged in the urine.
3. Immediately remove the test strip. Remove excess urine sample by tapping the ends of the strip into a paper towel or absorbent paper. Do not touch the test squares.
4. Align the test strip with the test squares on the label of the Dipstick container.
5. Match the colors of the test squares of the strip with the test squares on the Dipstick container.
6. Record the results.

Clinical Notes—Dialysis

When the kidneys are failing, dialysis can substitute for primary function of the kidneys in filtering waste products from the blood. There are two common lab tests that indicate when the kidneys are unable to filter the waste products from the blood: creatinine and blood urea nitrogen (BUN). Diabetes and hypertension can cause kidney failure, which requires dialysis. There are two types of dialysis: hemodialysis and peritoneal dialysis.

Hemodialysis is a machine with a special filter called dialysis membrane, where the blood of the patient is passed through to filter waste products, and return back to the patient through a special vascular tube inserted between an artery and a vein in the arm or leg.

Peritoneal dialysis uses a fluid that is placed into the patient's abdominal cavity through a plastic peritoneal dialysis catheter to remove excess waste products and fluid from the body. The patient's own tissues inside the abdominal cavity is used as a filter.

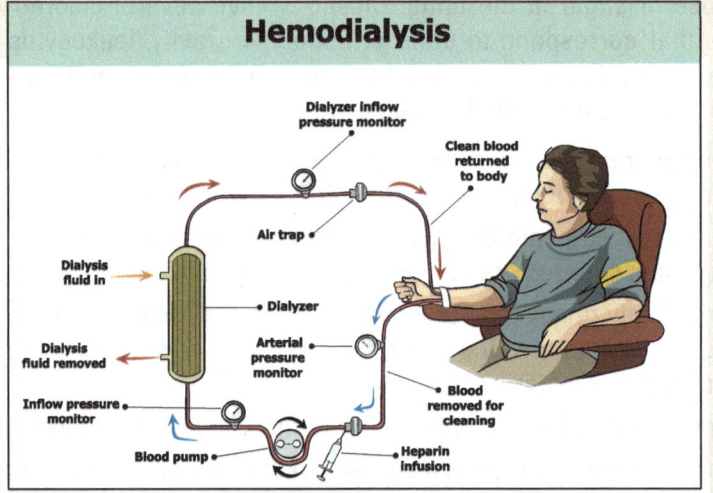

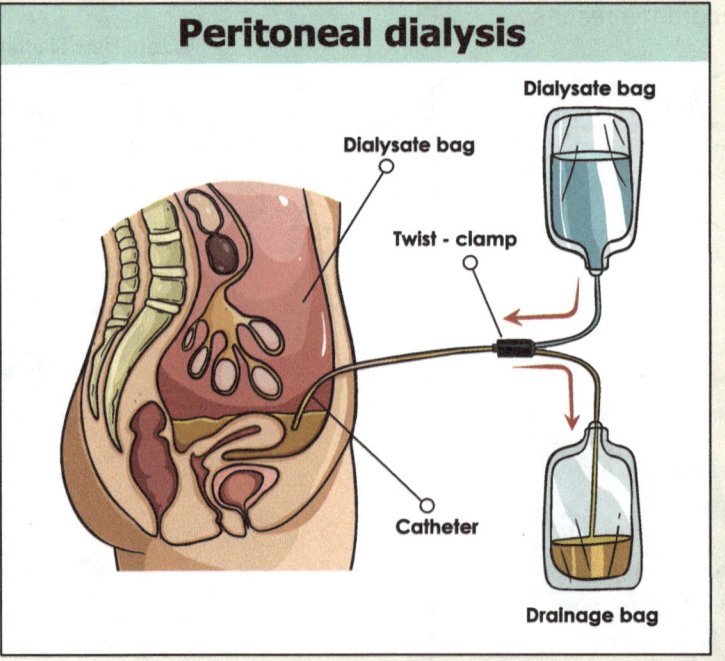

(continued)

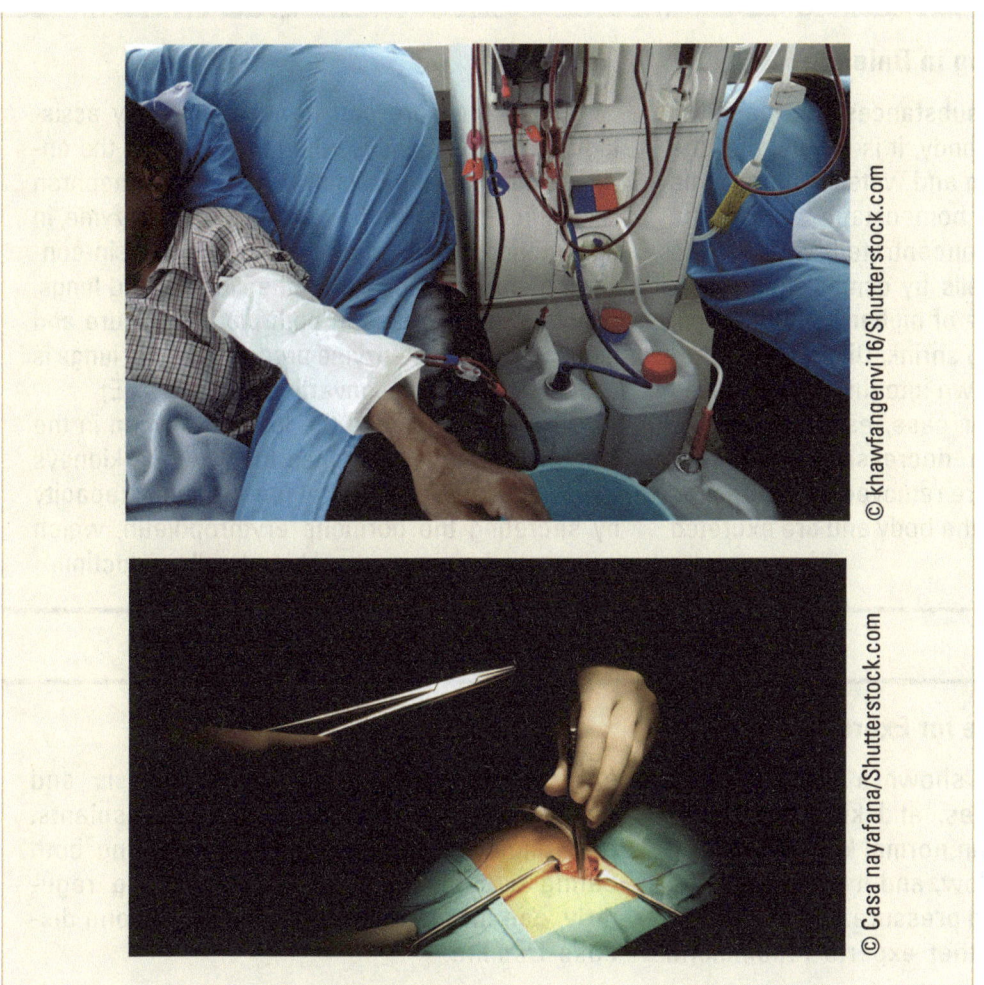

Diuretics Used for Hypertension and Congestive Heart Failure

One of the common methods used to treat congestive heart failure and some forms of hypertension is to decrease plasma volume through diuretics use. This group of drugs is known as *thiazide diuretics*. They inhibit the reabsorption of sodium from the renal tubule, causing additional water to also become urine. These drugs also result in an increased excretion of potassium. Compared with sodium, the homeostatic range of potassium is quite narrow. It is important for an individual taking thiazide diuretics (Lasix, for example) to take in additional potassium, either by eating high-potassium foods such as fruits and vegetables or by taking potassium supplements. It is highly unlikely that an individual could consume enough fruits and vegetables to overdose on potassium; however, ingestion of too many potassium supplements can cause serious or even fatal side effects.

Homeostasis—Holding in Balance

Because blood carries substances that reach virtually every cell in our body, it is critical that the concentration of sodium and water in the plasma be kept within a narrow homeostatic range. If the plasma becomes too concentrated, water could be drawn out of the cells by osmosis, as water moves to dilute the area of higher concentration, thereby causing them to shrink. Dilute plasma results in water being drawn into the cells, causing them to swell. In either case, especially in the brain, cellular function decreases. Other substances in the plasma are removed if in excess or if simply not needed by the body and are excreted in the urine.

Blood pressure can be maintained by assistance from the kidneys when they secrete the enzyme renin. If the flow of filtrate through the nephron is insufficient, the secretion of this renal enzyme, in conjunction with another enzyme, angiotensin-converting enzyme (ACE), from the kidneys and lungs, results in an increase in both blood pressure and blood volume. The enzyme produced by the lungs is called "angiotensin converting enzyme (ACE)".

The kidneys sense the level of oxygen in the blood. If the level becomes too low, the kidneys cause an increase in oxygen-carrying capacity by secreting the hormone erythropoietin, which causes an increase in red blood cell production.

There Is No Substitute for Exercise

Medical studies have shown a close link between obesity, diabetes, and kidney disease. Exercise helps maintain normal kidney function by increasing blood flow, and it decreases the incidence of high blood pressure. It is so important to overall health that experts recommend exercise for patients receiving dialysis and for those who have had kidney transplants. Maintaining a healthy lifestyle, including both eating a balanced diet and exercising regularly, can increase the possibility of a long disease-free life.

Clinical Scenario—Renal Failure

Mrs. Johnson is a 63-year-old female admitted to Green Meadow Hospital. Her problems were obvious. She was in renal failure (with no urine output for 12 hours) and had a left ventricular heart failure. She has a history of gastroesophageal reflux disease (GERD) and hyperlipidemia. Mrs. Johnson arrives at the emergency department via EMS complaining of a productive cough. At first, Mrs. Johnson thought her cough was a result of the GERD. But with other signs and symptoms, her daughters convinced her she needed to be seen in the local ED.

They went first to the local urgent care. After initial evaluation, it was obvious to the caregivers her problems and requirements were far more advanced than they could provide. The physician called the ED at Green Meadows Hospital to advise them of her imminent arrival and give the MD some background information. Mrs. Johnson was only slightly short of breath. She was complaining of an occasional flutter in her chest. A diagnostic ECG was requested. Once Mrs. Johnson was admitted to the ED, an IV was started. Prior to starting the infusion, blood was drawn for lab tests. These lab tests included a basic metabolic panel (BMP), electrolytes, and serial troponin levels.

(continued)

Her vital signs on presentation to the ED were:

BP	183/93
Temp	37.0
SpO_2	88%
HR	111
RR	20

Her room-air blood gas values were certainly abnormal, pH 7.28, $PaCO_2$ 31, HCO_3 16, and her PaO_2 was 60. The patient was placed on low-flow nasal oxygen. Her BUN, creatinine, potassium, and phosphate levels were all markedly elevated.

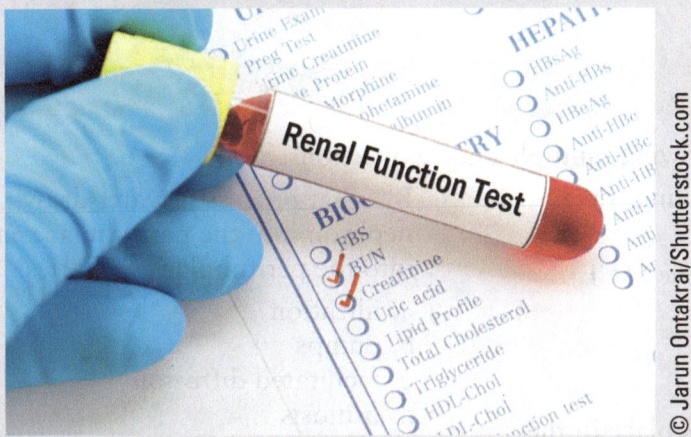

Chest pain is certainly one of the highest-risk chief complaints you will see in the Emergency Department setting. Because of this, the ED crew must be ready at any moment to move to a Code Blue situation if the patient would go into cardiac arrest. A chest x-ray is requested to evaluate for pneumonia, pneumothorax, heart size, esophageal rupture, and aortic dissection.

Mrs. Johnson was admitted to the medical floor. Her cardiologist and primary care physician (PCP) were contacted. The hospitalist assumed her in-hospital care. The hospitalist and her cardiologist started her on an aggressive cardiac regimen. Her renal failure, shortness of breath, and productive cough continued until day 3. At this point, her heart function began to return to her normal. As her heart function improved and quality blood flow was returned to her kidneys, her renal failure began to subside. She then started to produce urine. At this point, her productive cough resolved and she just felt better. Her heart flutter was gone. Discharge was planned for the following morning. Her BP was now her "normal," HR was 74, and her RR was 12 and normal. Her chest x-ray (CXR) had cleared, and she showed no evidence of heart damage. Her room-air ABGs were now normal. Mrs. Johnson was discharged home that morning. Her "at-home" care was considered minimal and that minimal care would be provided by her family.

Questions:
1. How does blood circulation affect the functions of both lungs and kidneys?
2. How could the patient's heart failure cause renal disease?
3. What is the significance of creatinine, BUN, electrolytes, and phosphate?

Contributed by Max Eskelson. © Kendall Hunt Publishing Company

CLINICAL TERMS TO REVIEW

Blood pH, filtration, urinary tract infection (UTI)
Kidney stones, urinalysis, dipstick,
Dialysis—hemodialysis, peritoneal dialysis
Kidneys, ureters, urinary bladder, urethra
Renal pyramid, glomeruli
Proximal convoluted tubule, distal convoluted tubule, Bowman's capsule, and ascending and descending limb
Antidiuretic hormone (ADH), aldosterone
Glomerular filtration, glomerular filtration rate (GFR), tubular reabsorption, tubular secretion
Urine, micturition

Test Yourself

Choose the best answer to the following multiple choice questions:

1. The chemical produced by the lungs that increases blood pressure is
   a. angiotensin-converting enzyme (ACE).
   b. renin.
   c. erythropoietin.
   d. antidiuretic hormone.

2. Reabsorption of nutrients takes place in the renal tubule in the
   a. glomerulus.
   b. distal convoluted tubule.
   c. collecting tubule.
   d. proximal convoluted tubule.

3. The speed at which fluid filters out of the glomerulus into the glomerular capsule is known as the
   a. nephron loading complex.
   b. glomerular filtration rate.
   c. glomerular capsule conductance factor.
   d. peritubular reabsorption rate.

4. All of the following are functions of the urinary system EXCEPT
   a. regulation of ions in the plasma.
   b. regulation of the production of red blood cells.
   c. activation of white blood cells.
   d. removal of wastes from the body.

5. Arteries in the renal columns are known as
   a. arcuate arteries.
   b. interlobular arteries.
   c. peritubular arteries.
   d. interlobar arteries.

6. By what process is water drawn out of the nephron into the peritubular space and back into the peritubular capillaries?
   a. filtration
   b. pumps
   c. facilitated diffusion
   d. osmosis

7. The kidneys are found in the
   a. upper posterior abdomen.
   b. pelvic cavity.
   c. inferior posterior thoracic cavity.
   d. anterior abdomen lateral to the navel.

8. How can the kidneys affect pH?
   a. by excretion or retention of carbon dioxide
   b. by excretion or retention of base and excretion of acid
   c. by the production of neutralizing buffers when told to do so by hormones from the posterior pituitary gland
   d. by parietal cells producing hydrochloric acid

9. Additional sodium or water can be reabsorbed from the only if hormones are present.
   a. glomerulus
   b. proximal convoluted tubule
   c. distal convoluted tubule
   d. minor calyces

10. The ability of the nephron to control the blood flow into its glomerulus is known as
    a. autoregulation.
    b. regurgitation.
    c. positive feedback.
    d. homeostatic compensation.

The Reproductive Systems

Chapter 13

LEARNING OBJECTIVES

Upon completion of this chapter, you will be able to:

1. Describe the components and functions of the male and female reproductive systems and how they propagate the human species.
2. Explain the female menstrual and reproductive cycles.
3. Describe male and female sexual function.
4. Describe the response of the reproductive systems to conception and to aging.

CHAPTER OUTLINE

Introduction
Functions of the Reproductive Systems
Structures of the Reproductive Systems
Undifferentiated Gonads
Male Reproductive System
- Anatomy of the Male Reproductive System
- Testes
- Epididymis
- Ductus Deferens
- Urethra
- Accessory Glands
- Sperm
- External Genitalia—The Penis
- Male Sexual Function

Female Reproductive System
- Anatomy of the Female Reproductive System

Female Sexual Function
- Physiology of the Follicle
- The Menstrual Cycle

Response of the Reproductive Systems to Conception and to Aging
- Pregnancy

Puberty
Menopause

INTRODUCTION

This is the last chapter in the anatomy and physiology class. There is a lot of material written on the reproductive system. The importance of learning the male and female reproductive system is not just on the anatomical and physiological aspects but should also include reproductive health. Reproductive health is about the total well-being or holistic approach to the proper functioning of the reproductive organs. Physical and mental well-being are essential to reproductive health.

Reproductive health is concerned with safe sexual practices, safety from sexually transmitted diseases (STD), early pregnancy, protection of the mother and the developing fetus, protection of the baby, fertility problems, and reproductive diseases.

STD rates continue to increase in the U.S. Amy Norton from *HealthDay Reporter*, October 2019, stated, "A new government report finds that combined cases of syphilis, gonorrhea and chlamydia reached an all-time high in 2018. Nearly 2.5 million cases of these sexually transmitted diseases (STDs) were reported to the U.S. Centers for Disease Control and Prevention for the year."

In order for the human race to survive, humans must be able to reproduce. Male and female reproductive systems complement each other anatomically, and each gender provides half the appropriate set of genes to produce viable offspring. The reproductive systems are also responsible for the physical and mental attributes of each sex (Figure 13.3).

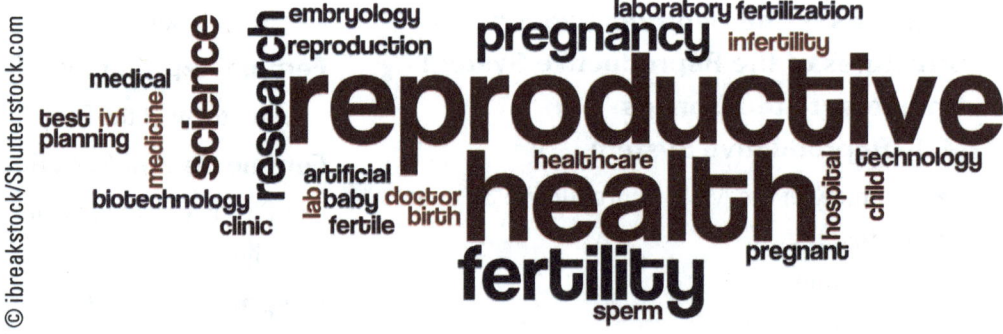

Figure 13.1

Figure 13.2

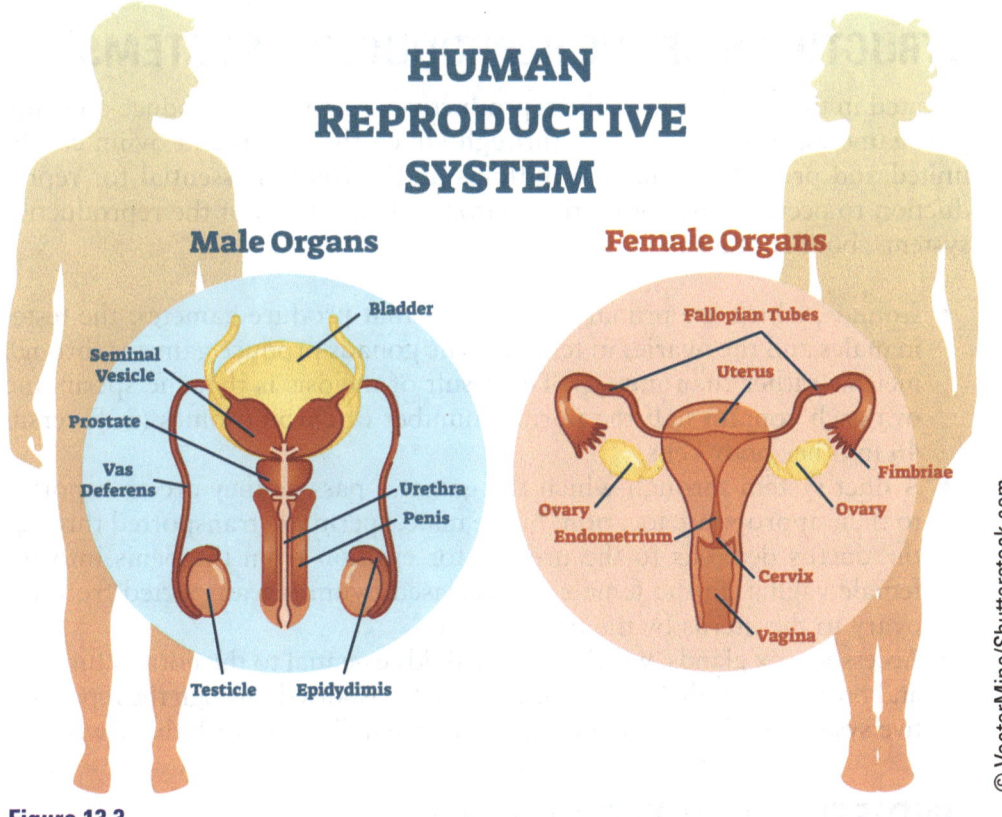

Figure 13.3

FUNCTIONS OF THE REPRODUCTIVE SYSTEMS

Reproduction requires the combination of genes from both parents to form offspring that have attributes from each. The reproductive system must provide a process of cell division by which half of the parent's chromosomes remain in the reproductive cells. When male and female reproductive cells are united, the result is a cell containing a complete set of chromosomes. The reproductive systems also support reproductive cells prior to conception and, in the case of the female, care for the fetus throughout gestation. During

Embryo the unborn human young for the first two months after conception.

Fetus the unborn human young from the third month after conception until birth.

Sperm cells produced by the male that possess the capacity to deliver 23 chromosomes from the male to the awaiting ovum.

Ova cells produced by the female that contain 23 chromosomes to provide the mother's genetic contribution to the offspring.

Gametes the collective term referring to sperm in the male or ova in the female.

Fertilization the process of one sperm uniting with one ovum to form a complete set of chromosomes from which the offspring develops.

Gonads the collective term for testes and ovaries, which produce sperm and ova, respectively, through the process of meiosis.

gestation, the unborn young is considered an embryo for the first 2 months after conception and a fetus from the third month until birth. Hormones produced by each reproductive system stimulate physiological changes that optimize the reproductive functions of the male and female. The functions of the reproductive system are as follows:

- To produce sperm in the male and ova (eggs) in the female. Collectively the sperm and ova are referred to as gametes. Included in this function is the ability to care for the gametes prior to their availability for fertilization.
- To provide a mechanism by which fertilization of the ovum (singular for *ova*) can occur. Once fertilization has taken place, it is essential that the female body provide a location for the protection, growth, and development of the fetus until it has matured enough to survive outside the mother's womb.
- To produce hormones that cause physical characteristics specific to each gender as well as sex drive. In the female, hormones also prepare the woman's body to receive a fertilized ovum and to nourish the fetus during gestation and the newborn after birth.

STRUCTURES OF THE REPRODUCTIVE SYSTEMS

Located in the pelvic region, the reproductive organs must produce and support gametes, provide a system through which the sperm and ovum can be united and protected, and supply fluids and hormones essential for reproduction to occur. Considering the anatomical structures of the reproductive systems, both contain the following:

- Gonads, which are primary sex organs that produce gametes: the testes in males and the ovaries in females. The gonads produce gametes through meiosis rather than mitosis. The result of meiosis is that the sperm and ova each contain half the normal number of chromosomes (23), versus 46 in other body cells.
- A duct system through which the gametes pass as they are transported to their appropriate location. In the male, sperm are transported through the ductus deferens to the urethra for ejection from the penis into the female vagina. In the female, the released ovum is transported from the ovary to the uterus by the uterine tube.
- Accessory sex glands, which produce fluids essential to the normal function and protection of the gametes as they are transported through the reproductive system. In the female they also produce milk to nourish the newborn.

UNDIFFERENTIATED GONADS

Early in the development of the fetus, the gonads of both males and females are in the same location—the developing pelvic cavity. At that point, it is impossible to physically determine whether the fetus is male or female. The developing placenta produces a hormone that instructs the male fetus to produce testicular-determining factor, a chemical that induces the formation of testes. As the testes form, a membrane also develops to draw the testes out of the abdomen into the scrotum (Figure 13.4). Sperm cannot survive at normal body temperature for extended periods, so the testes are moved into the scrotum to be away from body heat. As the testes descend into the scrotum, they draw their nerves and blood vessels with them. In females, the ovaries remain within the pelvic cavity throughout life.

Chapter 13 The Reproductive Systems 389

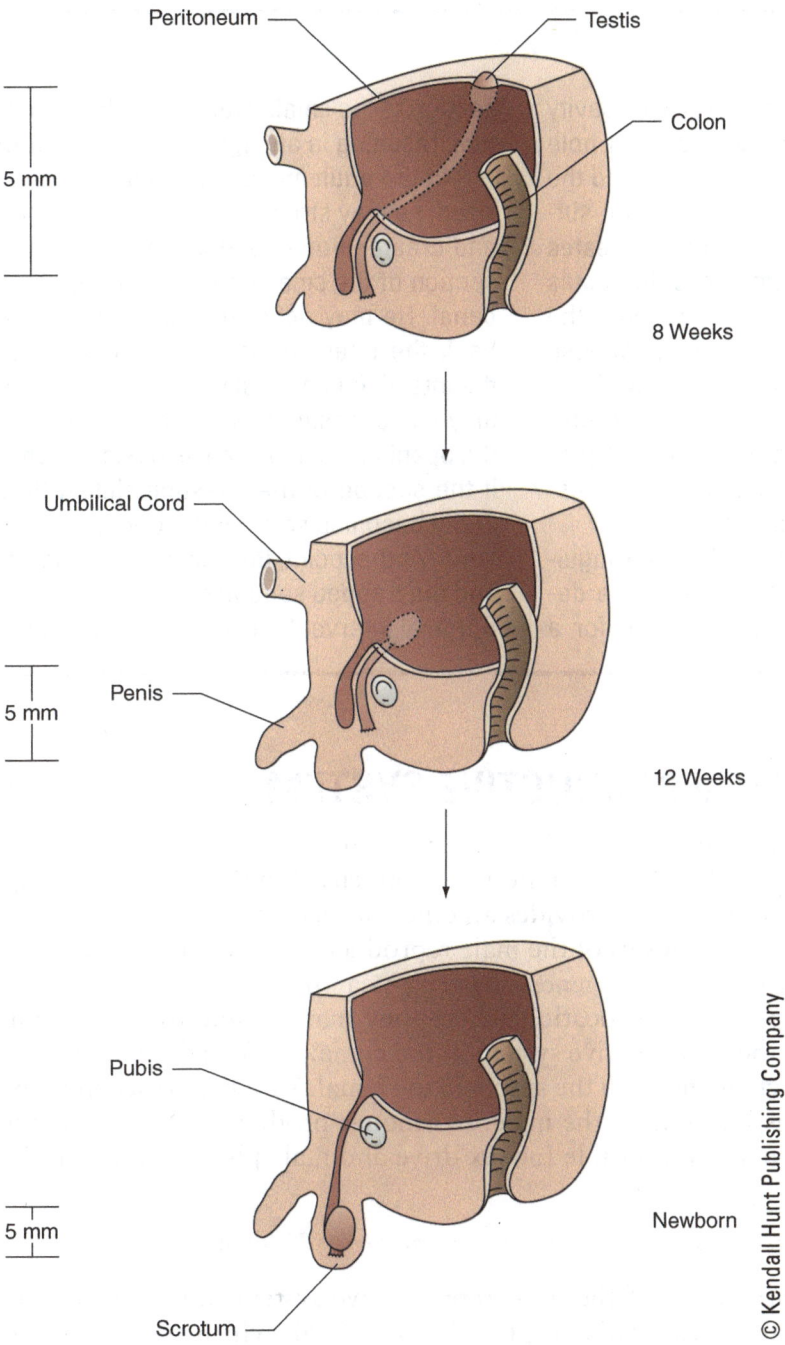

Figure 13.4 Descent of the Testes. The testes are drawn out of the abdomen into the cooler scrotum during fetal development.

Comprehension Check-up:

1. The ovaries and testes are collectively known as _____.
2. In the earliest stages of development, the ovaries and testes are both found in the _____.

1. gonads 2. lower abdomen

> **Descent of the Testes**
>
> For the testes to descend from the pelvic cavity into the scrotum, they must pass through a hole in the abdominal cavity. This opening is called the *inguinal canal*. If the size of the canal is not sufficient to allow passage of the testes, the testes remain in the pelvic cavity. Sometimes the testes are surgically moved through the canal into the scrotum shortly after birth. Some parents choose to wait to see if the process occurs naturally as the boy matures. If left in the pelvic cavity after puberty, the undescended testes would not produce sperm; in addition, the incidence of the testes becoming cancerous greatly increases.
>
> Conversely, there are times when the inguinal canal is too large. Once the testes have descended, the canal may be large enough for a section of the small intestine to slide into the scrotum, resulting in an inguinal hernia at birth.
>
> In the adult male, if the inguinal canal is oversized, he may strain to lift a heavy object or cough and create enough abdominal pressure to push a section of the small intestine through the inguinal canal. He may need to wear a supporter to hold back the intestine. If the canal is enlarged and the intestine continually slides through the area, it may be necessary to surgically reduce the size of the opening. There is cause for significant concern if the section of the intestine sliding through the canal becomes stuck in the opening and strangulated. At that point, the alimentary canal is blocked and the trapped section of small intestine may die. Surgical intervention is essential in these cases.

MALE REPRODUCTIVE SYSTEM

The male must provide a set of 23 chromosomes, his contribution to the offspring. His genetic material is condensed in the head of each sperm while the rest of the cell provides an efficient method of delivery by the movement of a tail. The testes of the male reproductive system form sperm by a process known as spermatogenesis (*sperm*, meaning "creation"). The sperm are then transported to a location where they mature and are stored until released from the reproductive system at the climax of sexual intercourse. Fluids that protect and nourish the sperm in the female's body also accompany the sperm as it is released by the male. Hormones produced by the male reproductive system are responsible for sex drive and male physical characteristics.

> Spermatogenesis the process of producing sperm.

Anatomy of the Male Reproductive System

The structures of the male reproductive system can be divided into the organs producing gametes, the ducts through which those reproductive cells travel, and the accessory glands that provide supporting fluid (Figure 13.5). They are:

- Testes, which produce sperm and androgens, the male hormones responsible for sex drive and male physical characteristics
- Ducts through which sperm pass
  - The epididymis, which serves to store, mature, and transport spermatozoa between the testis and the vas (vas deferens)
  - Ductus deferens, which transports sperm to the ejaculatory duct, which in turn receives the sperm from ductus deferens and seminal fluid from seminal vesicles, and drains them to the urethra
  - Urethra, which transports sperm and fluid through the penis to be delivered into the female vagina

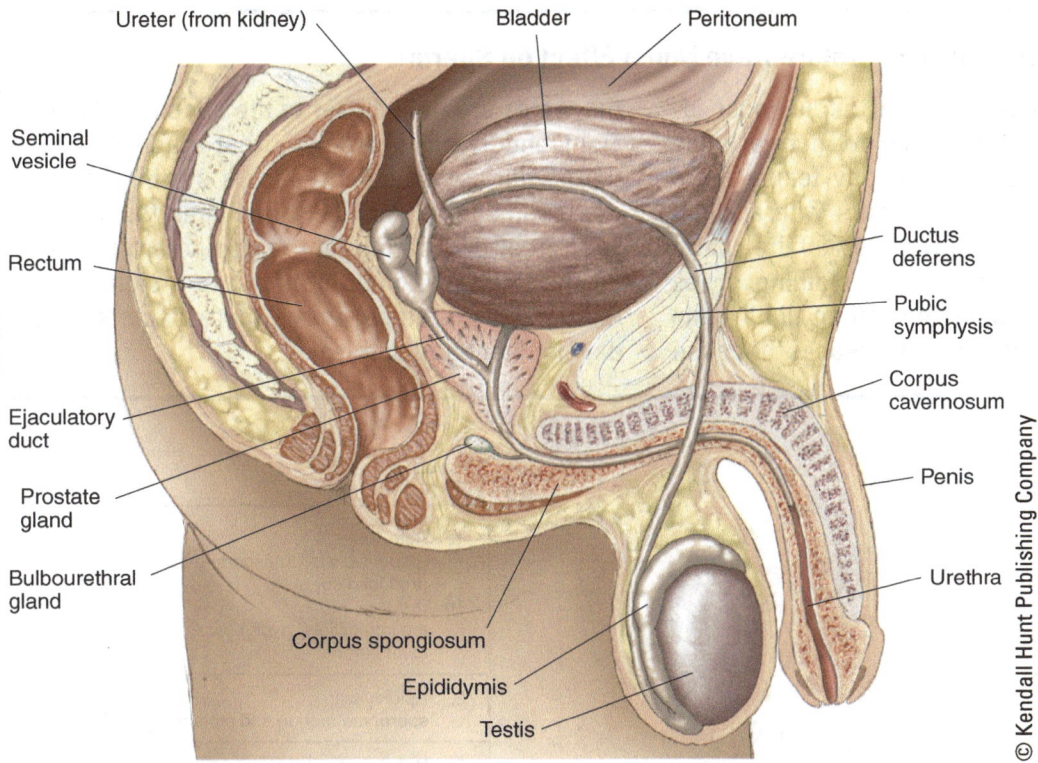

Figure 13.5 Male Reproductive System.

- Accessory glands, which produce fluid known as *semen* that contains the fluids secreted by the accessory glands, as well as the sperm and some fluid from the testes (testicular fluid) and epididymis
  - Prostate gland, which produces alkaline fluid to neutralize vaginal acid
  - Seminal vesicle, which secretes fluid high in fructose as an energy source for active sperm
  - Bulbourethral gland (also known as Cowper's gland), which produces a fluid containing mucus to lubricate the urethra as sperm passes through. This fluid is also alkaline, to assist in neutralizing acids that could harm sperm. About the size and shape of a pea, this gland is located inferior to the prostate gland.
- External genitalia
  - Penis, which becomes elongated and firm to provide a mechanism through which sperm can be delivered to the appropriate location in the female vagina
  - Scrotum, a tissue sac in which the testes are suspended outside the abdominal cavity away from body heat

Each of these structures is discussed individually.

Testes

The testes are in the shape of a small plum and are located away from body heat in the scrotum, the tissue sac inferior and posterior to the penis (Figure 13.6). Sperm cannot survive at normal body temperatures for more than 48 hours. Placement of the testes and epididymis outside the abdomen allows the developing sperm to complete their normal production. That said,

> **Scrotum** a tissue sac in the anterior pelvic region that holds the testes and epididymis away from normal body heat.

Does Increased Scrotal Temperature Have Much Effect on Sperm?

Sperm production and survival occur most effectively at 4° to 5°F (3°C) below normal body temperature. An increase in body temperature of one or two degrees, as occurs with a slight fever, may result in as many as 20% of the sperm being rendered defective. Even clothing styles can affect sperm production. When tight jeans are in style, the birth rate decreases because the testes are held closer to the abdominal heat. Tight underwear, especially that made of synthetic fabrics, may also cause a significant decrease in sperm viability. It has been found that even heat-dissipating laptop computers resting on the thigh of a male can create a significant rise in scrotal temperatures that affects sperm development and survival.

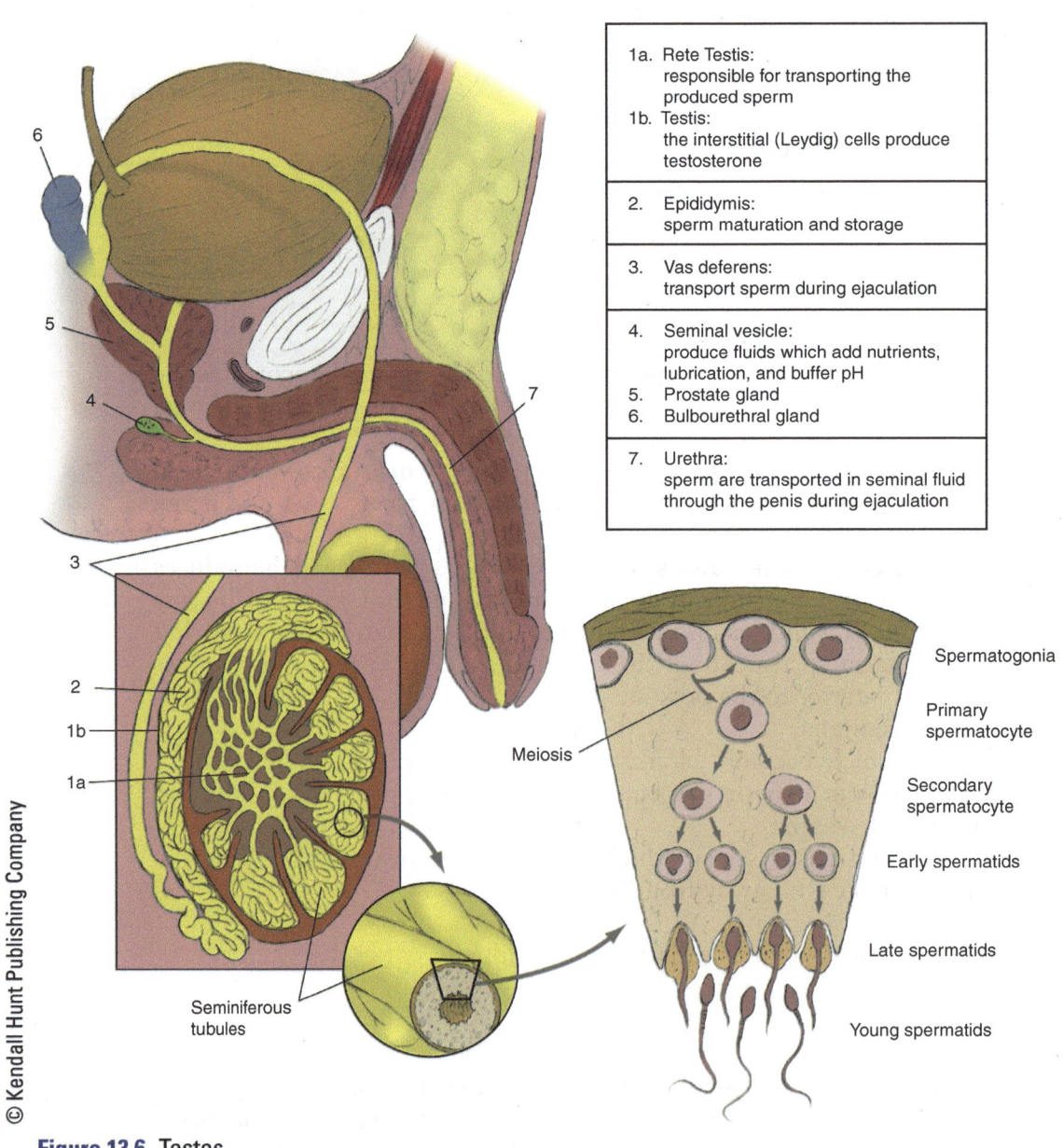

Figure 13.6 Testes.

increases in the temperature of this area can cause the deformation or destruction of increased numbers of sperm.

Inside the testes are roughly half a mile of *seminiferous tubules* actively producing sperm by meiosis. Between the seminiferous tubules are *interstitial cells* that produce testosterone. Testosterone production is controlled by interstitial cell-stimulating hormone (also known as luteinizing hormone), a hormone produced by the anterior pituitary gland. Testosterone is responsible for sex drive and for male physical characteristics such as facial, axillary, and pubic hair; lowering of the voice; and increased muscle mass and blood volume, as well as slowing the replacement of cartilage with bone, which results in males being generally taller than females.

Epididymis

Attached to the posterior surface of the testes is the epididymis, where they obtain the ability to swim and to fertilize the egg (Figure 13.6). The epididymis consists of tightly coiled tubes whose overall length is about 7 meters (23 feet). Sperm spend about 2 weeks passing through the epididymis. Phagocytes in the epididymis remove defective or diseased sperm to recycle their chemicals for new spermatogenesis.

> **Epididymis** a set of tightly coiled tubes attached to the testes, which store sperm until release or until they die.

Ductus Deferens

The ductus deferens, also known as the *vas deferens*, is a muscular duct that transports sperm by peristalsis from the epididymis to the urethra during ejaculation (ejection of sperm). The ductus deferens is attached to the proximal end of the epididymis in the scrotum and runs into the man's pelvic cavity through the inguinal canal, around the urinary bladder to connect with the ejaculatory duct inside the prostate gland. The section of the ductus deferens between the inguinal canal and epididymis also includes testicular arteries, veins, and nerves enclosed in a tissue sheath collectively known as the *spermatic cord*.

> **Ductus deferens** a muscular duct that transports sperm by peristalsis from the epididymis to the urethra during ejaculation.

Urethra

The urethra is a common tube for both urine and reproductive fluid. The ductus deferens connects to the ejaculatory duct inferior to the urinary bladder within the prostate gland. During erection of the penis, the sphincters on the urinary bladder remain tightly closed so that urine cannot escape.

Comprehension Check-up:

1. In the testes, sperm are produced in the _____.
2. Sperm are stored in the _____ until release.

1. seminiferous tubules   2. epididymis

Accessory Glands

Accessory glands provide secretions that are essential for the survival and motility of the sperm but that are not directly part of the duct through which the sperm pass on their way out of the body.

Prostate Gland

Inferior to the urinary bladder is the ping-pong ball-sized prostate gland that surrounds the intersection of the urethra and ductus deferens (Figure 13.7). It produces an alkaline fluid that accompanies the sperm as it is ejaculated from the man's body. The female vagina is kept exceedingly acidic as a protection against bacteria. To prevent the destruction of the sperm when they arrive in the vagina, alkaline fluid is also provided to temporarily neutralize the acid and protect the sperm. Nevertheless, a substantial portion of the sperm still die from the acid.

> **Prostate gland** located inferior to the urinary bladder, it surrounds the intersection of the urethra and ductus deferens. It produces an alkaline fluid that accompanies the sperm as it is ejaculated.

Seminal Vesicle

The seminal vesicles, located superior and posterior to the prostate gland, produce a fluid high in fructose, a nutrient source of energy for sperm on their way to the potentially waiting ovum (Figure 13.7). During ejaculation the seminal vesicles release their secretions, causing the sperm to become highly mobile.

> **Seminal vesicles** located superior and posterior to the prostate gland, the seminal vesicles produce a fluid high in fructose that accompanies sperm as it is ejaculated.

Semen

The alkaline fluid from the prostate gland and fructose secretions from the seminal vesicles along with some lubricating fluid blend to form semen, which accompanies the sperm as it is ejaculated from the man's body. If a man has had a vasectomy, he will still eject semen during sexual climax even though no sperm are present.

> **Semen** a fluid mixture containing sperm and secretions from accessory glands.

Sperm

The sole function of the sperm is to deliver 23 of the male's chromosomes to the awaiting ovum. Anatomically the sperm can be divided into three sections (Figure 13.8):

- The *head* of the sperm contains the male's genetic contribution to his offspring. At the forward tip of the head is the *acrosome*, which contains enzymes that become activated as the sperm travels through the woman's

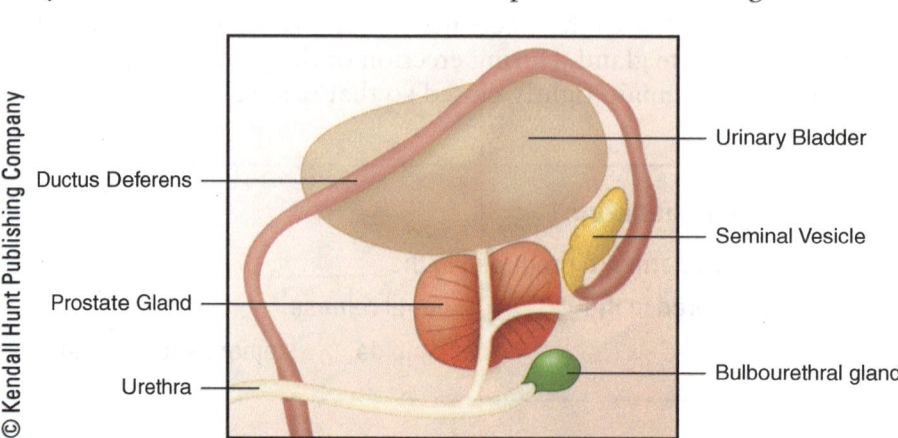

Figure 13.7 Seminal Vesicle, Prostate Gland, and Bulbourethral Gland.

The Prostate Gland and Age

As the male ages, the prostate gland tends to harden, making urination more difficult. The prostate gland may even become cancerous and swell enough to completely block the urethra. Prostate cancer, if caught early, is very treatable. If not treated early, it can rapidly become very serious. It is highly recommended that males older than age 50 have an annual checkup of their prostate glands.

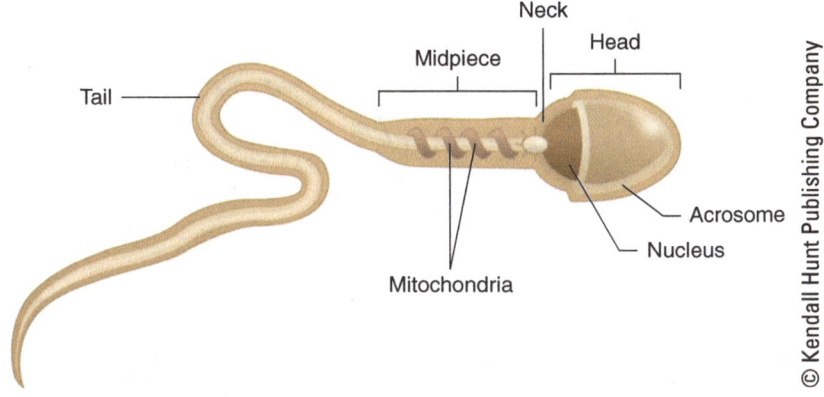

Figure 13.8 Sperm. Sperm can be divided into the head, midpiece, and tail.

re-productive system toward the ovum. The released egg is surrounded by two layers of protective material. To reach the ovum, the sperm must release their enzymes to dissolve the protective tissue to reach the surface of the egg.
- The *midpiece* is wrapped with mitochondria. These "powerhouses of the cell" produce the adenosine triphosphate (ATP) required to cause the tail to whip back and forth to move the sperm toward the ovum.
- The *tail* is the sperm's source of movement. This flagellum moves actively to allow the sperm to reach its destination.

The male produces approximately 100,000,000 (1×10^8) sperm per day. It would seem that the enormous quantity of sperm available to fertilize one ovum would be sufficient, but sperm have many barriers to cross. Sperm must be protected against vaginal acid, swim through downward current produced in the pathway to the ovum, and penetrate two layers of tissue surrounding the egg before reaching their ultimate destination.

External Genitalia—The Penis

The penis is a cylindrical organ that excretes urine or is to be able to deliver semen to the appropriate location in the female reproductive system. To explain the physiology of that process, it is essential to first consider the anatomy. The penis contains the following structures (Figure 13.9):

- The *urethra* is the tube transporting urine or semen out of the man's body.
- The *corpus spongiosum* is a spongy layer around the urethra that prevents the erect penis from blocking the urethra so that sperm and semen can be delivered.
- The *corpora cavernosa* are two areas containing erectile tissue that cause the penis to become elongated and firm so that sperm can be easily delivered.

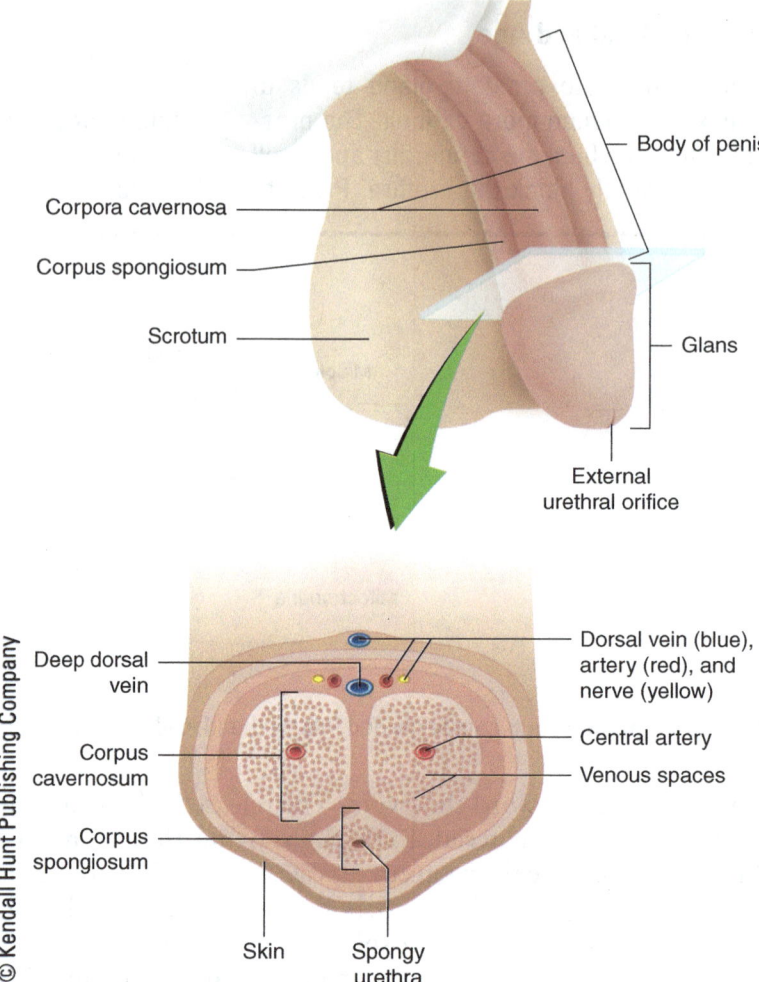

Figure 13.9 Penis. The penis consists of the urethra through which urine or semen pass. The urethra is surrounded by the corpus spongiosum. Two areas of erectile tissue, the corpora cavernosa, become engorged with blood to cause erection.

- The head of the penis is known as the *glans penis*. It contains the orifice of the urethra and is composed primarily of corpus spongiosum. It does not contain erectile tissue.
- There is a fold of skin that partially covers the glans penis, known as the *prepuce*. It is commonly referred to as the *foreskin*. Removal of the prepuce is termed *circumcision*.

Comprehension Check-up:

1. Fluid high in fructose that accompanies sperm is produced by the _____.

2. Enzymes needed to penetrate the protective tissue surrounding the ovum are found in the _____ in the head of the sperm.

1. seminal vesicle 2. acrosome

> **Circumcision**
>
> Circumcision originally became a common practice in the United States as a matter of hygiene. The prepuce often held chronic infection or irritating particles against the glans penis if the man's hygiene was poor. The constant irritation sometimes led to penile cancer, particularly in men who worked in jobs that involved a great deal of dust, such as coal miners or chimney sweeps. Improved hygiene of today's society in the United States has decreased the need for circumcision as a method for controlling penile cancer.

Male Sexual Function

The discussion of male reproductive physiology involves four areas. First is the hormone control of sperm production and the regulation of testosterone. Next is the production of sperm, called *spermatogenesis,* by the seminiferous tubules of the testes. Third involves the erection of the penis, and finally, there is ejaculation of semen.

Regulation of Hormones Affecting Male Reproduction

Two hormones, both produced by the anterior pituitary gland, affect male reproduction. Recall that hormones produced by the anterior lobe of the pituitary gland require releasing hormones from the hypothalamus, so the regulation of hormones affecting reproduction is under the control of the hypothalamus. Follicle-stimulating hormone increases meiosis in the seminiferous tubules in the testes to increase sperm production. Lutenizing hormone, also known as *interstitial cell-stimulating hormone,* causes interstitial cells in the testes to increase their testosterone output.

Spermatogenesis—Production of Sperm

Spermatogenesis begins during puberty, typically in the early teens, and continues throughout the male's adult life. In the testes, around the outer edges of the seminiferous tubules, are stem cells known as *spermatogonia* (Figure 13.10). They divide by mitosis into two cells. One of the two cells remains behind as a stem cell to divide again by the same process; the other continues on, dividing by meiosis, to become sperm. The cell destined to become sperm is known as a *primary spermatocyte.*

The primary spermatocyte proceeds through meiosis I and becomes what is known as a *secondary spermatocyte.* At the completion of meiosis II, the developing sperm has half the number of chromosomes (23 or N) found in other body cells (46 or 2N); these are known as *spermatids*. What remains is to streamline the spermatids into sperm. Their organelles and cytoplasm are stripped away as a tail develops, with the exception of the enzyme containing acrosome in the head. Even though the development of the sperm is complete before leaving the testes, it remains passive; that is, it is not yet mature enough to move by its own power.

Inside the seminiferous tubules are other cells called *Sertoli cells*. These cells care for the developing sperm by providing support and protection against potential attack from the immune system through the formation of the blood-testis barrier. Once sperm is complete, the Sertoli cells move them to the lumen of the seminiferous tubule, where they are flushed out of the tubule into the epididymis.

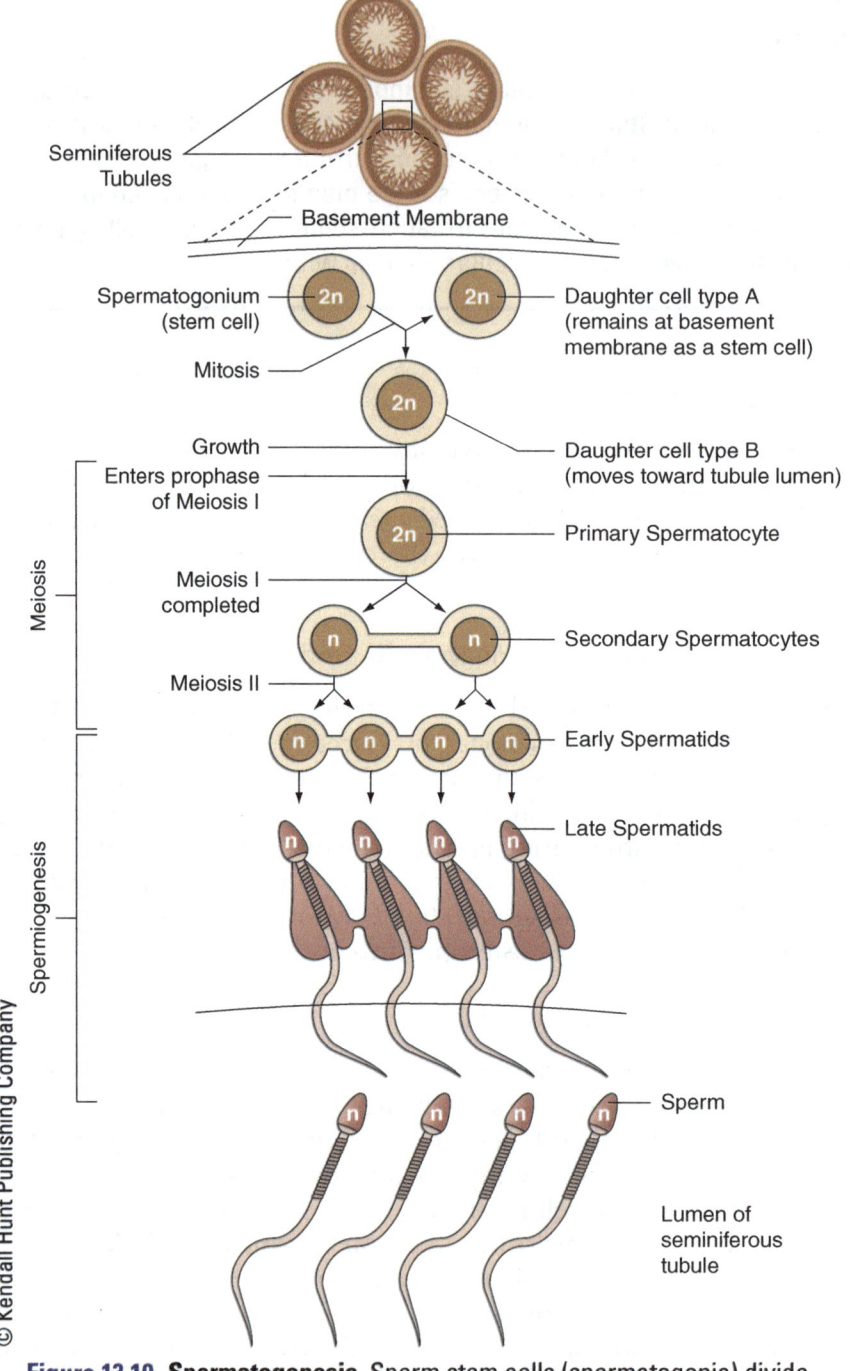

Figure 13.10 Spermatogenesis. Sperm stem cells (spermatogonia) divide by meiosis to become primary then secondary spermatocytes before becoming spermatids as they develop a tail and mature into sperm.

Erection of the Penis

For sperm to be delivered to the appropriate location inside the female vagina, the penis must elongate and become firm. The processes of sexual function require the participation of the autonomic nervous system.

As a result of sexual arousal, the parasympathetic nervous system causes blood flow to increase in the corpora cavernosa and they become engorged with blood. These erectile tissues balloon outward until they create pressure

> **Erectile Dysfunction**
>
> There are numerous causes for erectile dysfunction (ED), in which it becomes difficult for a male to achieve erection. Common causes may be physical or emotional factors, nerve damage, or a side effect of some medications. More than 50% of males between 40 and 70 years of age have ED, and this increases to about 70% after the age of 70. There are medications that do not themselves cause erection but rather lengthen the time of response of normal sexual arousal to result in erection.

against the tunica albuginia around the length of the penis, causing the penis to become erect and firm. The corpus spongiosum surrounding the urethra prevents the expanding corpora cavernosa from blocking the urethra and stopping the delivery of sperm.

Ejaculation

Ejaculation is the ejection of semen out of the penis through the urethra. At the climax of sexual stimulation, the sympathetic nervous system stimulates peristalsis of the ductus deferens and contraction of the prostate gland and seminal vesicles to release sperm and fluid. Between 2.5 and 5 ml of semen is released, containing approximately 300 to 400 million sperm.

> **Ejaculation** the ejection of sperm and semen out of the penis through the urethra.

> **Comprehension Check-up:**
>
> 1. The term for sperm production is _____.
> 2. The hormone responsible for increasing sperm production is _____.
>
> 1. spermatogenesis 2. follicle stimulating hormone

> **Clinical Notes—Male Reproductive Issues**
>
> Common male reproductive issues include enlarged prostate or known as benign prostatic hyperplasia (BPH), prostate cancer, testicular cancer, erectile dysfunction, male infertility, and testosterone deficiency.
>
> Prostate-specific antigen (PSA) levels increase when there is an enlarged prostate. However, elevated PSA levels can also be due to recent procedures, infection, surgery, or prostate cancer. PSA is a blood test.
>
> Male fertility testing initially starts with a physical exam and history conducted by a urologist. History includes questions on previous and type of surgeries performed, types of medications taken or currently taking, smoking, exercise habits, sex life, and sexually transmitted diseases (STD). A sample of semen may be required by the physician for analysis.
>
> Testing for erectile dysfunction (ED) includes medical and sex history, mental health and physical exam, lab tests, and imaging test by ultrasound.
>
> Testing for low testosterone levels includes blood tests for testosterone, luteinizing hormone (LH), follicle-stimulating hormone (FSH), and prolactin.
>
> *(continued)*

Lab tests are only ordered by the urologist when there are significant signs and symptoms of low testosterone levels such as low sex drive, erectile dysfunction, and infertility. Other symptoms may include depression, irritability, fatigue, delayed sexual development, obesity, poor concentration, and disturbed sleep.

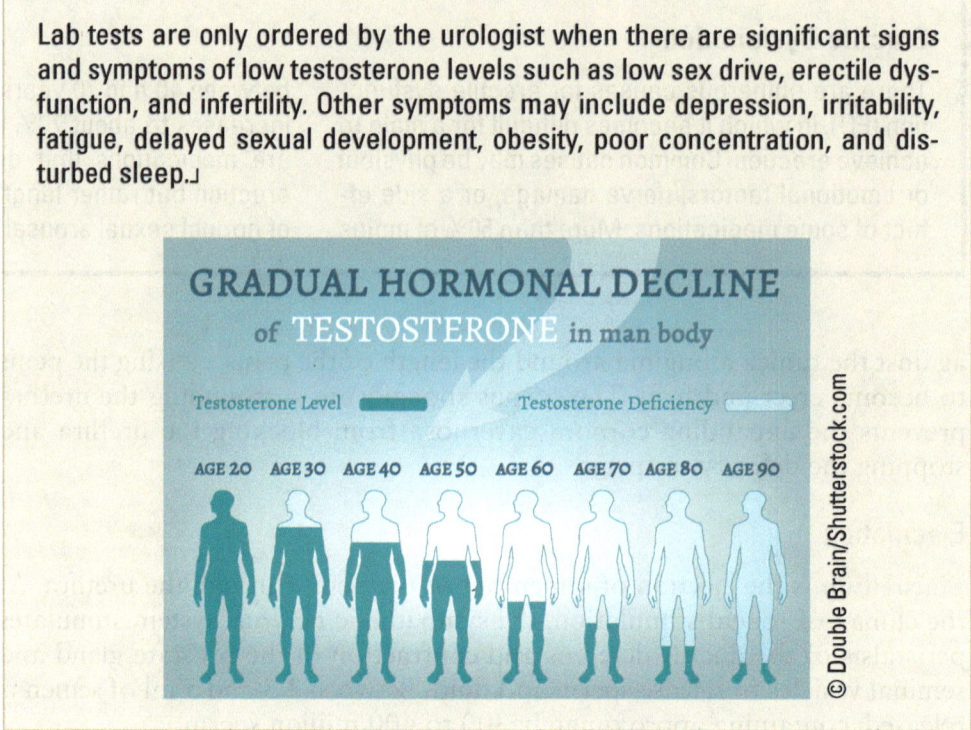

FEMALE REPRODUCTIVE SYSTEM

The female reproductive system produces ova by meiosis and hormones to assist in the control of reproduction. Each ovum is surrounded by a sphere of cells known as *follicular cells* or *granulose cells*. Every menstrual cycle, a follicle from one of the ovaries releases its ovum by a process known as ovulation. The released ovum travels through the uterine tube while it awaits the arrival of sperm. The uterus is also considered a duct specially designed to support the fetus during pregnancy. The female reproductive system is hormone dependent for cyclic communication between the ovaries and brain and between the developing fetus and the woman's body.

> Ovulation the release of an ovum from a follicle in the ovary.

Anatomy of the Female Reproductive System

The organs of the female reproductive system are located in the pelvic cavity (Figure 13.11). They produce and care for ova, provide a location for fertilization to occur, and provide nutrient-rich support for the development and birth of the fetus. Once birth has taken place, the system provides continued nourishment to the baby. They include:

- Ovaries produce and release ova and secrete the reproductive hormones estrogens and progesterone.
- Ducts provide transport of the ovum and receive sperm. They also assist in the development of the fetus:
  - Uterine tubes transport the released ovum to the uterus if ovulation occurs along the pathway.
  - The uterus provides a protective environment for the fetus and is where the placenta attaches.
  - The vagina provides a canal for the deposit of sperm.

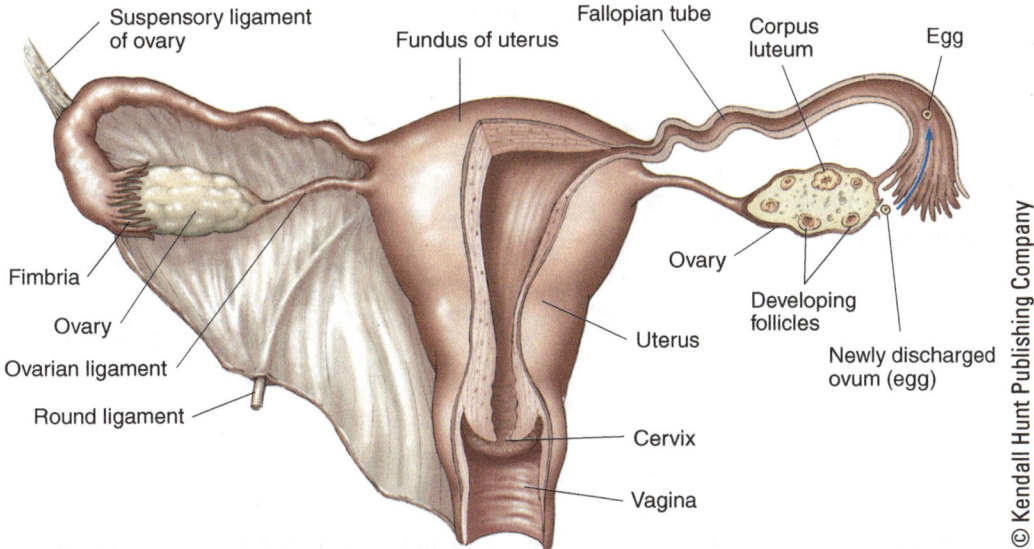

Figure 13.11 Female Reproductive System.

- External genitalia are protective tissues surrounding the vaginal orifice.
- The accessory gland consists of milk-producing glands, called *mammary glands,* to nourish the newborn child.

As with the male reproductive system, each of these organs is discussed individually.

Ovaries

The ovaries are roughly the size and shape of almonds and are found close to the lateral wall of the pelvic cavity suspended between the *suspensory ligament,* from the pelvic wall to the ovary, and the *ovarian ligament* attaching the ovary to the uterus (Figure 13.12). It is within the ovaries that meiosis takes place to form ova. Each ovum is surrounded by the follicular cells or granulose cells, which support and protect the egg until it is released from the follicle by a process known as *ovulation.* The ovum is released through the wall of the ovary, where it has resided throughout its existence. Ova are not released from a single outlet but can be released virtually anywhere on the ovarian surface. The vast majority of the potential ova are never released from their follicles. The process by which the ovary determines that one ovum is to be released during a menstrual cycle and others are not is not understood.

Uterine Tubes

The uterine tubes, also known as *fallopian tubes,* are muscular tubes where conception occurs and are where the earliest divisions of the embryo begin as it is moved to the uterus. The uterine tube is not attached to the ovary because the ovum may be released from anywhere on the ovary. On the distal end of each uterine tube is the *infundibulum,* a funnel-shaped structure with finger-like projections called *fimbriae* that surround the lip of the funnel (Figure 13.12). The infundibulum hovers over the follicle that is about to rupture. On the underside of the fimbriae are cilia that create a gentle current that moves the released ovum into the uterine tube. Because the ovum may be released from anywhere on the ovary, it is the function of the infundibulum to be close enough to the ovary to capture the ovum after it exits the follicle.

> **Uterine tubes** muscular tubes where conception occurs and the earliest divisions of the embryo begin as it is moved to the uterus.

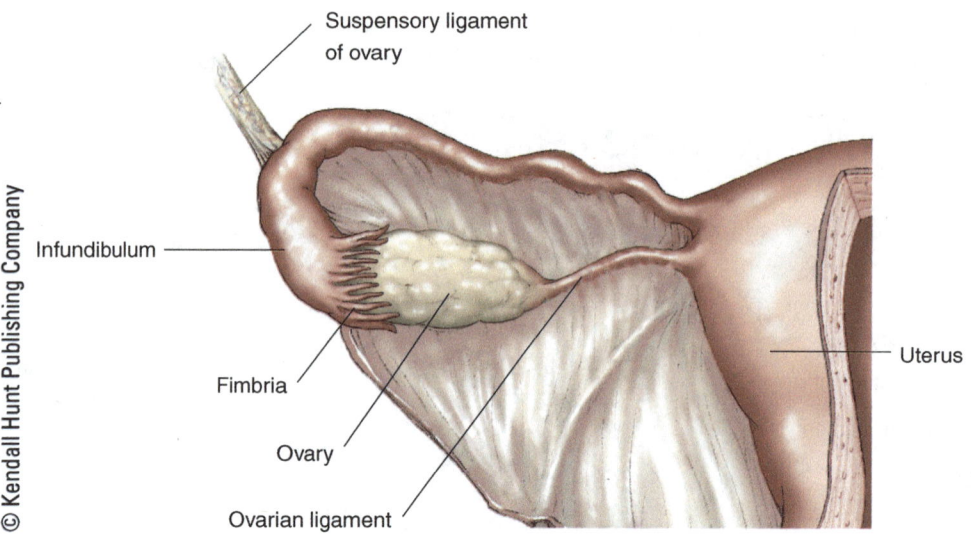

Figure 13.12 Ovary. The ovary releases ova "caught" by the distal end of the uterine tube, which forms a funnel-shaped infundibulum. The rim of the infundibulum possesses finger-like projections called *fimbriae* that hover over the follicle that is about to rupture.

Once the ovum is inside the uterine tube, movement of the egg is the result of the movement of cilia and exceedingly gentle peristaltic contractions of the smooth muscle in the wall of the tube. Movement of the ovum down the length of the tube, about 13 cm (5 inches) takes 3 or 4 days. It is during that time that sperm must be available to fertilize the ovum. The unfertilized ovum cannot survive outside the ovary for more than 24 hours. It is necessary for the sperm to travel up the uterine tube against the current created by the cilia lining the tube in order to reach the ovum. If conception does occur, the earliest divisions of the embryo take place as the fertilized ovum continues through the uterine tube on its way to the uterus. If conception does not occur, the dead ovum is removed by phagocytes.

> **Comprehension Check-up:**
> 1. The process of releasing an ovum from the ovary is known as _____.
> 2. At the end of the uterine tube are finger-like projections that assist in moving the released ovum from the ovary into the uterine tube. These projections are known as _____.
>
> 1. ovulation 2. fimbriae

Uterus

Uterus a pear-shaped organ that provides attachment for the placenta and support of the fetus through gestation.

The uterus is a pear-shaped organ that provides a location for the attachment of the placenta for exchange of nutrients, gases, and wastes; it also supports the fetus during pregnancy. During labor, the uterus contracts and abdominal pressure increases to eject the fetus. Several ligaments—the lateral cervical ligaments, uterosacral ligaments, and round ligaments—are needed to hold the uterus in place because of the substantial increase in weight as the fetus matures inside the womb. The area where the placenta attaches and the fetus resides throughout pregnancy is known as the *body of the uterus*.

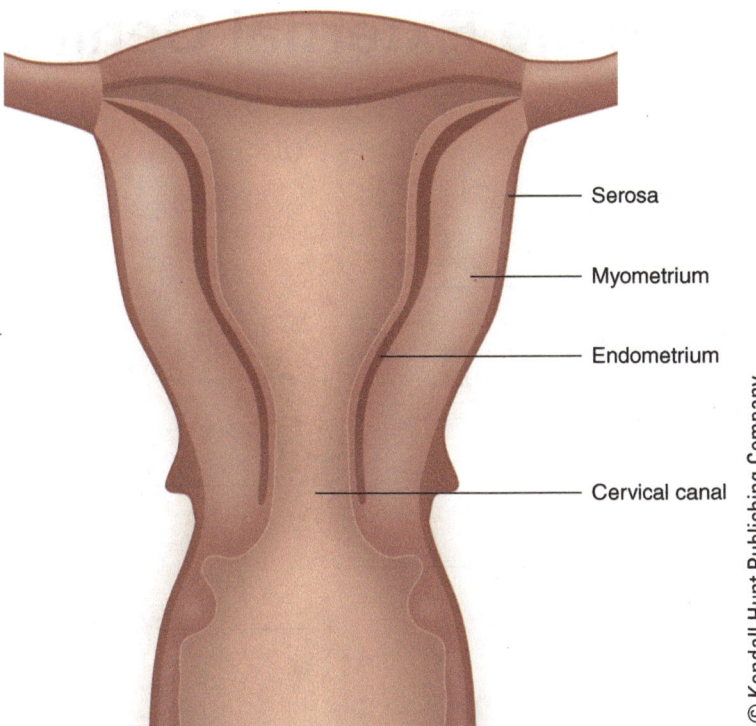

Figure 13.13 Uterus. The uterus is where the fetus develops during gestation. It is covered externally by the serosa. The myometrium, the middle layer, is composed of smooth muscle, which contracts during labor. The inner lining is the endometrium, to which the placenta attaches for exchange of nutrients, gases, and wastes.

The inferior cylindrical portion of the uterus is the cervix; the opening where sperm enter and through which the fetus exits is known as the *cervical canal*.

The wall of the uterus can be divided into three layers (Figure 13.13). The outermost covering is the *serosa*. There is a thick layer of smooth muscle that contracts during labor known as the *myometrium*. Lining the inside of the uterus is the endometrium. The placenta attaches to the endometrium to receive nourishment, for gas exchange, and for waste elimination.

Endometrium the inner lining of the uterus to which the placenta attaches during gestation for exchange of nutrients, gases, and wastes.

Vagina

The vagina is a muscular tube between the uterus and external genitalia that provides a canal for the deposit of sperm. As mentioned earlier, because the orifice of the vagina is on the surface of the body, it is possible for bacteria to migrate into the woman's body, potentially creating serious infection. As a defense against bacterial infection, the vagina is very acidic.

Vagina a muscular tube for the deposit of sperm. It is normally very acidic as a protection against bacterial invasion.

External Genitalia

The external genitalia provide protection and lubrication for the opening of the vagina, known as the *vestibule* (Figure 13.14):

- *Labia minora* are small tissue folds surrounding the vestibule.
- *Labia majora* are larger tissue folds lateral to the labia minora.
- *Clitoris* is the female equivalent to the penis. It contains erectile tissue that becomes engorged with blood during sexual arousal.

The external genitalia become sensitized during sexual arousal.

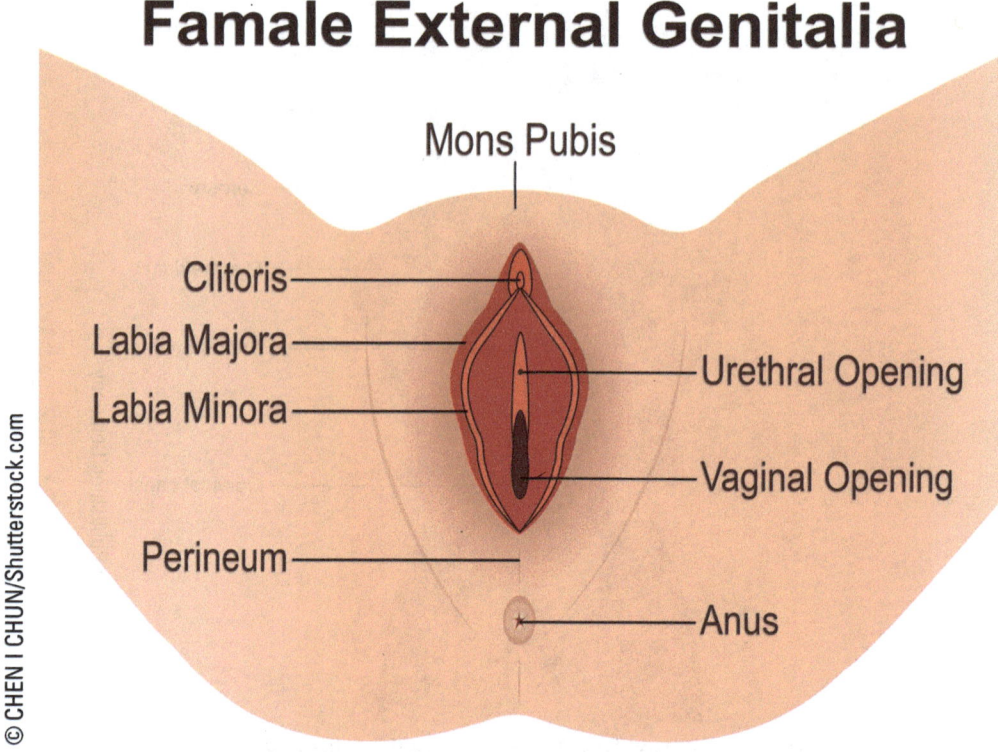

Figure 13.14

Accessory Glands—Mammary Glands

The function of the mammary glands is to produce milk to feed the newborn child. These glands are located in the anterior breast posterior to and radiating around the nipple (Figure 13.15). The mammary glands contain secretory lobules that produce milk. The milk flows from the lobules through an opening in the nipple. Two hormones are involved in the production and secretion of milk: prolactin and oxytocin. Prolactin, from the anterior pituitary gland, causes the lobules to produce milk. Oxytocin, from the posterior pituitary gland, causes the milk to be secreted. Sucking of the nipple after birth and during periods of breast-feeding stimulates the release of oxytocin. Fat deposits behind the mammary glands allow the breast to be easily manipulated to make feeding more comfortable for both mother and child. Suspensory ligaments hold the breasts in place when the woman is young, but they tend to lose their ability to support the breast with aging.

> **Mammary glands** located in the breasts for the production and secretion of milk to nourish the newborn.

> **Comprehension Check-up:**
> 1. The lining of the uterus to which the placenta attaches is known as the _____.
> 2. The hormone that stimulates lobules in the breast to produce milk is _____.
>
> 1. endometrium 2. prolactin

The Ovum

The ovum contains the female's contribution to the offspring. Whereas the head of the sperm is essentially stripped of any additional cellular components other than the enzymes contained in the acrosome, the ovum retains organelles to be available for the embryo.

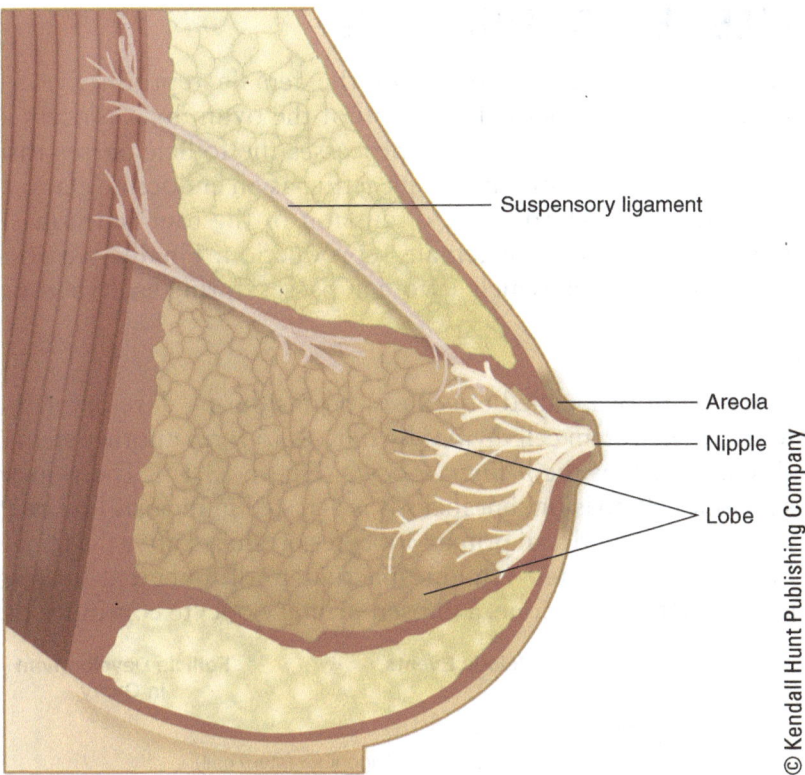

Figure 13.15 Anatomy of the Breast. Lobes of the mammary gland produce milk, which flows to the areola, where it is released through the nipple.

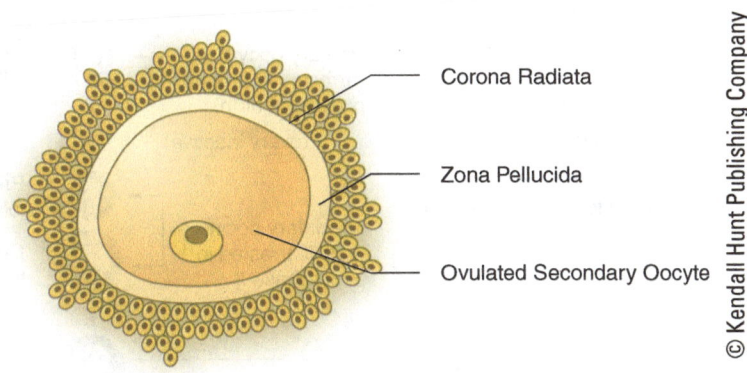

Figure 13.16 The Ovum Surrounded by Two Layers of Protective Tissue.

The released ovum is surrounded by two layers, the zona pelucida and corona radiate, which protect and nourish it (Figure 13.16). It is through these two layers that the enzymes in the sperm's head, the acrosome, must digest in order to reach the ovum.

Mitochondria Can Be Used to Trace Maternal Ancestry

The mitochondria found in the child originate from the mother because the mitochondria from the father were wrapped around the midpiece of the sperm and did not enter the ovum at conception.

Mitochondria contain a small amount of DNA. Since mitochondrial DNA is from the mother only, it is possible to use this DNA to trace maternal ancestry.

FEMALE SEXUAL FUNCTION

Our discussion of the physiology of the female reproductive system involves four areas. First is the formation of the ovum by a process known as oogenesis. Next is the study of the processes the follicle passes through not only in releasing the ovum but also its acting as an endocrine gland after ovulation has occurred. Third is the discussion of the four hormones involved with the rhythmic cycles of female reproduction. Finally, we consider the cyclic processes through which a woman passes during each menstrual cycle.

> **Oogenesis** the process of producing ova through meiosis.

Oogenesis

The formation of ova, called *oogenesis,* begins during development of the female fetus, when a stem cell divides by mitosis to form another stem cell and a second cell, known as the *primary oocyte* (Figure 13.17). The primary oocyte begins meiotic cell division, but meiosis is arrested in meiosis I by birth. Apparently there are no additional primary oocytes produced after the female is born.

The primary oocytes remain inactive until puberty. In response to follicle-stimulating hormone, a chosen primary oocyte continues the process

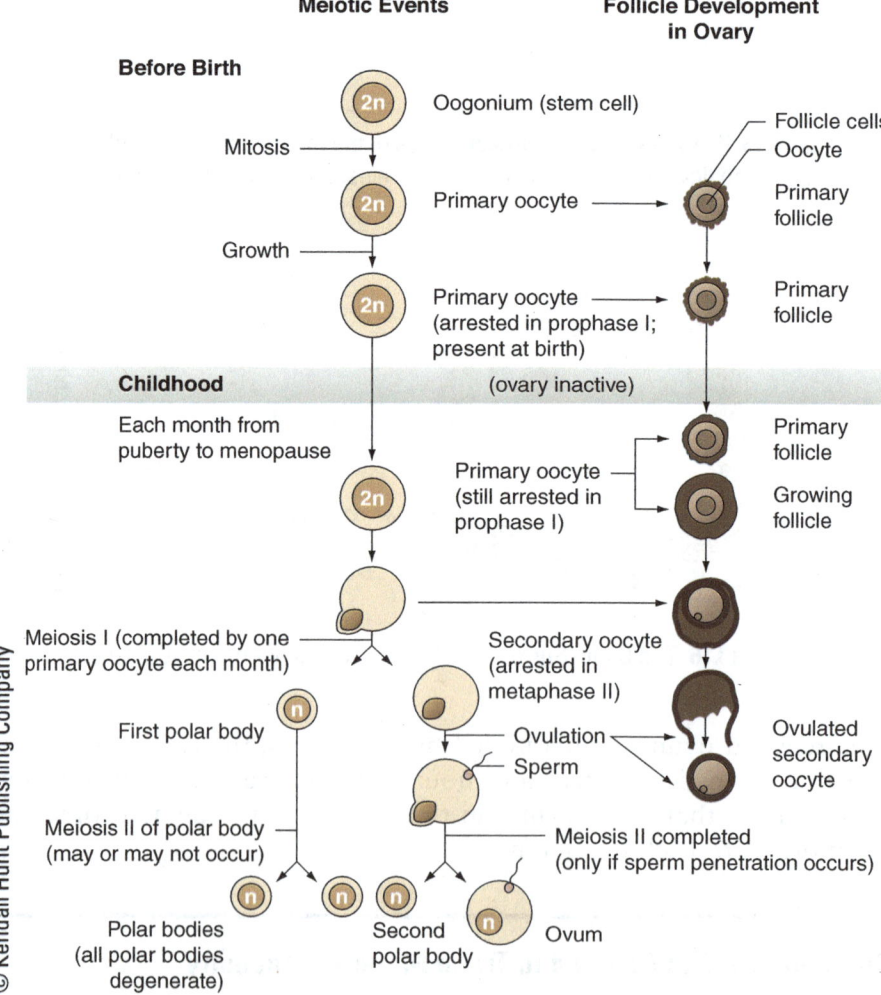

Figure 13.17 Oogenesis. Ova stem cells (oogonia) divide by meiosis prior to birth. Meiosis I is arrested from birth until after puberty. Prior to an ovum being released, it completes meiosis I; it then continues into meiosis II. Three of the four cells resulting from meiosis II become nonfunctional polar bodies, with only one cell completing meiosis II if it is fertilized by sperm.

of meiosis. How the ovary derives that choice is unknown. At the end of meiosis I, the two *secondary oocytes* each contain a set of 23 chromosomes. For some reason that is not clear, only one of the two secondary oocytes retains the bulk of the cytoplasm from the primary oocyte and remains a viable cell. The other cell gives up most of its cytoplam to the other oocyte to become what is known as a *polar body*. During meiosis II, the primary oocyte divides again, with one cell becoming the ovum and the other becoming a polar body. The polar body that resulted from meiosis I also divides into two polar bodies. In summary, the single primary oocyte passes through both stages of meiosis to end up with one ovum containing 23 individual chromosomes along with its cytoplasm and three polar bodies.

Physiology of the Follicle

As a result of follicle-stimulating hormone secretion from the anterior pituitary gland, a primary oocyte completes meiosis I as the follicle begins to accumulate fluid (Figure 13.18). As more and more fluid fills the follicle, it swells and bulges onto the surface of the ovary as meiosis II reaches completion. After 2 weeks, there is an increase in lutenizing hormone, also from the anterior pituitary gland, which causes the erosion of the outer surface of the follicle, causing it to rupture. The fluid within the follicle oozes out, carrying with it the ovum. The process of releasing an ovum from its follicle is known as *ovulation*. Lutenizing hormone converts the empty follicle into an endocrine gland known as the corpus luteum. The corpus luteum secretes the hormones progesterone and estrogens into the bloodstream to stimulate growth of the endometrium. If conception does not occur, about 10 to 12 days after ovulation, the corpus luteum becomes nonfunctional and eventually scars over. If conception did occur, the corpus luteum continues producing progesterone and estrogens until the placenta can take over their production.

> **Corpus luteum** once a follicle in the ovary has released its ovum, the empty follicle is converted into an endocrine gland known as the corpus luteum which secretes the hormone progesterone.

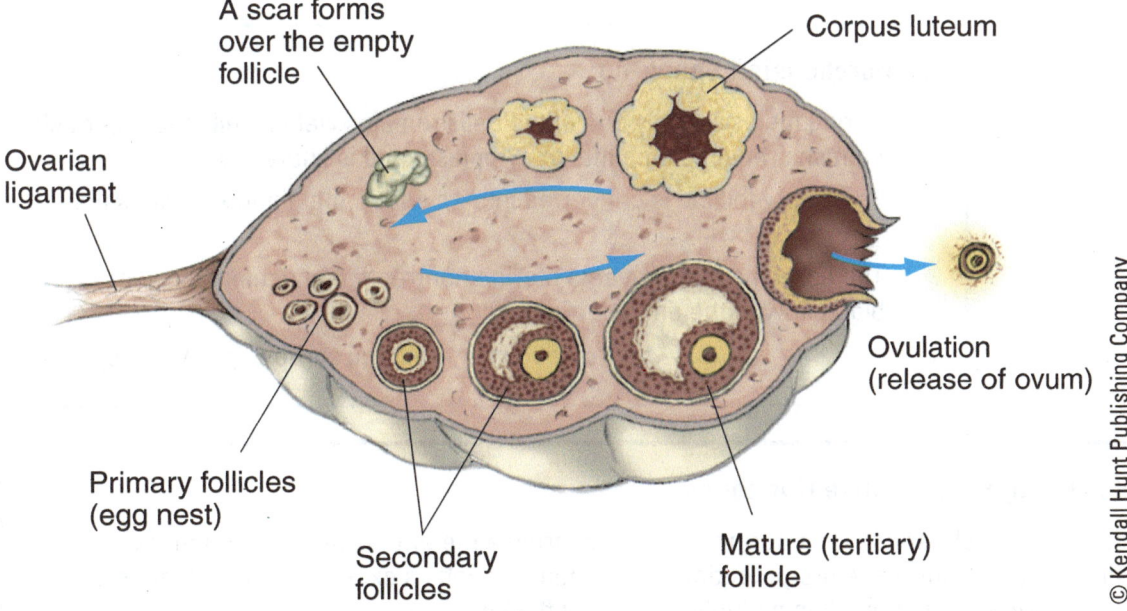

Figure 13.18 Follicle. A follicle releases its ovum after being stimulated by luteinizing hormone. The process begins as the follicle accumulates fluid, causing it to swell until the follicle ruptures. This releases the ovum during ovulation. The empty follicle is converted into the corpus luteum, which produces the hormones progesterone and estrogens for roughly a week, after which, if pregnancy does not occur, it dries up and scars over.

Hormones Affecting Female Reproduction

Four groups of hormones affect female reproduction. Two are produced by the anterior pituitary gland and two, by the ovaries. They are:

- *Follicle-stimulating hormone,* produced by the anterior pituitary gland, causes a follicle to prepare to release an ovum. It also assists in causing ovulation. If conception does not occur during one menstrual cycle, follicle-stimulating hormone increases at the end of this cycle to begin preparation of the next follicle for ovulation to occur in the subsequent cycle.
- *Lutenizing hormone,* from the anterior pituitary gland, stimulates the production of enzymes that cause the rupturing of the surface of the follicle, resulting in ovulation and causing the conversion of the empty follicle into the corpus luteum.
- *Estradiol (the primary estrogens)* is produced by the follicles and the corpus luteum in the ovary to communicate with the woman's body to prepare for the implantation of a fertilized ovum. It causes proliferation of the endometrium to provide a location for attachment of the placenta. It also alters the woman's blood flow, as well as causing other physiological changes, to prepare her body each cycle to carry the fetus throughout pregnancy.
- *Progesterone* is produced by the corpus luteum in the ovary and by the placenta during pregnancy. The corpus luteum maintains high blood flow to the endometrium so that the placenta has the ability to exchange nutrients, gases, and wastes. It also inhibits the contraction of the myometrium. Throughout pregnancy the contraction of the uterus must be inhibited until it is time for the fetus to be delivered. Progesterone also inhibits the release of follicle-stimulating hormone, stopping menstrual cycles while a fetus remains in the uterus.

The production of these hormones, especially follicle-stimulating hormone and estrogens, can be affected by stress, nutrition, and age.

Comprehension Check-up:

1. During meiosis I, only one of the cells remains viable while the other cell contains reduced amounts of cytoplasm and is known as a _____.
2. The empty follicle is converted into an endocrine gland known as the _____.
3. The hormone that prevents uterine contraction and maintains increased blood flow of the endometrium is _____.

1. polar body 2. corpus luteum 3. progesterone

Benefits of Altering Reproductive Hormones

- Taking additional follicle-stimulating hormone as a "fertility drug" may increase the potential for conception but may also result in multiple births.
- Estrogen therapy has become a common treatment for women during menopause to decrease the effects of this reproductive change, such as excessive mood swings and hot flashes.
- Progesterone can be taken as a birth control pill to falsely signal that the female is pregnant, preventing ovulation from occurring.

The Menstrual Cycle

The process a woman passes through from the start of several days of menstrual bleeding, known as *menses*, until the beginning of the next menses is referred to as the menstrual cycle (Figure 13.19). On average, the cycle lasts approximately 28 days. Generally, menses ends around day 4 or 5. Ovulation

> **Menstrual cycle** the process through which the female's reproductive system passes in preparation for the release of an ovum and its potential fertilization.

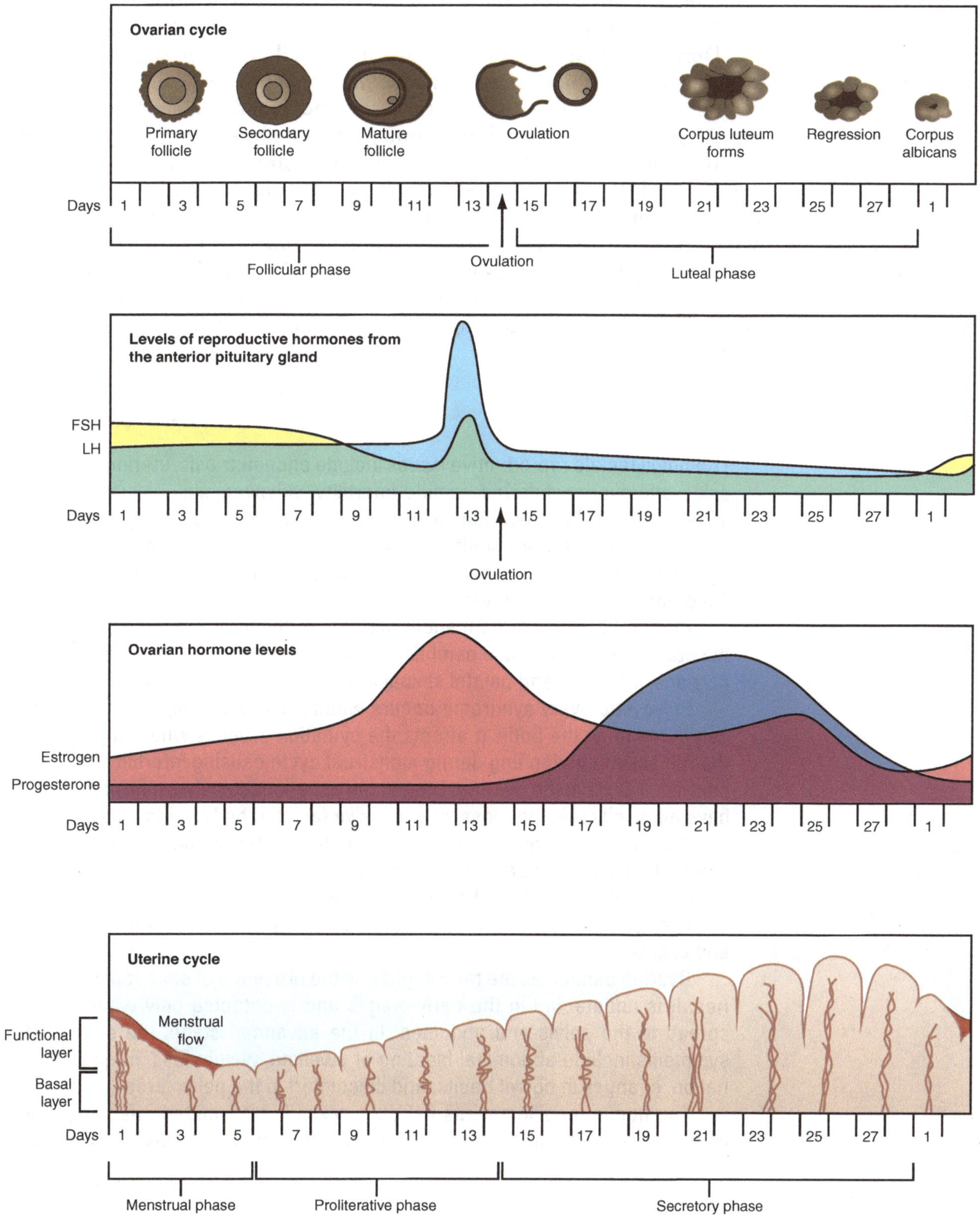

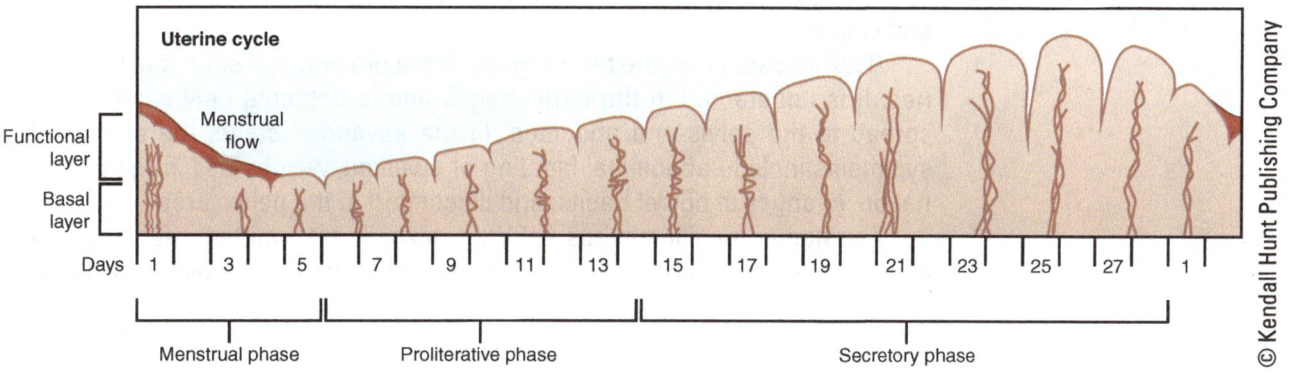

Figure 13.19 Events of the Menstrual Cycle.

occurs in the middle of the cycle, about day 14. Following that time, the corpus luteum is formed and conception is possible. On average, out of this 28-day period, the conception is possible during 3 days. Sperm can survive inside the female up to 48 hours before body heat overcomes them, so sperm can be available up to 2 days before ovulation. The ovum can survive unfertilized for approximately 1 day, so if the sperm arrive on the day of ovulation conception is highly probable. Outside of that 3-day window, it is unlikely conception will occur.

During the first 4 or 5 days of the menstrual cycle, the superficial layers of the endometrium, referred to as the *functional layer*, to which the placenta would attach, is being shed. Both tissue and blood are shed during menses. After that time, the endometrium, stimulated by estrogens, goes through rapid mitosis to rebuild the functional layer again for the next potential fetus. Toward the end of the menstrual cycle, if conception did not occur, the corpus luteum becomes nonfunctional. There is a drop in progesterone levels, which results in spasms in arteries supplying blood to the functional layer; consequently the tissue begins to die. By the end of the 28th day of the cycle, there is enough tissue death that the functional layer begins to tear away from the uterine wall, signaling the beginning of the next menstrual cycle.

Clinical Notes—Female Reproductive Issues

LCommon female reproductive issues include endometriosis, uterine fibroids, polycystic ovary syndrome, female infertility, vulvovaginitis, sexually transmitted disease (STD), ovarian cancer, cervical cancer, and vaginal cancer.

Endometriosis is a condition wherein the tissue that normally lines the uterus grows outside in other areas such as the ovaries, fallopian tubes, and the outer surface of the uterus.

Uterine fibroids are noncancerous tumors found in the muscle cells of the uterus. They can cause painful menstrual bleeding, lower back pain, miscarriage, infertility, and painful sexual intercourse.

Polycystic ovary syndrome occurs when there is an increase of androgen hormone in the body. It affects the ovulation process where it prevents the release of ovarian egg during menstrual cycle causing infertility. It forms fluid-filled cysts in the ovaries. The person experiences pelvic pain and would have acne, oily skin, hair loss, or excessive facial or body hair growth.

Vulvovaginitis is an excessive inflammation of the vagina or vulva tissues where the person experiences vaginal itching and inflammation, urinary discomfort, unpleasant vaginal odor, and vaginal discharge. It is caused by poor hygiene, sexually transmitted diseases, and infection by bacteria, viruses, and yeasts.

Ovarian cancer, as the term implies, is the presence of cancer in the ovaries. It is undetected in the early stages and is detected only when it has spread to the pelvis and abdomen. In the advanced stage, the signs and symptoms include abdominal bloating or swelling, weight loss, frequent urination, changes in bowel habits, and discomfort in the pelvic area.

The human papillomavirus (HPV), a sexually transmitted infection, is a common cause for most cervical cancers. The virus can survive for many

(continued)

years in the human body that can cause cervical cells to mutate into cancer cells. The cancer develops in the lower part of the uterus. An HPV vaccine is used to protect against the virus. Risk factors for cervical cancer includes multiple sex partners, early sexual activity, STDs, smoking, and weak immune system. The person experiences painful sexual intercourse, vaginal bleeding after intercourse, heavy bloody vaginal discharge with foul odor.

RESPONSE OF THE REPRODUCTIVE SYSTEMS TO CONCEPTION AND TO AGING

The physiological processes occurring in the female during pregnancy include support of the fetus; exchange of nutrients, gases, and wastes between maternal blood and the placenta; gestation and delivery of the fetus; and nursing of the newborn. We will also discuss the activation of the reproductive systems, puberty, and the end of the reproductive process, menopause.

> **Comprehension Check-up:**
> 1. Ovulation occurs on day _____ of the average menstrual cycle.
> 2. Blood pressure, heart rate, and muscle tone increase and sperm form into a mass during the _____ phase of intercourse.
>
> 1. 14 2. plateau

Pregnancy

A *zygote* is a "single cell" formed by the union of an ovum and a sperm cell. If fertilization occurs in the uterine tube, cell division by mitosis occurs within the first 24 hours (Figure 13.20). After the zygote has divided into a 2-cell, 4-cell, or 8-cell structure, it is called a *cleavage*. When it becomes a

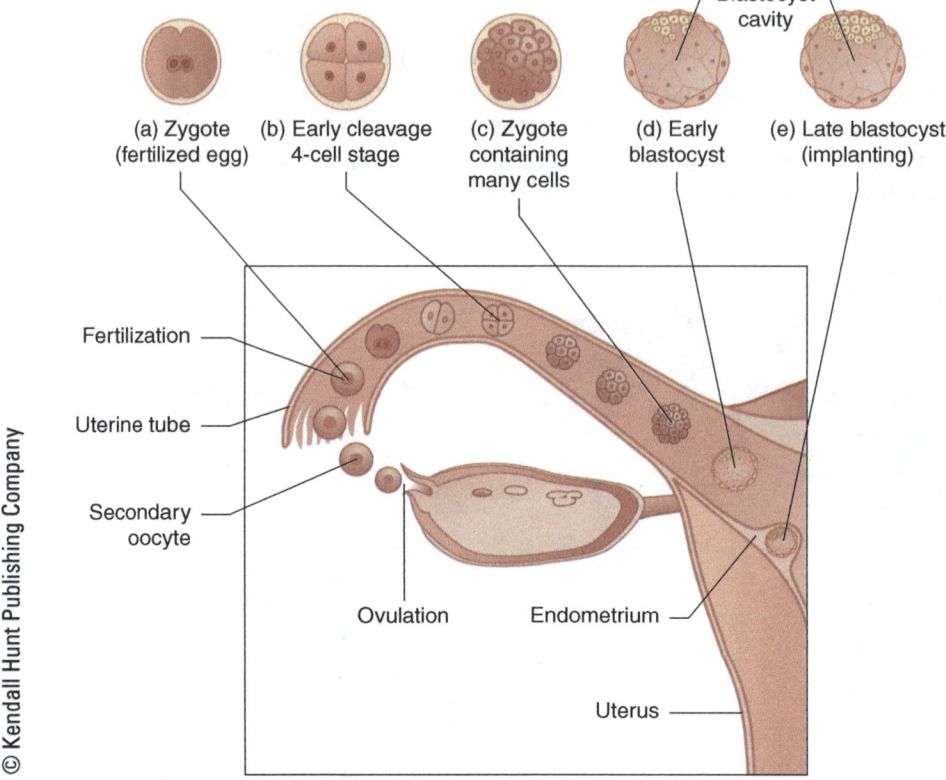

Figure 13.20 Conception and Implantation. Fertilization of the ovum occurs in the uterine tube. Once the fertilized ovum begins to divide, it is known as a *zygote*. While in this phase, subsequent cell divisions do not result in an overall increase in size. By the time the zygote reaches the uterus it is known as a *blastocyst*, which is a mass of cells and a cavity filled with amniotic fluid.

16-cell structure, it is called a *morula*. From day 4–16 after fertilization, it's called *blastocyte* (a multicellular preembryonic structure until 16 days after fertilization). After 16 days of fertilization, the multicellular structure turns into an *embryo*.

Once the blastocyte enters the uterus, it buries itself in the endometrium by a process known as *implantation*. Within a few days, a primitive placenta begins to form. When the developing placenta begins producing hormones, it is known as the *chorion*. One hormone it begins producing is *human chorionic gonadotropin (hCG)*. The hormone is significant for four reasons:

- It causes the male to begin producing testicular-determining factor, resulting in the formation of the male reproductive system.
- It informs the woman's body that she is pregnant. The corpus luteum will continue to produce progesterone and estrogens for an extended period until the placenta has developed sufficiently to produce this hormone throughout pregnancy.
- It causes morning sickness, although unpleasant, which is an early sign of pregnancy.
- Its presence can be determined with a home pregnancy test. The nonpregnant female does not normally produce hCG. Using an indicator that changes color in the presence of this hormone has become a reliable test for pregnancy.

Because the woman's body is now aware an embryo is growing in the uterus, she will increase and maintain a substantial blood flow to the endometrium to support the developing fetus. Throughout the entire gestation period, the mother's body will provide the fetus with nutrients from her own body. If a particular nutrient is in short supply, she develops cravings for foods containing those substances. Prenatal vitamins and calcium supplements are strongly recommended to supply both the mother's and fetus' needs. If too much calcium is removed from the mother's bones, there is an increased possibility of osteoporosis as she ages. A lack of sufficient folic acid has been linked to birth defects. It is strongly recommended that a woman who has a potential to become pregnant take additional folic acid because waiting until she determines she is pregnant may be too late to prevent defects.

Placenta

The placenta is the fetus' attachment to the mother and its source of nutrients, exchanges of oxygen and carbon dioxide, and disposal of wastes. The placenta attaches to the functional layer of the endometrium. Under the placenta, pools of the mother's blood forms. Capillaries from the placenta pass through these pools. There is no direct exchange of blood between the mother and the fetus. Rather, nutrients, gases, and wastes can diffuse across the placental capillaries, exchanging with the substances in the pools of the mother's blood.

The placenta is the only temporary endocrine gland found in the body. There are several hormones produced by the placenta. Progesterone and hCG are two of the hormones we have discussed. Once the baby has been delivered and the placenta detaches from the endometrium, these hormones are no longer available.

Gestation

The average developmental time for the fetus within the uterus is about 266 days (38 weeks). Typically the physician adds 14 days to the calculation, the time from the start of the last menstrual period to ovulation, to determine the approximate birth date after 280 days. This length may vary considerably. The large number of processes that occur in the development of the fetus between conception and birth is a study known as embryology. It is simply beyond the scope of this course to discuss this awe-inspiring series of events.

Labor

When it is time for labor to begin, oxytocin is released from the posterior pituitary gland to cause contractions of the myometrium. Normally, over a series of hours, the woman's body adjusts to prepare for the outward passage of the fetus. Contraction of the myometrium combined with increased abdominal pressure pushes the fetus and eventually the placenta through the cervix and vagina and out of the body.

Comprehension Check-up:

1. The attaching of the blastocyte to the lining of the uterus is referred to as _____.
2. The gestation period for the fetus is about _____.

1. implantation   2. 266 days – 38 weeks

Nursing

Recall from our earlier discussion that the production of milk by the mammary glands is due to the presence of the hormone prolactin. Although milk is produced, it is not released until nursing causes the posterior pituitary gland to secrete oxytocin.

During the first 3 days after giving birth, the mother's milk is especially high in protein and also contains the mother's antibodies. This milk is known as *colostrum*. Not only does this special milk provide extra protein for the baby, it also gives the newborn temporary passive immunity against diseases to which the mother has been previously exposed. After a few days, the mother's milk contains an increased level of fat and less protein and little in the way of protective antibodies.

PUBERTY

> **Puberty** the process of activation of the male or female reproductive system.

The reproductive systems of both males and females become activated by a process referred to as puberty. The activation of the reproductive systems is the result of stimulating hormones from the hypothalamus. In the female the hypothalamus becomes sensitive to estrogen from the ovaries. It is the intercommunication between these two organs that establishes the rhythmic patterns of the menstrual cycle. During this time, due to the periodic upsurges in hormone levels, rapid physiological and psychological changes may occur.

Because of the increased sensitivity of her hypothalamus to estrogen, hormones will be released in the female that cause the development of breasts and result in her experiencing the first periods of menstrual bleeding as her menstrual cycles increase in regularity. In males the hypothalamus stimulates activity of the testes, resulting in an increase in testosterone, which causes male physical characteristics to become apparent. Over the next several years, he will experience a deepening voice; increased growth of hair on his face, axilla, and pubic area; and enlargement of his penis. The changing levels of reproductive hormones during puberty also affect personality, including increased aggressiveness and irritability. Puberty typically occurs between 10 and 15 years of age for girls and in the early to mid-teens for boys.

MENOPAUSE

> **Menopause** the process of deactivation of the female reproductive system.

As a woman ages, her ovaries produce less estrogen. By the time she is 45 to 55, although there is much variation in this age, the hypothalamus has become less responsive to the lower levels of estrogen and her menstrual cycles become irregular and finally cease in a process known as menopause. It has also been found that the number of viable ova at menopause is virtually negligible. As the woman's body adjusts to the lack of menstrual cycles, there are often unpleasant side effects, such as sudden increases in blood flow to the skin causing hot flashes. She may have sudden mood swings or experience depression.

There does not appear to be an age at which men no longer produce sperm. After puberty, they are able to father offspring throughout life.

Hormone Replacement Therapy

In recent years, hormone replacement therapy, primarily estrogen, has eased the unpleasant effects of menopause. Recent studies, however, have determined abnormally high incidences of heart disease, breast cancer, stroke, and dementia in women taking hormone replacement therapy and its use has been greatly decreased.

Homeostasis—Holding in Balance

The reproductive system does not directly maintain homeostasis of other parameters in the body. There is, however, a direct correlation between the level of testosterone or estrogen and the control of bone and muscle density and red blood cell production. Testosterone increases the length of long bones in males and also stimulates the production of larger muscle mass and a higher percentage of red blood cells than in females. Estrogen also prevents the breakdown of bone in women during reproductive years. A postmenopausal woman, on the other hand, typically has lower levels of estrogen, resulting in bone loss that increases the probability of osteoporosis.

Comprehension Check-up:

1. The first 3 days after giving birth, the mother's milk contains her antibodies and is known as _____.
2. The onset of puberty in both males and females is the result of hormones from the _____.

1. colostrum 2. hypothalamus

CLINICAL TERMS TO REVIEW

Male reproductive issues, female reproductive issues
Embryo, fetus, sperm, ova, gametes, fertilization
Gonads, spermatogenesis
Testes, epididymis, ductus deferens, prostate gland, seminal vesicle, semen,
Cowper's gland, penis, scrotum
Ovaries, uterine tubes, uterus, vagina, mammary glands, ovum, oogenesis
Follicle-stimulating hormone, luteinizing hormone, estradiol
Menstrual cycle, menopause, hormone replacement therapy
Pregnancy, human chorionic gonadotropin (HCG), placenta, gestation, labor, contraction

Test Yourself

Choose the best answer to the following multiple choice questions:

1. The hormone responsible for sex drive and male physical characteristics is
   a. lutenizing hormone.
   b. follicle-stimulating hormone.
   c. estrogen.
   d. testosterone.

2. What is the function of the midpiece of the sperm?
   a. It contains mitochondria for ATP production to cause the tail to move.
   b. It contains chromosomes protected by the sperms head.
   c. It whips back and forth to cause the sperm to move.
   d. It contains acid-neutralizing chemicals.

3. The fold of skin on the penis removed during circumcision is the
   a. glans penis.
   b. prepuce.
   c. corpus spongiosum.
   d. corpora cavernosa.

4. The hormone responsible for an increase in sperm production is
   a. lutenizing hormone.
   b. testosterone.
   c. follicle-stimulating hormone.
   d. estrogen.

5. The process by which gametes are produced is
   a. mitosis.
   b. parthenogenesis.
   c. symbiosis.
   d. meiosis.

6. The funnel-shaped end of the uterine tube is called the
   a. infundibulum.
   b. cristae.
   c. cervix.
   d. frenulum.

7. Large tissue folds of the female external genitalia are known as
   a. clitoris.
   b. labia minora.
   c. corpus spongiosum.
   d. labia majora.

8. The hormone that prepares the ovary to release another ovum is
   a. lutenizing hormone.
   b. follicle-stimulating hormone.
   c. estrogen.
   d. testosterone.

9. The hormone responsible for causing uterine contraction during labor and milk release when nursing is
   a. prolactin.
   b. lactorferrin.
   c. oxytocin.
   d. colostrum.

10. The condition that results when the ovaries produce less estrogen and the hypothalamus becomes less sensitive to estrogen is known as
    a. hypertension.
    b. menopause.
    c. arteriosclerosis.
    d. puberty.

Appendix I

Physical Examination, History, and Interview

An overview of physical examination, history, and interview include the following:

1. Chief complaint and History of Present Illness (HPI)—the main the reason why the patient is seeking medical attention. It is what the patient tells you about his or her problem.
2. Past medical history—any illness, injury, diseases, seizures, surgery, medical procedures, and hospitalizations that the patient had in the past.
3. Family history—family members or relatives who experienced the same or similar illness or diseases, such as diabetes, hypertension, allergy, seizures, and so on.
4. Personal history—home environment, lifestyle, diet, eating habits, exercise, weight problems, smoking, alcohol, and social life, including sex life, that may have an effect on the current condition.
5. Occupational history—work environment, nature of work, exposure to smoke, paint, chemicals, factories, kitchen smoke, engines, and so on.
6. Review of Systems (ROS):
    - General symptoms—fever, malaise, fatigue, chills, night sweats, significant weight loss or weight gain, appetite, cough with phlegm, and so on.
    - General appearance—color, posture
    - Head and neck—physical exam
    - Chest and lungs—difficulty of breathing, coughing, phlegm, use of accessory muscles when breathing
    - Extremities—skin, color of nailbeds
    - Lymphatic—tenderness in armpit, neck, groin region
    - Musculoskeletal—pain in joints, difficulty in walking or standing
    - Gastrointestinal—nausea, vomiting, loss of appetite, abdominal pain, bloating, frequent passing of gas, bowel movement, diarrhea, constipation, blood in stools
    - Urinary—incontinence, pain during urination, cloudy urine, blood in urine
    - Mental health—mood swings, anxiety, depression, poor concentration
7. Vital signs:
    - Body temperature—96.7-100.5°F; above this value strongly indicates infection which needs further tests such as complete blood count
    - Pulse rate—60-100 beats per minute; above 100 is tachycardia
    - Respirations—12-20 breaths per minute; above 20 is tachypnea
    - Blood pressure—systolic of 100-120 mmHg; diastolic of 60-80 mmHg; hypertension—140-160 mmHg/90-100 mmHg

8. Techniques in examination
   - Interview, observing patient looking for signs and symptoms while talking
   - Inspection, observing the patient's reaction during inspection
   - Palpation—chest and abdomen
   - Percussion—chest and abdomen
   - Auscultation—chest, heart

Complete Blood Count (CBC), Basic Metabolic Panel (BMP), Kidney Panel, and Comprehensive Metabolic Panel (CMP), Liver Panel

Appendix II

General Count	Normal Values	Significance
Red Blood Cells (RBC)	3.5–5.5 million/cu mm	Low—anemia
White Blood Cells (WBC)	5.0–10.0 thousand/cu mm	High—infection
Hemoglobin (HB)	12–16 g/dl	Low—anemia
Hematocrit	36–47 vol%	Low—anemia
Platelet count	150–400 thousand/cu mm	Acute blood loss
Differential Count		
Neutrophils	50%–75%	High—bacterial infection
Lymphocytes	25%–45%	High—viral infection
Monocytes	0%–10%	High—chronic infection
Eosinophils	0%–4%	High—allergy, asthma
Basophils	0%–1%	High—inflammation
BMP—Includes Kidney Panel		
Glucose	70–110 mg/dl	High—diabetes
Calcium (Ca)	8.6–10.2 mg/dl	High—hyperthyroidism
Sodium (Na)	135–145 mEq/l	High—dehydration
Potassium (K)	3.5–5.0 mEq/l	High—renal failure
Chloride (Cl)	95–105 mEq/l	High—dehydration, starvation
Bicarbonate (HCO_3)	24–28 mEq/l	High—vomiting, alkalosis

(*continued*)

General Count	Normal Values	Significance
Kidney Panel		
Creatinine	0.5–1.2 mg/dl	Above 4 mg/dl—serious renal disease
Blood Urea Nitrogen (BUN)	10–20 mg/dl	High—kidney failure

CMP includes all tests under BMP and the following:

CMP—Includes Liver Panel	Normal Values	Significance
Albumin	3.5–5.4 g/dl	High levels—dehydration, inflammation, bone marrow disease, HIV/AIDS
Total protein	6–8 g/dl	High levels—dehydration, hepatitis, HIV/AIDS
Liver Panel		
Alkaline phosphatase (ALP)	30–120 units/l	High—hepatitis, liver cancer
Alanine aminotransferase (ALT)	4–36 units/l	High—acute hepatitis, liver cirrhosis
Aspartate aminotransferase (AST)	0–35 units/l	High—acute hepatitis; heart attack
Bilirubin	Total bilirubin: 0.3-1.0 mg/dl Indirect bilirubin: 0.2-08 mg/dl Direct bilirubin: 0.1-0.3 mg/dl	High—hepatitis, liver disease, bile duct blockage; patient appears jaundiced

Cardiac Panel, Lipid Panel, and Urinalysis

Appendix III

Cardiac Panel	Normal Values	Significance
Total Creatine kinase (CK), or Creatine Phosphokinase (CPK)	Male: 38–174 units/l Female: 96–140 units/l	Recent heart attack (myocardial infarction)
Creatine Kinase MB (CK-MB), or Creatine phosphokinase (CPK-MB)	5–251 units/l	Recent heart attack (myocardial infarction)
Troponin	Troponin I: 0–0.04 ng/ml Troponin T: 0–0.01 ng/ml	Recent heart attack (myocardial infarction)
Lipid Panel		
Total Cholesterol	Less than 200 mg/dl	Above 250—high risk for heart disease
Low-Density Lipoprotein (LDL)	Less than 100 mg/dl	Above 170—high risk for heart disease
High-Density Lipoprotein (HDL)	Above 60 mg/dl	Below 40—high risk for heart disease
Triglycerides	Less than 150 mg/dl	Above 200—high risk for heart disease

Urine Test	Normal Values	Significance
Specific gravity	1.002–1.035 Average is 1.010	Dehydration
Ph	4.5–8.0 ideal is 6.0	High pH—kidney stones, kidney failure Low pH—diabetes, diarrhea
Protein	Undetectable or 0	Positive—kidney damage
Glucose	Negative	Positive—diabetes
Ketones	Negative	Starvation, diabetes, high-protein diets, severe exercise
Leukocyte/WBC	Negative	Urinary tract infection (UTI)
Nitrite	Negative	UTI
Bilirubin	Negative	Liver/gallbladder disease
Urobilinogen	Negative	Liver/gallbladder disease
Red blood cells	Negative or trace	Kidney/ladder disease
White blood cells (WBC)	Negative or trace	UTI, inflammation
Epithelial cells	Negative or trace	UTI, inflammation
Microorganism	Negative	UTI

12-Lead ECG: Chest Electrodes and Placement

Appendix IV

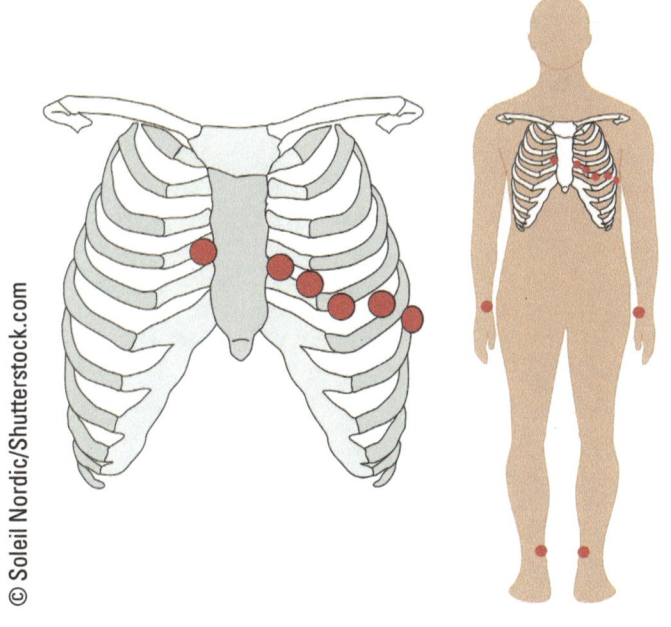

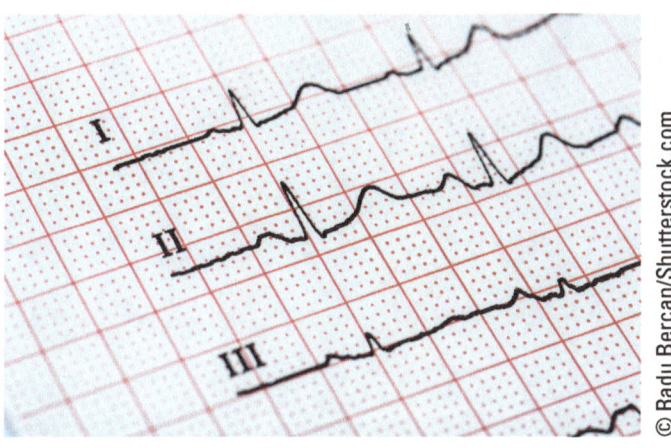

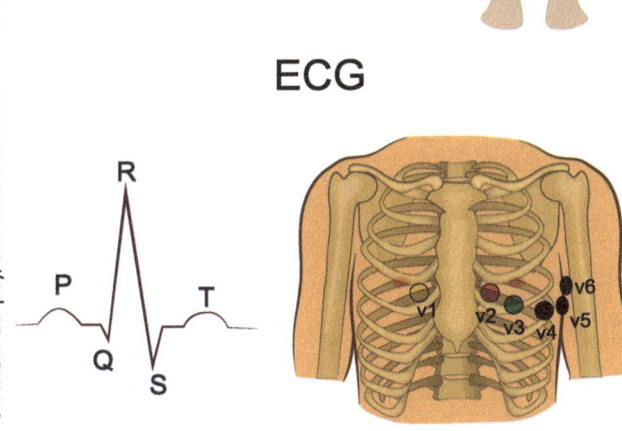

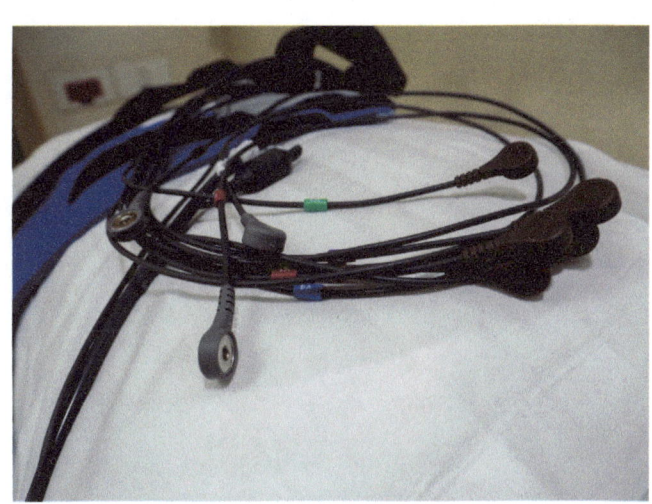

V1—Fourth intercostal space on the right sternum
V2—Fourth intercostal space at the left sternum
V3—Midway between placement of V2 and V4
V4—Fifth intercostal space at the midclavicular line
V5—Anterior axillary line on the same horizontal level as V4
V6—Mid-axillary line on the same horizontal level as V4 and V5

RA (Right Arm)—Anywhere between the right shoulder and right elbow
RL (Right Leg)—Anywhere below the right torso and above the right ankle
LA (Left Arm)—Anywhere between the left shoulder and the left elbow
LL (Left Leg)—Anywhere below the left torso and above the left ankle